ABOUT THE AUTHORS

*T*he authors of this book bring a wealth of professional and personal knowledge, training, and experience to the project. Among them, they have carried out geologic fieldwork on all of the Earth's continents. The diversity demonstrated in their own careers reflects the broad range of challenges that characterize Earth science and environmental geology today.

As an undergraduate *Barbara Murck* was a confirmed nonscientist, until an introductory geology course changed her plans. Since then her professional focus has ranged from igneous geochemistry and ore-deposit petrography to alternative energy sources and state-of-the-environment reporting. Her current work focuses primarily on environmental management training for decision makers in developing countries.

Throughout his career as a geologist, *Brian Skinner's* research has focused on the physical properties of minerals and on the genesis of base-metal deposits. He has worked extensively in Australia, Africa, and North America and with students in Asia and Europe. With Yale University colleagues, he has had the opportunity to explore a diversity of Earth science topics, including oceanography and climatic change, volcanic gases, economic models of resource depletion, and the geologic aspects of the space program.

Stephen Porter's professional career has largely been concerned with studies of glaciation in many of the world's major mountain systems and with the history of the climatic changes their deposits record. He has also studied the evolution of midocean and continental volcanoes and the products of their prehistoric eruptions and how volcanic eruptions may have influenced the Earth's climate. With colleagues from around the world, he has studied the hazards of large rockfalls in the Alps and the thick, extensive deposits of windblown dust in China that provide one of the longest continuous records of climatic change during the past several million years.

The authors' global perspective is reflected in this book by examples and illustrations from numerous foreign areas, for it is important to emphasize that geology is a global science—a science that recognizes no political boundaries.

Only by studying the Earth in its entirety can we hope to understand how our amazing planet works, how geologic processes affect our lives, and how, in turn, human activities affect the functioning of the Earth system.

PREFACE

*T*he study of the Earth has captured people's interest for as long as scientific enquiry has been conducted. To explore, to question, to seek an understanding of the natural processes that surround us is fundamental to human nature. In the Earth sciences today, more than ever before, we recognize what links the various parts of the Earth system to one another and what links all organisms—including human beings—to the geologic environment. Investigating the nature of these dynamic connections is a central goal of environmental geology and an important focus of this book.

Why We Wrote *Environmental Geology*

We love geology. We love to study, teach, and learn about geology. It is fun and exciting to help people understand the Earth beneath their feet and the geologic processes that affect their daily lives. It is rewarding when a student says, "I heard about that earthquake on the news last night, and I understood why it happened," or "On our holiday we saw a flood out the window of the airplane, and I explained to my family what was happening." By producing a first-rate text on environmental geology, we hope to facilitate such learning.

As a society (and, increasingly, as a global community) we want to continue to derive benefits from the Earth's resources without jeopardizing the future availability of those resources and without compromising the integrity of natural systems. We hope, also, to minimize the negative impacts of hazardous Earth processes. To achieve these goals we need to foster greater understanding of Earth systems and geologic processes. For those reasons we have chosen to explore environmental geology from a distinctly human perspective. Throughout the book we return to the themes of human-induced environmental change, the impacts of geology on human interests, and societal dependence on the Earth's resources. This book is an exploration of the planet. More precisely, it explores the human-planet relationship: how Earth processes influence human lives on a daily basis and how human actions, in turn, alter the functioning of Earth systems.

This is an environmental geology textbook. It is not a physical or historical geology textbook, although environmental geology does encompass some aspects of both physical and historical geology. The basic geology topics covered in the book were specifically selected to provide the foundation for concepts and terminology needed to understand the impacts of geologic processes on human interests. Students respond best to background material when they understand the reasons for learning it; hence, frequent explanations are offered. They might ask, for example, Why is it important to learn about the interior of the Earth? (Because internal Earth processes are fundamental in forming the landscape and causing hazardous Earth processes like earthquakes and volcanoes.) Why is it necessary to study the properties of rocks, minerals, and soils? (Because we need these materials as resources and because the properties of Earth materials can affect human interests in a wide variety of ways—in their ability to resist mass wasting, to absorb wastes, or to hold and transmit oil or water, for example.) Students who read this book will gain the basic geologic background they need to interpret geologic events.

To the Reader

Most likely, you are reading this book in the context of an introductory course on environmental geology. People take such courses for a variety of reasons; consequently, this book may mean different things to different people. If you are planning to become a geologist or an environmental geologist, you can consider the book an introduction to the many fascinating possibilities and challenges that await you in your career. If you are studying environmental geology out of personal interest or perhaps to fulfill a degree requirement, we hope it will help you become more aware of the geologic nature of our environment and the role of geoscientists in events of public concern. And we hope that all of our readers will emerge better prepared to make informed decisions about the natural processes that affect our lives on a daily basis.

Organization

The book is organized in four main parts. In *Part I: Geologic Framework* we provide a brief background in Earth system science and physical and historical geology. In this part you will find basic concepts and terminology concerning the structure and materials of the Earth and the functioning of Earth systems and cycles. The topics covered are not overly technical; they were carefully chosen to provide just the background needed to be able to understand the material presented in subsequent chapters. *Part II: Hazardous Geologic Processes* covers the broad range of geologic events that are damaging to human interests, including Earthquakes, volcanic eruptions, landslides, and floods. In this section we look at both the impacts of such events on humans and the geologic processes that underlie the events. In *Part III: Using and Caring for Earth Resources* we investigate the nature of the Earth's resources, their importance for modern society, the geologic processes that control their formation and occurrence, and some of the environmental implications of their use and management (or mismanagement). *Part IV: Human Impacts on the Environment* focuses on contaminants—waste and pollution, where they come from, how they are transported in the geologic environment, and how they can be managed. This part of the book ends with a look at the impacts of human activity on the atmosphere and at the potential impacts of global atmospheric change.

Special Features

The following special features of content, organization, graphical presentation, and pedagogy set this book apart from others in similar subject areas.

Part Opening Essays

Each of the four main parts of the book opens with an essay. These essays are not add-ons; they are integral to the text. They set the context and provide an overview of the fundamental concepts that recur throughout the ensuing chapters. The essay for Part I, entitled *The Home Planet,* puts Earth in perspective by placing it in the context of the solar system and by examining some of the Earth-forming processes that have made this planet hospitable to life. *Assessing Geologic Hazards and Risks,* the essay for Part II, introduces natural, geologic, technological, and anthropogenic hazards and some of the approaches used to assess human vulnerability to these hazards. In the essay for Part III, *The Nature of Earth Resources,* renewable and nonrenewable resources are introduced, and the finite nature of some Earth resources and the need for appropriate management are discussed. The Part IV

opening essay, *Managing Wastes,* examines waste and contaminants in the context of a "throwaway" society.

Chapter-Opening Vignettes

Each chapter opens with a short vignette, or anecdote, illustrated with a chapter-opening photograph on the facing page. The purpose of these vignettes is to provide a glimpse into some aspect of the geologic environment and how that environment affects humans (and vice versa). To a certain extent the opening comments are designed to provide a context or rationale for the chapters. For example, students often wonder why the study of minerals is relevant to environmental geology. The vignette for chapter 2, concerning the properties of asbestos minerals and how they affect human interests, provides an answer.

Boxes: *The Human Perspective* and *Focus On . . .*

Scattered throughout the book you will find material set aside in shaded boxes; at least one box appears in each chapter. Boxes entitled *The Human Perspective* are intended to highlight particular aspects of the human–planet relationship, such as human impacts on the environment, the impacts of geologic processes on human interests, or human institutions (projects, programs, etc.) devoted to some aspect of environmental geology. Boxes entitled *Focus On . . .* are intended to provide an in-depth look at some of the more technical aspects of environmental geology and related sciences. These boxes provide depth and detail for a course that goes beyond the introductory level or is designed specifically for geology or environmental geology majors. If desired, however, the *Focus On . . .* boxes can be skipped without affecting the integrity of the material presented in the rest of the chapter.

Artwork

The artwork in this book has been carefully crafted to complement the text and to facilitate learning. For the line drawings, the authors worked closely with illustrators who are familiar with the subject matter and are highly skilled at translating concepts into accurate full-color representations. Intensive research has allowed us to carry on the tradition of beautiful, up-to-date, instructive photographic images for which Wiley's Earth science textbooks are known.

Pedagogical Material

Each chapter includes a list of *Important Terms to Remember.* A page reference is given for each of the important terms, and a glossary is provided at the end of the book. The chapters also include a *Summary* of key points presented in a numbered list. The *Questions and Activities* that end each chapter are meant to stimulate independent thought and study and critical thinking. Some of the activities are appropriate for group discussions, projects, or field trips; others lend themselves to term-paper or essay assignments. The *Appendices* provided at the end of the book contain useful reference information for students (and instructors) on units and conversions, the chemical elements, and the geologic time scale.

Supplements

A full range of supplementary material to accompany *Environmental Geology* is available to assist both instructors and students. The ***Study Guide,*** prepared by Barbara Murck, provides an inexpensive way for students to derive full benefit from this textbook. It includes chapter summaries; brief discussions of the most important terms and key points in each chapter; study pointers and guidelines; and practice questions to help students review and apply concepts and to prepare for examinations.

The ***Instructor's Manual and Test Bank*** was also prepared by Barbara Murck. It includes chapter synopses and lecture lead-ins; sample syllabi and options for course organization; suggestions for further reading; a full description of supplementary materials; and additional written, audiovisual, and computer resources. The *Test Bank* is also available in computerized format for both IBM and Macintosh applications. The computerized format features an easy-to-use test-generating program that allows instructors to select test questions from the printed test bank, print the completed tests for use in the classroom, and save the tests for later use or modification.

Environmental Science Activities, by Dorothy B. Rosenthal, provides forty-six interdisciplinary student activities that require a hands-on approach and inspire critical thinking. Through the use of readily available materials, this supplement allows students to investigate, analyze, and appreciate the world around them in a class or on their own.

The ***Wiley Geology Transparency Set*** includes 150 full-color textbook illustrations, resized and edited for maximum effectiveness in large lecture halls. The ***Wiley Geology Slide Set*** comprises the 150 images provided as transparency acetates in 35-mm slide form. The ***Environmental Geology Overhead Transparency Set*** consists of full-color line drawings from this book, resized for maximum visibility in large lecture halls. These images are also available as 35-mm slides in the ***Environmental Geology Slide Set.***

The ***Wiley Geosciences CD–ROM*** contains many images from *Environmental Geology,* as well as from *Dynamic Earth* and *The Blue Planet,* both by Brian J. Skinner and Stephen C. Porter. Available for Macintosh or IBM and compatible computers, the CD–ROM includes an image manager that allows instructors to customize their presentations and eight animations that illustrate key geological concepts.

ACKNOWLEDGMENTS

It has been a pleasure to work with the talented, efficient, and ever patient professionals at John Wiley & Sons and the freelance experts associated with them. Thanks to everyone involved with this project; you deserve a gold medal (and a vacation) for consistent good humor in dealing with chronically truant authors.

The concept of an environmental geology text originated with Barry Harmon, then Earth sciences editor, and publisher Kaye Pace and continued under the editorial guidance of Chris Rogers. Developmental editor Rachel Nelson prodded, advised, commiserated, and otherwise guided the project to a successful conclusion. Others who contributed their considerable talents to the success of the project include (in no particular order): Bonnie Cabot, who oversaw an extremely tight production schedule; Stella Kupferberg, Alexandra Truitt, Kim Khatchatourian, and Michelle Orlans, who found just the right photographs for the text; Anna Melhorn, who managed (or juggled) the illustration program; Karin Kincheloe, the designer of the book; John Woolsey and his staff, who exhibited amazing skill at translating crude sketches into accurate full-color representations; Eric Stano, who coordinated the supplementary materials; Carolyn Smith, whose ruthless green pen greatly improved the readability of the text; Catherine Faduska, who developed the marketing program for the book; Diane Kraut, permissions editor; and Beth Brooks and Pui Szeto, who provided editorial assistance. Special mention is due to Judith Peatross, who contributed valuable suggestions on organization, pedagogy, and scientific content throughout the early stages of manuscript development. Judith also wrote a number of the *Human Perspective* and *Focus On . . .* boxes; thank you for these contributions.

Geology in general—and environmental geology in particular—is an interdisciplinary endeavor, one that encompasses many areas of expertise. For this reason we sought the input of colleagues who collectively represent a wide range of experience in all aspects of research and teaching related to environmental geology. The careful reading and extensive commentary by these colleagues improved the book immeasurably. Their thoughtful suggestions touched on every aspect of the volume, from the overall organization to the tiniest details. They helped keep us up-to-date in a science that is constantly changing. More importantly, through their comments the reviewers made available to us their many years of collective experience in conveying this material to beginning students of geology and environmental geology. Thank you to those who assisted us by reviewing all or part of the manuscript. They are:

REVIEWERS

Yemane Asmerom
University of New Mexico

James Bell
Linn Benton Community College

Jane Boger
State University of New York at Geneseo

Phillip Boger
State University of New York at Geneseo

James Bugh
State University of New York at Cortland

John Busby
Hardin-Simmons University

Ellen Cowan
Appalachian State University

Stanley Dart
University of Nebraska at Kearney

Pascal de Caprariis
Indiana University–Purdue University

Jeremy Fein
McGill University

Robert Furlong
Wayne State University

Malcolm Hill
Northeastern University

Jerry Horne
San Bernadino Valley College

Larry Kodosky
Oakland Community College

Joan Licari
Cerritos College

Barbara Manner
Duquesne University

George Meyer
College of the Desert

Kevin Mickus
Southwest Missouri State University

Michael Nielson
University of Alabama at Birmingham

Nathaniel Ostrom
Michigan State University

Donald Palmer
Kent State University

Eugene Perry
Northern Illinois University

Frank Revetta
Potsdam College

Glenn Roquemore
Irvine Valley Community College

Laura Sanders
Northern Illinois University

Catherine Shrady
St. Lawrence University

Frank Simpson
University of Windsor

Lonnie Thompson
Ohio State University

John Vitek
Oklahoma State University

Daniel Zarin
University of Pennsylvania

BRIEF TABLE OF CONTENTS

CONTENTS

ENVIRONMENTAL GEOLOGY

OUR PLACE IN THE ENVIRONMENT

Geology controls our lives.
• Jack Oliver

*E*gypt has always been dominated by its geology. Upper Egypt consists of a long valley carved by the Nile River into the rocky desert plateau of the eastern Sahara. Lower Egypt is a vast delta built outward into the Mediterranean Sea by the deposition of sediment from the Nile. Ancient Egyptians called the Nile Valley *Kemit,* "the Black Land," referring to the dark, fertile soil along the banks of the river. The deserts bordering the valley were known as "the Red Land," also named for their distinctive soil. This region was harsh and forbidding, yet it was the source of the great mineral wealth that made Egypt a powerful center of civilization.

The role of geology in Egyptian civilization is reflected in mythology, especially in the figure of Osiris. Osiris was killed by his jealous brother Set, who cut his body into 14 pieces and threw them into the Nile. Isis, the wife of Osiris, found the pieces and, with the help of the gods, restored Osiris to life. The power of resurrection vested in Osiris was associated with the fertilizing effects of the annual flooding of the Nile, which watered and enriched the valley. Each year for a period of about 100 days the river

swelled as if by magic, since Egypt experienced no great rains. The flooding left behind deposits rich in potash and soft silts that were easily tilled. These rich muds formed a natural fertilizer that, when warmed by the spring sun, allowed crops to flourish. Yet as soon as one crop was harvested, the nutrients in the sediment were essentially exhausted. Only the next flooding of the Nile would render the field suitable for cultivation again.

The Nile was also a destructive force for the ancient Egyptians. Only through the coordination of human efforts could its power be brought under control. The success of Egyptian agriculture required not only local control of the river but a united effort throughout the two lands. Regions were divided into political districts, and taxes were based on the height of flood waters. The river thus became the impetus for agricultural innovation, technologic advancement, and the development of governmental and legal systems.

For modern Egyptians the Nile remains both a provider and a destroyer. The heritage of technologic control over geologic processes survives in the form of the Aswân High Dam, which generates more than half the electricity produced in Egypt. The construction of the dam and the creation of Lake Nasser added significantly to the amount of land fit for cultivation. At the same time, however, the dam has interfered with the distribution of sediment to agricultural areas downstream, forcing them to become dependent on chemical fertilizers. The lack of nutrient-laden sediments reaching the Mediterranean has affected the offshore fishing industry. Other unexpected effects include salinization of irrigated areas, erosion of sediment-starved coastlines, saltwater intrusion of coastal freshwater supplies, and the spread of water-borne diseases such as schistosomiasis. The Egyptian people have exerted a massive amount of effort to control the force of the Nile and to draw resources from it, but these activities have had some unanticipated consequences.

◄ —————————————————————

Egypt and the River Nile viewed from space. This landsat image was made in the infrared so vegetation shows up red. The red ribbon is the floodplain of the Nile and it has the color it does because it is farmed intensely. The river is the thin, wavy blue line inside the red ribbon. The direction of river flow is northward, from the bottom of the page to the top; the triangular region at the top is the delta, where the river empties into the Mediterranean Sea. Cairo is the blue-grey spot at the apex of the delta. The Aswân Dam and Lake Nasser are visible at the bottom of the image.

WHAT IS ENVIRONMENTAL GEOLOGY?

This book is an exploration of the planet Earth. More precisely, it explores the human-planet relationship—how Earth processes influence human lives on a daily basis, and how human actions, in turn, alter the functioning of Earth systems.

Geologic processes affect every inhabitant of the Earth every day. Some of these processes, such as earthquakes, landslides, and floods, are obvious. Others are subtle. They include the role of mountains in shaping climatic zones; the influence of volcanism on the chemical evolution of the atmosphere; the formation of rich mineral deposits; and the contribution of floodwaters to the creation of fertile agricultural soils, as in the example of the Nile Valley (Fig. In.1).

The human residents of this planet are also fundamentally dependent on the material resources of the Earth, from metals and minerals to water, air, soil, and even energy. In extracting these resources we become part of the *geologic cycle,* in that we move materials, create wastes, and alter natural biogeochemical cycles. The changes we make in the natural systems around us can cause changes in our surroundings, eventually affecting our quality of life and our health, as well as those of other species.

The study of all these things—the functioning of Earth systems and how they affect and are affected by human activities—is the domain of **environmental geology.** Let's briefly examine the components of the term "environmental geology" in order to clarify its meaning.

Geology: The Study of the Earth

Geology is the scientific study of the Earth. The word comes from two Greek roots: *geo-,* meaning "of the Earth," and *-logis,* meaning "study" or "science." Scientists who study the Earth are called **geologists.** The science of geology encompasses the study of the planet, its formation, its internal structure, its materials and their properties, its chemical and physical processes, and its history—physical, chemical, and biologic.

Traditionally, geology has been divided into two broad topic areas with related but differing aims. **Physical geol-**

▲ F I G U R E In.1
Water and river-born silt make the Nile valley bloom. Looking from the arid desert cliffs near Luxor, on the west bank of the Nile, across the fertile valley. The river, which can be seen in the upper third of the photo, flows from right to left.

ogy is concerned with understanding the *processes* that operate at or beneath the surface of the Earth and the *materials* on which those processes operate. The causes of volcanic eruptions, earthquakes, landslides, and floods are processes. Materials include soils, sediments, rocks, air, and seawater. **Historical geology** is concerned with the chronology of *events,* both physical and biologic, that have occurred in the past. It seeks to resolve questions such as when the oceans formed, when dinosaurs first appeared, when the Rocky Mountains rose, and when and where the first trees appeared.

Environmental geology encompasses some of the goals of both historical and physical geology. It emphasizes the understanding of Earth processes and materials, both of which affect our lives in obvious ways. But the human–planet relationship is a dynamic one, characterized by change. The planet itself is ever changing, and humans contribute to such changes on a massive scale. Historical geology not only helps us understand the past but also establishes a context within which to understand the magnitude, scope, and direction of planetary changes.

Environment: Our Surroundings

The United Nations Environment Programme has defined **environment** as the outer biophysical system in which people and other organisms exist. Other definitions abound; many are more specific, identifying individual components of the physical environment, such as water, air, and the solid Earth. In a broad sense, however, the word *environ-ment* can be used to refer to anything, living or nonliving, that surrounds and influences living organisms.

In considering the environment, it is important to note that the world is not made up of separate, isolated components—rock, soil, sediments, water, air, and organisms. Instead, these components have evolved interactively in the course of our planet's 4.6 billion-year history, forming inseparable parts of the Earth's biophysical system. In other words, the distinction between organisms and their surroundings is artificial. What "environment" really represents is the interconnections, the dynamic relationships between organisms and their physical and biologic surroundings.

HUMANS AS AGENTS OF GEOLOGIC CHANGE

Human actions can perturb natural systems, sometimes in an unpredictable manner, as we saw in the example of the Aswân High Dam. Over many millennia humans slowly changed the Earth's natural landscapes as they built villages and cities, converted forests to agricultural land, and dammed and diverted streams (Fig. In.2). Beginning in the nineteenth century, however, ever-greater amounts of mineral and energy resources were needed to fuel the industrial technologies of increasingly populous societies. In particular, the exploitation of fossil fuels helped raise the standard of living for most people well beyond that of their forebears.

◄ F I G U R E In.2
The human capacity to change the landscape is enormous. The Uribante-Caparo hydroelectric dam project, on the Rio Caparo, Venezuela, has created a vast lake and restructured the valley walls.

▲ F I G U R E In.3
An extreme case of environmental deterioration—deforestation in Rondonia, Brazil. The formerly luxuriant rainforest has been cut and burned; accelerated runoff and soil erosion follow. After a few years of cropping or grazing, the little soil that remains loses its natural fertility; all that is left is a degraded landscape of little value.

In spite of the benefits involved, the exploitation of our planet's rich natural resources has not been without cost. In many parts of the world, environmental deterioration is epidemic (Fig. In.3). In addition to scarring and poisoning the Earth's land surface, we have unwittingly polluted the oceans and groundwater, changed the composition of the atmosphere, and caused the extinction of species at a rate unmatched in the past 65 million years. Today natural and human activities are inextricably intertwined, and it is increasingly apparent that *Homo sapiens* has become a major factor—a global factor—in geologic change.

The massive impact of humanity on the Earth's surface is partly a result of consumption-oriented lifestyles and partly due to the sheer numbers in the "human herd." We are only just beginning to realize how immense our collective activities are. Consider the quantity of mineral resources we take from the ground and use to make automobiles, build roads, and produce the myriad other objects we use. On the average, 10 metric tons of material a year are dug up and utilized for every man, woman, and child on the Earth. In 1994 there were about 5.5 billion people on the Earth, which means that the total amount of Earth material dug up that year was 55 billion tons (not including material moved for such purposes as construction or dredging). Compare this enormous figure with the 16.5 billion tons of dissolved and suspended matter carried to the sea each year by the world's rivers and you will get an idea of the magnitude of the geologic impact of human activities. It is also worth noting that even this figure—the 16.5 billion tons of sediment carried to the sea by rivers each year—is estimated to be two to three times larger than it was before humans began altering the Earth's surface through land-clearing activities such as agriculture and mining.

In the pages of this text we will compare natural causes of geologic change to human causes. We will see how natural processes affect and shape our daily lives. And we will see how human activities have brought about unprecedented changes that will present major challenges in coming decades.

FUNDAMENTAL CONCEPTS OF ENVIRONMENTAL GEOLOGY

A number of themes are woven through the fabric of this book. They occur in a variety of contexts because they are fundamental concepts of environmental geology. These concepts are summarized here so that you can watch for them as you read.

Earth Systems

The Earth is unique. A remarkable combination of geologic and astronomic processes has led to the development of a habitable planet. As far as we know, the Earth is unique among planets as a home base for life. That this should be so is a constant source of awe and respect for those who study the workings of the Earth. It also serves as a reminder of how precious are the natural life-support systems on which we depend. It is important that we strive to maintain the integrity of these natural systems.

The Earth is a closed system. This is perhaps the most fundamental theme of the book; its implications will be explored in more detail in Chapter 1. For now, let's just say that this concept implies that the different parts of the Earth's biophysical system are intricately interrelated and dynamically balanced. If you make a change in one part of the system, its impacts are likely to be felt elsewhere in the system. As a science, geology is moving toward the study of the Earth as a unified system rather than focusing on separate parts of the system.

Materials and energy tend to cycle from one reservoir to another. Within the Earth system, materials and energy are constantly moving from one reservoir to another. For example, water may reside in the atmospheric reservoir, in the oceans, in soil moisture, as groundwater, bound up tightly in mineral structures, on the surface in rivers or lakes, or in living organisms. An individual molecule of water may spend 100,000 years in a glacier, 1000 years in an underground reservoir, 7 years in a lake, 10 days in a cloud, or a few hours in an animal's body. Water is transferred from one of these reservoirs to another via any of innumerable pathways (Fig. In.4). In many respects, environmental geology is the study of these pathways and reservoirs: how materials and energy move around within the Earth system, how they interact along the way, and what controls their behavior while they are residing in a given reservoir or moving along a particular pathway.

The physical structure and chemical composition of the Earth affect our lives in many different ways. Although it may not seem evident, the structure and composition of the Earth, inside and out, are important to us in a variety of ways. The internal structure of the Earth plays a significant part in shaping our landscape and causing geologic events that may be hazardous for people. We rely on the materials of the solid Earth as resources. We need to learn as much as possible about how they are formed, how to find them, and how to use them wisely. The materials of the Earth also have distinct physical and chemical properties that can affect their ability to hold and transmit such fluids as water or oil; their capacity to absorb waste or stop its migration; their tendency to flow or to fail; and other important characteristics.

Geologic processes and human beings operate on different time scales. People are newcomers on the geologic scene, to say the least. The history of our species spans less than a half million years. (For comparison, the dinosaurs were

▲ F I G U R E In.4
Movement of water between natural reservoirs. The sun heats the sea and causes water to evaporate; water vapor in the atmosphere condenses to form clouds, rain, or snow. Compacted snow forms the ice in a glacier; the glacier flows slowly down to the sea where huge bergs calve off and slowly melt. This photograph shows icebergs calving from the Hubbard Glacier in Alaska.

around for about 100 million years.) The things that are important to us socially and historically are measured on time scales of years, decades, or centuries (Fig. In.5). Yet the processes that affect our lives on a daily basis operate on time scales ranging from a few seconds (e.g., an earthquake) to a few millennia (e.g., soil formation) to millions, even billions of years (e.g., the formation of mineral deposits and the evolution of the atmosphere).

Different time scales make for difficult management problems, especially since people tend to perceive events in terms of human time scales. For example, in some parts of the world huge agricultural enterprises are based on the pumping of groundwater, which, once depleted, may not be replenished for centuries (Fig. In.6). The key to the sustainable use and effective management of Earth resources is to learn to function within the constraints of the time scales imposed by natural processes.

Hazardous Processes

Hazardous geologic processes have always existed. Earth processes that we term "hazardous"—such as floods, earthquakes, and volcanic eruptions—are natural geologic processes. They have operated, for the most part, since early in Earth history, although in some cases the magnitude or timing of the events may have changed. We are *anthropocentric* (human-centered) in that we define Earth processes as hazardous only when they have a direct negative impact on human interests. Some hazardous processes, such as massive volcanic eruptions, are beyond human influence. However, many processes can be influenced, either positively or negatively, by human activities. In fact, human actions have generated a whole new category of hazardous processes—*technologic hazards*—that encompasses the hazards involved in the use of Earth materials in the built environment.

Risk is characteristic of the human–planet relationship. Geologic processes challenge us ruthlessly and continuously with events that seem, to us, hazardous or even catastrophic. Floods and volcanic eruptions routinely strike the most fertile land; earthquakes level cities; and the inexorable processes of climate and weathering wear away at the solid Earth around us. One of the main goals of environmental geology is to understand such processes so that we

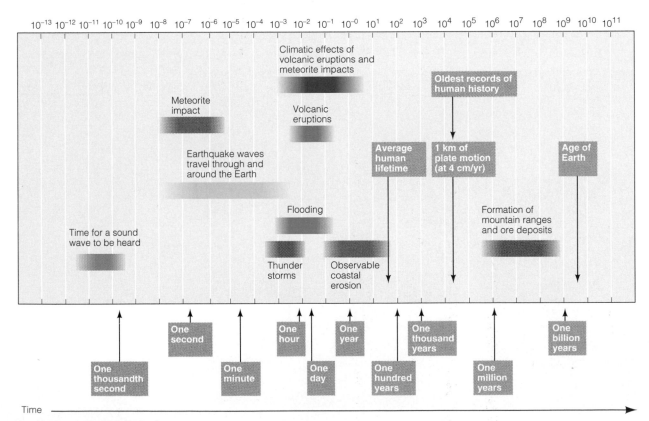

▲ **F I G U R E In.5**
Human beings and geologic processes operate on different time scales. The things that are important to us socially and politically are measured on a scale of days to years. Geologic processes, meanwhile, operate on scales ranging from a few seconds (e.g., an earthquake) to a few millennia (e.g., soil formation) to a few hundred million years (e.g., the formation of an ore deposit or the uplifting of a mountain range).

◀ FIGURE In.6
Potatoes grown by irrigation in the Negev Desert, Israel. When groundwater is pumped and used for irrigation at a faster rate than the underground reservoir is replenished, the water resource will eventually run out and irrigation will no longer be possible.

can predict and control them if possible, or at least mitigate their effects.

Earth Resources

We are fundamentally dependent on Earth resources for the conduct of modern society. One of the main themes of this book is the ultimate dependence of humans on the Earth. We depend on this small planet for all of our material resources—the minerals, rocks, and metals with which we construct our built environment; the energy with which we run it; the soil that sustains agriculture and other plant life; and the air and water that sustain life itself (Fig. In.7). Geologic processes control our lives by determining the abundance, availability, and distribution of these resources.

Earth resources are limited. Another consequence of the fact that the Earth is a closed system is the finite nature of the planet's material resources. Some resources, such as groundwater, are renewable on a human time scale—*if* they are managed properly. Others, such as fossil fuels and min-

◀ FIGURE In.7
Practically everything we do depends directly or indirectly on nonrenewable fossil fuels and mineral resources. This family holiday would be very different without such resources. Their automobile, bicycle, radio, and camper stove are made of metal; the inflatable raft, backpack, beach balls and even the clothes they are wearing are made from chemicals derived from fossil fuel.

eral resources, are nonrenewable on any humanly accessible time scale. Once these resources have been used up, we can't wait around for hundreds of millions of years while new ones are created.

Earth resources can be managed properly. We are not likely to run out of either energy or mineral resources in the near future. However, mineral and energy resources that are affordable and environmentally acceptable may become harder to find. We rarely pay the full cost, including the cost of environmental impacts, for the resources we consume. If we are to continue to use Earth resources and ensure their availability for future generations, we must manage them more wisely. This includes taking into account the real environmental costs of extraction, production, and use, as well as the possibility of depletion or exhaustion of resources.

Waste

Human activities generate waste. In extracting and utilizing Earth resources humans generate leftovers or waste. It is thermodynamically impossible to eliminate all waste, as

▲ **FIGURE In.8**
Because we use a lot of material resources we also generate a lot of "used" resources, the material we call "waste." In this garbage dump in Edmonton, Alberta, the "wastes" range from building and garden materials to appliances, food scraps, and old packing materials. If such materials are not buried properly they can become long-term sources of ground-water pollution.

long as we keep using resources. But we can (and must) minimize our production of waste through more efficient industrial methods and reuse or recycling of materials.

There is no 'away' to throw things to. Another corollary of the concept of the Earth as a closed system is that the planet must eventually resorb all waste materials. Wastes must be handled properly and disposed of in an appropriate geologic environment, or they will cause trouble down the road (Fig. In.8).

Humans have become a geologic force to be reckoned with. Through the amount of material we move each year, through our interference in the functioning of Earth processes, and through our influence on biogeochemical cycles, we have become an enormous and ever-growing agent of geologic change. Some of the changes caused by human actions manifest themselves not just on local or regional scales but on a global scale. It appears likely that some of the changes may be irreversible.

The Human–Planet Relationship

Managing the environment means managing human behavior. It is futile to attempt to manage the environment without managing human behavior, because the two are so closely intertwined. Managing human behavior means, for one thing, dealing with exponential population growth and its environmental implications. It means reexamining our consumption-oriented lifestyles and minimizing the impacts of the "footprint" we are leaving on the Earth's surface. Because some crucial Earth resources, such as the oceans and the atmosphere, are "owned" in common by all the Earth's peoples, managing human behavior also means finding a way to reach an international consensus on the best course of action for the protection and maintenance of such resources.

Restoration and preservation are also part of the human–planet relationship. The human–planet relationship does not have to be an adversarial or a controlling one. Humans can have positive impacts on the environment, as well as negative ones. Many human activities have made the world a nicer place in which to live. No one could deny that building cities and clearing land for farms causes large changes in the environment. But who would argue that a beautiful city such as Paris is not a proud achievement (Fig. In.9)? Think, too, of the abundant food produced by modern agriculture. Our ancestors had a much harder time feeding themselves than we do today. It is also becoming increasingly clear that economic development and care for the environment—once thought to be polar opposites— can be mutually compatible goals. It *is* possible to continue to make use of Earth resources and manipulate our physical surroundings, but to do it in such a way that natural systems can continue to function unimpaired and undepleted.

We still have much to learn about the functioning of Earth systems. We are only beginning to understand the complexi-

▲ FIGURE In.9
Paris, one of the world's most beautiful cities, is built along the River Seine. This photograph is taken looking in a southerly direction across the ancient center of Paris. The Eiffel Tower is visible center left.

ties and interrelationships of such systems as the Earth's climate, oceans, and shifting continents. Sophisticated modeling tools and advanced technologies for detecting and monitoring small changes in the environment have helped our understanding considerably. Still, uncertainty is a fact of life when dealing with systems as intricate as these.

GOALS AND DIRECTIONS

Geology has always been an interdisciplinary science because Earth processes involve interacting biologic, physical, and chemical processes. Yet we are discovering that the interactions are more complex and dynamic than we would have believed as recently as a few decades ago. We now know more about how the chemicals that are essential for life move from reservoir to reservoir in biogeochemical cycles; how the Earth's climate is intimately connected with oceanic and atmospheric circulation and with the distribution of land masses; how volcanoes contribute to processes as diverse as the evolution of the atmosphere and the shift-

ing of continents. We also have a more profound appreciation of the role of humans in geologic change and the need to study the Earth's biophysical system as a whole rather than in individual fragments.

In the future we need to refine and deepen this understanding. We need to understand how to continue to derive practical benefits from Earth resources without jeopardizing their future availability. We need to distinguish between the effects of human-induced change and those of natural change and to be able to predict the impacts of both. Above all, in the years to come we must improve the relevance of scientific knowledge to those who make decisions concerning land use, resource management, and environmental protection. Scientific understanding must be combined with a commitment to take action on behalf of the environment.

IMPORTANT TERMS TO REMEMBER

environment (p. 3)   geology (p. 2)
environmental geology (p. 2)   historical geology (p. 3)
geologist (p. 2)   physical geology (p. 2)

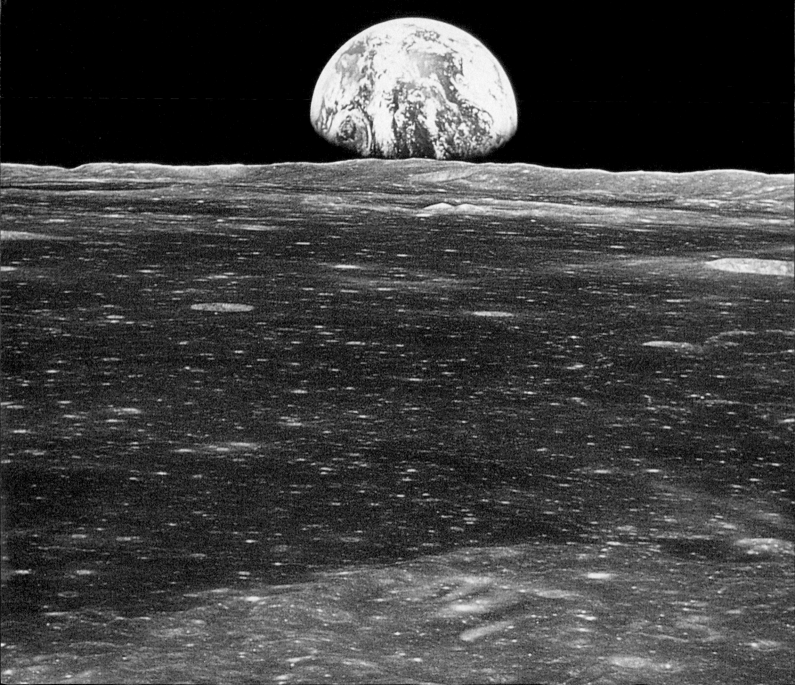

GEOLOGIC FRAMEWORK

★ E·S·S·A·Y ★

THE HOME PLANET

Perhaps the single most powerful image of all times is a photograph of the Earth taken from space (Fig. I.1). We first saw the image in the 1960s, at the dawn of the "space age." There it was, the entire planet in one sweeping view. We could see everything at a glance—the clouds, the oceans, polar ice caps, and continents—all at the same time and in their proper scale. As never before it was possible to see and understand that the Earth really is just a small planetary body—"Spaceship Earth"—in orbit around an ordinary, medium-sized star. This understanding has fueled a global outpouring of concern for the environment. It has been fundamental in shaping societal views about how the resources available on Spaceship Earth should be managed and utilized for the expanding population and at the same time nurtured and protected for generations to come. To fully appreciate Spaceship Earth, to comprehend its limits and vulnerabilities, it is helpful to look first at the Earth's place among its neighbors in space.

THE SOLAR SYSTEM

Studies of the Earth in the context of the **solar system**—the group of objects in orbit around the Sun—reveal a planet that is at once akin to and distinct from other planets in the system. The Earth is one of nine planets in the solar system, which also includes the Sun, at least 61 moons, a vast number of asteroids, millions of comets, and innumerable small fragments of rock and dust called *mete-*

oroids. All the objects in the solar system move through space in smooth, regular orbits, held in place by gravitational attraction. The planets, asteroids, and meteoroids all circle the Sun, whereas the moons circle the planets.

The planets can be separated into two groups on the basis of their physical characteristics, especially density, and

▲ FIGURE I.1
The Earth from space. The blue ocean, the thin, oxygen rich layer of atmosphere, and the green, plant-covered land areas make it clear that the Earth is a unique planet.

The Earth rising over the Moon.

▲ F I G U R E I.2
The planets can be divided into two groups—the terrestrial planets, which are the small, rocky planets closest to the Sun—and the jovian planets, which are the large gaseous bodies distant from the Sun.

their closeness to the Sun (Fig. I.2). The innermost planets—Mercury, Venus, Earth, and Mars—are small, rocky, and relatively dense. Each has an overall density of 3 g/cm³ or more (which is about the density of the average rock). They are similar in size and composition and are called **terrestrial planets** because as a group they resemble *Terra* (the Latin word for the Earth). With the exception of Pluto, the planets farther from the Sun than Mars are much larger than the terrestrial planets yet much less dense. The masses of Jupiter and Saturn, for example, are 317 and 95 times the mass of the Earth, but their densities are only 1.3 and 0.7 g/cm³, respectively. These **jovian planets**—Jupiter, Saturn, Uranus, Neptune, and Pluto—take their name from *Jove,* an alternate designation for the Roman god Jupiter. They all probably have small solid centers that resemble terrestrial planets, but (again with the exception of Pluto) most of their planetary mass is contained in thick atmospheres of hydrogen, helium, and other gases. These atmospheres (which are what we actually see when we observe these planets) are the reason for the low densities of the jovian planets (Fig. I.3).

▲ F I G U R E I.3
Jupiter, the largest planet, is a gas-shrouded giant that conceals its presumably solid interior. This image of Jupiter was taken February 5, 1979 from a distance of 29 million km by the spacecraft *Voyager 2*. The banded pattern is due to turbulence and Jupiter's rotation. A particularly violent storm is visible in the lower left-hand corner.

*T*HE ORIGIN OF THE SOLAR SYSTEM

How did the solar system form? We may never know the precise answer to this question, but we can discern the outlines of the process from evidence obtained by astronomers and from the laws of physics and chemistry.

The process began with space that was not entirely empty because earlier suns had exploded in what as-

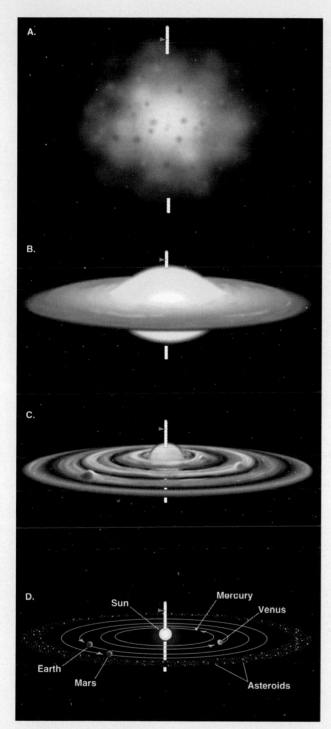

▲ F I G U R E 1.4
Birth of a solar system. A. The gathering of atoms in space created a rotating cloud of dense gas. B. The center of the gas cloud eventually became the Sun; the planets formed by condensation of the outer portions of the gas cloud (C, D).

tronomers call supernovas. The explosions scattered atoms of various elements throughout a huge volume of space. Most of the atoms were hydrogen and helium, but small percentages of all the other chemical elements were also present. It is from these thinly spread atoms that everything in the solar system was eventually constructed. In a sense, the Earth and everything on it is made of star dust. Even though they were thinly spread, the atoms of star dust formed a turbulent, swirling cloud of cosmic gas. Over a very long period the gas thickened as a result of a slow gathering of the atoms in response to mutual gravitational attraction. As the atoms moved closer together, the gas became hotter and denser. Near the center of the cloud of gas, hydrogen and helium atoms eventually became so tightly pressed and so hot that they began to fuse and form heavier elements. At that point, estimated at about 6 billion years ago, the Sun was born.

At some stage the cool outer portions of the cosmic gas cloud became compacted enough to allow solid objects to condense, in the same way that ice condenses from water vapor to form snow (Fig. 1.4). The solid condensates eventually formed the planets, moons, and other solid objects of the solar system.

Condensation of the gas cloud is only the first part of the planetary birth story. Condensation formed innumerable small rocky fragments, but the fragments still had to be joined together to form a planet. This happened as a result of impacts between fragments drawn together by gravitational attraction. The largest masses slowly swept up more and more of the condensed rocky fragments, grew larger, and became the planets. Meteorites, such as the ones in Fig. 1.5, still fall on the Earth, proving that even now

▲ F I G U R E 1.5
Messengers from space carrying some of the history of the earliest days of the solar system. These stony meteorites fell to the Earth at Pueblo de Allende, Mexico, in 1969.

▲ F I G U R E I.6

Meteor Crater, near Flagstaff, Arizona. The crater was created by the impact of a meteorite about 50,000 years ago. It is 1.2 km in diameter and 200-m deep. Note the raised rim and the blanket of broken rock debris thrown out of the crater. Many impacts larger than the Meteor Crater event have occurred during the Earth's long history, and some have caused major disruptions to living creatures.

some ancient rocky fragments still exist in space. Meteorites and the scars of ancient impacts (Fig. I.6) provide evidence of the way the terrestrial planets grew to their present sizes. The growth process—the gathering of more and more bits of solid matter from surrounding space—is called *planetary accretion.* The formation of the Earth, the terrestrial planets, and the other large planetary bodies through the processes of condensation and planetary accretion was essentially complete approximately 4.6 billion years ago.

Brothers and Sisters: A Comparison of the Terrestrial Planets

The nine planets and the other planetary bodies are like siblings; all were born of the same processes that gave rise to the whole solar system family. Some of the siblings are more alike than others. The Earth and its close neighbor Venus, for example, are so much alike in size, density, and overall composition that they are often called "sister" planets.

As a group, the terrestrial planets have many things in common beyond their small size and rocky composition. They have all been hot and, indeed, partially melted at some time early in their histories. All of them appear to have dense metallic iron cores, which probably separated from the rocky outer layers during this period of partial melt. The planets have all experienced volcanic activity, dominated by one type of rock, *basalt,* which is common to all the terrestrial planets. They have all passed through a period of intense cratering and surface modification by meteorite impacts, a process that continues today. Unlike the jovian planets, all the terrestrial planets have lost their envelopes of gaseous material, which were swept away by the solar wind. The terrestrial planets that ended up with atmospheres (Earth, Mars, and Venus) had to evolve their own secondary gaseous envelopes.

In sum, when we look at the solar system we see a group of planets and other objects that are related by birth and by their association with our Sun. Within this system is a smaller group, the terrestrial planets, linked even more closely by their similar geologic histories and planetary characteristics. The Earth is what it is and the environment works the way it does because of all the things that have happened during its long history. The Earth's history is just sufficiently different that planet Earth is habitable but the other terrestrial planets are not.

THE HABITABLE PLANET

Although the terrestrial planets are alike in composition and density, they differ greatly in the compositions of their atmospheres and the presence or absence of water and life. The space image of our blue planet reveals just what makes the Earth unique and different from the other planetary bodies.

Spheres of the Earth

Atmosphere and Hydrosphere

The Earth has an overall blue and white hue because it is surrounded by an **atmosphere** of gases, predominantly nitrogen, oxygen, carbon dioxide, and water vapor. No other planet in the solar system has such an atmosphere. The Earth's atmosphere contains white clouds of condensed water vapor. The clouds form because water evaporates from the **hydrosphere** ("water sphere"), which consists of the oceans, lakes, and streams; underground water; and snow and ice, including glaciers. The hydrosphere is another unique characteristic of the Earth. Planets farther from the Sun are too cold for water to exist as both a liquid and a gas (water vapor), and planets closer to the Sun are too hot. Other planets have hydrospheres, but only the Earth has a hydrosphere consisting of water, ice, and water vapor.

Biosphere and Regolith

Another unique feature of the Earth is the **biosphere** ("life sphere")—the totality of the Earth's living matter. When the Earth is viewed from space, the biosphere is most dramatically revealed by blankets of green plants on some of the land masses (Fig. I.7). The biosphere embraces innumerable living things, large and small, which are grouped into millions of different species. It also includes dead plants and animals that have not yet been completely decomposed.

A fourth reason the Earth is special concerns its solid surface. Regions that are not green because of dense plant cover appear brown and weather-beaten because of *weathering*—the chemical alteration and mechanical breakdown of rock during exposure to the atmosphere, hydrosphere, and biosphere. As a result of weathering, the Earth is covered by an irregular blanket of loose rock debris, or **regolith.** Soils, muds in river valleys, sands in the desert, and all the other friable rock debris are part of the regolith. Other planets and planetary bodies with rocky surfaces have regolith too, but in those cases the regolith has formed primarily as a result of endless meteorite impact cratering. The Earth's regolith is unique because it is formed by complex interactions among physical, chemical, and biologic processes, usually involving water, and because it is teeming with life forms. Most of the world's plants and animals live on or in the regolith or in the hydrosphere.

▲ F I G U R E I.7

The biosphere determines the color of the land surface viewed from space. A composite of numerous satellite images shows densely vegetated regions in green, dry deserts in yellow or brown, and ice-covered regions in white. The variations in color reflect climatic variations and their influence on the biosphere.

Lithosphere

Finally, the Earth differs from all other planets we know of in the unique relationship that exists between the outermost rocky layer, the **lithosphere,** and the hotter, more plastic material in the planet's interior. The Earth is the largest of the terrestrial planets; for this reason, it has cooled off more slowly than the others. Because the Earth's lithosphere is a thin, cold, brittle shell, it has broken up into a series of enormous rocky plates. These plates move around and jostle one another as a result of movements in the hot, mobile material underneath.

Plate Tectonics

The result of the movement of lithospheric plates is a set of processes that we refer to as **plate tectonics.** Plate tectonic activity—that is, the motion and consequent interactions of lithospheric plates—has been a force throughout much of geologic history (at least 2 billion years) and has shaped the landscape. It is responsible for the creation of mountains (Fig. I.8), volcanoes, and deep ocean basins, and has influenced the formation of the atmosphere, the development of climatic zones, and the evolution of life. As we will see in later chapters, plate tectonics has also provided geologists with a concept that enables many years' worth of ob-servations of natural processes to be understood in a unified context.

We know of no other planet where plate tectonics has played, and continues to play, such an important role in forming the environment. We know of no other planet, anywhere, where the temperature permits water to exist near the surface in solid, liquid, and gaseous forms. No other planet that we now know of would have been hospitable to the origin and evolution of life as we know it. There are billions upon billions of suns in the universe, so it is almost inevitable that there are billions of planets too; surely a few of these planets must be Earthlike and therefore may support life. However, if a relatively advanced civilization does exist on a planet somewhere out in space, so far we haven't heard or seen any sign of it. The Earth might well be unique in the universe.

*T*HE RELATIONSHIP BETWEEN LIFE AND THE EARTH

A fundamental principle of environmental geology is that the relationship between life and the Earth is one of mutual influences. Nowhere is this delicately balanced relationship better portrayed than in the story of the atmosphere.

◄ F I G U R E I.8
The Appalachian mountain range in Pennsylvania viewed from space. Layers of rock, once horizontal, were twisted and contorted as a result of a collision between two plates of lithosphere several hundred million years ago. The hills are the eroded roots of a once much grander mountain range.

Life on the Earth

Life originated on the Earth because of this planet's special characteristics, especially the presence of liquid water. In turn, life itself has had a profound influence on the chemical evolution of the Earth's atmosphere. For example, carbonate rocks called *limestones,* which are formed from discarded shells of billions of tiny sea creatures, provide a home, or "sink," for carbon dioxide (CO_2) (Fig. I.9). Since relatively early in the Earth's history, these limestones—which would never have existed if life itself had not first evolved—have kept huge quantities of carbon dioxide bound up at the bottoms of shallow seas where oceans meet continents. If all the carbon dioxide in limestones were released, we would have a carbon dioxide–dominated atmosphere like that of Venus. The concentrations of carbon dioxide in Venus' atmosphere trap so much of the Sun's warmth that the planet's surface temperature is hot enough to melt lead! Carbon dioxide performs the same function in the Earth's atmosphere, but because so much carbon dioxide is bound up in limestone there is only a small amount of it in the atmosphere, and the planet's surface stays just warm enough for life to continue to exist.

▲ FIGURE I.9
The Dolomites, a mountain range in northern Italy, are an example of the huge carbonate masses in which carbon dioxide is locked up in rocks.

The process of *photosynthesis,* whereby green plants utilize the energy of the Sun to combine water and carbon dioxide to make carbohydrates and oxygen, represents another fundamental interaction between life and the Earth. The Earth's early atmosphere was probably much more reducing (i.e., oxygen-poor) than the present atmosphere. We can trace the development of an oxygen-bearing atmosphere, as well as the oxygenation of the hydrosphere, by looking at weathered materials preserved in the regolith of 1 billion years ago and more. These materials are a record of the interaction between the oxygen-bearing atmosphere, the hydrosphere, and the lithosphere. In them we can actually observe the transition from oxygen-poor to oxygen-rich minerals. This transition became possible only after the advent of photosynthetic plants, especially algae.

Expanding Horizons for a Shrinking Earth

We, who belong to the present generation, have inherited the fruits of countless thousands of inquiring minds. We are heirs to a vast legacy of knowledge about the Earth—our environment, our home. Anyone can, if he or she will study this legacy, understand how the Earth works, how the natural activities that shape it are interwoven in complex ways, how the environment of which each of us is a part was created. Anyone can understand the threat that the environment can be changed, distorted, and even shattered by human activities.

Being heirs to a body of knowledge is nothing new. Each generation enjoys that privilege. This generation, however, is not merely another in the long line of generations; it is unique because it is the first to face the reality that humans are now so powerful and so numerous that collectively we rival the other forces of nature that nurture, shape, and change our fragile environment. Therefore, we have a crisis on our hands—one that came to a head so recently and so quickly that earlier generations, confused by long-accustomed ways of doing and thinking, have failed to join battle with it. But we also have something that prior generations lacked; we have a point of view toward the human place in the natural world. We must now use our collective understanding of the Earth to impel this generation, and future ones, to make the hard decisions, to sort out and choose a list of priorities by which the quality of life can be maintained and enhanced, and to accept the discipline that our decisions and priorities will require.

IMPORTANT TERMS TO REMEMBER

atmosphere (p. 15)
biosphere (p. 15)
hydrosphere (p. 15)
jovian planets (p. 12)
lithosphere (p. 16)

plate tectonics (p. 16)
regolith (p. 15)
solar system (p. 11)
terrestrial planets (p. 12)

CHAPTER 1

EARTH SYSTEMS AND CYCLES

We shall not cease from exploration.
And the end of all our exploring
Will be to arrive where we started
And know the place for the first time.

• T.S. Eliot *Four Quartets*

A farmer in Indonesia tends his rice fields. The soil is rich, full of minerals and nutrients. But the slopes are steep, and the farmer must terrace his fields carefully to avoid losing the soil through erosion. Further up the mountainside, near a steaming vent by the shore of a volcanic lake, a worker carries a basket full of yellow sulfur, the day's harvest. The sulfur will be sold in town, to be processed and combined with phosphate to make fertilizer. Downslope, women are busy scooping baskets of volcanic sand from the bottom of a river, to be used in making cement for road construction.

Volcanoes are a fundamental part of life in Indonesia, a chain of more than 400 volcanic islands along the southwestern edge of the Pacific Ocean. Even the earliest human inhabitants of the region were affected by volcanic eruptions as far back as 70,000 years ago; the tools and weapons of prehistoric hunters have been found buried in volcanic ash nearby. Tales of catastrophic eruptions are common in Indonesian folklore, such as the legend of the eruption of Tangkubanprahu. Many years ago, according to the legend, there lived a prince who sought to marry his own mother. But the mother, a beautiful princess, feared the wrath of the gods over such a union and demanded of her son a seemingly impossible task: to create a lake in the mountains nearby and build a suitable nuptial canoe, all before the sun went down that day. The prince set to with a will, damming a river gorge and hewing a canoe from an enormous tree. The wedding took place, and the gods were not pleased. They sent down a holocaust of fire and thunder that overturned the canoe and drowned the participants. To this day, it is said, the upturned hull may be seen

◄ ──────────────────────────

Terraced rice field on the steep volcanic slopes of the island of Bali, Indonesia.

in the elongated shape of Tangkubanprahu, which means "capsized canoe."

Indonesians depend on the mineral-rich soils and other resources provided by the volcanoes. But the impacts of volcanic eruptions on Indonesian history and civilization represent only a small part of the story. The geologic turmoil that characterizes the Pacific Ocean rim causes the volcanoes to erupt. The largest eruptions cause changes in regional and global climate, sometimes lasting for years. The eruptions also form mountains, change the landscape, replenish the soils, and deposit sulfur and volcanic ash. While volcanic activity replenishes the land, the rains come and wash away the minerals, eventually carrying them to the sea. Volcanoes erupt, rain falls, water flows, and the cycle continues. In Indonesia, the interconnections among various parts of the Earth system and their impacts on human activities are very apparent.

THE EARTH SYSTEM

A new approach is taking hold in the Earth sciences. The traditional way to study the Earth has been to focus on separate units—the atmosphere, the oceans, or even a single mountain range—in isolation from the others. In the new approach, the Earth is studied as a whole and viewed as a unified system. In particular, Earth scientists are now focusing on interactions and interrelationships among the various parts of the Earth system.

It is sometimes said that the Earth is a *closed system*. This is not just a catchy saying; it has a specific scientific meaning and is rich in connotations that are particularly applicable to the study of environmental geology. In this chapter we will take a close look at the meaning of the word *system*, at some of the characteristics of the Earth system and its component parts, at the interactions that characterize the system, and at how those interactions combine to produce

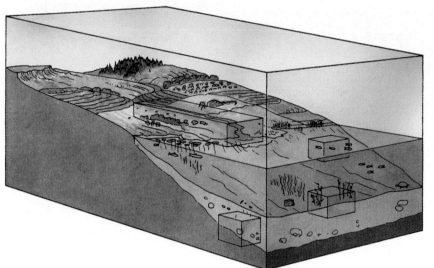

◀ F I G U R E 1.1
The system concept. The river is a system, as is the lake it flows into. Together they form a larger system—the watershed. The small volumes of water and sediment indicated by boxes are examples of smaller systems.

the environment in which humans have evolved. It will be useful to begin our discussion by considering the system concept in greater detail.

The System Concept

The system concept is a helpful way to break down a large, complex problem into smaller, more easily studied pieces. A **system** can be defined as any portion of the universe that can be isolated from the rest of the universe for the purpose of observing changes. By saying that *a system is any portion of the universe,* we mean that the system can be whatever

the observer defines it to be. That's why a system is only a concept; you choose its limits for the convenience of your study. It can be large or small, simple or complex (Fig. 1.1). You could choose to observe the contents of a beaker in a laboratory experiment. Or you might study a lake, a hand sample of rock, an ocean, a volcano, a mountain range, a continent, or even an entire planet. A leaf is a system, but it is also part of a larger system (a tree), which in turn is part of an even larger system (a forest).

The first step in viewing the Earth as a system is to identify the smaller systems that are its component parts. There are four principal systems within the larger Earth system:

◀ F I G U R E 1.2
The four parts of the Earth system that most directly concern environmental geology: lithosphere, biosphere, atmosphere, and hydrosphere.

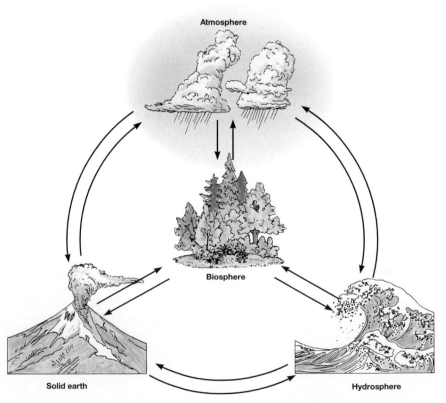

Atmosphere

Biosphere

Solid earth

Hydrosphere

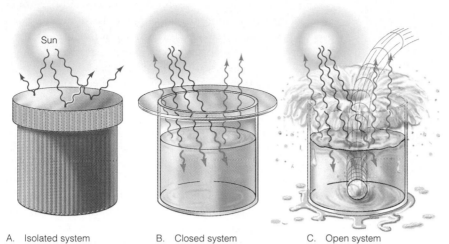

◄ F I G U R E 1.3
The three basic types of systems:
A. An isolated system. B. A closed
system. C. An open system.

A. Isolated system   B. Closed system   C. Open system

the atmosphere, the hydrosphere, the biosphere, and the lithosphere (Fig. 1.2). Each of these can be further divided into smaller, more manageable study units. We can divide the hydrosphere into the oceans, glacier ice, streams, and groundwater, for example.

The fact that a system has been *isolated from the rest of the universe* means that it must have boundaries that set it apart from its surroundings. The nature of those boundaries is one of the most important defining characteristics of a system, leading to three basic kinds of systems, as shown in Fig. 1.3. The simplest type of system to understand is an *isolated system;* in this case the boundaries are such that they prevent the system from exchanging either matter or energy with its surroundings. The concept of an isolated system is easy to understand, but such a system is imaginary because although it is possible to have boundaries that prevent the passage of matter, in the real world it is impossible for any boundary to be so perfectly insulating that energy can neither enter nor escape.

The nearest thing to an isolated system in the real world is a **closed system;** such a system has boundaries that permit the exchange of energy, but not matter, with its surroundings. An example of a closed system is an oven, which allows the material inside to be heated but does not allow any of that material to escape. The third kind of system, an *open system,* is one that can exchange *both* matter and energy across its boundaries. Rain falling on an island is a simple example of an open system: some of the water runs off via streams and groundwater while some evaporates back to the atmosphere (Fig. 1.4).

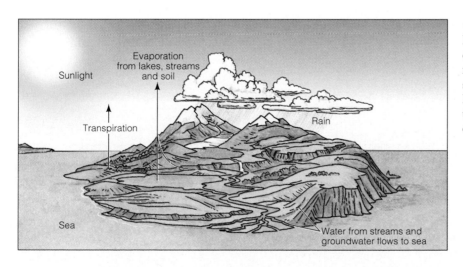

◄ F I G U R E 1.4
Example of an open system. Energy (sunlight) and water (rainfall)
reach an island from external
sources. The energy leaves the island as long wavelength radiation;
the water either evaporates or
drains into the sea.

Box Models

Systems are commonly depicted in the form of box models, as shown in Fig. 1.3. The advantages of doing so are simplicity and convenience. A box model can be used to show the following essential features of a system:

1. The rates at which material and/or energy enter and leave the system.
2. The amount of matter or energy in the system.

The island system mentioned above can be depicted in this way (Fig. 1.5). Water stays on the island for some time before it flows off or is evaporated. The island thus is a **reservoir,** or storage tank, for water in this system. The average length of time water spends in the reservoir is called the **residence time.** The essential features of a box model of the island's water budget are, therefore, the rate at which water falls as rain (*input*), the rate at which water leaves the island (*output*), and the average amount of water on the island at any given time (the size of the reservoir). The size of the reservoir is a function of input and output. If input increases or output decreases, the size of the reservoir must increase to compensate. If input decreases or output increases, the reservoir's size must decrease. The main challenge in using box models to study systems lies in identifying and measuring all the inputs and outputs in order to understand how the system reaches a balance.

Living in a Closed System

As you may have realized, the Earth is a natural closed system—or at least a very close approach to such a system (Fig. 1.6). Energy reaches the Earth in abundance in the form of solar radiation. Energy also leaves the system in the form of longer wavelength infrared radiation. It is not quite correct to say that no matter crosses the boundaries of the Earth system, because we lose a small but steady stream of hydrogen atoms from the upper part of the atmosphere and we gain some extraterrestrial material in the form of meteorites. However, the amount of matter that enters or leaves the Earth system is so minuscule compared with the mass of the system as a whole that for all practical purposes the Earth is a closed system.

The fact that the Earth is a closed system has two important implications for environmental geology:

1. *The amount of matter in a closed system is fixed and finite.* This means that the mineral resources on this planet are all we have and, for the foreseeable future, all we will ever have. Someday it may be possible to visit an asteroid for the purpose of mining nickel and iron; there may even be a mining space station on the Moon or Mars at some time in the future. But for now it is realistic to think of the Earth's resources as being finite and therefore limited. This means that we must treat them with respect and use them wisely and cautiously.

Another consequence of living in a closed system is that material wastes must remain within the confines of the Earth system. As environmentalists are fond of saying, "There is no *away* to throw things to."

2. *When changes are made in one part of a closed system, the results of those changes will eventually affect other parts of the system.* Even though the Earth system is closed, its innumerable smaller parts are open systems. These systems are in a dynamic and sometimes delicate state of balance. When something disturbs one of the smaller systems, the rest also change as they seek to reestablish a state of balance or *equilibrium.* Sometimes an entire chain of events may ensue; for example, a volcanic eruption in Indonesia could throw so much dust into the atmosphere that it could initiate climatic changes leading to floods in South America and droughts in California, eventually affecting the price of grain in west Africa.

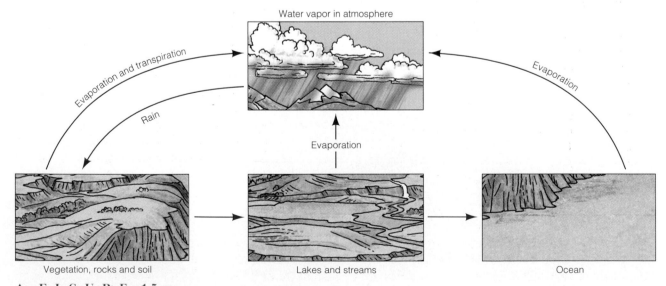

▲ **F I G U R E 1.5**
Depiction of the open system in Figure 1.4 by a box model.

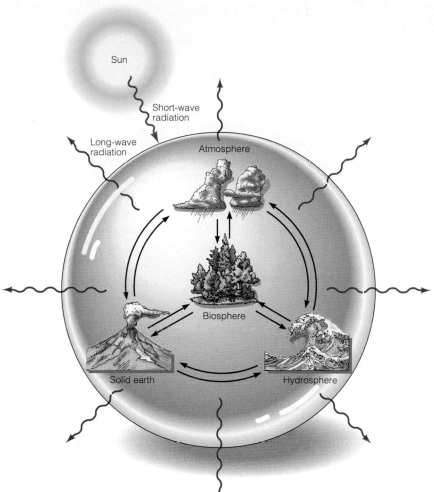

Sun

Short-wave radiation

Long-wave radiation

Atmosphere

Biosphere

Solid earth

Hydrosphere

◀ F I G U R E 1.6
The Earth is a closed system. Energy reaches the Earth from an external source and eventually returns to space as long wavelength radiation. Smaller systems within the Earth, such as the atmosphere, biosphere, hydrosphere, and lithosphere, are open systems.

DYNAMIC INTERACTIONS AMONG SYSTEMS

The causes and effects of disturbances in a complex closed system are very difficult to predict. Consider the anomalously warm ocean tide called El Niño, which occurs every few years off the west coast of South America. El Niños (discussed in greater detail in Chapter 9) are characterized by weakening of trade winds; suppression of upwelling cold ocean currents; worldwide abnormalities in weather and climatic patterns; and widespread incursions of biologic communities into areas where they do not normally occur. These features of El Niños are reasonably well known; what is *not* known is the triggering event. In other words, the interactions among processes in the atmosphere, hydrosphere, and biosphere are so complex, and these subsystems are so closely interrelated, that scientists can't pinpoint exactly what it is that initiates the whole El Niño process. It has even been suggested that changes originating in the lithosphere—in the form of localized heating of ocean water resulting from submarine volcanic activity—may create enough of an imbalance to trigger an El Niño.

From an environmental point of view, the significance of interconnectedness is obvious: When human activities produce changes in one part of the Earth system, their effects—often unanticipated—will eventually be felt elsewhere. When sulfur dioxide is generated by a coal-fired power plant in Ohio, it can combine with moisture in the atmosphere and fall as acid rain in northern Ontario. When pesticides are used in the cotton fields of India, the chemicals can find their way to the waters of the Ganges River and thence to the sea, where some may end up in whale blubber. The chemicals also may be ingested by fish, which in turn may be caught and eaten. In this way, pesticides sometimes end up in the breast milk of mothers halfway around the world from the place where they were applied. Such processes can take a long time to happen, and that is why they have been all too easy to overlook in the past.

Cycling and Recycling

Since material is constantly being transferred from one of the Earth's spheres to another, you may wonder why those systems seem so stable. Why should the composition of the atmosphere be constant? Why doesn't the sea become saltier, or fresher? Why does rock 2 billion years old have

INTERNATIONAL GEOSPHERE–BIOSPHERE PROGRAM

*T*oday a growing number of international, interdisciplinary research programs focus on developing and refining our understanding of Earth processes, the interrelationships between them, and how they are affected by human actions. One such research effort is the International Geosphere–Biosphere Program (IGBP), a far-reaching program operating under the aegis of the International Council of Scientific Unions (ICSU). The goals of the IGBP are to examine in detail some of the unknown links in global Earth cycles, to collect data and facilitate communications among researchers throughout the world, to promote the development and use of new research technologies, and to integrate the findings into a clearer and more directly applicable understanding of global processes and global change.

It's quite a tall order! But the task has been narrowed somewhat by focusing the program's international research efforts on a set of relatively specific questions, such as "How do ocean biogeochemical processes influence and respond to climatic change?" and "How do changes in land use affect the resources of the coastal zone, and will changes in sea level and climate alter coastal ecosystems?" Each of the core projects and participating research agencies focuses their efforts on one or more of these research priority questions. Information systems and data-management methodologies are also being developed to support the research efforts.

The IGBP has obtained funding from a variety of sources for the establishment of 10 regional centers for global change studies in developing countries. The regional centers are intended to facilitate collaboration and research on global change, emphasizing the processes of particular importance in each region. The IGBP is probably the most ambitious research program ever mounted on an international scale, and the list of its supporters is impressive: UNEP (the United Nations Environment Program), UNESCO (the United Nations Educational, Social, and Cultural Organization), the Third World Academy of Sciences, the Commission of the European Communities, the African Biosciences Network, the Organization of American States, and many other organizations and national governments. Most of the involvement to date has been through governmental, international, and scientific agencies. However, the IGBP is particularly anxious to promote the involvement of the private sector in the program because "In the next decades, with possibly great environmental changes, industry will need to be fully apprised of the best available scientific knowledge concerning environmental changes and the research that is planned to increase our understanding and predictive ability. Industry must know, on global as well as regional scales, the anticipated impacts of environmental change."

Source: Adapted from *On Common Ground* by Ranjit Kumar and Barbara Murck, John Wiley & Sons Canada Ltd., Toronto, 1992, p. 75. Quotation from: International Geosphere–Biosphere Program, "A Study of Global Change" (Initial Core Projects), IGBP Report 12 (June 1990) in Overview, pp. 1–3.

the same composition as rock only 2 million years old? The answers to these questions are the same: The Earth's natural processes follow cyclic paths. Materials flow from one system to another, but the systems themselves don't change much because the different parts of the flow paths balance each other: The amounts added equal the amounts removed. This cycling and recycling of materials and dynamic interaction among subsystems has been going on since the Earth first formed and it continues today.

The Cycling of Carbon

Carbon is a familiar example of a material that constantly cycles from one reservoir to another. Carbon can find a home in the biosphere (where it is the fundamental building block in virtually all molecules that make up living creatures); in the lithosphere (existing in rocks such as coal or limestone, which are made from the remains of living creatures); in the hydrosphere (where it is held in solution by a number of complex mechanisms); or in the atmosphere (as part of carbon dioxide gas, which helps keep the planet warm enough for life to continue). The complex set of interactions involving carbon can be depicted using a box model, as shown in Fig. 1.7.

As shown in the model, the atmosphere contains about 750 billion tons of carbon in the form of carbon dioxide. Photosynthesis by plants removes about 120 billion tons of

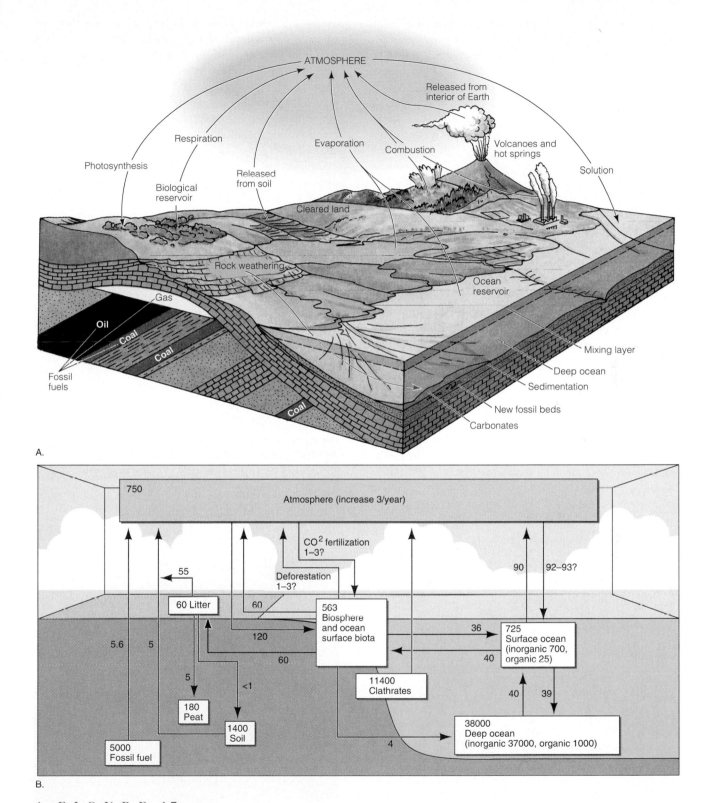

▲ FIGURE 1.7
A. The global carbon cycle. B. Box model for carbon cycle. Each box represents a different reservoir; the numbers indicate the mass of carbon, in billions of tons, in that reservoir. Arrows and numbers between boxes represent exchanges (flows or fluxes) of carbon in billions of tons per year.

carbon per year from the atmosphere, but plant respiration plus the decay of dead plants and animals returns about the same amount. Living plants and animals (mostly plants) contain about 560 billion tons of carbon; plant detritus that has not yet decayed, together with organic matter buried in soils, amounts to 1400 billion tons. Approximately 11,000 billion tons of carbon are trapped in chemical compounds called *clathrates* [complexes of water and methane (CH_4)] found in sediments in the ocean floor and in some onshore regions of permafrost. The oceans contain 38,000 billion tons of carbon, most of it in the form of dissolved carbon dioxide.

By affecting such factors as the growth and death rates of plant and animals and the exchange of carbon dioxide between the oceans and the atmosphere, climatic changes cause the distribution of carbon among the reservoirs to vary. Thus, the atmospheric concentration of carbon dioxide will fluctuate naturally as climatic conditions cause more of it to be drawn into the oceans or the biosphere or, alternatively, as more is released from these sources back into the atmosphere. One of the striking features of past climatic variations is the remarkably close correlation between temperature and atmospheric concentrations of carbon dioxide (Fig. 1.8). What, then, are the potential effects on climate if human interventions cause imbalances in the global carbon cycle? We will consider some possible answers to this question in more detail in Chapter 18, but first let us consider some of the ways in which human activities are changing the natural carbon cycle.

With the advent of the Industrial Revolution 200 years ago, humans began burning massive and ever-increasing quantities of fossil fuels (oil, gas, and coal) for energy, thereby unlocking the vast amounts of carbon stored in these substances and releasing it to the atmosphere. At the same time, a rapidly expanding human population began to consume more and more of the world's forests as the demand for fuel, building materials, and—most of all—agricultural land grew. The burning of fossil fuels adds nearly 22 billion tons of carbon dioxide (about 6 billion tons of carbon) to the atmosphere every year. Deforestation is estimated to add a further 1.6 to 2.7 billion tons of carbon a year (Fig. 1.9). These amounts may seem small in comparison to the large fluxes of the natural carbon cycle. However,

▲ FIGURE 1.9
Deforestation reduces the carbon reservoir of the forest. If the trees are burned or left to decay, their carbon content will be released into the atmosphere. The photograph shows the destruction of a luxuriant rain forest in the Amazon basin near Maraba, Brazil.

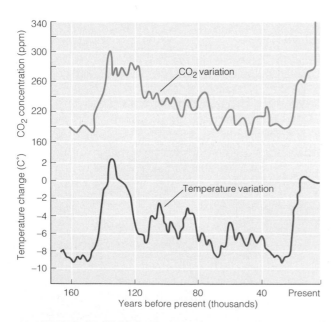

▲ FIGURE 1.8
Variation of temperature over Antarctica and of global atmospheric carbon dioxide concentration during the last 160,000 years. High temperatures are correlated with high CO_2 levels in the atmosphere.

long-term imbalances in fluxes to and from the atmosphere resulting from the natural cycle generally amount to less than 1 billion tons a year. Compared to this small number, the fluxes of carbon resulting from human activities are, in fact, quite large. As a result, humankind is radically altering a delicately balanced natural system.

Cycles in the Earth System

As we have just seen in the discussion of the carbon cycle, it is useful to envisage interactions within the Earth system as a series of interrelated cycles. In the carbon example we discussed the movement of material between reservoirs. The movement of energy can be similarly treated. Both materials and energy can be stored in reservoirs, and the storage times can differ greatly. For example, carbon stored in plants may have a residence time of a few months or years, whereas carbon buried in the rock reservoir may have a residence time of millions of years. This means that a single cycle may include processes that operate on several different time scales.

A few basic cycles can serve to illustrate most of the Earth processes that are of importance in environmental geology. These include the *energy cycle,* the *hydrologic cycle,* the *rock cycle,* and *biogeochemical cycles* (of which the carbon cycle is an example). In the discussions that follow we will briefly consider each of these cycles. It is also possible to extend the concept of cycles to include human-controlled cycles that involve or affect natural processes; examples of such cycles will be introduced at appropriate places throughout this book.

THE ENERGY CYCLE

The **energy cycle** (Fig. 1.10) encompasses the great "engines"—the external and internal energy sources—that drive the Earth system and all its cycles. We can think of the Earth's energy cycle as a "budget": energy may be added to or subtracted from the budget and may be transferred from one storage place to another, but overall the additions and subtractions and transfers must balance each other. If a balance did not exist, the Earth would either heat up or cool down until a balance was reached.

Energy Inputs

The total amount of energy flowing into the Earth's energy budget is more than 174,000 terawatts (or $174,000 \times 10^{12}$ watts). This quantity completely dwarfs the 10 terawatts of energy that humans use per year. There are three main sources from which energy flows into the Earth system.

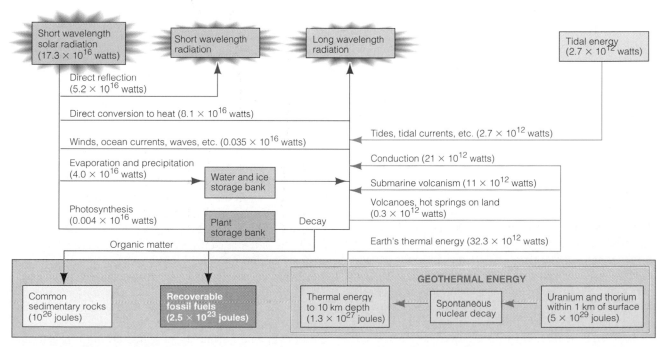

▲ F I G U R E 1.10
The energy cycle. There are three main sources of energy in the cycle: solar radiation, geothermal energy, and tidal energy. Energy is lost from the system through reflection and through degradation and reradiation.

Solar Radiation

Incoming short-wavelength solar radiation overwhelmingly dominates the flow of energy in the Earth's energy budget, accounting for about 99.986 percent of the total. An estimated 174,000 terawatts of solar radiation is intercepted by the Earth. Some of this vast influx powers the winds, rainfall, ocean currents, waves, and other processes in the hydrologic cycle. Some is used for photosynthesis and is temporarily stored in the biosphere in the form of plant and animal life. When plants die and are buried, some of the solar energy is stored in rocks; when we burn coal, oil, or natural gas, we release stored solar energy.

Geothermal Energy

The second most powerful source of energy, at 23 terawatts or 0.013 percent of the total, is **geothermal energy,** the Earth's internal heat energy. Geothermal energy eventually finds its way to the surface of the Earth, primarily via volcanic pathways. It drives the rock cycle (discussed below) and is therefore the source of the energy that uplifts mountains, causes earthquakes and volcanic eruptions, and generally shapes the face of the Earth.

Tidal Energy

The smallest source of energy for the Earth is the kinetic energy of the Earth's rotation. The Moon's gravitational pull lifts a tidal bulge in the ocean; as the Earth spins on its axis, this bulge remains essentially stationary. As the Earth rotates, the tidal bulge runs into the coastlines of continents and islands, causing high tides. The force of the tidal bulge "piling up" against land masses acts as a very slow brake, actually causing the Earth's rate of rotation to decrease slightly. The transfer of tidal energy accounts for approximately 3 terawatts, or 0.002 percent of the total energy budget.

Energy Loss

The Earth loses energy from the cycle in two main ways: reflection, and degradation and reradiation.

Reflection

About 40 percent of incoming solar radiation is simply reflected, unchanged, back into space, by the clouds, the sea, and other surfaces. For any planetary body, the percentage of incoming radiation that is reflected is called the *albedo*.

Each different material has a characteristic reflectivity. For example, ice is more reflectant than rocks or pavement; water is more highly reflectant than vegetation; and forested land reflects light differently than agricultural land. Thus, if large expanses of land are converted from forest to plowed land, or from forest to city, the actual reflectivity of the Earth's surface, and hence its albedo, may be altered.

Any change in albedo will of course have an effect on the Earth's energy budget.

Degradation and Reradiation

The portion of incoming solar energy that is not reflected back into space, along with tidal and geothermal energy, is absorbed by materials at the surface of the Earth, in particular the atmosphere and hydrosphere. This energy undergoes a series of irreversible degradations in which it is transferred from one reservoir to another and converted from one form to another. The energy that is absorbed, utilized, transferred, and degraded eventually ends up as heat, in which form it is reradiated back into space as long-wavelength (infrared) radiation. Weather patterns are a manifestation of energy transfer and degradation.

THE HYDROLOGIC CYCLE

For most people, the most familiar cycle is the **hydrologic cycle,** which describes the fluxes of water between the various reservoirs of the hydrosphere. We are familiar with these fluxes because we experience them as rain, snow, and running streams (Fig. 1.11). Like all the cycles in the Earth system, the hydrologic cycle is composed of pathways, the various processes by which water is cycled around in the outer part of the Earth, and reservoirs, or "storage tanks," where water may be held for varying lengths of time. The hydrologic cycle maintains a mass balance, which means that the total amount of water in the system is fixed and the cycle is in a state of dynamic equilibrium. There are fluctuations on a local scale—sometimes quite large fluctuations, such as those that cause floods in one area and droughts in another—but on a global scale these fluctuations balance each other out.

Pathways

The movement of water in the hydrologic cycle is powered by heat from the Sun, which causes *evaporation* of water from the ocean and land surfaces. The water vapor thus produced enters the atmosphere and moves with the flowing air. Some of the water vapor condenses and falls as *precipitation* (either rain or snow) on the land or ocean.

Rain falling on land may be evaporated directly or it may be intercepted by vegetation, eventually being returned to the atmosphere through their leaves by a process called *transpiration*. Or it may drain off into stream channels, becoming *surface runoff*. Or it may *infiltrate* the soil, eventually percolating down into the ground to become part of the vast reservoir of *groundwater*. Snow may remain on the ground for one or more seasons until it melts and the meltwater flows away. Snow that nourishes glaciers remains locked up much longer, perhaps for thousands of

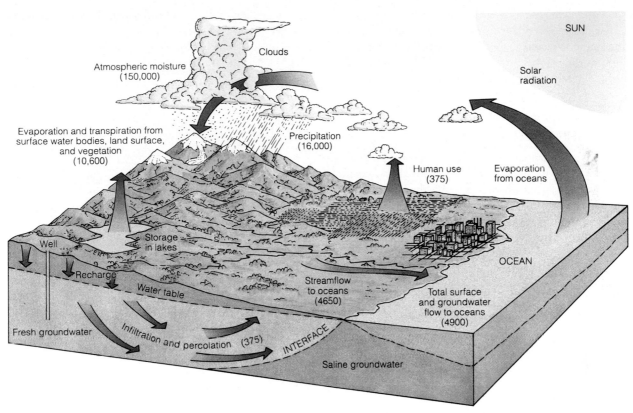

▲ FIGURE 1.11
The hydrologic cycle.

years, but eventually it too melts or evaporates and returns to the oceans.

Reservoirs

The largest reservoir for water in the hydrologic cycle is the ocean, which contains more than 97.5 percent of all the water in the system. This means that most of the water in the cycle is saline, not fresh water—a fact that has important implications for humans because we are so dependent on fresh water as a resource for drinking, agriculture, and industrial uses. Surprisingly, the largest reservoir of fresh water is the permanently frozen polar ice sheets, which contain almost 74 percent of all fresh water. The ice sheets represent a long-term holding facility; water may be stored there for thousands of years before it is recycled. Of the remaining unfrozen fresh water, almost 98.5 percent resides in the next largest reservoir, groundwater. Only a very small fraction of the water passing through the hydrologic cycle resides in the atmosphere or in surface freshwater bodies such as streams and lakes.

There is a correlation between the size of a reservoir and the residence time of water in that reservoir: residence time

in the large-volume reservoirs, such as the oceans and the ice caps, is many thousands of years; in groundwater it is tens to hundreds of years, whereas in the small-volume reservoirs it is short—a few days in the atmosphere, a few weeks in streams and rivers.

BIOGEOCHEMICAL CYCLES

A **biogeochemical cycle** describes the movement of any chemical element or chemical compound among interrelated biologic and geologic systems. This means that biologic processes such as respiration, photosynthesis, and decomposition act alongside and in association with such nonbiologic processes as weathering, soil formation, and sedimentation in the cycling of chemical elements or compounds. It also means that living organisms can be important storage reservoirs for some elements. The carbon cycle is an important biogeochemical cycle; so are the nitrogen, sulfur, and phosphorus cycles, because each of these elements is critical for the maintenance of life.

It is difficult to produce a box model, even a highly simplified one, that accurately describes the biogeochemical behavior of an element as it cycles through the Earth system. These cycles potentially involve a wide variety of reservoirs and processes, and elements often change their chemical form as they move through the cycle. This complexity is illustrated by the nitrogen cycle, which we discuss briefly here.

The Nitrogen Cycle

Amino acids are essential components of all living organisms. They are given the name *amino* because they contain amine groups (NH_2), in which nitrogen is the key element. Nitrogen therefore is essential for all forms of life. The key to understanding the nitrogen cycle is understanding how nitrogen moves among the four major reservoirs of the Earth system—the atmosphere, biosphere, oceans, and soil and sediment. Figure 1.12 shows the reservoirs, the estimated number of grams of nitrogen in each reservoir, and the paths by which nitrogen moves among the reservoirs.

Nitrogen exists in three forms in nature. In the atmosphere it is present in the elemental form (N_2); reduced forms such as ammonia (NH_3) and oxidized forms such as nitrate (NO_3) also exist. Only in the reduced forms can nitrogen participate in biochemical reactions; N_2 cannot be used directly by organisms.

Nitrogen is removed from the atmosphere and/or made accessible to the biosphere in three ways:

1. Solution of N_2 in the ocean.
2. Oxidation of N_2 by lightening discharges to create NO_3, which is rained out of the atmosphere and into the soil and sea. Plants can reduce NO_3 to NH_3, thereby making nitrogen available to the biosphere.

3. Reduction of N_2 to NH_3 through the action of nitrogen-fixing bacteria in the soil or sea. The reduced nitrogen is quickly assimilated by the biosphere.

Once reduced, nitrogen tends to stay reduced, remain in the biosphere, and either be reused by other organisms or oxidized back to N_2 and returned to the atmosphere. The main route by which nitrogen returns to the atmosphere, however, is the reduction of nitrate. This route is kept open by bacteria that use the oxygen in nitrate during metabolism.

THE ROCK CYCLE

The hydrologic and biogeochemical cycles are driven by energy from the Sun. The most important cycle driven by geothermal energy is the **rock cycle.** Before we discuss this complex cycle it is necessary to consider some aspects of the internal structure of the Earth and the phenomenon of plate tectonics.

Heat Transfer in the Earth

Some of the Earth's internal heat makes its way slowly to the surface through the process of *conduction.* However, conduction (which basically works by passing thermal energy from one atom to the next) is a slow way to transfer heat. It is faster and more efficient for an entire packet of hot material to be transported, heat and all, from the hot part of the Earth's interior to the surface. This is essentially what happens when a fluid boils on a stovetop. If you watch a fluid such as pudding or spaghetti sauce as it boils, you will see that it turns over and over as packets of hot material rise from the bottom of the pot to the top. When it reaches the surface, the packet of hot fluid releases its heat and is swept back down to the bottom of the pot. The cycle of motion from bottom to top and back is called a *convection cell,* and this entire mode of heat transfer is called *convection.*

If you resume watching your pot of boiling pudding or sauce, you will see that a thin, hard film or skin forms on top of the fluid, where it is coolest. This film tends to ride around on the convecting fluid underneath. The same is true of the Earth: the lithosphere, or outer 100 km (approximately) of the Earth, is a cold, thin outer layer lying on top of hot, convecting material (Fig. 1.13). The thickness of the lithosphere relative to that of the Earth as a whole is about the same as that of the skin of an apple relative to the whole apple, or the glass sphere of a lightbulb relative to the whole bulb. Heat reaches the bottom of the lithosphere by convection; it passes through the lithosphere by conduction.

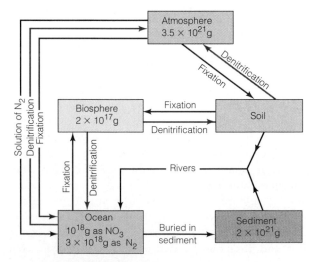

▲ F I G U R E 1.12
The nitrogen cycle.

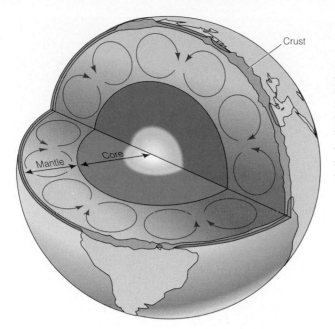

◄ F I G U R E   1.13
A sliced view of the Earth reveals a layered internal structure. The compositional layers, starting from the inside, are the core, the mantle, and the crust. Note that the crust is thicker beneath the continents than under the oceans. The outermost rocky layer, comprising the crust and the very top of the mantle, is called the lithosphere. The lithosphere "rides" around on the hot, convecting mantle beneath.

Think about this for a moment: If an exceedingly thin layer of cold, brittle material is riding and jostling around on top of a hot, mobile, convecting fluid, what can happen to the cold boundary layer? It can break, of course, and that is exactly what has happened to the rocky outer layer of the Earth. The lithosphere has broken into a number of jagged, rocky pieces called *plates,* which range from several hundred to several thousand kilometers in width (Fig. 1.14).

The lithospheric plates are riding around on an underlying layer of hot, ductile, easily deformed material called the *asthenosphere,* or "weak layer."

Some of the lithospheric plates are composed primarily of oceanic crustal material, whereas others are composed primarily of continental material. If it were possible to remove all the water from the ocean and view the dry Earth from a spaceship, we would see that the continents stand,

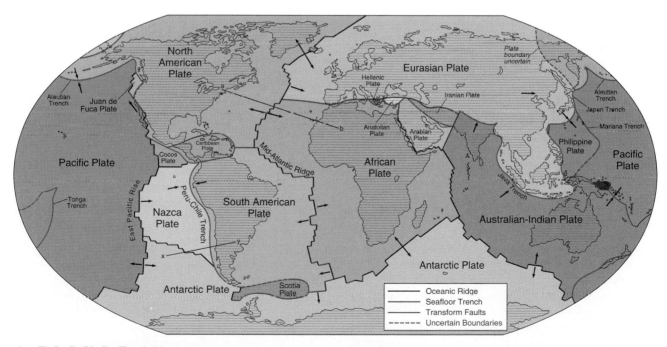

▲ F I G U R E   1.14
Six large plates and several smaller ones cover the Earth's surface and move steadily in the directions shown by the arrows.

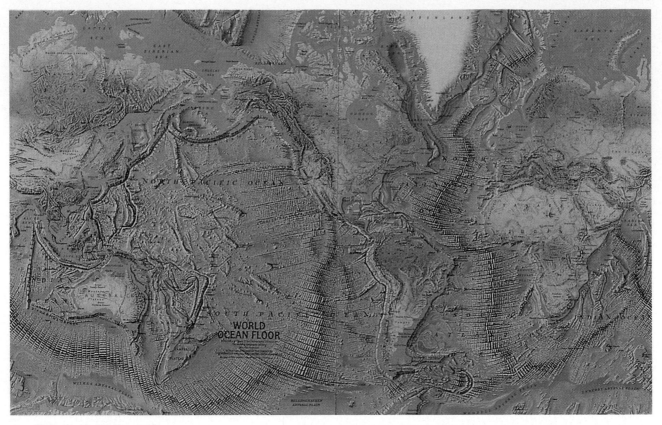

▲ FIGURE 1.15
The topography of the continents and ocean floor. Note the difference in elevation between the continents and the oceans.

on average, about 4.5 km above the floor of the ocean basins (Fig. 1.15). Continental crust is relatively light (density 2.7 g/cm³), whereas oceanic crust is relatively heavy (density close to 3.2 g/cm³). Because the lithosphere is floating on the weak asthenosphere, the plates capped by light continental crust stand high while those capped by heavy oceanic crust sit lower.

Plate Tectonics and the Earth's External Structure

Convection within the Earth is constantly moving the plates of lithosphere and slowly changing the Earth's surface. Such mountains as the Alps or Appalachians that seem changeless to us are only transient wrinkles when viewed from the perspective of geologic time. Mountain ranges grow when fragments of moving lithosphere collide and heave masses of twisted and deformed rock upward; then the ranges are slowly worn away, leaving only the eroded roots of an old mountain range to record the ancient collision (Fig. 1.16). The continents are still slowly moving at rates up to 10 cm a year, sometimes bumping into each other and creating a new mountain range and sometimes splitting apart so that a new ocean basin forms. The Himalaya is a range of geologically young mountains that began to form when the Indian subcontinent collided with Asia about 45 million years ago. The Red Sea is a young ocean that started forming about 30 million years ago

when a split developed between the Arabian Peninsula and Africa as the two land masses began to move apart.

But it is not just the continents that move, it is the entire lithosphere. The continents, the ocean basins, and everything else on the surface of the Earth are moving along like passengers on large rafts; the rafts are huge plates of lithosphere that float on the underlying convecting material. As a result, all the major features on the Earth's surface, whether submerged beneath the sea or exposed on land, arise as either a direct or an indirect result of the motion of lithospheric plates.

Such motions involve complicated events, both seen and unseen, all of which are embraced by the term *tectonics,* derived from the Greek word, *tekton,* which means carpenter or builder. Tectonics is the study of the movement and deformation of the lithosphere. The special branch of tectonics that deals with the processes by which the lithospheric plates move and interact with one another is called **plate tectonics.**

Plate tectonics provides a unifying theory that can be used to explain hundreds of years of independent observations of the processes of rock formation, mountain building, and terrain modification. It also provides an effective framework for our discussion of geologic processes that affect people and form the environment in which we live.

Today the lithosphere is broken into six large plates and numerous small ones (Fig. 1.14), all moving at speeds ranging from 1 to 10 cm a year. As a plate moves, everything on it moves too. If the plate is capped partly by oceanic crust and partly by continental crust, then both the ocean floor and the continent move at the same speed and in the same direction. The term *continental drift* is sometimes used to describe continental movement, but it must be remembered that everything on a plate moves, not just the continents.

Plate Margins

Plates move as individual units, and interactions between plates occur along their edges. The most pronounced manifestations of these interactions are earthquakes and volcanoes, which occur primarily along plate margins. Through studies of these phenomena, particularly earthquakes, geologists have been able to decipher the shapes of the plates. Plate margins are of special interest in environmental geology because so many hazardous natural events tend to occur in these regions.

Plates have three kinds of margins (Fig. 1.17):

1. **Divergent margins,** which are also called *rifting* or *spreading centers* because they are fractures in the lithosphere where two plates move apart. Divergent margins are marked by submarine volcanism and frequent but weak earthquakes.

2. **Convergent margins,** where two plates move toward each other. Along convergent margins, one plate must either sink beneath the other, in which case we refer to

▲ F I G U R E 1.16
The scar of an ancient collision. Layers of rock, once horizontal, were twisted and contorted as a result of a collision between two plates. These eroded roots of an ancient mountain range north of Adelaide, South Australia, were recorded in a Landsat image in September 1983.

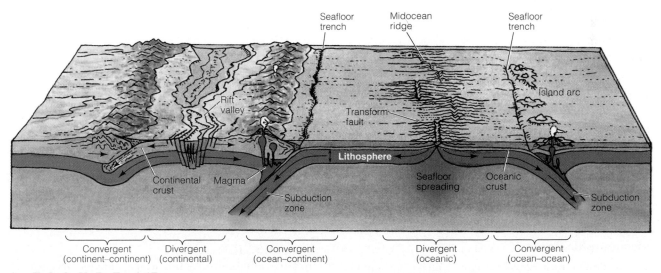

▲ F I G U R E 1.17
The various types of plate margins: divergent (spreading or rifting); convergent (subduction and collision zones); and transform (lateral motion). In an oceanic setting, divergent margins are marked by midocean ridges, like the Mid-Atlantic Ridge. In a continental setting, divergent margins are marked by rift valleys, like the East African rift. Ocean-ocean and ocean-continent convergent margins where subduction occurs are marked by deep trenches and lines of volcanoes. Continental collision zones, as in the Himalaya, are marked by mountain ranges. Transform margins vary in their topographic features; they are sometimes marked by a long linear valley.

the margin as a *subduction zone,* or the two plates must collide, in which case we refer to the margin as a *collision zone.* When oceanic crust is involved in the interaction (i.e., in an ocean–ocean or ocean–continent convergence), a subduction zone will occur. When only crustal material is involved, a collision zone forms. Convergent margins are often marked by explosive volcanism and powerful earthquakes.

3. **Transform fault margins** are fractures in the lithosphere where two plates slide past each other, grinding and abrading their edges as they do so. Transform fault margins are frequent sites of powerful earthquakes but are not associated with volcanism.

The Three Rock Families

An understanding of the rock cycle also requires familiarity with the major kinds of rocks. There are three large "families" of rocks, each defined by the processes that form them.

Igneous Rocks

The first major rock family consists of **igneous rocks** (named from the Latin *igneus,* meaning fire). Igneous rocks are formed by the cooling and consolidation of *magma,* or molten rock. Magma that cools and crystallizes underneath the ground becomes *plutonic rock* (after Pluto, the Greek god of the underworld). If the magma finds its way to the Earth's surface, erupting through volcanic conduits, we refer to the molten rock as *lava,* and when it solidifies it is called *volcanic rock.* Most igneous rock is formed along the spreading edges and convergent margins of plates.

Sedimentary Rocks

Rocks at the surface of the Earth are exposed to water, ice, and air, with which they interact both physically and chemically. The interactions cause rocks to break down into smaller particles, or *weather.* Some products of weathering are soluble in water and are carried away in solution by streams and rivers, but most are loose particles in the regolith. Loose particles that are transported by water, wind, or ice, and then deposited, are called **sediment.** Sediment eventually becomes **sedimentary rock,** a term that refers to any rock formed by chemical precipitation or by cementation of sediment. Sedimentary rocks constitute the second rock family and can be found anywhere on the Earth.

Metamorphic Rocks

The final major rock family is metamorphic rock (from the Greek *meta,* meaning change, and *morphe,* meaning form: hence, change of form). **Metamorphic rocks** are rocks whose original form has been changed as a result of high temperature, high pressure, or both. *Metamorphism*—the process that forms metamorphic rocks from sedimentary or igneous rocks—is analogous to the process that occurs when a potter fires a clay pot in an oven. The mineral grains in the clay undergo a series of chemical reactions as a result of the increased temperature; new compounds form, and the formerly soft clay becomes hard and rigid. Metamorphism occurs most noticeably along plate collision margins.

Rocks in the Crust

The Earth's crust is 95 percent igneous rock or metamorphic rock derived from igneous rock. However, as shown in Fig. 1.18, most of the rock that we actually see at the surface of the Earth is sedimentary. Sediments are products of weathering, and as a result they are draped as a thin veneer over the largely igneous crust. This distribution of rock types is one consequence of the rock cycle and the interactions between internal and external processes in the cycle.

The internal processes that form magma and in turn lead to the formation of igneous rock interact with external processes through weathering and erosion. When rock weathers, the particles form sediment. The sediment is transported and deposited. It may eventually become cemented, usually by substances carried in water moving through the ground, and in this way they are converted into new sedimentary rock. In places where such sedimentary rock forms, it can reach depths at which pressure and heat cause new compounds to form, with the result that the sedimentary rock becomes metamorphic rock. Sometimes metamorphic rock settles so deep that the high temperatures of the Earth's interior melt it and magma is formed. The new magma can then move upward through the crust, where it can cool and form another body of igneous rock. Eventually the new body of igneous rock can be uncovered

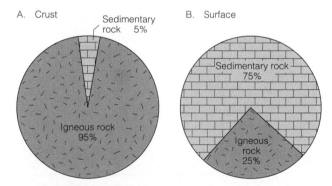

▲ F I G U R E 1.18
Relative amounts of sedimentary and igneous rock. (Metamorphic rocks are considered to be either sedimentary or igneous, depending on their origin.) A. The great bulk of the crust consists of igneous rock (95 percent) but sedimentary rock (5 percent) forms a thin covering at and near the surface. B. The extent of sedimentary rock cropping out at the surface is much larger than that of igneous rock, so 75 percent of all rock seen at the surface is sedimentary , and only 25 percent is igneous.

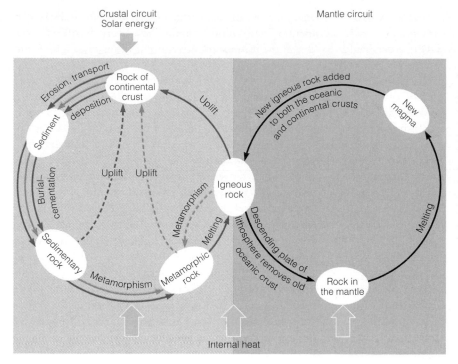

Crustal circuit
Solar energy

Mantle circuit

Internal heat

◀ FIGURE 1.19
The rock cycle—an interplay of
internal and external processes.
Rock material in the continental
crust can follow any of the arrows
from one phase to another. At one
time or another it has followed
all of them. Within the mantle
circuit, magma rises from below
and forms new igneous rock in the
lithosphere. The old lithosphere
descends again to the mantle,
where it is eventually remixed.

and subjected to erosion, the eroded particles start once more on their way to the sea, sediment is laid down, and the cycle is repeated.

Rock Cycles and Circuits

The cycle involving igneous and sedimentary rock has occurred again and again throughout the Earth's long history. It is not the only possible cycle, however; it is just one circuit among many that occur in the continental crust. As the dashed lines in Fig. 1.19 show, other circuits involve bodies of sedimentary rock that are neither metamorphosed nor melted before they are uplifted and eroded. Whether the circuits are long or short, the continental crust is being endlessly recycled as a result of erosion, on the one hand, and plate tectonics, on the other. Because the mass of the continental crust is large, the average time a rock takes to complete the cycle is long. Estimates of the length of the cycle vary, but the average age of all rock in the continental crust seems to be about 650 million years.

The rock cycle of oceanic crust is faster than that of continental crust. When sinking lithosphere carries old oceanic crust back down into the mantle, some of the crust melts and rises to form volcanoes, and the rest is eventually remixed into the mantle. Thus, the most ancient crust of the ocean basins is only about 180 million years old, and the average age of all oceanic crust is only 60 million years.

The magma that rises to form new oceanic crust forms hot igneous rocks that react with seawater. In this reaction, some constituents in the hot rock, such as calcium, are dis-

solved in the seawater and constituents already in the seawater, such as magnesium, are deposited in the igneous rock. Because the magma that forms oceanic crust comes from the mantle, the reactions between hot crust and seawater are one way in which the mantle plays a role in determining the composition of seawater. They are also an important example of the interaction between the rock and hydrologic cycles.

UNIFORMITARIANISM AND EARTH CYCLES

Among the many important questions that have faced geologists who study Earth processes is one that concerns the importance of small, slow changes, such as erosion caused by a single rainstorm, as opposed to large-scale changes, such as earthquakes and floods, that are infrequent but cause dramatic changes in the landscape. During the seventeenth and eighteenth centuries, before geology became the scientific discipline it is today, people believed that all the Earth's features had been produced by a few great catastrophes. Those catastrophes were thought to be so huge that they could not be explained by ordinary processes but must have supernatural causes. This concept came to be called **catastrophism.** The catastrophes were thought to be gigantic and sudden, and they were also thought to have occurred relatively recently and to fit a chronology of catastrophic events recorded in the Bible.

During the late eighteenth century the concept of catastrophism was reexamined, compared with geologic evidence, and found wanting. The person who assembled much of the evidence and proposed a countertheory was James Hutton (1726–1797). Hutton, a Scottish physician and gentleman farmer, was intrigued by what he saw in the environment around him. He observed the slow but steady effects of erosion: the transport of rock particles by running water and their ultimate deposition in the sea. He reasoned that mountains must slowly but surely be eroded away, that rocks must form from the debris of erosion, and that those rocks in turn must be slowly thrust up to form mountains. Hutton didn't know the source of the energy that caused mountains to be thrust up, but he argued that everything moved slowly along in a repetitive, continuous cycle.

Hutton's ideas evolved into what we now call the principle of **uniformitarianism,** which states that the same external and internal processes that we recognize in action today have been operating throughout the Earth's history. The principle of uniformitarianism provides a first and very significant step toward understanding the Earth's history. We can examine any rock, however old, and compare its characteristics with those of similar rocks that are forming today in a particular environment. We can then infer that the ancient rock very likely formed in the same sort of environment. For example, in many deserts today we can see gigantic sand dunes formed from sand grains transported by the wind. Because of the way they form, the dunes have a distinctive internal structure (Fig. 1.20A). Using the principle of uniformitarianism, we can safely infer that any rock composed of cemented grains of sand and having the same distinctive internal structure as modern dunes (Fig. 1.20B) is the remains of an ancient dune.

Geologists since Hutton's time have explained the Earth's features in a logical manner by using the principle of uniformitarianism. But in so doing they have made an outstanding discovery—the Earth is incredibly old. An enormously long time is needed to erode a mountain range, or for huge quantities of sand and mud to be transported by streams, deposited in the ocean, and cemented into new rocks, and for the new rocks to be deformed and uplifted to form a new mountain. Yet, slow though it is, the cycle of erosion, formation of new rock, uplift, and more erosion has been repeated many times during the Earth's long history.

Uniformitarianism and the Rates of Cycles

During the nineteenth century, geologists tried to estimate the duration of the rock cycle by estimating the thickness of all the sediments that have been laid down through geologic time. They assumed that the principle of uniformitarianism applied to the rates at which processes occur as well as to the processes themselves, and hence that rates of deposition of sediment have always been constant and equal to today's rates. Thus, they thought, it would be a simple calculation to estimate the time needed to produce all the sediments. The results, we now know, were greatly in error. One of the reasons for the error was the assumption of constancy of geologic rates.

The more we learn about the Earth's history and the more accurately we determine the timing of past events through *radiometric dating* (using rates of decay of naturally occurring radioactive atoms to determine the ages of rocks), the clearer it becomes that rock cycle rates have not always been the same. Some rates were once more rapid, others much slower. This means that the relative importance of different geologic processes has probably differed in the past. For example, just because glaciation is an important process today, we cannot assume that it has been equally important throughout geologic time. But we *can* assume that when glaciation did affect the Earth in geologically remote times, the processes and effects were the same as those we observe in glaciated regions today.

Uniformitarianism says that "the present is the key to the past"—that we can study present Earth processes in order to understand the processes that have shaped our environment in the past. Now we are discovering that the reverse is also true—the past holds important keys for understanding the present. For example, scientists are documenting changes in the chemical composition of the atmosphere that may signal major global climatic changes. But the Earth's climate system is highly complex, with cyclical variations that are as yet poorly understood. How can we be sure that we really understand the magnitude and significance of the changes we are witnessing? Studies of past climatic changes, in the scientific field of study known as *paleoclimatology,* are providing much-needed clues and a geologic baseline against which we can assess the significance of present changes.

Neo-Catastrophism

Uniformitarianism is a powerful principle, especially when it is constrained by radiometric dating and other geologic techniques, but should we abandon catastrophism as a totally incorrect hypothesis? Recent discoveries suggest not. For example, there is a growing body of evidence that at least once, and perhaps several times, the Earth has been struck by a meteorite large enough to wipe out many life forms. The impact of a large meteorite may have been responsible for the extinction of the dinosaurs 66 million years ago. (The effects of such an event are discussed more fully in Chapter 10.) Even more dramatic mass extinctions have occurred at other times in the past. The geologic record indicates that about 245 million years ago almost 90 percent of all living plants and creatures became extinct. We still have not discovered the cause of that occurrence, but we can be sure that it was a catastrophic disaster of some sort.

Such infrequent massive events fall somewhere between uniformitarianism and catastrophism. When we view the Earth's history as a series of such repeated but sporadic

A.

◀ F I G U R E 1.20
The internal structure of sand dunes, ancient and modern. A. A distinctive pattern of wind-deposited sand grains can be seen in a hole dug in this dune near Yuma, Arizona. B. The same distinctive pattern in rocks in Zion National Park, Utah, lets us infer that these rocks were once sand dunes too.

B.

events, there is no evidence to suggest that similar events will not occur in the future. Nor is there any evidence to suggest when another such event might occur. This new theory of catastrophism—based on the geologic record rather than on the Biblical record—is sometimes called *neo-catastrophism*. It recognizes uniformitarianism as the guiding principle in understanding Earth processes while acknowledging the role of infrequent catastrophic events in generating massive, far-reaching environmental changes on a very short time scale.

A fascinating but frightening possibility is that a catastrophe of a different kind may already be happening. It has been suggested that the collective activities of humans may be changing the Earth so rapidly, and so massively, that they may cause a catastrophe similar in magnitude to some of the major ones in the geologic record. This hypothesis—whether or not it proves true—emphasizes an important fact, and a major theme throughout the remainder of this book: Geology and the welfare of the human race are indissolubly linked.

SUMMARY

1. The concept of a system—a portion of the universe that can be isolated from the rest for the purpose of observing changes—can be a useful way to approach the study of complex problems. The first step in studying the Earth as a system is to identify the smaller systems that are its component parts: the atmosphere, hydrosphere, biosphere, and lithosphere.

2. Box models are often used to portray the rates at which material and/or energy enter or leave a system; the amount of matter or energy in the system; the various storage reservoirs for material and/or energy within the system; and the pathways by which matter and/or energy are transferred from one part of the system to another.

3. The Earth is a natural example of a closed system. This has two important implications for environmental geology: (1) The amount of matter in the system is fixed and finite. (2) When changes are made in one part of the system, the results will eventually affect other parts of the system.

4. Even though material is constantly being transferred from one of the Earth's systems to another, the systems themselves don't tend to change much because the different parts of the flow paths tend to balance one another. This is shown by the cycling of carbon, which is constantly transferred among the biosphere, hydrosphere, atmosphere, and lithosphere. The natural carbon cycle is balanced, but human activities since the Industrial Revolution have radically altered this state of balance.

5. The Earth cycles that are of most importance in environmental geology are the energy cycle, the hydrologic cycle, biogeochemical cycles, and the rock cycle.

6. The energy cycle encompasses the external and internal energy sources that drive the Earth system and all its cycles. The three sources of inputs into the energy cycle are solar radiation, geothermal energy, and tidal energy. The two sources of loss from the cycle are (1) reflection and (2) degradation and reradiation.

7. The hydrologic cycle describes the fluxes of water between the various reservoirs of the hydrosphere. The processes by which water is cycled around the Earth (such as evaporation and precipitation) are driven by energy from the Sun. The ocean is the largest reservoir in the hydrologic cycle. The polar ice caps are the largest reservoir for fresh water, and groundwater is the largest reservoir for unfrozen fresh water.

8. A biogeochemical cycle describes the movement of any chemical element or compound among interrelated biologic and geologic systems. In biogeochemical cycles, such as the nitrogen cycle, biologic processes (such as respiration, photosynthesis, and decomposition) act alongside and in association with nonbiologic processes (such as weathering, soil formation, and sedimentation).

9. Unlike the hydrologic cycle and biogeochemical cycles, which are driven by energy from the Sun, the rock cycle is driven by the Earth's internal energy—especially by the release of heat through the process of convection.

10. The Earth's lithosphere is a cold, brittle outer layer "floating" on hot, convecting material. As a result, the lithosphere has broken into large, jagged plates that range from several hundred to several thousand kilometers in width.

11. Plate tectonics refers to processes by which the Earth's lithospheric plates move around and interact with one another. Where plates interact, three types of plate margins can form: divergent margins, convergent margins (subduction zones and collision zones), and transform fault margins. Plate margins are of special interest in environmental geology because many hazardous natural events occur in these regions.

12. There are three major rock families: igneous rocks, which form by cooling and consolidation of molten rock (magma); sedimentary rocks, which form by chemical precipitation or by cementation of sediment; and metamorphic rocks, which form when rocks

change as a result of high temperature, high pressure, or both.

13. In the various circuits of the rock cycle, internal processes that form magma interact with external processes through erosion and sedimentation. Rocks do not always follow the same pathway through the rock cycle.

14. The principle of uniformitarianism states that the same external and internal processes that we recognize in action today have been operating throughout the Earth's history. Thus, scientists can observe present Earth processes and draw conclusions about rocks that formed in similar environments long ago. Our understanding of past processes—climatic changes, for example—also can provide a baseline against which we can assess the magnitude and significance of current changes in the Earth system.

IMPORTANT TERMS TO REMEMBER

biogeochemical cycles (p. 29)
catastrophism (p. 35)
closed system (p. 21)
convergent margin (p. 33)
divergent margin (p. 33)
energy cycle (p. 27)
geothermal energy (p. 28)

hydrologic cycle (p. 28)
igneous rock (p. 34)
metamorphic rock (p. 34)
plate tectonics (p. 32)
reservoir (p. 22)
residence time (p. 22)
rock cycle (p. 30)

sediment (p. 34)
sedimentary rock (p. 34)
system (p. 20)
transform fault margin (p. 34)
uniformitarianism (p. 36)

QUESTIONS AND ACTIVITIES

1. The most important biogeochemical cycles are the carbon, nitrogen, phosphorus, sulfur, and oxygen cycles. (Some people consider the cycle of water—the hydrologic cycle—to be a biogeochemical cycle, too, because of the importance of water in supporting life.) Choose one of these cycles to investigate in detail. What are the main reservoirs and pathways in the natural cycle? Where do humans fit into the cycle, and what is the extent to which human actions have affected the natural balance? Try to draw a pre- and postindustrial box model of your biogeochemical cycle to highlight the impacts of human activities.

2. Our planet operates as a closed, dynamic system. In what ways do human actions demonstrate an understanding of this? Can you think of examples of human actions that seem to show ignorance of the implications of living in a closed system? How might we reorganize human activities and structures to better reflect this closed system and the finite nature of most Earth resources? These might be good topics for an essay or a class discussion.

3. Based on your own knowledge, see if you can construct a rough box model for a cycle that interfaces with natural Earth systems but is primarily controlled by human actions. Possibilities include the cycles of mineral resources or fossil fuels (from its extraction to processes, consumption, and waste disposal), or the cycle of water use and wastewater treatment and disposal at a factory.

4. Figure 1.1 uses the example of a river flowing into a lake to illustrate the system concept. How many smaller systems can you think of that are component parts of the river–lake system? (The examples of a small volume of water or sediment are shown in the figure; try to think of as many others as you can.) Now, think big—what are some of the larger systems of which the river–lake system is a component part?

5. What kind of rock underlies your house (igneous, sedimentary, or metamorphic)? What about your school? Do you live near a plate margin? If so, what kind (divergent, convergent, or transform fault)? You can investigate the geology of your area through library research or by contacting the state or provincial geological survey.

6. Find out more about the International Geosphere–Biosphere Program. Are there any researchers at your college or university who participate in the program? You can receive the *IGBP Global Change Newsletter* by writing to the IGBP Secretariat, The Royal Swedish Academy of Sciences, Box 50005, S-104 05, Stockholm, Sweden. Can you find any other examples of scientific research programs that take an interdisciplinary approach to the study of the Earth system?

EARTH STRUCTURE AND MATERIALS

It isn't the size that counts so much as the way things are arranged.

• E. M. Forster, 1910

Sometimes the way a material is put together is its most important quality. This appears to be true for the mineral asbestos. "Asbestos" is not actually a true mineral name; rather, it is an industrial or commercial term applied to a group of minerals with different chemical compositions and crystal structures. A mineral is called *asbestos* (or "asbestiform") if it occurs in the form of fine, flexible, hairlike fibers. The fibrous form gives these minerals strength and flexibility as well as other qualities, such as incombustibility and low thermal conductivity. As a result, asbestos is a convenient material to use in any part of a building or machinery that requires insulation—the roofs and floors of building, brake linings, and many other applications.

Asbestos was widely used for such purposes in the period following the Second World War. Then in the 1960s evidence emerged that some people had suffered devastating health consequences as a result of exposure to asbestos. Incidents of lung cancer, asbestosis, and mesothelioma (a cancer of the lining of the chest or abdomen that is nearly always fatal) were undeniably linked to the inhalation of airborne fibers by miners and others who were exposed to asbestos, principally in occupational settings.

◀ _____

Removal of chrysotile asbestos from a public building. To avoid ingesting or inhaling fibers, the workman is completely covered and is breathing from a contained air supply.

The medical and epidemiologic evidence ignited an intense debate about the extent to which asbestos fibers in the environment should be regulated and exactly how they should be handled. Occupational hazard legislation now protects miners, firefighters, construction workers, and others who are routinely exposed to asbestos in the workplace. By 1973, regulations were passed limiting the spraying of asbestos insulation. In 1982, the United States passed legislation banning the use of asbestos in schools and mandating inspections of school premises (although the legislation does not specify any required corrective action).

One particular type of asbestos—crocidolite, which is infrequently encountered—has been shown to be more carcinogenic than others. Some studies have concluded that the fibers of the chrysotile form of asbestos, which accounts for about 95 percent of the asbestos currently in use, are as much as 10 times less hazardous than those of the crocidolite form. This finding has intensified the debate about how to deal with asbestos in the built environment. If the most commonly used form of asbestos is the safest, is the high cost of removal warranted? Should regulations (and funding) be focused on the most hazardous varieties? Or should all asbestos be removed, no matter what form it is? An additional problem is that the removal itself may contribute to the hazard. Asbestos causes health problems when it is inhaled; removing old, asbestos-bearing construction materials may generate dust with a high concentration of fibers.

The properties of asbestos minerals affect human lives on a daily basis. There are many unresolved questions—geologic, medical, political, and financial questions—about the hazards of asbestos and how they should be managed.

THE COMPOSITION OF THE EARTH

Environmental geologists study the physical structure and chemical composition of the Earth for three main reasons: (1) The structure of the Earth, particularly its internal structure, has a lot to do with shaping our landscape and causing events that may be hazardous for humans. (2) Humans are dependent on the materials of the solid Earth as resources. Because the Earth is a closed system, these resources are limited; we need to learn as much as possible about how they form, how to find them, and how to use them wisely. And (3) the materials of the Earth have distinct physical and chemical properties; we need to understand these materials because they can affect human lives in many different ways. In this chapter, therefore, we explore several aspects of the Earth's composition and structure.

What Is the Earth Made Of?

If you were given the task of figuring out what the Earth is made of, how would you go about it? The Earth is almost 13,000 km in diameter; even in the deepest hole ever drilled (to a depth of more than 12 km in Russia) we have direct access to only the top few kilometers of the Earth's outermost layer. Those few kilometers of material are so variable and inhomogeneous in composition, and some parts are so remote, that it is extraordinarily difficult to document the composition of even this outermost layer.

As it turns out, scientists have been able to compile information from a wide variety of sources—including not only sampling and direct observation but also experimentation and theoretical modeling—and have used that information to build up and refine our understanding of the composition and distribution of materials throughout the Earth. In the first part of this chapter, we examine the internal structure and the different compositional zones within the Earth, and some of the tools and approaches that Earth scientists have used to discover more about them. We then take a look at the Earth's basic building blocks, minerals and rocks, and learn something about the properties and characteristics of different Earth materials.

THE EARTH: INSIDE AND OUT

The Earth is a *differentiated* body. This means that material is not distributed evenly or homogeneously throughout the planet. *Differentiation* is the process whereby planetary objects develop concentric layers or zones that differ in their compositions. It is common to all of the terrestrial planets as well as a number of smaller bodies, such as the Moon. Planetary differentiation occurred early in the history of the solar system, when newly formed bodies like the Earth were very hot and partially or completely molten. The least dense of the melted materials, dominated by the elements silicon, aluminum, sodium, potassium, and oxygen, rose toward the surface. The rocks that form the outermost part of the Earth are still rich in these elements. The densest melted materials, principally molten iron and nickel, sank to the center of the planet. Between these two groups of materials is a vast mass of intermediate-density rocks that are rich in elements such as magnesium, calcium, and titanium. Escaping gases, mainly water vapor, carbon dioxide, methane, and ammonia, gave rise to the Earth's early atmosphere.

The processes of partial melting and differentiation turned the Earth into a layered planet (Fig. 2.1). Like other differentiated bodies, the Earth contains three major *compositional layers* (Fig. 2.2). At the center is the densest layer, the **core.** The core is a spherical mass, composed largely of metallic iron, with lesser amounts of nickel and traces of other elements. The thick shell of intermediate-density, rocky matter that surrounds the core is called the **mantle.** The mantle is less dense than the core. Above the mantle lies the thinnest and least dense layer, the **crust.**

The core and the mantle have nearly constant thicknesses. The crust is far from uniform, though, and may dif-

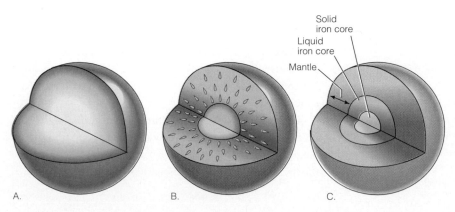

Solid
iron core

Liquid
iron core

Mantle

A.

B.

C.

◀ F I G U R E 2.1
Compositional differentiation of the early Earth. Early in Earth history, when the planet was partially molten, (A) the densest materials sank to the center and the lightest materials rose to the surface (B). This process, known as differentiation, made the Earth into a compositionally layered body (C).

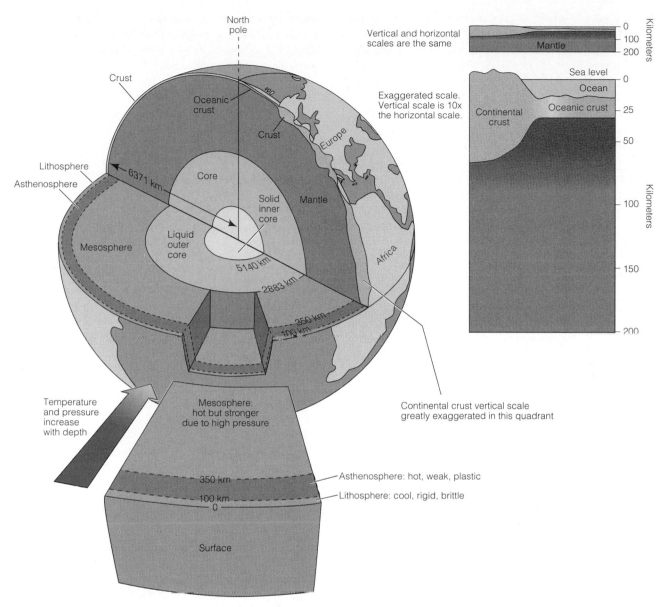

▲ F I G U R E 2.2
The Earth's internal structure, including compositional layers and layers of differing rock properties. Note that boundaries between zones of differing rock strength—such as the lithosphere and the underlying, weaker asthenosphere—do not coincide with compositional boundaries.

fer in thickness from place to place by a factor of nine. Also, as discussed in Chapter 1, there are two different types of crust, each with its own average thickness, composition, and properties. *Oceanic crust* has an average thickness of about 8 km, a density of 3.2 g/cm^3, and a composition rich in calcium, magnesium, and iron. In contrast, *continental crust* (average density 2.7 g/cm^3) ranges in thickness from 30 to 70 km and has a composition richer in light elements, such as silicon, aluminum, sodium, and potassium.

Layers of Differing Physical Properties

It is important to understand the difference between compositional layers and other types of layering within the Earth. Physical properties such as the strength of rock vary with depth in the Earth. These changes are not due to compositional changes but are controlled mainly by the interplay between temperature and pressure. Strange as it may seem, the composition of a rock plays a much smaller role than temperature and pressure in determining its strength.

When a solid is heated, it loses strength. When it is compressed, it gains strength. The result of this interplay between temperature and pressure in the Earth is that the places where physical properties change do not coincide with the compositional boundaries between the crust, mantle, and core (Fig. 2.2).

The most profound change in physical properties is found deep within the Earth's core, where pressures are so great that iron is solid despite its high temperature. The solid center of the Earth is the *inner core*. Surrounding the inner core is a zone where temperature and pressure are so balanced that the iron is molten and exists as a liquid. This is the *outer core*. The difference between the inner and outer cores is not one of composition. Instead, the difference lies in the physical states of the two: one is a solid, the other a liquid.

Differences in temperature and pressure divide the rocky layers above the core into three distinct regions. In the lower part of the mantle, the rock is so highly compressed that it has considerable strength even though the temperature is very high. Thus, a solid region of high temperature but also relatively high strength extends from the core–mantle boundary (at 2883 km depth) to a depth of about 350 km and is called the *mesosphere* ("intermediate," or "middle, sphere") (Fig. 2.2). Above the mesosphere, from 350 to about 100 km below the Earth's surface, is the *asthenosphere* ("weak sphere"), where the balance between temperature and pressure is such that rocks have very little strength. Rock in the asthenosphere is weak and easily deformed, like butter or warm tar. Above the asthenosphere, and corresponding approximately to the outermost 100 km of the Earth, is a region where rocks are cooler, stronger, and more rigid than those in the plastic asthenosphere. This hard outer region, which includes the uppermost mantle and all of the crust, is the *lithosphere* ("rock sphere").

Keep in mind that even though the crust and mantle differ in composition, it is strength, not composition, that differentiates the lithosphere from the asthenosphere. Rocks in the lithosphere are strong and brittle; rocks in the asthenosphere are weaker and can be easily deformed. As we saw in Chapter 1, the lithosphere is broken into a number of plates. The lithospheric plates move about on the hot, convecting asthenosphere and interact with each other along plate boundaries. The rock cycle and plate tectonics are among the consequences of the differences in physical properties between the lithosphere and the asthenosphere.

Layers of Differing Chemical Composition

The composition of the Earth's crust is of particular interest to environmental geologists for a number of reasons. The crust is the only layer to which we have direct access. We live on it and are affected on a daily basis by the properties of crustal materials. We derive mineral and energy resources from the crust, and we draw water from it. Determining

the composition of the crust has been challenging enough, but first let's take a quick look at how scientists have deduced the compositions of the layers to which we have no direct access—the mantle and core.

Composition of the Mantle and Core

Because we cannot see and sample either the core or the mantle, indirect measurements are used to find out about their composition and physical properties. One way to determine composition is to measure how the density of rock changes with increasing depth below the Earth's surface. We can do this by measuring the velocities with which earthquake waves pass through the Earth. The velocities of these waves are a function of the elastic properties and density of rocks. From such measurements, we discover that density increases with depth, but not smoothly. At some depths, abrupt increases in velocity indicate sudden increases in density. From this we infer that the solid Earth does not have a uniform composition but instead consists of distinct layers with different densities. Knowing these densities, we can estimate more precisely what the compositions of the different layers must be.

We can confirm the estimates of the composition of the mantle by comparing them with the composition of materials that have been brought to the surface from great depth by volcanoes. Molten rock that erupts as lava sometimes carries along fragments of rocks, called *xenoliths* (from the Greek for "foreign rock"), which have been torn away from the region deep in the Earth where the volcano originated. We find these xenoliths encased in volcanic rocks that have cooled and solidified at the Earth's surface (Fig. 2.3). However, the minerals in the xenoliths tell us that they once existed at extremely high temperatures and pressures. The composition of most xenoliths is consistent with estimates of mantle composition based on studies of earthquake waves. Taken together, all of the various sources of knowledge about the composition of the mantle suggest that it is composed of chemical compounds made primarily of the elements magnesium, silicon, iron, and oxygen.

Determining the composition of the core presents the greatest difficulty. The Earth's overall density is 5.52 g/cm^3. The density of rocks at the surface is no more than 3.2 g/cm^3. Even rock that is compressed at the bottom of the mesosphere is not as dense as 5.52 g/cm^3. Scientists have therefore deduced that the Earth must have a very dense core. Some of the best evidence for the composition of the core comes from iron meteorites. Such meteorites are believed to be fragments from the core of a small differentiated planetary body that was shattered by a gigantic impact early in the history of the solar system. Scientists presume that this object must have had compositional layers similar to those of the Earth and the other terrestrial planets. Additional evidence comes from the Earth's magnetic field. The presence of a dipolar magnetic field requires the presence of a convecting, electrically conducting fluid in its interior; as in the other terrestrial planets, partially molten metallic

◄ F I G U R E 2.3
Kimberlite, a volcanic rock from the mantle which has carried up mineral fragments and rounded xenoliths of mantle rocks. One of the minerals in the sample is a large diamond (indicated by red arrow). Diamonds can only form under very high-temperature, high-pressure conditions in the mantle.

iron is the most likely material. Studies of the behavior of earthquake waves near the Earth's core confirm that the outer core is, indeed, in a liquid state.

Composition of the Crust

Slight compositional variations probably exist within the mantle and core, but it is difficult to determine very much about them. We can see and sample the crust, however. The samples reveal that the crust's overall composition and density are very different from those of the mantle, and that the boundary between them is distinct.

As we learned in our consideration of the rock cycle (Chapter 1), the composition of the crust is quite varied and the rocks that make up the crust are unevenly distributed. Those that are most common at the surface of the Earth, the sedimentary rocks, are not representative of the most common crustal rocks, the igneous rocks. Even among the igneous rocks, chemical compositions vary widely. If you were a geochemist, how might you go about estimating the overall composition of such a varied assortment of materials?

One early approach was to assume that the overall composition of the crust was essentially the same as the average composition of igneous rocks, given that igneous rocks were predominant in the crust (Fig. 1.18). But this did not produce a very refined estimate of crustal composition. Another approach, used in the 1920s by the Norwegian geochemist V. M. Goldschmidt, was to analyze the composition of fine glacial sediments deposited in lakes by vast ice sheets during the last glaciation. Goldschmidt reasoned that the moving glaciers had scoured fine rock powder from an extremely large area of igneous rock, blending the powder into a homogeneous mixture that was representative of the wide region of crustal material sampled by the glaciers.

Modern estimates of the composition of the crust usually take into account different tectonic environments, such as the deep ocean basins, young mountain chains, continental margins, and stable continental interiors. The relative proportions and average compositions of rock types in each of these regions are estimated, then weighted and added together according to the importance of each environment in the crust as a whole. Table 2.1 gives a modern estimate of the composition of the crust of the Earth.

T A B L E 2.1 • **The Most Abundant Chemical Elements in the Continental Crust**

Element	Percentage by Weight
Oxygen (O)	45.20
Silicon (Si)	27.20
Aluminum (Al)	8.00
Iron (Fe)	5.80
Calcium (Ca)	5.06
Magnesium (Mg)	2.77
Sodium (Na)	2.32
Potassium (K)	1.68
Titanium (Ti)	0.86
Hydrogen (H)	0.14
Manganese (Mn)	0.10
Phosphorus (P)	0.10
All other elements	0.77
Total	100.00

Compared to the Earth as a whole, the rocks that make up the crust of the Earth are rich in relatively light elements (Fig. 2.4). As shown in Table 2.1 and Figs. 2.4 and 2.5, the crust is overwhelmingly dominated by oxygen—over 45 percent by weight, over 60 percent by atomic proportions and over 90 percent by volume. The dominance of oxygen in the crust prompted Goldschmidt to comment that the crust of the Earth should be called the "oxysphere." Other relatively light elements, notably silicon and aluminum, make up the remainder of the material in the crust.

It may seem odd that so much oxygen—an element that we normally associate with the atmosphere—should be

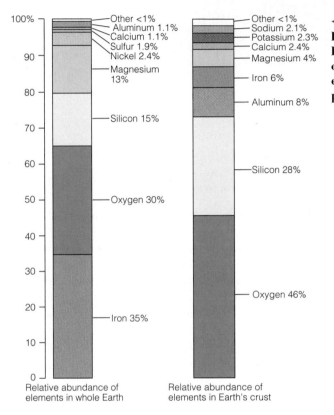

◄ F I G U R E 2.4
Relative abundance by weight of elements in the whole Earth and in the Earth's crust. Differentiation has created a light crust depleted in iron and magnesium and enriched in oxygen, silicon, aluminum, calcium, and potassium.

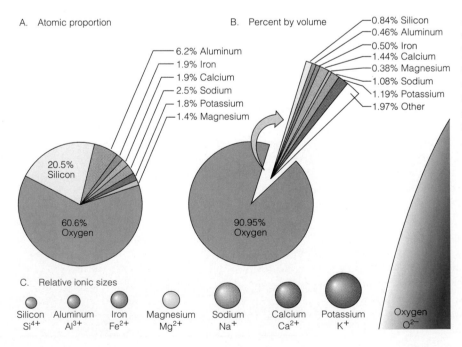

◄ F I G U R E 2.5
The abundances of elements in the Earth's crust. A. Abundances by atomic proportions. B. Percentages of elements by volume; oxygen makes up most of the volume because it has a large atomic radius. C. Relative ionic sizes of the most abundant chemical elements.

found in the solid crust of the Earth. But the oxygen in the crust differs significantly from that in the atmosphere in that crustal oxygen is very tightly linked or *bonded* to other elements in the form of naturally occurring chemical compounds called *minerals*.

MINERALS

The word **mineral** has a specific connotation in geology; it refers to any naturally formed, solid, inorganic chemical substance having a specific composition and a characteristic crystal structure. Minerals are chemical compounds, and as such they are made of chemical *elements*. Let's briefly review these before going on to discuss minerals and their properties in more detail.

Chemical Elements and Compounds

If you were a chemist and were asked to analyze a sample of mineral or rock, your report would list the kinds and amounts of chemical elements present in the sample. An **element** is the most fundamental substance into which matter can be separated by chemical means. For example, table salt (the chemical compound sodium chloride) is not an element because it can be separated into sodium and chlorine. But neither sodium nor chlorine can be further broken down chemically; thus, each is an element.

Every element is identified by a symbol, such as H for hydrogen and Si for silicon. Some symbols, such as that for hydrogen, come from the element's name in English. Other symbols come from other languages. For example, the symbol for iron is Fe, from the Latin *ferrum;* the symbol for copper is Cu, from the Latin *cuprum,* which in turn comes from the Greek *kyprios;* and the symbol for sodium is Na, from the Latin *natrium.* The 92 naturally occurring elements and their symbols are listed in Appendix B. The fundamental building blocks of the elements are discussed in further detail in Box 2.1.

The binding force between elements in a chemical compound is called *bonding.* The properties of compounds are quite different from those of their constituent elements. For example, the elements sodium (Na) and chlorine (Cl) are highly toxic, but the compound sodium chloride (NaCl, the mineral *halite* or table salt) is essential for human health. The character and strength of the bonds in a mineral help determine the specific physical and chemical properties of the mineral, as we will see shortly.

Definition of a Mineral

To be classified as a *mineral,* a substance generally must meet five requirements:

1. It must be *naturally formed.*
2. It must be a *solid.*
3. It must be *inorganic.*
4. It must have a *specific chemical composition.*
5. It must have a *characteristic crystal structure.*

Let's take a closer look at each of these requirements and its implications.

Naturally formed. The requirement that minerals be naturally formed excludes the vast numbers of substances produced in laboratories, such as, for example, synthetic ruby, steel, glass, or plastic. Technically, none of these substances is a mineral.

Solid. All liquids and gases—including naturally occurring ones such as oil and natural gas—are also excluded because minerals are solids. This requirement is based on the physical state of the material, not on its composition. Thus, for example, ice in a glacier is a mineral but water in the ocean and water vapor in the atmosphere are not, even though all three compounds have the same chemical formula (H_2O).

Inorganic. Materials that are derived from living organisms and contain organic compounds, such as leaves, are not minerals. This means that coal, for example, is not a mineral because it is derived from the remains of plant material and contains organic compounds. The teeth and bony parts of dead animals and the shells of sea creatures, which are preserved as fossils, present a slightly trickier case; in principle, such materials are not minerals because they are formed by organic processes. However, over the many millennia during which fossilization occurs, the organic remains are usually replaced by minerals, a process called *mineralization.* If you were to perform a chemical analysis of a dinosaur bone, for example, you would most likely find few or no organic compounds, only inorganic minerals, even though the fine internal structures of the bone may have been preserved by the mineralization process.

Specific chemical composition. The requirement that a mineral must have a specific chemical composition has several implications. Quartz, for example, has the chemical formula SiO_2. To a very limited extent, elements may be added to or substituted for the silicon and oxygen in quartz; however, if the chemical mixture strays too far from this simple formula, it will no longer be quartz. It won't look like quartz; its internal atomic structure will have changed; and it will no longer have the physical properties whereby we identify a mineral as "quartz."

However, the fact that the chemical compositions of minerals have specific limits doesn't necessarily mean that their chemical formulas are all simple ones like the formula for quartz. For example, the chemical formula of the mineral phlogopite, a common form of mica, is $K_2Mg_6Si_6Al_2O_{20}(OH)_4$, and many other minerals have even more complicated formulas. Nevertheless, the same thing is true for phlogopite as for quartz: if the chemical compound strays too far from this specific formula, the material will no longer have the same characteristics and will no longer be identifiable as phlogopite.

ATOMS AND IONS

A piece of a pure element, even a tiny piece no bigger than the head of a pin, consists of a vast number of identical particles called *atoms*. An **atom** is the smallest individual particle that retains all the properties of a given chemical element. Atoms are so tiny that they can be seen only with the most powerful microscopes ever invented, and even then the image is imperfect because individual atoms are only about 10^{-10} m in diameter. As you probably recall from high school chemistry, atoms are built up from *protons* (which have positive electrical charges), *neutrons* (which are electrically neutral), and *electrons* (which have negative electrical charges) (Fig. B1.1). The number of protons in an atom is what gives it its special physical characteristics, which, in turn, identify it as a specific element.

An atom that has an excess positive or negative electrical charge caused by the loss or addition of an electron is called an **ion.** When the charge is positive (meaning that the atom gives up electrons), the ion is called a *cation;* when the charge is negative (meaning that the atom adds electrons), it is an *anion.* The convenient way to indicate ionic charges is to record them as superscripts. For example, Li^+ is a cation (lithium) that has given up an electron, while F^- is an anion (fluorine) that has accepted an electron.

Chemical compounds form when one or more anions combine with one or more cations in a specific ratio. For example, lithium and fluoride combine to form the compound lithium fluoride, which is written LiF to indicate that for every Li atom there is a counterbalancing F ion. Similarly, two cations of H^+ combine with one anion of O^{2-} to make the compound H_2O. The smallest unit that retains all the properties of a compound is called a **molecule.**

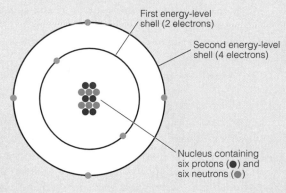

First energy-level
shell (2 electrons)

Second energy-level
shell (4 electrons)

Nucleus containing
six protons (●) and
six neutrons (●)

(●) Proton   (●) Neutron   (●) Electron

◄   F I G U R E   B1.1
Schematic diagram of an atom of carbon. At the center is the nucleus, containing protons and neutrons (six of each in the case of carbon). Six electrons orbit the nucleus.

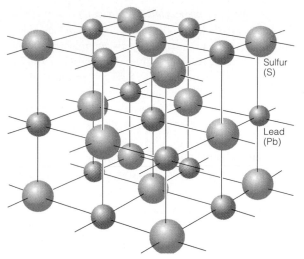

▲ FIGURE 2.6
Arrangement of atoms in galena (PbS), the most common mineral containing lead. The orderly packing arrangement is repeated continuously throughout the crystal, and the atoms are so small that a cube of galena 1 cm on its edge contains 10^{22} atoms each of lead and sulfur.

The requirement of specific limits on the chemical compositions of minerals also serves to exclude materials, like glass, that vary in composition within a range that cannot be expressed by an exact chemical formula.

Characteristic crystal structure. Glass—even when it is naturally formed volcanic glass—is further excluded from being a mineral by the requirement that minerals must have characteristic crystal structures. The atoms in gases, liquids, and glasses are randomly jumbled, but the atoms in many solids are organized in regular, repetitious geometric patterns, as shown in Fig. 2.6. The geometric pattern that atoms assume in a solid is called the *crystal structure,* and solids that have a crystal structure are said to be **crystalline.** Solids that lack crystal structures are called *amorphous* solids (from the Greek for "without form") and are not minerals. All minerals are crystalline, and the crystal structure of a mineral is a unique property of that mineral. All specimens of a given mineral have an identical crystal structure. Extremely powerful, ultra-high-resolution microscopes enable scientists to look into the crystal structures of minerals and actually see the orderly arrangement of atoms in the mineral, which resembles that of eggs lined up in a carton (Fig. 2.7).

Common Minerals

Scientists have identified approximately 3000 minerals. Most occur in the Earth's crust, but a few have been identified in meteorites, and two new ones were discovered in the rocks brought back from the Moon by astronauts. The total number of minerals may seem large, but it is tiny when compared with the astronomically large number of ways in which a chemist can combine naturally occurring elements to form compounds. The reason for the disparity becomes apparent when we consider the relative abundance of the chemical elements. As Table 2.1 shows, only 12 elements—oxygen, silicon, aluminum, iron, calcium, magnesium, sodium, potassium, titanium, hydrogen, manganese, and phosphorus—occur in the crust in amounts greater than 0.1 percent. Together these 12 make up 99.23 percent of the crust's mass. The crust is constructed, therefore, of a limited number of minerals in which one or more of the 12 abundant elements is an essential ingredient. Minerals containing the scarce elements certainly do occur, but only in small amounts and only under special and restricted circumstances.

As you have already learned, two elements—oxygen and silicon—make up more than 70 percent of the crust by weight. Oxygen forms a simple anion, O^{2-} (see Box 2.1); compounds that contain the O^{2-} anion are called *oxides.* Oxygen and silicon together form an exceedingly strong grouping called a silicate anion $(SiO_4)^{4-}$. Minerals that contain the silicate anion are called *silicates,* or **silicate minerals.** These are the most abundant of all naturally occurring inorganic compounds; oxides are the second most abundant group. There are other mineral groups based on different anions—for example, carbonates $(CO_3)^{2-}$ and phosphates $(PO_4)^{3-}$—but they are much less common than the silicates and oxides.

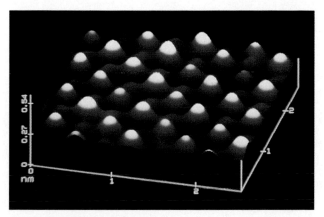

▲ FIGURE 2.7
Atoms can be seen through special kinds of microscopes. Sulfur atoms (large) and lead atoms (small) at the surface of a galena (PbS) crystal are revealed with a scanning-tunneling microscope.

Silicate Minerals

The four oxygen atoms in a silicate anion are tightly bonded to the single silicon cation. As shown in Fig. 2.8A, the four oxygens sit at the corners of a polyhedral shape called a *tetrahedron,* and the small silicon cation sits in the space between the oxygens at the center of the tetrahedron. The structures and properties of silicate minerals are all determined by the way the $(SiO_4)^{4-}$ silicate tetrahedra are packed together in the crystal structure.

Two silica tetrahedra can link together by sharing an oxygen atom at one apex (Fig. 2.8B). In this way, the tetrahedra link to form double tetrahedra, rings, chains, sheets, or three-dimensional frameworks (Fig. 2.9). The linking process is called *polymerization.* Different common cations (such as Ca^{2+}, Al^{3+}, Mg^{2+}, Fe^{2+}, Na^+, and others) are able to fit into the spaces or *interstices* between the linked silica tetrahedra. What finally determines the identity of a silicate mineral is a combination of (1) how the silica tetrahedra are linked together; (2) which cations are present in the interstices; and (3) how the cations are distributed throughout the crystal structure.

Oxides and Other Mineral Groups

As just noted, the oxides represent the second most abundant group of minerals in the crust. Oxides are compounds based on the O^{2-} anion, which is bonded in various

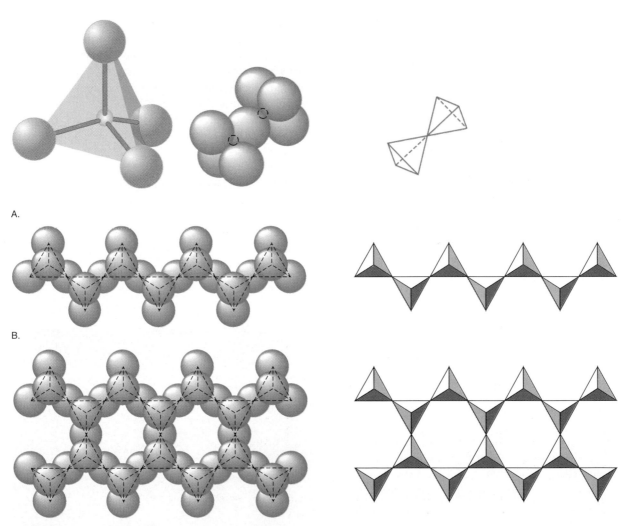

A.

B.

▲ FIGURE 2.8

The silica tetrahedron and silicate mineral structures. A. The structure of the silica tetrahedron. The large oxygen anions are at the four corners, equidistant from a small silicon cation at the center. B. Silica tetrahedra can link together in pairs, chains, sheets, and other geometric structures. The properties of silicate minerals are determined, to a great extent, by the types of linkages between the silica tetrahedra.

	Arrangement of silica tetrahedra	Formula of the complex ions	Typical mineral	
			Name	Composition
Isolated tetrahedra	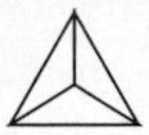	$(SiO_4)^{4-}$	Olivine	$(Mg, Fe)_2SiO_4$
Isolated polymerized groups	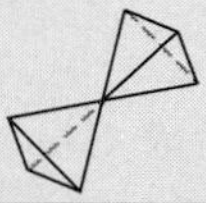	$(Si_2O_7)^{6-}$	Epidote	$Ca_2Fe_2Al_2O[Si_2O_7][SiO_4](OH)$
	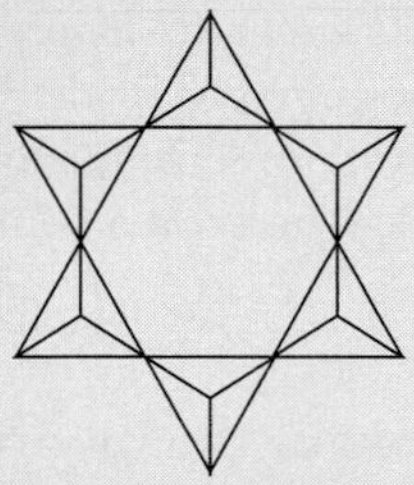	$(Si_6O_{18})^{12-}$	Beryl	$Be_3Al_2Si_6O_{18}$
Continuous chains	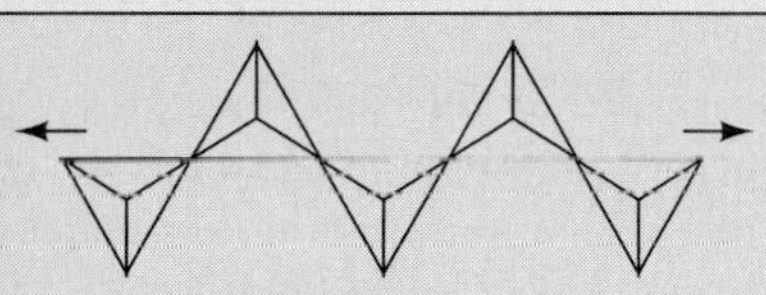	$(SiO_3)_n^{2-}$	Pyroxene	$CaMg(SiO_3)_2$ (Variety; diopside)
	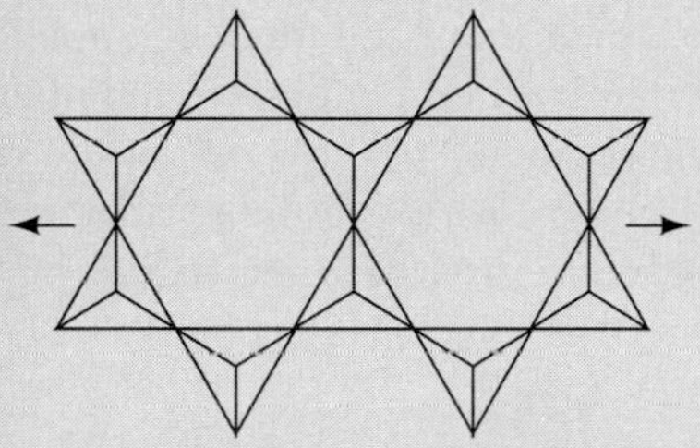	$(Si_4O_{11})_n^{6-}$	Amphibole	$Ca_2Mg_5(Si_4O_{11})_2(OH)_2$ (Variety; tremolite)
Continuous sheets	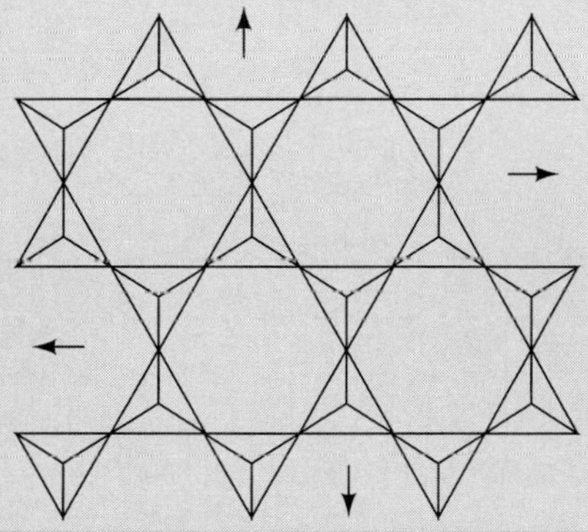	$(Si_4O_{10})_n^{4-}$	Mica	$KAl_2(Si_3Al)O_{10}(OH)_2$ (Variety; muscovite)
Three-dimensional networks	Too complex to be shown by a simple two-dimensional drawing	(SiO_2)	Quartz	SiO_2

▲ FIGURE 2.9
Summary of the ways silica tetrahedra link together to form the common silicate minerals. Linkages other than those shown are known but do not occur in common minerals.

arrangements with the common cations. Because iron is one of the most abundant elements in the crust, the iron oxides magnetite (Fe_3O_4) and hematite (Fe_2O_3) are the two most common oxide minerals. Magnetite takes its name from the Greek word *Magnetis,* meaning "stone of Magnesia," an ancient town in Asia Minor. Magnetis had the power to attract iron particles, and hence is the source of our words *magnet* and *magnetite.* The word hematite is derived from the red color of the mineral when powdered, the Greek word for red blood being *haima.* Magnetite and hematite are the main minerals from which the metal iron is derived. Other important oxides are uraninite (U_3O_8), the main source of uranium; cassiterite (SnO_2), from which tin is derived; and rutile (TiO_2), a widely used constituent of paints.

Another important mineral group is the *sulfides,* all of which contain the simple anion S^{2-} in combination with metal cations. Sulfide minerals are typically dense and have a metallic appearance. The two most common are pyrite (FeS_2) and pyrrhotite (FeS). Many of the sulfides are ore minerals (Fig. 2.10), which means that they are sought and processed for their valuable metal content. For example, most of the world's lead is won from galena (PbS), most of the zinc from sphalerite (ZnS), and most of the copper from chalcopyrite ($CuFeS_2$).

Like the silicates, oxides, and sulfides, each of the other mineral groups is based on combinations of cations with a particular anionic complex. The *carbonates* (Fig. 2.11)

contain the CO_3^{2-} anion, sulfates contain the SO_4^{2-} anion, and *phosphates* are based on the PO_4^{3-} anion. Each of these groups includes minerals that are important for one reason or another. For example, the phosphate mineral apatite [$Ca_5(PO_4)_3(F,OH)$] is the substance from which bones and teeth are made. Gypsum ($CaSO_4 \cdot 2H_2O$), a sulfate, is the raw material from which plaster is made. And the carbonate mineral calcite ($CaCO_3$), from which organisms such as molluscs construct their shells, is the principal constituent of marble.

Finally, some materials occur in nature as *native elements,* that is, not in combination with other elements. Minerals that occur in this form include some metals, such as gold (Au) and silver (Ag), and some nonmetals, such as sulfur (S) and graphite (C).

Minerals as Indicators of the Environment of Their Formation

Finally, it is worth noting that minerals are not merely objects of beauty or economically valuable resources. Their makeup offers clues to the conditions under which they are formed. The study of minerals, therefore, can provide insight into the chemical and physical conditions in regions of the Earth that we cannot observe and measure directly.

Our understanding of the conditions in which minerals are formed has come largely from laboratory studies. For example, scientists have been able to determine the temper-

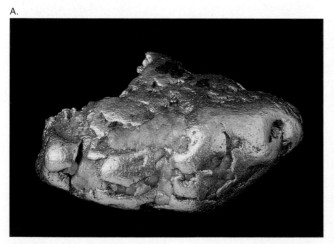 

▲ **FIGURE 2.10**
Examples of ore minerals. A. A nugget of metallic gold, from White Flats, Yukon River, Alaska. The nugget is 3 cm in length. B. A mixture of galena (gray) and chalcopyrite (yellow) from a richly mineralized vein in Peru. The specimen is 5 cm across. Galena (PbS) is the main ore mineral of lead, chalcopyrite ($CuFeS_2$) is the main ore mineral of copper. The two minerals must be separated into pure concentrates prior to treatment for recovery of the contained metals.

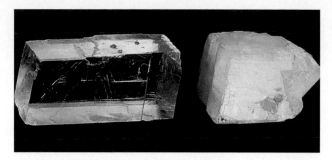

◀ F I G U R E 2.11
The two most common carbonate minerals, calcite
($CaCO_3$) on the left, dolomite ($CaMg(CO_3)_2$) on the right.

atures and pressures at which diamond forms instead of graphite. Because we know that diamonds are stable only at very high temperatures and pressures, and we understand how temperature and pressure increase at great depths within the Earth, we can state with certainty that rocks in which diamonds are found were formed in the mantle at least 150 km below the Earth's surface.

But studies of minerals are not limited to laboratory experiments. Field observations are also valuable. For example, the minerals that form in the regolith during weathering are controlled by the climate. Past climates can therefore be deciphered from the kinds of minerals preserved in sedimentary rocks. The composition of seawater in past ages can also be determined from the minerals formed when the seawater evaporated and deposited its salts.

ROCKS

In Chapter 1 we introduced the concept of the rock cycle and the three families of rocks: (1) *igneous rocks,* which have crystallized from molten rock, either deep underground or at the surface of the Earth; (2) *sedimentary rocks,* which have formed under low-temperature conditions near the Earth's surface, through chemical precipitation or cementation of sediment; and (3) *metamorphic rocks,* previously existing rocks that have been changed by the action of heat and pressure. But we did not present a precise definition of the term *rock.* Now that you have learned more about minerals—the basic components of rocks—we can be more specific.

A **rock** is a naturally formed, nonliving, firm, and coherent aggregate mass of solid matter that constitutes part of a planet. A pile of loose sand grains is not a rock because the grains are not locked together—that is, they are not coherent. A tree is not a rock, even though it is solid, because it is living. But coal, which is a compressed and coherent aggregate of twigs, leaves, and other bits of dead plant matter, is a rock.

To a nongeologist, the distinction between a rock and a mineral may not seem very clear, but it becomes clearer if one remembers that rocks are *aggregates.* This means that rocks are composed of lots of grains stuck together. The grains may be coarse or fine; they may all be the same type of mineral grain, or they may represent a variety of minerals; they may be tightly or loosely stuck together. All of these possible differences contribute to the description of different types of rock.

Describing Rocks

At first glance rocks seem confusingly varied. Some are either distinctly layered and have pronounced, flat surfaces covered with mica. Others are coarse and evenly grained and are not layered, yet they may contain the same kinds of minerals. Studying a large number of rock specimens soon makes it clear that no matter what kind of rock is being examined—sedimentary, metamorphic, or igneous—the differences between samples can be described in terms of two kinds of features: *mineral assemblage* and *texture.*

Mineral Assemblage

The first obvious feature of a rock is the kinds of minerals that are present. A few kinds of rock contain only one type of mineral, but most contain two or more types. The variety and the relative abundance of minerals present in rock, commonly called the *mineral assemblage* of the rock, are important pieces of information for interpreting how a rock was formed.

Texture

The second obvious feature is the rock's *texture*—the overall appearance produced by the size, shape, and arrangement of its constituent mineral grains. For example, the mineral grains may be flat and parallel to each other, giving the rock a platy or flaky texture like that of a deck of cards (Fig. 2.12A). In addition, the various minerals may be unevenly distributed and concentrated into specific layers, producing a texture that is both layered and platy. Alternatively, the

BOX 2.2
•
FOCUS ON. . .

PROPERTIES OF MINERALS

The properties of minerals are determined by their composition and crystal structure. Once we know which properties are characteristic of particular minerals, we can use them to identify those minerals. It is not necessary, therefore, to analyze a mineral chemically or to determine its crystal structure to discover its identity. The properties most often used to identify minerals are obvious ones, such as color, external shape (the crystal form or *habit*), and hardness. Less obvious properties, such as *luster* (the quality and intensity of light reflected from the mineral), *cleavage* (the tendency of the mineral to break in preferred directions), and *density* (i.e., the "heaviness" of the mineral), are also used in identifying minerals.

The fibrous form of asbestos minerals results in properties that are both useful and hazardous for humans, as we saw in the opening paragraphs of this chapter. Here are a few more examples of the ways in which the composition and arrangement of atoms can dictate the physical properties of minerals. The mineral quartz (SiO_2) is normally colorless, with a glassy-looking luster (called *vitreous*) and elongate, six-sided crystals (Fig. B2.1A). When a tiny amount of manganese (as little as 0.01 percent) is introduced into the crystal structure of quartz, it takes on a deep purple color and we refer to it as amethyst (Fig. B2.1B). If the quartz occurs in minute grains and is associated with iron (Fe^{3+}), it looks completely different and we know it as chalcedony (Fig. B2.1C). Similarly, the mineral corundum (Al_2O_3) is normally white or grayish. When small amounts of Cr^{3+} are introduced, the corundum takes on a blood-red color and we call it ruby; if Fe and Ti are

A.

B.

C.

▲ FIGURE B2.1
The variable habits of quartz (SiO_2). A. Clear, six-sided, glassy-looking crystals of quartz. Individual crystals in this sample are about 5 cm in length. B. Beautiful purple-colored quartz crystals (variety *amethyst*). The color is caused by trace amounts of manganese. C. Colored bands of quartz comprised of grains so small they can only be seen with powerful microscopes. The colors are due to trace amounts of chemical elements such as iron and manganese. Banded, fine-grained quartz is called *chalcedony.*

present, the corundum becomes deep blue, and another prized gem, sapphire, is the result.

The hardness of a mineral and the way the mineral breaks are determined primarily by the way the atoms are linked together and the strength of the bonds. For example, diamond (Fig. B2.2A) has the same chemical composition as graphite (pure carbon), but in a diamond the carbon atoms are linked together by an extremely strong type of bond, a covalent bond. In graphite (Fig. B2.2B), the carbon atoms are tightly bonded together in flat sheets, but the Van der Waals bonding between the sheets is very weak. The result is a series of very strong, flexible sheets that separate easily. Graphite (used in pencil leads) feels slippery when it is rubbed between the fingers because the rubbing breaks the weak bonds and the sheets slide past each other. Other minerals with a sheetlike structure include talc, the mineral in "talcum" powder, micas, and a variety of clay minerals.

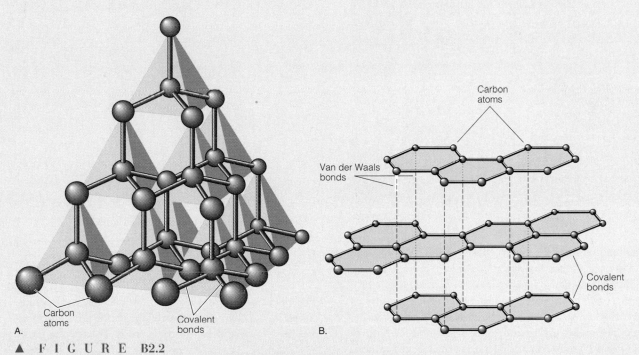

▲ FIGURE B2.2
A. The three-dimensional geometric arrangement of carbon atoms in diamond. Note that each atom is surrounded by four others. B. The geometric arrangement of carbon atoms in graphite. Bonding within sheets is strong, but binding between sheets it is weak.

A.

B.

◄ F I G U R E 2.12
Examples of rock textures. A. A distinct layered texture produced by parallel, flattened mineral grains is characteristic of many metamorphic rocks. This sample, called a *gneiss*, consists of the minerals feldspar (pink), quartz (gray), and the mica (black). The sample is 6 cm in width. B. Oolitic limestone, a sedimentary rock consisting of small, rounded grains of calcite ($CaCO_3$) and broken shell fragments. All of the grains are about the same size (0.3 mm in diameter). The texture is homogeneous.

rock's texture may be homogeneous, made up of grains that are more or less the same size (Fig. 2.12B).

The mineral grains in some kinds of rock are firmly held together, whereas in other kinds the grains are easily broken apart. Rocks with a small proportion of empty space between the grains are igneous and metamorphic; both of these types of rock contain intricately interlocked mineral grains. During the formation of rocks of these types, the growing mineral grains crowd against each other, filling all spaces and forming an intricate three-dimensional jigsaw puzzle (Fig. 2.13A). A similar interlocking of grains holds together steel, ceramics, and bricks.

The forces that hold together the grains of sedimentary rocks are less obvious (Fig. 2.13B). Sediment is a loose aggregate of particles. It must be transformed into sedimentary rock in either of two ways. In one, water circulates slowly through the open spaces between sediment grains and deposits new materials that cement the grains together. In the other, sediment becomes deeply buried and the temperature rises, causing mineral grains to recrystallize; the growing grains interlock and form strong aggregates. The latter process is the same as that which occurs when ice crystals in a pile of snow recrystallize to form a compact mass of ice.

EARTH MATERIALS IN THE ENVIRONMENT

The characteristics of rocks and minerals that we have discussed in this chapter ultimately determine their properties and behavior in different settings in the crust of the Earth. For example, a rock composed primarily of olivine (Mg_2SiO_4), a mineral that is stable in rocks found in the mantle, will weather and eventually crumble into sediment much more readily than a rock composed of quartz, which is stable in rocks found in the crust. A rock in which the grains are loosely held together with lots of space between them is more likely to allow the passage of fluids than a rock with tightly interlocking grains. A rock that is strongly layered is more likely to split along the layering, which may lead to movement or, eventually, landsliding along the split.

The chemical and physical properties of a rock or a mineral also determine whether or not it is of interest as an economic resource. Does the rock contain gold? Is the gold concentrated enough, and in a form that can be extracted using current technologies? Is the rock packed loosely enough to permit oil to be stored in the spaces? Or water?

In addition to rocks and minerals, we discussed a variety of other Earth materials in the context of the rock cycle

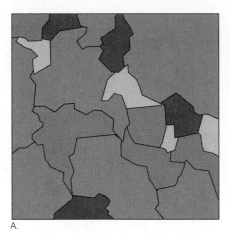
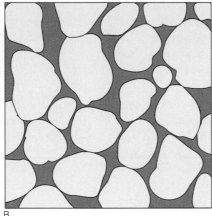

A.　　　　　　　　　　　　　　　　B.

▲ FIGURE 2.13

A. The close interlocking of grains in an igneous or metamorphic rock leads to a rock with few open spaces. Each color represents a different mineral species. B. Sediments and sedimentary rocks may have a significant amount of open space between grains. Some of the spaces may become filled through the processes of cementation and recrystallization. Sedimentary grains are shown in yellow, cement in brown.

(Chapter 1), including magma, lava, and sediment. Each of these materials has properties and characteristics with potential impacts on the environment or on other conditions that concern humans. For example, certain types of magma don't flow easily, so they tend to erupt explosively. Some sediments amplify seismic waves, leading to increased damage due to large earthquakes. Some sediments crack when they dry out, permitting radioactive gas to leak out. Some clay minerals chemically attract ions and therefore are useful for stopping leaks of toxic or nuclear waste. Other clay minerals chemically attract water, which may make them unstable as supporting foundations for buildings. And so on; each of these examples and many more will come up at appropriate places in later chapters. Instead of continuing with an inventory of the many ways in which the characteristics of Earth materials can affect our lives, let's go on to the next part of the book, in which we look at hazardous Earth processes in detail.

SUMMARY

1. Environmental geologists study the physical structure and chemical composition of the Earth because: (1) The internal structure of the Earth has a lot to do with shaping our landscape and causing events that may be hazardous for humans. (2) Humans depend on the materials of the solid Earth as resources. (3) The materials of the Earth have distinct physical and chemical properties that can affect human lives in many different ways.

2. The Earth is a differentiated body, consisting of three major compositional layers: the core, the mantle, and the crust.

3. In addition to being layered compositionally, the Earth has layers of differing rock properties. Most important are the layers of differing rock strength, which are determined primarily by the interplay between temperature and pressure at depth within the Earth. The Earth has a solid inner core; a liquid outer core; an intermediate zone called the mesosphere; a zone called the asthenosphere (from about 350 to 100 km below the Earth's surface), in which rocks have very little strength; and a stronger, more rigid outer layer, the lithosphere.

4. The core and mantle are not directly accessible, so their compositions have been determined by inference from indirect evidence such as the velocities of earthquake waves. The core is composed primarily of iron and nickel, whereas the mantle consists mainly of silicon, oxygen, magnesium, and iron.

5. The Earth's crust is highly variable in composition. Modern estimates of the composition of the crust are based on determinations of the abundances of different rock types in various tectonic environments, weighted with respect to the relative importance of each environment in the crust as a whole.

6. As a result of differentiation, in which the lightest elements rose to the top and the densest elements sank to the core, the Earth's crust is rich in light elements (such as silicon, aluminum, sodium, and potassium) compared to the Earth as a whole. The crust is particularly rich in oxygen (90 percent by volume); however, the oxygen in the crust differs significantly from that in the atmosphere in that it is tightly bonded to other elements.

7. A mineral is a naturally formed, solid, inorganic chemical substance having a specific composition and a characteristic crystal structure. Scientists have identified approximately 3000 minerals.

8. The most common mineral type is the silicate group, which is based on a grouping of silicon and oxygen atoms in the form of a tetrahedron. The silica tetrahedra link together in different ways to form the different minerals of the silicate group. The type of linkage determines the physical properties of the minerals. Other important mineral groups are the oxides, carbonates, sulfides, phosphates, and sulfates.

9. A rock is a naturally formed, nonliving, firm, and coherent aggregate mass of solid matter that constitutes part of a planet. The differences among rocks can be described in terms of their differing mineral assemblages and textures.

10. The chemical and physical characteristics of rocks and minerals determine their properties and, thus, their behavior in different settings in the crust of the Earth and their potential as economic resources.

IMPORTANT TERMS TO REMEMBER

atom (p. 48)
core (p. 42)
crust (p. 42)
crystalline (p. 49)

element (p. 47)
ion (p. 48)
mantle (p. 42)
mineral (p. 47)

molecule (p. 48)
rock (p. 53)
silicate minerals (p. 49)

QUESTIONS AND ACTIVITIES

1. Find out more about some of the minerals you encounter in your daily life. You could choose to investigate hazardous minerals, such as asbestos; or ore minerals, such as galena; or gems, such as emerald; or industrial minerals, such as diamond or garnet. Find out about the chemical and physical properties of the mineral that make it useful (or hazardous, or beautiful, or valuable).

2. Go on a walking tour of building stones in your town. Notice the types of rock that have been used to construct floors, countertops, and building facades. Office buildings, banks, and government buildings—old and new—are usually good places to start looking. Notice the different ways in which building stones respond to environmental forces, such as wind, water, or acid rain. Find out where the building stones used in your area have come from—are they from local quarries or imported from far away? Headstones in cemeteries can also provide interesting clues about rocks that have been used commercially in your area.

3. Do some research to find out more about how scientists use indirect techniques to study the interior of the Earth. Have some of these techniques also been applied to the study of other planetary interiors?

4. Grow your own crystals. Stir into a glass of warm water as much sugar as you can without allowing the sugar to collect on the bottom of the glass. Tie a string to a pencil and lay the pencil across the top of the glass, with the string hanging down into the water. Let the glass of water cool for a few days in a quiet place and then check the string for crystals. What is the shape of the crystals that formed on the string? Are these crystals actually minerals? Why or why not?

HAZARDOUS GEOLOGIC PROCESSES

"Civilization exists by geologic consent—subject to changes without notice."

• **Will Durant**

★ E • S • S • A • Y ★

ASSESSING GEOLOGIC HAZARDS AND RISKS

Geologic processes that we refer to as "hazardous" have existed throughout Earth history. Even the most destructive events are part of the normal functioning of this dynamic planet; to a great extent, they make the planet habitable. Earthquakes and volcanic eruptions, for example, are among the processes that have formed the continents, shaped the landscape, determined the positions of climatic zones, and allowed for the creation and stabilization of the atmosphere and oceans. The action of wind and water causes flooding, landslides, and windstorms but also replenishes soil and sustains life.

In other words, geologic processes affect our lives every day in ways that are both subtle and conspicuous, beneficial and harmful. Knowledge about those processes and the hazards associated with them should play an integral role in the planning of human activities. But before we address some of the challenges inherent in assessing geologic hazards and risks, we need to clarify the meanings of these and other terms.

Types of Hazards

The terms *natural hazard* and *geologic hazard* denote a wide range of geologic circumstances, materials, processes, and events. **Geologic hazards** include earthquakes, volcanic eruptions, floods, landslides, and other processes and occurrences. They are included in the broader concept of **natural hazards,** which encompasses processes or events such

◀ ─────────────

Collapse of the Antelope Valley Freeway, California, as a result of the Northridge earthquake on January 18, 1994.

as locust infestations, wildfires, and tornadoes in addition to strictly geologic hazards.

Some natural hazards are catastrophic events—occurrences that strike quickly but with devastating consequences. For example, there is a calculable, though small, risk that a large comet or meteorite might strike the Earth, with potentially disastrous effects. Events that strike quickly and with little warning, such as earthquakes, flash floods, or sudden windstorms, are called **rapid onset hazards.** Other hazardous processes operate more slowly. Droughts, for instance, can last 10 years or more. The socioeconomic impacts of extended droughts are caused by the cumulative effects of season after season of below-average rainfall.

In general, natural processes are labeled "hazardous" only when they present a threat to human life, health, or interests, either directly or indirectly. In other words, we tend to take an anthropocentric approach to the study and management of natural and geologic hazards. This is human nature, of course; we are justifiably concerned with the protection of human life and property. But this approach has important implications because it can lead to an adversarial style of hazard management in which geologic processes are cast as the "enemy" and efforts are made to manipulate the environment into submission. A somewhat different approach, which is currently receiving a lot of attention, focuses on improving scientific understanding of natural processes and their triggering mechanisms in order to provide a foundation for better management.

A different category of hazard, sometimes referred to as **technological hazards,** is associated with everyday exposure to naturally occurring hazardous substances, such as radon, mercury, asbestos fibers, or coal dust, usually through some aspect of the use of these substances in our built environment. Still another type of hazard arises from pollution and degradation of the natural environment, which have led to problems such as acid rain, contamination of surface and underground water bodies, depletion of the ozone layer, and global climatic warming; we might refer to these as primarily **anthropogenic hazards,** or human-generated hazards.

Primary, Secondary, and Tertiary Effects

Geoscientists and other specialists often refer to hazardous events or processes as having primary, secondary, and tertiary effects. *Primary effects* result from the event itself: water damage resulting from a flood; wind damage caused by a cyclone; the collapse of a building as a result of ground motion during an earthquake. *Secondary effects* result from hazardous processes that are associated with, but not directly caused by, the main event. Examples include forest fires touched off by lava flows, house fires caused by gas lines breaking during an earthquake, and disruption of water and sewage services as a result of a flood. *Tertiary effects* and higher order effects are long term or even permanent. These might include the loss of wildlife habitat or permanent changes in a river channel as the result of flooding; regional or global climatic changes and resulting crop losses after a major volcanic eruption; or changes in topography or land elevation as a result of an earthquake. For example, the 1964 "Good Friday" earthquake near Anchorage, Alaska, caused permanent topographic changes along an extensive section of the Alaskan coastline.

VULNERABILITY AND SUSCEPTIBILITY

Exactly how vulnerable are we to the damaging effects of these and other natural hazards? During the past two decades as many as 3 million lives have been lost as a direct result of hazardous events, and at least 800 million people have suffered adverse effects such as loss of property or health. A United Nations committee has estimated that during the 1990s the Earth will experience tens of thousands of landslides and earthquakes; 1 million thunderstorms; 100,000 floods; and many thousands of tropical cyclones and hurricanes, tsunamis, droughts, and volcanic eruptions. According to researchers at the World Bank, natural disasters cause about US$40 billion each year in physical damage; windstorms, floods, and earthquakes alone cost about US$18.8 million *per day!*

The concept of **vulnerability** has a very specific meaning to hazard specialists. It encompasses not only the physical effects of a natural hazard but also the status of people and property in the area. A large number of factors can increase one's vulnerability to natural hazards, especially catastrophic events. Aside from the simple fact of living in a hazardous area, vulnerability depends on population density, scientific understanding of the area, public education and awareness of hazards, the existence of an early-warning system and effective lines of communication, the availability and readiness of emergency personnel, construction styles and building codes, and cultural factors that influence public response to warnings.

Many of these factors help explain the fact that less developed countries are much more vulnerable to natural hazards than are industrialized countries. Whereas the actual dollar value of property damage from an event such as an earthquake or a flood may be higher in an industrialized country, the *relative* value of monetary losses is *much* greater, on average, in developing countries. There have even been occurrences in which single disasters have caused economic losses equal to a country's entire gross national product.

Poverty itself is a contributor to increased vulnerability. Inadequate housing and high population densities on sensitive lands contribute to much higher losses of life in natural disasters. Something as simple as an inoperative telephone line—an everyday reality in many countries—could cause an early-warning system to fail.

Human intervention in the functioning of natural systems can increase vulnerability in two ways: (1) through the development and habitation of lands that are sensitive or susceptible to hazards (e.g., floodplains or deltas), and (2) by increasing the severity or frequency of natural hazards (e.g., overintensive agriculture leading to increased erosion; mining of groundwater leading to subsidence; or global climatic change leading to increased intensity of tropical cyclones).

It is interesting to note that *either* poverty *or* affluence can cause these types of pressures on the environment. For example, extensive deforestation of lands bordering the Sahara Desert by poverty-stricken residents in desperate need of fuelwood has almost certainly accelerated the process of desertification, contributing to drought and famine in those areas. In contrast, in San Francisco economic pressures have led to the development and urbanization of large areas of land reclaimed from the Bay; these lands are particularly susceptible to failure through liquefaction (a quicksand-like condition) in a major earthquake.

ASSESSING HAZARDS AND RISKS

In order to incorporate knowledge about natural processes into the planning of human activities, we have to assess the hazards and risks associated with them. Although the terms *hazard assessment* and *risk assessment* are often used interchangeably, they are not synonymous.

Hazard Assessment

Hazard assessment involves asking questions like, "How often can we expect a hazardous event to occur?" and "When such an event does occur, what effects is it likely to have?" Specifically, hazard assessment consists of the following activities:

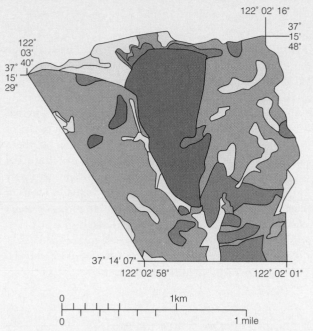

Relative stability	Map area	Geologic conditions	Recommended land use		
			Houses	Roads	
				Public	Private
Most stable ↑		Flat to gentle slopes; subject to local shallow sliding, soil creep, and settlement	Yes	Yes	Yes
		Gentle to moderately steep slopes in older stabilized landslide debris; subject to settlement, soil creep, and shallow and deep landsliding	Yes	Yes	Yes
		Steep to very steep slopes; subject to mass-wasting by soil creep, slumping and rock fall	Yes	Yes	Yes
		Gentle to very steep slopes in unstable material subject to sliding, slumping and soil creep	No	No	No
		Moving shallow (>10 ft) landslide	No	No	No
↓ Least stable		Moving, deep landslide, subject to rapid failure	No	No	No

▲ F I G U R E II.1
A landslide susceptibility map and recommended land use policies for the Congress Springs area near San Francisco. The results of hazard assessment are often portrayed in map format so they are understandable and useful for local planners and decision makers.

- Determining when and where hazardous events have occurred in the past.
- Determining the severity of the physical effects of past events of a given *magnitude* or size.
- Determining how frequently we can expect events that are strong enough to generate severe physical effects.
- Determining what a particular event would be like if it were to occur now, in terms of the type of effects it would have.
- Portraying all this information in a form that can be used by planners and decision makers.

The results of hazard assessment are (or should be) used by political leaders who have to make decisions concerning evacuation or contingency funding; by emergency personnel, who have to make decisions concerning levels of response and readiness; by planners and engineers, who have to make decisions concerning land use and zoning or building regulations; and by local scientific experts, who may be the first to sound an alert on the basis of such information. Often the results of hazard assessment are portrayed in the form of maps (as shown in Figs. II.1 and II.2) in order to make the information as accessible and understandable as possible.

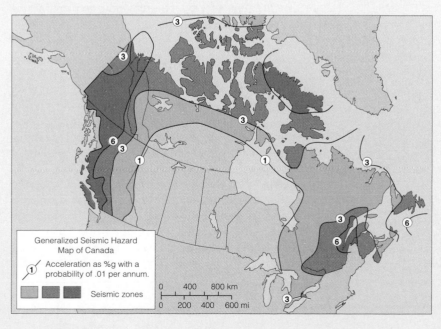

Generalized Seismic Hazard Map of Canada

① Acceleration as %g with a probability of .01 per annum.

Seismic zones

◀ F I G U R E II.2
Earthquake probability map of Canada. The map clearly identifies those areas with the highest probability of exceeding a certain magnitude of ground motion during an earthquake.

Risk Assessment

Risk assessment differs from hazard assessment in some important respects. In the context of natural hazards, risk is a statement of the economic losses, injuries and deaths, and loss of functioning of urban support systems expected when a specific physical effect, such as ground shaking triggered by an earthquake, strikes a given area. Risk assessment starts by establishing the *probability* that a hazardous event of a particular magnitude will occur within a given period. It then goes on to take into account factors such as the following:

- The locations of buildings, facilities, and emergency systems in the community.

- Their potential exposure to the physical effects of the hazardous situation or event.

- The community's vulnerability—that is, potential loss of life, injury, or loss in value—when subjected to those physical effects (Fig. II.3).

Thus, risk assessment incorporates social and economic considerations in addition to the scientific factors involved in hazard assessment. Whereas hazard assessment focuses on characterizing the physical effects of a particular event, risk assessment typically focuses on the extent of the damage anticipated, control or mitigation of the damage, and actions that might reduce vulnerability. Risk assessment involves asking such questions as, "When a situation of this type exists or an event of this type occurs, what kinds of

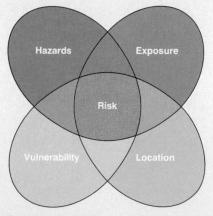

ELEMENTS OF RISK

▼ F I G U R E II.3
The elements of risk. Risk assessment involves establishing the probability and characterizing the physical effects of natural hazards as well as incorporating societal and economic considerations such as location, exposure to effects, and vulnerability.

damages can the community expect?" and "Even though the likelihood of an event of this magnitude occurring is small, would the consequences be unacceptably severe?"

Risk is sometimes stated in terms of probabilities. For example, it has been estimated that smoking 1.4 cigarettes, drinking 0.5 liter of wine, having a single chest x-ray, or being exposed to earthquake hazards by living in Southern California for seven months all carry the same statistical risk: They all increase the chance of death by approximately 1 in a million. Alternatively, risk can be stated in terms of the cost (damages and injuries expressed in terms of dollar value lost) if a hazardous event were to occur. In either case, a risk assessment can help both decision makers and scientists compare and evaluate hazards, set priorities, and decide where to focus attention and resources.

PREDICTION AND WARNING

Predicting hazardous events and advising decision makers about their expected severity are other important aspects of dealing with geologic hazards. A **prediction** is a statement of probability based on scientific observation. Accurate prediction requires the continuous monitoring of geologic processes. Monitoring usually focuses on identifying anomalies that might be **precursors**—that is, small physical changes leading up to a catastrophic event. When observations of precursor phenomena have accumulated to the point at which they signal the imminent occurrence of a hazardous event, a prediction can be made. For example, before Mt. Pinatubo in the Philippines erupted violently in 1991, scientists observed a variety of precursor events, of which the most important were increases in the number and intensity of earthquakes in the area and changes in their locations, as well as changes in the quantity and composition of gases emitted by the volcano. Careful monitoring of trends in precursor phenomena allowed scientists to predict the time of eruption with great accuracy, saving many thousands of lives.

Forecasting

Sometimes the term **forecast** is used synonymously with *prediction;* in other contexts, it is used quite differently. For example, in the prediction of floods or hurricanes, forecasting generally refers to short-term prediction of the specific magnitude and time of occurrence of an event (days or hours ahead of time, rather than months or years). In the prediction of earthquakes, on the other hand, the term *forecast* is generally used to refer to a long-term, nonspecific statement of probability. Thus, before the Loma Prieta earthquake of October 17, 1989 (the "World Series earthquake"), the U.S. Geological Survey had issued a forecast indicating a 50 percent probability of a large earthquake occurring along the San Andreas fault in the region of Santa Cruz within 30 years. This was a relatively nonspecific long-range forecast based on general scientific understanding of seismicity and the geology of the area, rather than on the observation of specific precursor phenomena. (The threat of a major earthquake in the San Francisco-Santa Cruz area is still very real; in fact, after the Loma Prieta earthquake the USGS issued a report forecasting a 67 percent probability of a major earthquake occurring in the area within the next 30 years.)

Early Warning

The final step in preparing a community to deal with a hazardous event is the issuance of an **early warning.** A warning is quite different from a prediction or forecast and carries many implications. As one expert put it, a warning is "a public declaration that the normal routines of life should be altered for a period of time to deal with the danger posed by the imminent event." Warnings depend heavily on timeliness, effective communications and public information systems, and credible sources. If a warning is issued prematurely or irresponsibly and the event does not occur as predicted, the results can be disastrous. Like the boy who cried "Wolf!" the scientist may be unable to regain credibility, and the public may be slow to respond to a *real* threat.

RESPONSE AND THE ROLE OF GEOSCIENTISTS

Some natural hazards, such as meteorite impacts, are impossible to prevent and very difficult to predict within any useful time frame. Although we know that such an event might occur, there is virtually nothing we can do with current technologies to decrease the risk. However, each day we are faced with a wide range of natural hazards to which we *can* adapt by making certain choices and taking conscious actions to prepare ourselves or decrease our vulnerability. Many of these adaptive actions fall outside the realm of science; they involve economic, legal, political, or lifestyle choices. People who live with risk have adopted widely varying methods of responding to the risk, ranging from denial to acceptance to panic. What, then, should the role of the geoscientist be in shaping and informing the public's response to natural hazards?

There is a strong need on the part of the general public, especially decision makers, for knowledge concerning geologic hazards. Unfortunately, there are many gaps between the acquisition of scientific knowledge about hazardous processes and the effective transfer of this information to people who can use it in formulating plans and adopting policies for the reduction of hazards and risks (Fig. II.4). Scientists tend to have different priorities from those of

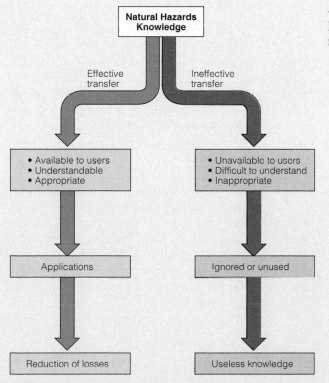

◀ F I G U R E II.4
Effective and ineffective transfer of knowledge about natural hazards.

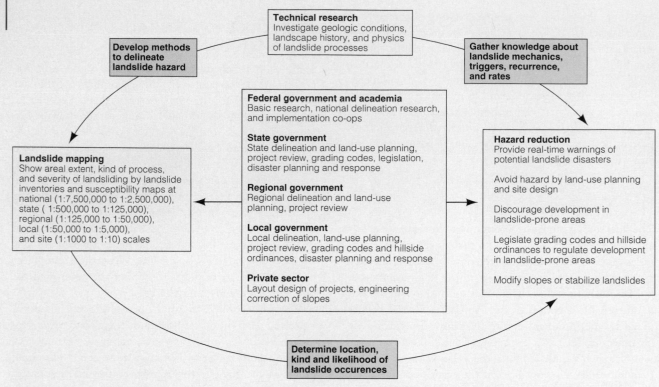

▲ FIGURE II.5
An integrated system designed to reduce landslide hazards. Geologic information can contribute to the establishment of an integrated system designed to reduce or mitigate the damage caused by natural hazards.

other members of the community and to use different modes of communication. They are concerned primarily with observing and understanding natural phenomena; in contrast, a government official is likely to be more concerned with the socioeconomic implications or with the feasibility and cost of hazard reduction strategies. Scientists typically communicate through technical papers, which are published in scientific journals and use specialized terminology; few government officials have the training or the time to keep up to date in the areas of study that are relevant to the management of natural hazards. These differing sets of priorities and communication styles can lead to gaps in communication and, thus, to the misdirection, misunderstanding, or misuse of important information.

Ideally, scientific understanding can contribute to the establishment of an integrated system in which geoscientists cooperate with government and private-sector decision makers to apply scientific and technical knowledge to the reduction of natural hazards. An example of such a system

is presented in Figure II.5. In this integrated approach to the management of landslide hazards, scientific and technical research contribute to the understanding of landslide mechanisms and the temporal and spatial delineation of the hazard. This understanding, in turn, leads to the development of hazard and risk reduction strategies and landslide hazard maps. All levels of government, academic institutions, and private-sector organizations would be involved in a balanced program of this type.

Geoscientists have a particular responsibility to contribute to such efforts. Because so many natural disasters are associated with geologic processes, geoscientists play an important role in furthering our understanding of hazardous Earth processes, assessing the hazards and risks involved, accurately predicting hazardous events, and assisting in prevention or the mitigation of impacts. All of these tasks depend on effective communication of scientific understanding about geologic processes.

IMPORTANT TERMS TO REMEMBER

anthropogenic hazard (p. 62)

early warning (p. 65)

forecast (p. 64)

geologic hazard (p. 61)

hazard assessment (p. 62)

natural hazard (p. 61)

precursors (p. 64)

prediction (p. 64)

rapid onset hazard (p. 61)

risk assessment (p. 64)

technological hazard (p. 61)

vulnerability (p. 62)

EARTHQUAKES

We rode on a sea of mountains and jungles, sinking in rubble and drowning in the foam of wood and rock. The Earth was boiling under our feet...making bells ring, the towers, spires, temples, palaces, houses, and even the humblest huts fall; it would not forgive either one for being high or the other for being low.

• **Survivor of an earthquake in Guatemala, 1773**

*I*n October 1989 climatologist Iben Browning predicted that a devastating earthquake would occur in the area of New Madrid, Missouri on December 2 or 3, 1990, plus or minus 2 days. His prediction was based on the idea that earthquakes may be triggered by peaks in the Earth's tides, a theory that has often been discussed in scientific journals but has never been substantiated. Scientists at the U.S. Geological Survey were quick to discredit Browning's prediction, stating that there was only a 1 in 60,000 chance that such an earthquake would occur in that location on that date.

In spite of the prediction's lack of scientific credibility, reporters across the United States jumped on the New Madrid story. Newspaper articles and radio talk shows about earthquakes proliferated. NBC aired a made-for-TV movie titled "The Big One." Government agencies in the central states, not wanting to be caught unaware, issued pamphlets about earthquake preparedness. Schools rehearsed emergency procedures, and most were scheduled to close on the day the quake was expected to occur. T-shirts and other quake memorabilia began to appear.

◄ ——————————————

On January 17, 1995, an earthquake occurred in southern Honshu, the main island of Japan. The city most severely affected by the quake was Kobe. More than 5000 people were killed and a much larger number were injured when buildings and roadways collapsed. This high-speed elevated roadway through downtown Kobe buckled and collapsed without warning.

New Yorker correspondent Sue Hubbell, who owns a farm near New Madrid, reported with some irony on conditions in the area as "E day" approached:

When I got to my farm, I felt as though I had stumbled into the countdown for Armageddon. Everywhere, I heard stories of people buying gas cans, batteries, even emergency power generators that cost thousands of dollars, stocking their pickups with emergency food and driving them out to the middle of their fields. I heard of white roses blooming blue out of season. The Mississippi River was said to be bubbling sulfur. A kindly, grandmotherly woman told me of the Bag 'Em and Tag 'Em Project: an "outfit" (unspecified) was said to have offered plastic bags and Magic Markers to the school in New Madrid so that overbusy undertakers could enlist the help of children to go out, pick up the dead, and label them as best they could. When I reacted with disgust and disbelief, she said, "Well, it is just awful, and they turned down the plastic bags. Took the Magic Markers, though."

The predicted catastrophe did not materialize; even the most sophisticated techniques do not yet permit scientists to predict earthquakes with such precision. But the episode may have had some positive outcomes. New Madrid does in fact lie within an earthquake-prone region. In the winter of 1811–1812, the area was shaken by a series of three great quakes that caused tremendous damage and were felt over two thirds of the contiguous United States. Geologists believe that it is still possible for a major earthquake to occur in the region. As a result of Browning's prediction, residents and emergency personnel may be better prepared for such an event should it occur.

WHAT CAUSES EARTHQUAKES?

Try this experiment. Ask a friend to hit one end of a wooden plank or the top of a wooden table with a hammer while you press your hand on the other end. You will feel vibrations set up by the energy of the blow. The harder the blow, the stronger the vibrations. The reason you can feel those vibrations is that some of the energy imparted by the hammer is transferred to your hand by vibrations traveling through the solid wood. In the Earth, a bomb blast or a violent volcanic explosion will serve as an energy source. So, too, will the sudden slipping of rock masses along fault surfaces, which results in an earthquake.

When an earthquake occurs, it is as if someone has struck the Earth with a huge hammer. Energy that has been built up and stored in rocks, sometimes over a long period, is suddenly released. The more energy released, the stronger the quake. Just how the stored energy is built up, how it is released, and the impacts on people in the area are the subjects of this chapter.

Fractures and Faults

Most earthquakes are caused by the sudden movement of blocks of the Earth's crust. Such movement involves the fracture of brittle rocks and the movement of rock along the fractures. A fracture in a rock along which movement occurs is called a **fault.** Tectonic forces produce *stress,* or directional pressure, which causes large blocks of rock on either side of a fault to move past one another. But that movement is not a smooth, continuous slippage. Instead, stress builds up slowly until the friction between the two blocks is overcome; then the blocks slip abruptly again.

If stresses persist, the cycle of slow buildup followed by abrupt movement repeats itself many times. Although movement along a large fault may eventually total many kilometers, this distance is the sum of numerous small, sudden slips. Each of those slips may cause an earthquake and, if the movement occurs near the Earth's surface, disrupt and displace surface features (Fig. 3.1).

Figure 3.1 illustrates abrupt horizontal movement, but vertical movements along faults are also well documented. The largest abrupt vertical displacement ever observed occurred in 1899 at Yakutat Bay, Alaska. During a major earthquake, a stretch of the Alaskan shore was suddenly lifted as much as 15 m above sea level. The visible vertical displacement may be less than the total amount, because the fault is located offshore and the block of crust on the other side of it, lying beneath the sea, may have moved downward, thus adding to the total displacement of the stretch of beach.

Movement along a fault is usually, but not always, abrupt. Detailed measurements along the San Andreas Fault in California reveal places where gradual slipping occurs, sometimes by as much as 5 cm a year. Geologic studies reveal that movement has been occurring along the San Andreas Fault for at least 65 million years. The total extent of that movement is not known, but there is evidence suggesting that it amounts to more than 600 km.

It may be that no spot on the Earth is completely stationary. Measurements by surveyors over the past 100 years reveal large areas of the United States where the land is slowly sinking and other places where it is slowly rising. The causes of these vast, slow vertical movements are not well understood, but the movements prove that the Earth is not as rigid as it seems and that great internal forces are continuously deforming its crust.

◄ F I G U R E 3.1
An orange grove planted across the San Andreas Fault in southern California. Movement on the fault has displaced the rows of trees. The direction of plate motion is such that trees in the background have moved from left to right relative to the trees in the foreground.

Classification of Faults

Faults are classified according to (1) the inclination, or slope, of the fault surface (called its *dip*) and (2) the direction of relative movement along the fault. The common

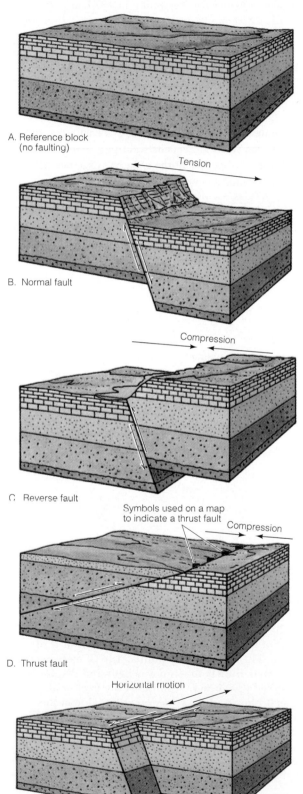

A. Reference block (no faulting)

Tension

B. Normal fault

Compression

C. Reverse fault

Symbols used on a map to indicate a thrust fault

Compression

D. Thrust fault

Horizontal motion

E. Slip-strike fault

classes of faults are shown in Fig. 3.2. Movement along many faults is entirely vertical or entirely horizontal, but along some faults combined vertical and horizontal movement occurs.

Normal faults are caused primarily by *tension,* that is, by movement that tends to pull crustal blocks apart. The fault shown in Fig. 3.2B is a normal fault. Normal faulting tends to stretch and thin the crust. **Reverse faults,** in contrast, arise from *compression,* that is, movement that tends to push crustal blocks together. The relative movement on a reverse fault is such that one block of crustal material moves upward relative to the other. These relationships are illustrated in Fig. 3.2C. Reverse fault movement shortens and thickens the crust. A special class of reverse faults, called *thrust faults,* are low-angle (i.e., shallow, rather than steep) reverse faults (Fig. 3.2D). Such faults, which are common in great mountain chains, are noteworthy because along some of them the relative movement of crustal blocks may amount to many kilometers.

Strike-slip faults are those in which the principal movement is horizontal (Fig. 3.2E). Such faults arise from stresses that lead to horizontal, or *translational,* motions of the fault blocks; in other words, the two crustal blocks are sliding past one another. The San Andreas Fault is an example of a strike-slip fault. In Fig. 3.1, it is strike-slip movement on the San Andreas Fault that is offsetting the rows of orange trees.

Horizontal fault movement is designated as follows: To an observer standing on either block, the movement of the other block is *left-lateral* if it is to the left and *right-lateral* if it is to the right. The San Andreas Fault, for example, is a right-lateral strike-slip fault. If you are standing on the east side of the fault and looking across toward the ocean, the block on the other side (the Pacific Ocean plate) is moving in a northwesterly direction relative to where you are standing (i.e., toward your right), carrying a little bit of the California coast along with it. The relative motion in a right- or left-lateral fault is the same regardless of which block the observer is standing on. In other words, if you were to stand on the Pacific Ocean side of the San Andreas Fault and look across toward the continental United States, you would be looking at a block (the North American plate) that is moving toward the southeast relative to where you are standing (i.e., still moving toward your right).

Earthquake Mechanisms

Sudden movement along faults causes earthquakes, but some earthquakes are millions of times stronger than others. The reason for this is that in some cases stored energy is released by thousands of tiny slips and small earthquakes,

◀ F I G U R E 3.2
The principal kinds of faults.

while in other cases it is released in a single immense quake. In this section we take a closer look at the mechanisms involved in earthquakes.

The Elastic Rebound Theory

The most widely accepted explanation of the origins of earthquakes is the **elastic rebound theory.** It is based on the mechanics of *elastic deformation* of rocks: reversible changes in the volume or shape of a rock (or other material) that is subjected to stress. When the stress is removed, the elastically deformed material returns to its original size and shape. You can demonstrate the storage of energy in an elastically deformed material with a heavy steel spring or a long metal ruler. When you compress the spring or bend the ruler across your knee, the material undergoes elastic deformation. When you suddenly release the spring or ruler, it bounces back to its original shape, releasing the built-up energy. Similarly, the elastic rebound theory suggests that energy can be stored in elastically deformed bodies of rock when they are subjected to stress along a fault. Eventually the stored energy is sufficient to overcome the

friction between the blocks. The energy is suddenly released in the form of an earthquake, and the elastically strained bodies of rock rebound to their original shapes.

The first evidence supporting the elastic rebound theory came from studies of the San Andreas Fault. During long-term field observations beginning in 1874, scientists from the U.S. Coast and Geodetic Survey determined the precise position of many points both adjacent to and distant from the fault (Fig. 3.3). As time passed, movement of the points revealed that the crust was slowly being bent. Near San Francisco, however, the fault was locked and did not slip. On April 18, 1906, the two sides of this locked fault shifted abruptly. The elastically stored energy was released as the fault moved and the bent crust snapped back, creating a violent earthquake. Subsequent repetition of the survey revealed that the bending of the crust had disappeared.

Most earthquakes occur in the brittle rock of the lithosphere. *Brittleness* is the tendency of a solid material to fracture when the deforming stress exceeds the material's *elastic limit,* that is, the limit beyond which deformation becomes permanent and the material will not be able to rebound to

A.

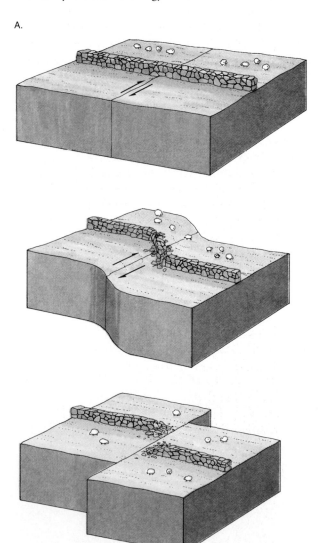

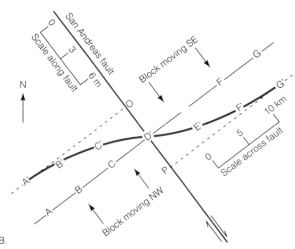

B.

◄ **F I G U R E   3.3**
An earthquake, caused by the sudden release of energy. A. Sketch based on detailed surveys near the San Andreas Fault, California, before and after the abrupt movement that caused the earthquake of 1906. A fence crosses the fault and is slowly bent as rock is elastically deformed. After the earthquake, the two segments of the fence are offset by 7 m. B. Results of the survey. The seven survey points, *A* to *G*, were originally aligned. Slowly the line was bent and displaced to the curved line *A'* to *G'*. Suddenly the frictional lock was broken and the rocks on either side of the fault rebounded. The surveyed points lay along the lines *A'O* and *PG'*.

A.

B.

▲ FIGURE 3.4
**Examples of rock deformation. A. Fracture of strata by brittle deformation. B. Bending of
rock layers by ductile deformation.**

its original shape or volume. At great depths below the Earth's surface, temperatures and pressures are so high that rocks will bend and fold but not break; this is called *ductile deformation* (Fig. 3.4). Under such conditions rocks can neither fracture nor store elastic energy. Instead, like putty, they undergo changes of shape that remain after the deforming forces have been removed. Earthquakes, therefore, are phenomena of the brittle, outer, cooler portion of the Earth.

HOW EARTHQUAKES ARE STUDIED

The study of earthquakes is known as **seismology,** from the ancient Greek word for earthquake, *seismos.* Scientists who study earthquakes are called *seismologists,* and the device used to record the shocks and vibrations caused by earthquakes is a **seismograph.**

Seismographs

An ideal way to record the vibrations and motions of the Earth would be to put a seismograph on a stable platform that is not affected by the vibrations. But a seismograph must stand on the Earth's vibrating surface, and it will therefore vibrate along with that surface. This means that there is no fixed frame of reference for making measurements. The problem is the same one that a sailor in a small boat faces when attempting to measure waves at sea. Because the boat moves up and down with each wave, there is no "platform" for measuring the height of the waves.

To overcome this problem, most seismographs make use of *inertia,* the resistance of a large stationary mass to sudden movement. If you suspend a heavy mass, such as a block of iron, from a light spring and suddenly lift the spring, you will notice that because of inertia the block remains almost stationary while the spring stretches (Fig. 3.5). This is the principle used in *inertial seismographs*

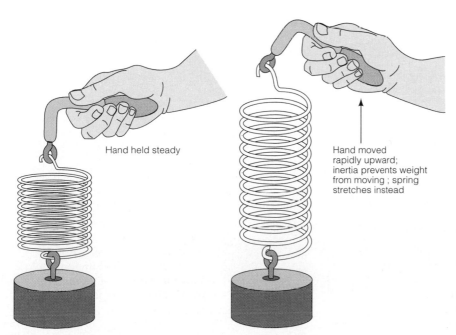

Hand held steady

Hand moved rapidly upward; inertia prevents weight from moving ; spring stretches instead

◀ FIGURE 3.5
The principle of the inertial seismograph.

(Fig. 3.6). Vertical motion is measured by the device shown in Fig. 3.6A, in which a heavy mass is supported by a spring and the spring is connected to a support that in turn is connected to the ground. When the ground vibrates, the spring expands and contracts but the mass remains almost stationary. The distance between the ground and the mass can be used to sense the vertical displacement of the ground surface.

Horizontal displacement can be measured by suspending a heavy mass from a string to make a pendulum (Figure 3.6B). Because of its inertia, the mass does not keep up with the horizontal motion of the ground, and the difference between the movement of the pendulum and that of the ground serves as a measure of the ground motion. Inertial seismographs are commonly used in groups so that up–down, east–west, and north–south motions can be measured simultaneously.

Another device, a *strain seismograph*, employs two concrete piers about 35 m apart (Fig. 3.6C). When the Earth vibrates, the two piers move independently of each other

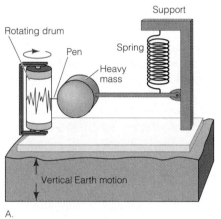

A.

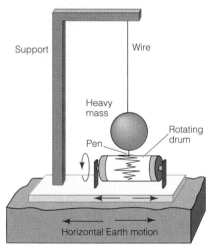

B.

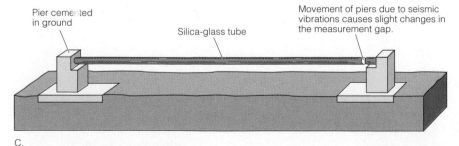

C.

◀ F I G U R E 3.6
Seismographs measure vibrations generated by earthquakes. A. An inertial seismograph for measuring vertical motion. B. An inertial seismograph for measuring horizontal motion. C. A strain seismograph.

and the distance between them changes. Attached to one pier is a long, rigid silica-glass tube. The other pier carries a detector that can measure even the slightest movement in the end of the silica-glass tube. Strain seismographs are usually installed in mines, tunnels, and other places where temperature is relatively constant and wind disturbance minimal.

Modern seismographs are incredibly sensitive because any movement is amplified electronically. Vibrational movements as tiny as one hundred millionth (10^{-8}) of a centimeter can be detected. Indeed, many instruments are so sensitive that they can detect the ground depression caused by a moving automobile several blocks away.

Seismic Waves

The elastically stored energy released by an earthquake is transmitted to other parts of the Earth. As with any vibrating body, waves (vibrations) spread outward from the earthquake's point of origin. These waves, called **seismic waves**, spread in all directions, just as sound waves spread in all directions when a gun is fired. Seismic waves are elastic disturbances, so the rocks through which they pass return to their original shapes after the passage of the waves. The waves therefore must be measured and recorded while the rock is still vibrating. For this reason, continuously recording seismograph stations have been installed around the world. Wherever and whenever an earthquake occurs, the characteristic signatures and arrival times of each seismic wave are recorded by many seismographs; the records obtained in this way are called **seismograms.**

Seismic waves are of two main types. **Body waves** travel outward from the point of origin and have the capacity to travel through the Earth. **Surface waves,** on the other hand, are guided by and restricted to the Earth's surface. Body waves are analogous to light and sound waves, which travel outward in all directions from their points of origin. Surface waves are analogous to ocean waves because they are restricted to the vicinity of a free surface, such as the Earth's surface both where it meets the atmosphere and where it meets the ocean.

Body Waves

Rocks can be elastically deformed by a change in either volume or shape. One kind of body wave, **compressional waves,** deforms rocks through a change in volume. A compressional wave consists of alternating pulses of compression and expansion acting in the direction in which the wave is traveling (Fig. 3.7A). Sound waves are also compressional waves. When a sound wave passes through the air, it does so by alternating compression and expansion of the air. Compression and expansion produce changes in the volume and density of a medium. Compressional waves can pass through solids, liquids, or gases because all three can sustain changes in density. When a compressional wave passes through a medium, the compression pushes atoms closer together. Expansion, on the other hand, is an elastic response to compression that increases the distance between atoms. A solid subjected to compressional waves moves back and forth in the line of the wave's motion. Compressional waves have the greatest velocity of all seismic waves—6 km/s is a typical value for the uppermost

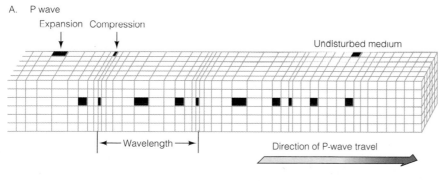

A. P wave

Expansion   Compression

Undisturbed medium

←— Wavelength —→

Direction of P-wave travel

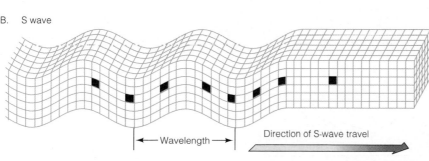

B. S wave

←— Wavelength —→

Direction of S-wave travel

◀ F I G U R E 3.7
Seismic body waves of the P (compressional) and S (shear) types. A. P waves cause alternating compression and expansion in the rock the wave is passing through. An individual point in a rock will move back and forth parallel to the direction of P-wave propagation. As wave after wave passes through, a square will repeatedly expand to a rectangle, return to a square, contract to a rectangle, return to a square, and so on. B. S waves cause a shearing motion. An individual point in a rock will move up and down perpendicular to the direction of S-wave propagation. A square will repeatedly change to a parallelogram and back to a square.

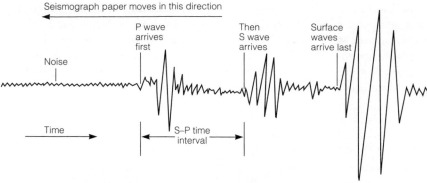

Seismograph paper moves in this direction

Noise

P wave arrives first

Then S wave arrives

Surface waves arrive last

Time

S–P time interval

▲ F I G U R E 3.8

Different travel times of P, S, and surface waves as shown on a typical seismogram. The P and S waves leave the point of origin at the same instant. The fast-moving P waves reach the seismograph first, and some time later the slower moving S waves arrive. The delay in arrival times is proportional to the distance traveled by the waves. The surface waves travel more slowly than either P or S waves.

portion of the crust—and they are the first waves to be recorded by a seismograph after an earthquake. They are therefore called **P** (for *primary*) **waves.**

Another kind of body wave, **shear waves,** deform materials through a change in shape. Because gases and liquids do not have the elasticity to rebound to their original shape, shear waves can be transmitted only by solids. Shear waves consist of an alternating series of sidewise movements, each particle in the deformed solid being displaced in a direction perpendicular to the direction of wave travel (Fig. 3.7B). A typical velocity for a shear wave in the upper crust is 3.5 km/s. Because shear waves are slower than P waves and reach a seismograph some time later, they are called **S** (for *secondary*) **waves.**

Surface Waves

Surface waves travel along or near the surface of the Earth like waves along the surface of a body of water. They travel more slowly than P and S waves, and they pass around the Earth rather than through it. Thus, surface waves are the

last to be detected by a seismograph. Figure 3.8 shows a typical seismogram, in which the P wave's arrival is seen first, followed by the arrival of the S wave and finally by that of the surface waves. Surface waves are important for planners and builders because they cause much of the ground shaking that causes damage to structures during large earthquakes.

Focus and Epicenter

The place where energy is first released to cause an earthquake is called the **focus** (plural: *foci*). Because most earthquakes are caused by movement on a fault, the focus is not a single point but a region that may extend for several kilometers. Earthquake foci lie below the surface, ranging from shallow (just below the surface) to deep (up to 700 km below the surface). Because of the variations in depth, it is more convenient to identify the site of an earthquake by its **epicenter,** the point on the Earth's surface lying vertically above the focus (Fig. 3.9). A good way to describe the loca-

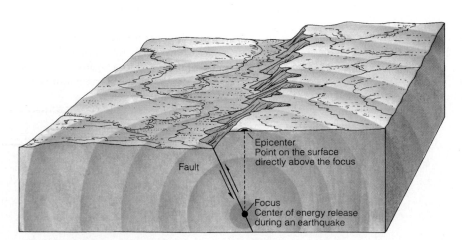

Epicenter
Point on the surface directly above the focus

Fault

Focus
Center of energy release during an earthquake

◀ F I G U R E 3.9
The focus of an earthquake is the site of first movement on a fault and the center of energy release. The epicenter of an earthquake is the point on the Earth's surface that lies vertically above the focus.

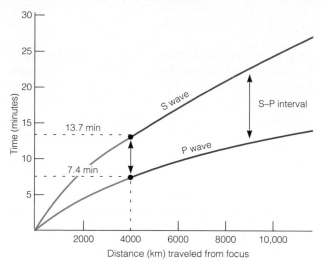

▲ F I G U R E 3.10

Average travel-time curves for P and S waves in the Earth, used to locate an epicenter. For example, when seismologists determine the S–P interval to be 13.7 min - 7.4 min = 6.3 min, they know that the epicenter is 4000 km away from their station.

tion of an earthquake's focus is to state the location of its epicenter and its *focal depth,* that is, how far below the surface it lies.

Locating the Epicenter

If an earthquake's waves have been recorded by three or more seismographs, its epicenter can be determined through simple calculations. The first step is to determine how far the seismograph is from the epicenter. This is done by comparing the arrival times of the P and S waves (Fig. 3.10). The greater the difference between the arrival times, the greater the distance from the epicenter. After using a graph like the one shown in Fig. 3.10 to determine the distance from the seismograph to the epicenter, the seismologist draws a circle with a radius equal to the calculated distance, with the seismic station at the center of the circle. When similar information is plotted for three or more seismographs, the exact location of the epicenter can be determined by triangulation (Fig. 3.11).

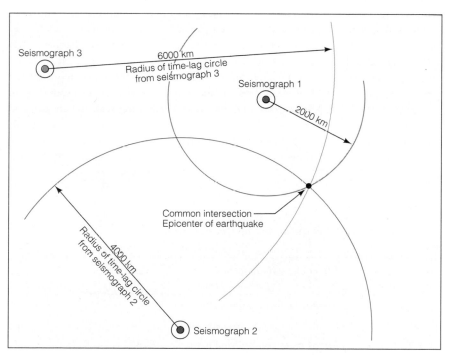

▲ F I G U R E 3.11

Locating an epicenter. The effects of an earthquake are felt at three stations. The time differences between the first arrival of the P and S waves depend on the distance of a station from the epicenter. The following distances are calculated by using the curves in Fig. 3.10.

	Time Difference	*Calculated Distance*
Seismograph 1	**8.8 min – 4.7 min = 4.1 min**	**2000 km**
Seismograph 2	**13.7 min – 7.4 min = 6.3 min**	**4000 km**
Seismograph 3	**17.5 min – 9.8 min = 7.7 min**	**6000 km**

On a map, a circle of appropriate radius is drawn around each station. The epicenter is located at the point where the three circles intersect.

Measuring Earthquakes

Very large earthquakes, such as the one that destroyed much of San Francisco in 1906, are relatively infrequent (Table 3.1). Such quakes occur more frequently in some areas than in others. In earthquake-prone regions such as San Francisco and the surrounding area, very large earthquakes occur, on average, about once a century. This means that it takes roughly 100 years to build up energy to the point at which the frictional locking of a fault is overcome. Small earthquakes may occur during this time as a result of local slippage, but even so, stored energy may be accumulating because parts of the fault remain locked.

The Richter Magnitude Scale

Measurements of elastically deformed rocks before an earthquake, and of undeformed rocks after an earthquake, can provide an accurate indication of the amount of energy released by the quake. But this task is time consuming, and the pre-earthquake measurements often are not available. Therefore, seismologists have developed a way to estimate the energy released by a quake by measuring the recorded heights, or *amplitudes,* of the resulting seismic waves. The **Richter magnitude scale,** named after seismologist Charles Richter, who developed the scale in 1935, is calculated from the maximum heights of P and S waves as recorded on a seismogram. Because wave signals vary in strength by factors of a hundred million or more, the Richter scale is *logarithmic;* each unit increase in magnitude corresponds to a 10-fold increase in the amplitude of the wave signal. Thus, a magnitude 6 signal has an amplitude 10 times larger than a magnitude 5 signal, but a magnitude 7 signal has an amplitude 100 times larger than that of a magnitude 5 signal (10×10).

Richter magnitudes are corrected for distance from the epicenter. This means that the magnitude calculated for a given earthquake is the same whether you are standing at the epicenter, where the effects of the earthquake would feel strongest, or 1000 km away, where the effects would feel less intense.

Figure 3.12 illustrates how a Richter magnitude is calculated. The energy of a seismic wave is a function of its amplitude (i.e., the height of the wave on a seismogram) and the duration of a single oscillation, denoted T. Divide the maximum amplitude, X (measured in steps of 10^{-4} cm on a suitably adjusted seismograph) by T (measured in seconds). Then add the distance correction factor, Y (determined from the S–P interval). X/T is a measure of the maximum energy reaching the seismograph. The formula is:

Richter magnitude, $M = \log X/T + Y$

As just noted, each step in the Richter magnitude scale corresponds to a 10-fold increase in amplitude (X). However, the increase in energy released is proportional to X^2, that is, a hundredfold. The duration of a single oscillation differs greatly from one earthquake to another. The most energetic earthquakes have a higher proportion of long-duration waves. As a result, the increase in energy corresponding to an increase of one unit on the Richter scale, when summed over the whole range of waves in a wave record, is about 30-fold. Thus, a magnitude 6 earthquake releases almost 1000 times as much energy as a magnitude 4 quake (a 30-fold in-

T A B L E 3.1 • Earthquake Magnitudes and Frequencies of Occurrence, with Characteristic Damaging Effects

Richter Magnitude	*Number per Year*	*Modified Mercalli Intensity Scale[a]*	*Characteristic Effects of Shocks in Populated Areas*
<3.4	800,000	I	Recorded only by seismographs
3.5–4.2	30,000	II and III	Felt by some people who are indoors
4.3–4.8	4,800	IV	Felt by many people; windows rattle
4.9–5.4	1,400	V	Felt by everyone; dishes break, doors swing
5.5–6.1	500	VI and VII	Slight building damage; plaster cracks, bricks fall
6.2–6.9	100	VII and IX	Much building damage; chimneys fall; houses move on foundations
7.0–7.3	15	X	Serious damage, bridges twisted, walls fractured; many masonry buildings collapse
7.4–7.9	4	XI	Great damage; most buildings collapse
>8.0	One every 5–10 years	XII	Total damage, waves seen on ground surface, objects thrown in the air

[a]Mercalli numbers are determined by the amount of damage to structures and the degree to which ground motions are felt. These depend on the magnatude of the earthquake, the distance of the observer from the epicenter, and whether an observer is in or out of doors.

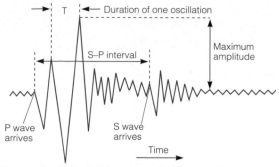

T — Duration of one oscillation

S–P interval

Maximum amplitude

P wave arrives

S wave arrives

Time

Y is correction factor that depends on the distance of the seismograph from the epicenter. It is calculated from the S-P interval.

◀ F I G U R E 3.12
Measurements used for determining Richter magnitude (M) from a seismogram. Y is a correction factor that depends on the distance of the seismograph from the epicenter. It is calculated from the S–P interval.

crease for each step in the scale, or $30 \times 30 = 900$). This means that a magnitude 8 earthquake, such as the New Madrid earthquakes of 1811–1812 mentioned at the beginning of the chapter, is not twice as large as a magnitude 4 quake but almost 1 million times as large ($30 \times 30 \times 30 \times 30 = 810,000$)! So even though it may happen infrequently, a single very large earthquake can release as much stored energy as many thousands of smaller quakes.

The Richter scale is open ended, which means that theoretically there is no upper limit on the possible size of an earthquake. The largest earthquakes recorded to date have Richter magnitudes of about 8.6. Many of the greatest earthquakes ever recorded occurred in subduction zones; they include the Chilean earthquake of 1960 (magnitude 8.5) and the Alaskan "Good Friday" earthquake of 1960 (magnitude 8.6). It is possible that earthquakes cannot be any larger because rocks cannot store more elastic energy. Before they are deformed further, they fracture and release the energy.

Through careful measurement of rocks along the San Andreas Fault, seismologists have found that about 100 J (joules) of energy can be accumulated in 1 m^3 of elastically deformed rock. This is not very much—it is equivalent to only about 25 calories of heat energy—but when billions or trillions of cubic meters of rock are strained, the total amount of stored energy can be enormous. When the frictional lock is broken and an earthquake occurs, the elastic energy is released during a few terrible seconds. (Even near the epicenter of a very large earthquake, the duration of strong ground motion is rarely more than about 30 seconds.) The amount of elastically stored energy released during the 1989 Loma Prieta earthquake (the San Francisco "World Series quake") was about 10^{15} J, and the 1906 San Francisco earthquake released at least 10^{17} J. For purposes of comparison, a hydrogen bomb blast also releases about 10^{17} J of energy.

The Modified Mercalli Intensity Scale

Another scale besides the Richter scale is sometimes used to measure the intensity of an earthquake. The **modified Mercalli intensity scale,** developed in the late 1800s and later modified by Father Giuseppi Mercalli and others, is based on the amount of vibration people feel during low-magnitude quakes and the extent of damage to buildings during high-magnitude quakes. The scale ranges from I (not felt except under favorable circumstances) to XII (waves seen on ground surface, practically all works of construction destroyed or greatly damaged). The correspondence among Mercalli intensity, Richter magnitude, and the approximate frequency of earthquakes is shown in Table 3.1.

Note that the Mercalli scale is not corrected for distance from the epicenter. This means that a single earthquake could have a Mercalli magnitude (i.e., a felt intensity) of IX or X near the epicenter, where the intensity is greatest, whereas 400 or 500 km away its intensity would be only I or II. The distance over which the effects are felt is partly a function of the size of the earthquake—that is, the intensity of the seismic waves—and partly a function of the efficiency with which the waves are transmitted by different materials (Fig. 3.13A).

The Mercalli scale has been particularly useful in the study of earthquakes that occurred before the development of modern measurement equipment. For example, the exact sizes of the series of earthquakes that hit New Madrid, Missouri, in 1811–1812 are unknown. However, historical eyewitness accounts indicate that all three quakes had Mercalli intensities of XI near the epicenter. Combining this with information about how far away the effects were felt, researchers have determined that the three shocks had Richter magnitudes of about 7.5, 7.3, and 7.8, respectively. Another great earthquake struck Charleston, South Carolina, on August 31, 1886, killing 60 people and de-

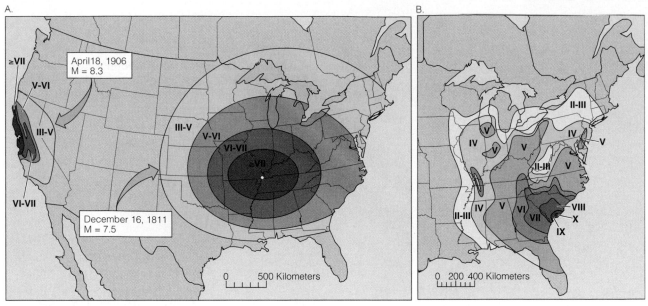

▲ F I G U R E 3.13
Intensities of great earthquakes on the modified Mercalli scale. A. Areas affected by the 1811–1812 New Madrid earthquakes (Richter magnitudes 7.5–7.8) and the 1906 San Francisco earthquake (Richter magnitude 8.3). The larger area around the New Madrid epicenter results from the greater efficiency of eastern crustal rocks in transmitting seismic waves. B. The earthquake that occurred on August 31, 1886, near Charleston, South Carolina, was felt throughout the eastern United States. The map shows intensities on the modified Mercalli scale. Estimates of Richter magnitude based on these observations range from 6.6 to 7.1.

stroying 90 percent of the city. The highest intensity of this quake on the Mercalli scale was X and was felt just northwest of Charleston (Fig. 3.13B). The effects of the earthquake were felt as far away as Minneapolis, Bermuda, and Cuba. The Richter magnitude estimated from these observations was in the range of 6.6–7.1.

Some Good News About Earthquakes

Before we turn our attention to the considerable hazards associated with earthquakes, it is worth noting that not everything about earthquakes is bad. Through measurements of earthquake vibrations, we can find out about parts of the Earth that are inaccessible to direct study. Used in this way, earthquake vibrations are like the X-rays a doctor uses to study the inside of a human body; they are the probes we use to sense and measure the world beneath our feet. The arrival times of seismic waves at seismographs around the world provide records of waves that have traveled along many different paths. From such records, it is possible to calculate how rock properties change with increasing depth below the Earth's surface. We can also identify the boundaries between layers with sharply different properties. For example, Danish seismologist Inge Lehmann used observations of the behavior of P waves passing through the center of the Earth to determine that the Earth has a solid inner core within its outer, liquid core. Her discovery was first published in a scientific paper entitled "P," which may qualify as the shortest article title ever published!

EARTHQUAKE HAZARDS AND RISKS

Each year there are many hundreds of thousands of earthquakes. Fortunately, only one or two are large enough, or close enough to major population centers, to cause loss of life. As discussed in the essay for Part II, population distribution, building codes, and emergency preparedness and response are significant factors in determining the risks associated with natural phenomena such as earthquakes. In this section we examine some of the risks and hazards associated with earthquakes and consider how scientific research can assist in addressing them.

Quantifying and Addressing the Risks

Seventeen earthquakes are known to have caused 50,000 or more deaths apiece (Table 3.2). The most disastrous one on record occurred in 1556 in Shaanxi Province, China. Many of the estimated 830,000 people who died in that quake lived in cave dwellings excavated from cliffs made of *loess* (fine, wind-deposited silt), which collapsed as a result of the quake. The worst earthquake disaster of the twentieth century also occurred in China. At 3:42 A.M. on July 28, 1976, while the 1 million inhabitants of T'ang Shan were asleep, a 7.8 magnitude quake leveled the city. Hardly a building was left standing, and the few that withstood the first quake were destroyed by a second one (*M* 7.1) that struck at 6:45 P.M. the same day. When the wreckage was cleared, 240,000 people were dead. The losses were so high because most of the buildings had unreinforced brick walls. When the ground started to shake, the walls collapsed, the roofs caved in, and the sleeping inhabitants were crushed.

Sometimes very large earthquakes occur in sparsely populated areas without causing significant damage. For example, in August 1993 a magnitude 8.1 quake occurred in the Pacific Ocean off the southeast coast of Guam. There were no deaths and only minor injuries, although some buildings sustained significant damage. The loss of life and property damage caused by the great Alaska earthquake (M 8.6) of 1964 were also much lower than they would have been had the quake struck in a more densely populated area.

Scientists have produced computer-simulated scenarios of what might happen if a great earthquake were to strike an area like Los Angeles or San Francisco. In one such model, a major earthquake (M >8.0) in a heavily populated area along the San Andreas Fault is projected to cause between 3000 and 13,000 deaths. The difference between the two extremes is simply the difference between day and night; the lower figure pertains to a scenario in which the quake occurs at 2:30 in the morning, whereas the higher one applies if the quake were to strike at 4:30 on a weekday afternoon. The reason for the difference is that wood-frame, single-family buildings are more earthquake resistant than the medium-rise, unreinforced masonry build-

ings that are common in many California towns. If Californians are at home in their beds when the "Big One" occurs, they will be less vulnerable overall than if they are at work or commuting to or from work.

Architecture and Building Codes

Certain areas, some of them heavily populated, are known to be earthquake prone, and building codes in those areas require structures to be as resistant as possible to earthquake damage. California, for example, imposes strict building codes and requires that special studies be conducted before any skyscraper, public building, or large construction project can be located near an active fault. Yet all too often earthquakes occur in areas in which buildings have not been constructed to withstand the effects of ground shaking. Examples include the T'ang Shan earthquake of 1976 and the magnitude 6.8 quake that struck Armenia in 1988. The Armenian earthquake occurred in a heavily populated area with poorly constructed buildings, killing an estimated 25,000 people. Another tragic example is the 1993 earthquake in Latur, India (M 6.4), in which more than 11,000 people were killed when their masonry homes collapsed into rubble. The horrendous casualties from these earthquakes belie the fact that their magnitudes were lower than those of both the 1989 Loma Prieta (San

T A B L E 3.2 • Earthquakes Occurring During the Past 800 Years That Have Caused 50,000 or More Deaths

Place	Year	Estimated Number of Deaths
Silicia, Turkey	1268	60,000
Chihli, China	1290	100,000
Naples, Italy	1456	60,000
Shaanxi, China	1556	830,000
Shemaka, USSR	1667	80,000
Naples, Italy	1693	93,000
Catalina, Italy	1693	60,000
Beijing, China	1731	100,000
Calcutta, India	1737	300,000
Lisbon, Portugal	1755	60,000
Calabria, Italy	1783	50,000
Messina, Italy	1908	160,000
Gansu, China	1920	180,000
Tokyo and Yokohama, Japan	1923	143,000
Gansu, China	1932	70,000
Quetta, Pakistan	1935	60,000
T'ang Shan, China	1976	240,000
Iran	1990	52,000

▲ **F I G U R E 3.14**
When a magnitude 7.0 earthquake struck Northridge, California, on January 20, 1994, an apartment house collapsed and crushed the automobiles parked beneath it.

BOX 3.1

•

THE HUMAN PERSPECTIVE

SEISMIC VERIFICATION OF NUCLEAR TESTING

*S*eismology is a global science by its very nature. Seismic data are collected and exchanged by scientists who collaborate with their international colleagues. Earthquakes generate seismic waves that are transmitted through the Earth and across its surface. The frequency, period, and amplitude of those waves create characteristic signatures that are received by an established global network of 125 monitoring stations known as the World-Wide Standard Seismological Network (WWSSN). The seismic waves of underground nuclear explosions detonated by countries testing nuclear devices are also received at the WWSSN stations. Scientists can interpret the seismic signatures of these waves to establish the location of a specific seismic event and determine whether it has a natural or nuclear source.

Since the period when only the United States and the former Soviet Union were nuclear opponents, decades of

research have gone into developing the long-range verification system used by the WWSSN. The first completely underground test, RAINIER, was conducted in 1957 in Nevada. This test was also the first for which seismology was employed as a diagnostic tool. Following RAINIER, a global seismology program and network evolved between the academic institutions and government agencies of the United States and 30 other countries.

The Soviet Union and the United States had a bilateral seismic monitoring network throughout the Cold War period, with monitoring stations at predetermined distances from suspected underground nuclear testing sites. Seismic monitoring technology was designed and calibrated to detect seismic waves that traveled long distances through the Earth's deep interior. Now that more

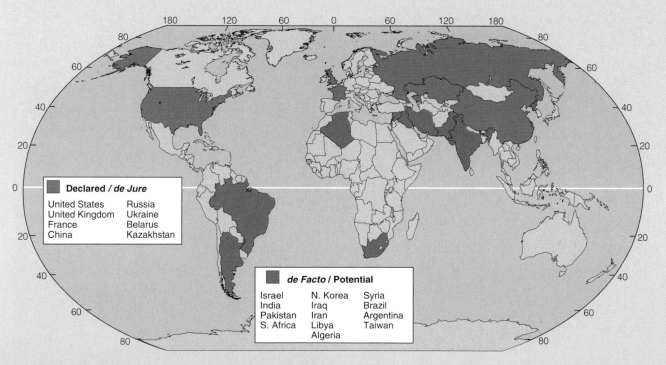

▲ F I G U R E B1.1
The new seismic monitoring task. The map shows countries of the world with declared (blue) and potential (orange) nuclear capabilities.

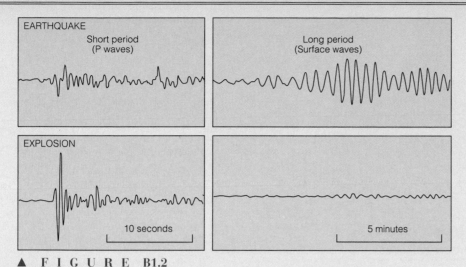

▲ F I G U R E B1.2
A comparison of seismic waves from explosions and earthquakes. Both
events took place in October 1984, in the former Soviet Union. The most
important difference between the signals from the two types of events is
the much smaller surface waves produced by the explosion.

than 20 countries must be monitored for nuclear testing
(Figure B1.1), the emphasis for the global monitoring network must be on detecting a variety of seismic signals from
short-distance regional waves.

Earthquakes produce a more complicated set of seismic
waves than do underground nuclear explosions. Earthquake waves emanate from a wide area over an interval of
time, as plates move past one another deep within the
Earth. Therefore, the seismic signal of earthquakes contains many long-period wavelengths. An underground nuclear explosion, by contrast, is focused within a relatively
small area for a short time and produces more short-period
wavelengths. For seismic events with large magnitudes
(>4.5), the nuclear and earthquake signals are readily dis-
tinguishable (Figure B1.2). For events of smaller magnitude, a seismic signal of natural origin is more difficult to
discriminate from one generated by an underground nuclear test.

Some countries have used these seismic principles to
evade detection by the global monitoring states. For exam-

ple, the testing of small-magnitude tactical weapons creates
high-frequency waves that are virtually impossible to detect over long distances. Nuclear devices have also been
tested during natural earthquakes, which masks the seismic
signal from the explosion. In other cases, countries have
detonated a series of tests in order to mimic an earthquake
signal, and sited tests in a large underground cavern in
order to muffle the seismic signal. This last method,
known as *decoupling*, blocks seismic waves emanating from
the explosion because the stresses from the explosion on
the rock will not exceed the elastic limit if the cavern is
large enough.

The development of innovative seismic instrumentation
and the ongoing cooperation among nations in the World-
Wide Standard Seismological Network have led to a
greater ability to discriminate between underground nuclear explosions and other seismic signals. The existence of
this seismic network has critical importance today, both
for the scientific exploration of the Earth's interior and for
the deterrence of nuclear proliferation.

◀ F I G U R E 3.15
Collapse of the Nimitz Freeway, San Francisco, as a result of the Loma Prieta earthquake on October 21, 1989.

Francisco) earthquake (M 7.1), in which 62 people were killed, and the 1994 Northridge (Los Angeles) earthquake (M 6.9), in which 51 died (Fig. 3.14). The comparatively low death tolls in the California earthquakes can be attributed in part to structural engineering and geoscientific research into the effects of intense ground shaking on buildings.

Even where special building codes are in effect, however, unanticipated problems may arise, such as the collapse of the Nimitz Freeway and the Bay Bridge during the Loma Prieta earthquake (Fig. 3.15) and the collapse of Interstate Highway 5 during the Northridge quake. In 1985, an earthquake destroyed parts of Mexico City; many of the buildings that collapsed had been built to meet "California standards,"—the most exacting standards in effect at that time (Fig. 3.16). Further reminders were seen in the twisted

rubble and collapsed buildings of Kobe, Japan, a city well known for earthquake preparedness, following a magnitude 7.2 earthquake in January 1995.

Structural engineers are continually refining their understanding of the factors that make tall buildings and other structures more resistant to collapse (Fig. 3.17). However, much of this understanding is based on theory and laboratory work; there are few real-life situations in which the response of a building to exceptionally intense ground motion can be tested and the standards verified.

Seismic Hazard and Risk Mapping

As discussed in the Part II essay, describing the nature and intensity of an expected event is only one part of risk and hazard assessment. Another important step is determining

◀ F I G U R E 3.16
When an earthquake struck Mexico City in September, 1985, this apartment house collapsed and sandwiched occupants between floors. The rescuer is searching the wreckage for survivors.

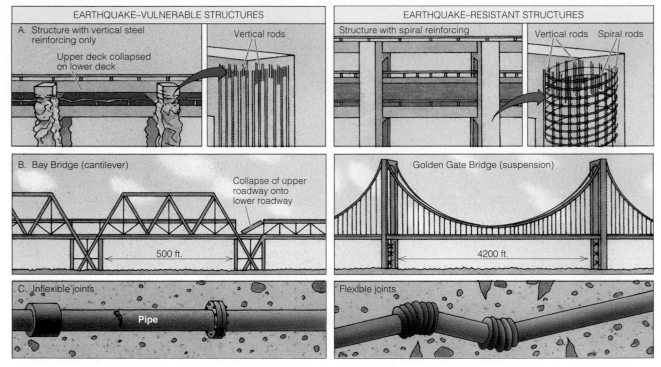

▲ F I G U R E 3.17
The way structures are designed and built can determine how they respond to ground shaking during an earthquake. A. The worst structural failure during the 1989 Loma Prieta earthquake was the collapse of the Nimitz Freeway, where a number of concrete columns gave way. The columns had been built in the 1950s with vertical steel rods inside, but they lacked the spiral reinforcing rods used in modern construction. B. During the Loma Prieta earthquake the Bay Bridge, with its inflexible cantilever-style design, collapsed, while the Golden Gate Bridge merely swayed. C. Flexible joints between gas and water pipes that give the pipes some flexibility during ground shaking can prevent gas leaks, water damage, and water shortages following an earthquake.

exactly where such an event is most likely to occur and comparing the expected event with human factors such as population distribution, building codes and zoning laws, and emergency preparedness in the area. This is often done with seismic hazard maps like those shown in Figs. 3.18 and 3.19.

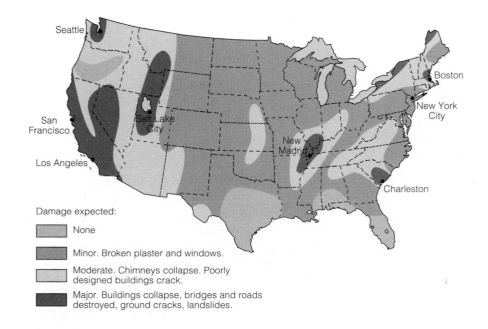

Damage expected:

◻ None

◼ Minor. Broken plaster and windows.

◻ Moderate. Chimneys collapse. Poorly designed buildings crack.

◼ Major. Buildings collapse, bridges and roads destroyed, ground cracks, landslides.

◀ F I G U R E 3.18
A seismic-risk map of the United States based on quake intensity. Zones refer to maximum earthquake intensity and therefore to maximum possible destruction. The map does not indicate the frequency of earthquakes. For example, frequency is high in southern California but low in eastern Massachusetts. Nevertheless, when earthquakes occur in eastern Massachusetts they can be as severe as the more frequent quakes in southern California.

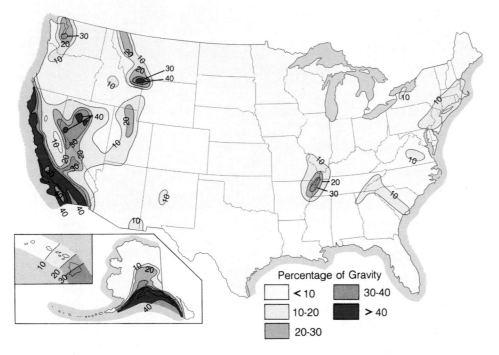

▲ F I G U R E 3.19

A seismic-risk map based on horizontal acceleration. Because acceleration during
ground movement is the most important factor in the design of an earthquake-stable
building, this type of risk map is preferred by builders and city planners. Numbers
on contours refer to the maximum horizontal acceleration during an earthquake,
expressed as a percentage of the acceleration due to gravity (980 cm/s^2). The
probability that an acceleration of the amount indicated will occur in any given
50-year period is 1 chance in 10.

Active plate margins are a logical starting point for attempts to quantify the hazards and risks associated with
earthquakes. For example, when earthquakes are mentioned, most people in the United States immediately think
of California and the San Andreas Fault, which marks the
active boundary separating the North American plate from
the Pacific plate. However, some of the most intense earthquakes to jolt North America in the past 200 years were the
ones centered near New Madrid, Missouri and Charleston,
South Carolina, which are located far from active plate
boundaries. Thus, the possibility that a major earthquake
might occur in North America east of the Mississippi River
is a topic of concern to seismologists and planners alike.

On the basis of known geologic structures (mainly
faults) and the location and intensity of past earthquakes,
the U.S. National Oceanographic and Atmospheric Administration (NOAA) prepared the seismic-risk map shown
in Fig. 3.18. Although such a map is very informative, it is
not particularly useful to people who plan and build roadways and public buildings or calculate insurance premiums.

Planners need to know just how strongly the ground is
likely to shake and what the frequency of large earthquakes
might be. To meet this need, the U.S. Geological Survey
publishes seismic-risk maps like the one shown in Fig.
3.19. In this type of map the strength of a possible earthquake is compared to the acceleration from gravity, which
is 980 cm/s^2 (1.0 G). Scientists have calculated that there is
1 chance in 10 that a given acceleration will be exceeded in
a 50-year period. The maximum acceleration is likely to
occur in California, reaching 80 percent of gravity (0.8 G).
Damage begins at about 0.1 G.

Hazards Associated with Earthquakes

A variety of hazards are associated with earthquakes. The
first two discussed here—ground motion and faulting—are
primary effects; they cause damage directly. The other effects are secondary; they cause damage indirectly as a result
of processes set in motion by the earthquake. Tertiary and
longer term effects—such as displacement of people; dis-

ruption of services; loss of jobs; destruction of wildlife habitat; and permanent changes in groundwater levels, waterways, or coastlines—are also caused by large earthquakes.

Ground Motion

Ground motion results from the movement of seismic waves, especially surface waves, through surface rock layers and regolith. In the most intense earthquakes ($M > 8.0$) the surface of the ground can sometimes be observed moving in waves. Ground motion is the most significant primary cause of damage from earthquakes. It can damage and sometimes completely destroy buildings. As discussed earlier, proper design of buildings and other structures can do much to prevent these effects, but in a very strong earthquake even the best designed buildings may suffer some damage.

The wide variety of effects caused by earthquakes results partly from differences in the way Earth materials transmit seismic waves. For example, the New Madrid earthquakes of 1811–1812 were felt over a much wider area than the 1906 San Francisco earthquake of similar magnitude because seismic body waves are transmitted more efficiently by the crustal rocks in the central United States, as shown in Fig. 3.13A.

Sometimes local geologic factors can amplify the effects of ground motion. For example, the damage due to ground shaking during the Loma Prieta earthquake was of two types: that caused by normal ground shaking and that caused by enhanced (i.e., higher than expected) ground shaking. Enhanced ground motion is often attributable to the presence of soft sediments, like the landfill in the San Francisco Bay area. It has been calculated that enhanced shaking may have been responsible for as much as 70 percent of the losses incurred during the quake, whereas nor-

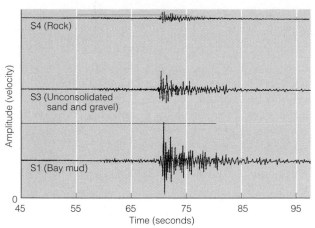

▲ **F I G U R E   3.20**

Seismograms recorded for a magnitude 4.1 aftershock to the Loma Prieta earthquake, on mud near the collapsed Nimitz Freeway (S1), on unconsolidated material (sand, gravel, etc.) near an uncollapsed portion of the freeway (S2), and on bedrock in the Oakland Hills (S3). All three seismograms are plotted on the same scale. The horizontal component of ground motion is amplified in the seismogram recorded at the mud site.

mal ground shaking caused only about 28 percent of total losses (Table 3.3). Enhanced ground shaking resulting from the presence of soft sediments may also be implicated in the collapse of the Nimitz Freeway during that event. As shown in Fig. 3.20, seismic signals recorded during a magnitude 4.1 aftershock show that ground motions were amplified by the soft bay muds, compared to other types of Earth materials in nearby sites.

T A B L E   3.3   •   Losses Resulting from the Loma Prieta Earthquake, by Type of Hazard

Earthquake Hazard	Total Damages (millions)	Loss (% of Total)
Ground shaking		
Normal	$1635	28
Enhanced	$4170	70
Liquefaction	$97	1.5
Landslides	$30	0.5
Ground rupture	$4	0.0
Tsunami	$0	0.0
Total	$5936	100.00

From Holzer, Thomas L. *EOS*, vol. 75, no. 26, June 28, 1994, p. 300.

Faulting and Ground Rupture

Where a fault breaks the ground surface, buildings can be split, roads disrupted, and any feature that lies on or that crosses the fault broken apart. Large cracks and fissures can open in the ground (Fig. 3.21). Even the numerous small earthquakes resulting from slow, continuous movements along faults can contribute to structural damage. Such movement can be seen in Carrizo Plains, California, which sits astride the San Andreas Fault (Fig. 3.22).

Aftershocks

Earthquake crises are often exacerbated by **aftershocks,** the (usually) smaller earthquakes that occur shortly after a major quake. In the four months following the 1964 Alaska earthquake, for example, 1260 aftershocks were recorded. In some cases large earthquakes trigger aftershocks at locations distant from the original epicenter. The Landers (near Los Angeles) earthquake of 1992 (*M* 7.3) triggered secondary events at 14 locations, some of them as far as 1250 km away.

Fires

A secondary effect, but one that can pose a greater hazard than ground motion, is fire. Ground movement displaces stoves, breaks gas lines, and loosens electrical wires, thereby starting fires. Because ground motion also breaks water mains, there often is no water available to put out the fires.

In the earthquakes that struck San Francisco in 1906 and Tokyo and Yokohama in 1923, as much as 90 percent of the damage to buildings was caused by fire. In the 1906 San Francisco earthquake the fire destroyed 12 km², or 521 city blocks, in 3 days. For many years thereafter the quake was referred to as the "Great Fire." In 1989 fires caused by the Loma Prieta earthquake again ravaged the Marina district of downtown San Francisco (Fig. 3.23).

Landslides

In regions with steep slopes, earthquake vibrations may cause slipping of regolith, collapse of cliffs, and other rapid downslope movements of Earth material. This is particu-

▲ F I G U R E 3.22
Slow movement of the San Andreas Fault, Carrizo Plains, California, has detached the headwaters of a stream (upper right) from the downstream portion (lower left). The fault runs across the center of the photo. The land to the rear is moving right relative to the land in the foreground. Because motion on the fault is slow and continual, the two halves of the stream system remain in contact as water flows down the depression caused by the fault.

▲ F I G U R E 3.21
Fissures opened in Santa Cruz, California, as a result of the Loma Prieta earthquake, October 21, 1989.

◄ F I G U R E 3.23
Fire caused by gas lines that broke as a result of the Loma Prieta earthquake in 1989. This shot shows a fire in the Marina district of San Francisco.

larly true in Alaska, parts of southern California, China, and hilly places such as Iran and Turkey. Houses, roads, and other structures are destroyed by rapidly moving regolith. In 1970, for example, a devastating landslide in Yungay, Peru, in which at least 18,000 people were killed, was triggered by an earthquake of magnitude 7.75.

Liquefaction

The sudden shaking and disturbance of water-saturated sediment and regolith can turn seemingly solid ground into a liquidlike mass of quicksand. This process, called **liquefaction,** occurs when vibration causes sediment grains to lose contact with one another, allowing interstitial water to bubble through. Soil liquefaction and the resulting landslides were major causes of damage during the earthquake that destroyed much of Anchorage, Alaska, in 1964 (Fig. 3.24). In the same year liquefaction and uneven ground settling resulting from an earthquake caused apartment houses to sink and collapse in Niigata, Japan. Many of the buildings were not structurally damaged; they simply keeled over onto their sides. Apartment dwellers who were later permitted to enter the buildings retrieved their belongings by rolling wheelbarrows up the walls and lowering themselves through the windows.

◄ F I G U R E 3.24
Liquefaction of clay layers as a result of the Alaskan earthquake of March 27, 1964 destroyed these houses in Turnagain Heights.

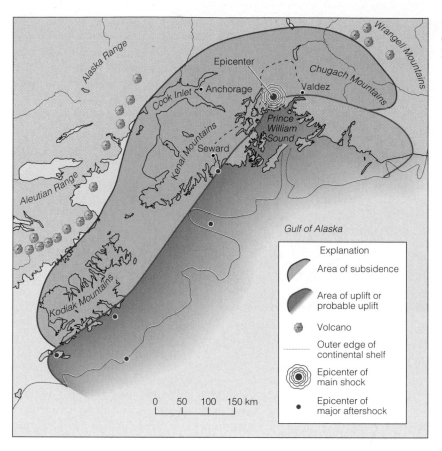

◄ FIGURE 3.25
Changes in the crust after the Good Friday, 1964, earthquake in Alaska.

Changes in Ground Level

Sometimes the level of the ground changes over vast areas as a result of a very large earthquake (Fig. 3.25). As noted earlier, in the 1964 Alaska earthquake vertical displacements occurred along almost 1000 km of the coastline from Kodiak Island to Prince William Sound. The ground-level changes resulting from that earthquake included both uplift and subsidence. Subsidence, a lowering of ground level, was as much as 2 m, whereas vertical uplift reached 11 m in some places.

Tsunamis

Another secondary effect of earthquakes is **seismic sea waves,** also called *tsunamis.* Submarine earthquakes are the main cause of these waves, which are particularly destructive around the Pacific Ocean rim. A well-known example is the tsunami generated by a severe submarine earthquake near Unimak Island, Alaska, in 1946. The wave traveled across the Pacific Ocean at a velocity of 800 km/h, striking Hilo, Hawaii, about 4.5 hours later. Although the amplitude (height) of the wave in the open ocean was less than 1 m, it increased dramatically as the wave approached land. When it hit Hawaii, the wave had a crest 18 m higher than normal high tide. It demolished nearly 500 houses, damaged 1000 more, and killed 159 people. Another devastat-

ing tsunami was generated by an earthquake off the coast of Portugal in 1755. It caused the deaths of 60,000 people in Lisbon alone. The waves were observed a few hours later as far away as the West Indies. (Tsunamis are discussed in more detail in Chapter 5.)

Flooding

Flooding is a secondary or tertiary effect of earthquakes, usually resulting from ground subsidence (as in the 1964

▼ FIGURE 3.26
Flooding in Portage, Alaska, due to ground subsidence, following the earthquake of March 27, 1964.

NORTHRIDGE EARTHQUAKE: A PERSONAL ENCOUNTER

I used to think that earthquakes were fun and exciting. I used to tell my friends and students that I wanted to be in Los Angeles when the oft-predicted BIG ONE visits us. I no longer feel that way. I realize now that this attitude toward earthquakes stemmed from having been safely distant from the epicenters of past tremors. Earthquakes were fun, exciting, and made great lecture material for my class.

January 17, 1994, changed that perspective. At 4:31 A.M., I and several million of my closest neighbors were jarred from deep sleep. I knew instantly that the demonic shaking of my Valley condominium was earthquake induced, and I knew it was going to be bad. In total citywide darkness, my three-storey building shook and flexed and rolled with a ferocity so unimaginable to me that I was sure it would fall. A horrendously loud and ugly cacophony—the dissonant, overprinted sounds of furniture toppling, glass breaking, wood flexing, people screaming, and car alarms blaring—accompanied the intense lurching and vibrating of the building around me. With a flashlight, I ran down my swaying stairs to save a cabinet filled with prized art glass. That cabinet was my focus in the midst of a situation in which I had no control. The shaking and noises lessened, than ceased. With heart pounding, I stood in the almost quiet darkness beside the rescued cabinet and surveyed my littered living room. I remember my wonderment then—a feeling that continues three weeks later—that my home was still standing and that it was structurally undamaged. Just two blocks away and 10 miles from the epicenter, a major freeway interchange had suffered heavy damage and had nearly collapsed.

The Northridge earthquake. 25–30 seconds duration. 57 deaths. 9,300 injuries. 10,000 jobs lost. 5,700 mobile homes knocked off their foundation and another 240 burned when ruptured gas lines were ignited. 13,000 buildings destroyed or severely damaged. 22,000 dwelling units ordered vacated. 51,600 residences damaged. 55,000 persons homeless, at least temporarily. 304,000 people seeking financial aid. More than $10 billion in damage, and several million bruised states of mind.

The BIG ONE? Still wanna be here when it comes, Greg? All two or three minutes of it? Thanks, but I'll pass. Earthquakes just aren't fun anymore. And now I know, the big ones never were.

[Source: From *Geotimes,* May 1994. Contributed by Greg Davis. Editor's note from *Geotimes:* This piece was written three weeks after the Northridge earthquake, and some of the information cited has been officially updated since then.]

Alaska earthquake, Fig 3.26), the rupture of dams, or tsunamis. Reelfoot Lake, across from New Madrid on the Tennessee side of the Mississippi River, was created by flooding following ground subsidence during the series of 1811–1812 earthquakes.

PREDICTION AND CONTROL OF EARTHQUAKES

Charles Richter once said, "Only fools, charlatans, and liars predict earthquakes." Today, however, seismologists use sensitive instruments and sophisticated techniques to monitor seismically active zones. We still cannot predict the exact magnitude and time of occurrence of an earthquake, but our knowledge about seismic hazards and mechanisms and our awareness of the tectonic environments in which earthquakes occur have improved greatly since Richter's time. Long-term earthquake forecasting is based primarily on our understanding of the tectonic cycle and the geologic settings in which earthquakes occur. Short-term earthquake predictions, on the other hand, are usually based on observations of precursor phenomena, physical anomalies that may be early warning signs of earthquake activity.

The World Distribution of Earthquakes

The first step in earthquake prediction is to examine the world distribution of earthquakes and establish connections between the plate tectonic cycle and the mechanisms by which earthquakes are generated. Although no part of the Earth's surface is exempt from earthquakes, several well-defined **seismic belts** are subject to frequent shocks (Fig. 3.27). Of these, the most obvious is the *circum-Pacific belt*, for it is here that about 80 percent of all recorded earthquakes originate. The belt follows the mountain chains in the western Americas from Cape Horn to Alaska and crosses to Asia, where it extends southward down the coast through Japan, the Philippines, New Guinea, and Fiji and finally loops far southward to New Zealand. Next in prominence, giving rise to 15 percent of all earthquakes, is the Mediterranean-Himalayan belt, which extends from Gibraltar to Southeast Asia. Lesser belts follow the midocean ridges.

A lot of the Earth's internal energy is released in these seismic belts. One might therefore expect other manifestations of internal energy also to appear in these regions, and indeed some of them do. Midocean ridges, deep-sea trenches, volcanoes, and many other features that outline the active margins of lithospheric plates either coincide with or closely parallel those margins. If you compare Fig. 3.27 with Fig. 1.14 you will see that seismic belts outline the plate boundaries.

Seismicity and Plate Tectonics

We learned in Chapter 1 that there are three kinds of plate boundaries: (1) *divergent boundaries,* or spreading centers, which coincide with rift valleys and midocean ridges; (2) *transform fault boundaries,* where two plates or portions of plates are sliding past each other; and (3) *convergent bound-*

aries, which coincide with oceanic trenches and/or continental collision zones. Each kind of boundary is characterized by earthquakes with distinctive fault motions and depth of foci. In addition, as we have seen, very large earthquakes occasionally occur in the interiors of plates, in the normally stable continental intraplate environment.

Divergent Boundaries

Along a divergent plate boundary or spreading center, two plates move apart from each other and the lithosphere is stretched by tensional stresses (Fig. 3.28A). The faults associated with tensional stresses are normal faults. Earthquakes at spreading centers tend to have low Richter magnitudes and foci that are invariably less than 100 km and usually less than 20 km deep. This indicates that the lithosphere is thin beneath a spreading center and that the ductile material underneath must come close to the surface.

Transform Fault Boundaries

Transform faults are huge, vertical strike-slip faults that cut down through the lithosphere. They mark the boundaries where two plates slide past each other (Fig. 3.28B). Earthquakes along transform faults always have shallow foci, no deeper than 100 km, and often have high Richter magnitudes. The locations of the earthquake foci suggest that when a transform fault cuts the continental crust, a system of parallel faults rather than a single fracture can develop. This seems to be true in the case of the San Andreas Fault, as shown in Fig. 3.29.

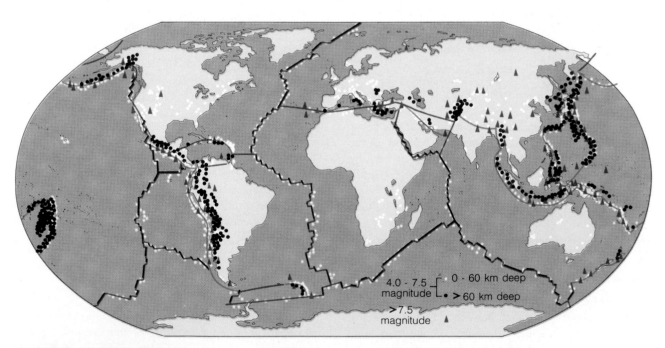

▲ F I G U R E 3.27
The Earth's seismicity outlines plate margins. This map shows earthquakes of magnitude 4.0 or greater from 1960 to 1989.

Convergent Boundaries

As discussed in Chapter 1, convergent plate boundaries are of two kinds: (1) *subduction boundaries,* where lithosphere capped by oceanic crust is subducted into the asthenos-

phere and mesosphere; and (2) *collision boundaries,* where two continents collide. Each kind of convergent boundary has a characteristic pattern of seismic activity.

When oceanic lithosphere is subducted, it is subjected to complex stresses that can result in different kinds of earthquakes (Fig. 3.28C). Lithosphere bends downward as it is subducted, and the bending causes normal faults in the upper part of the plate. Earthquakes associated with such faults have very shallow foci and small Richter magnitudes. Subduction involves one plate sliding beneath the other, so the boundary between the two plates is a thrust fault. Above a depth of at least 100 km (the region where the two plates of lithosphere are in contact) earthquakes often have large Richter magnitudes and are invariably associated with thrust faulting. Below 100 km, where the subducted lithosphere is sinking through the asthenosphere, earthquakes occur *in* the subducted slab. Some earthquakes indicate

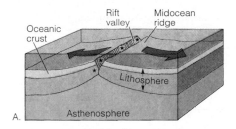
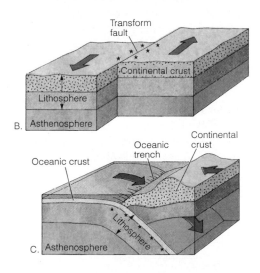
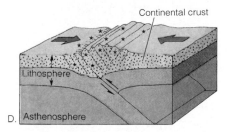

▲ F I G U R E 3.28
The relationship between seismicity and plate boundaries. Stars indicate earthquake foci. A. Divergent boundary. Earthquakes have shallow foci, low Richter magnitudes, and normal fault motions. B. Transform fault boundaries. Earthquakes have foci down to 100 km, sometimes high Richter magnitudes, and strike-slip motions. C. Subduction boundary. Earthquakes are complex. Earthquakes with low Richter magnitudes, shallow foci, and normal fault motions are observed seaward of the oceanic trench. Deeper earthquakes can have high Richter magnitudes and thrust fault movement. D. Continental collision boundary. Earthquakes have foci down to 300 km, thrust fault motions, and sometimes high Richter magnitudes.

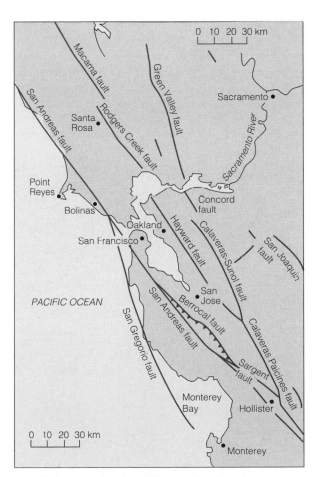

▲ F I G U R E 3.29
Map showing faults in the region of San Francisco, along which displacement has occurred within the last 10,000 years. The San Andreas Fault is not a simple fault but a complex assemblage of fractures marking the transform boundary between two plates.

tensional stresses (normal faults), whereas others indicate compressional stresses (reverse faults).

A few earthquakes originate at depths as great as 700 km. Exactly why and how such deep earthquakes occur in subducted slabs remains something of a puzzle. A very large (*M* 8.2), deep-seated (640 km) earthquake, the largest deep-focus earthquake ever measured, occurred in Bolivia in 1994. This event afforded scientists a rare opportunity to study the mechanisms associated with earthquakes that originate deep within the Earth, where high temperatures and pressures normally cause rocks to flow and deform ductilely under stress, rather than fracture.

Collision boundaries are the places where two continents collide. The Caucasus mountain chain between Russia and Asia Minor is a present-day collision boundary; the devastating 1988 earthquake in Armenia was a manifestation of this collision. A zone of collision tends to be several hundred kilometers wide, and within it rocks are intensely compressed and thrust-faulted, as shown in Fig. 3.28D. Within a collision zone, the crust is locally thickened. Earthquakes may have foci as deep as 300 km and high Richter magnitudes.

Intraplate Earthquakes

Whereas active plate boundaries are zones of frequent and intense earthquake activity, the stable interiors of continents tend to be seismically very quiet by comparison. But intraplate environments also can be the sites of large earthquakes. In some cases, such as that of the 1811–1812 New Madrid earthquakes, the seismic activity is believed to be related to ancient, deep-seated fault structures that are reactivated from time to time. In other cases, the earthquakes may be related to incipient continental rifting. In still others, such as the Latur earthquake of 1993, which occurred in the ancient, stable heart of the Indian subcontinent, the causes and mechanisms are simply not understood.

Approaches to Earthquake Prediction

As we have seen, some of the most dreadful natural disasters in history have been caused by earthquakes. It is hardly surprising, therefore, that a great deal of research around the world focuses on earthquake prediction. The hope is that through this research seismologists will be able to improve their forecasting ability to the point at which effective and accurate early warnings can be issued.

Long-Term Forecasting: Paleoseismology and Seismic Gaps

As noted earlier, the long-term prediction of earthquakes is based on knowledge of the tectonic cycle. In places where earthquakes are known to occur repeatedly, such as along plate boundaries, it is sometimes possible to detect a regular pattern in the recurrence interval of large quakes. To do so,

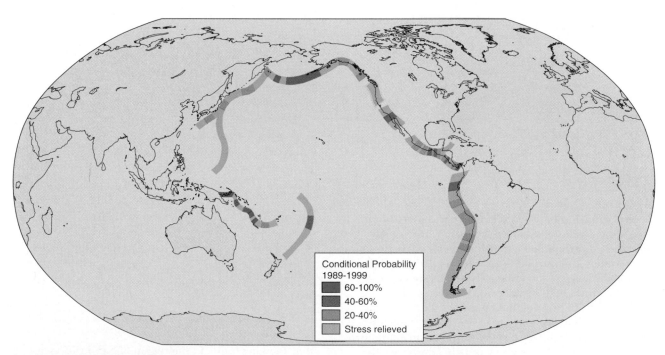

▲ F I G U R E 3.30
Seismic gaps in the circum-Pacific belt. In the areas indicated, earthquakes of magnitude 7.0 or greater are known to occur but have not done so in recent times. Strain is now building up in each seismic gap, raising the probability that a large quake will occur.

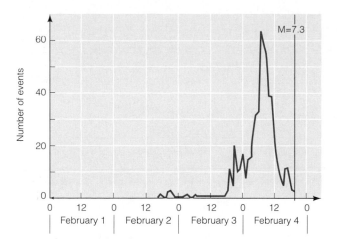

▲ F I G U R E 3.31

Frequency of foreshocks before the 1975 earthquake (*M* 7.3) at Haicheng, People's Republic of China. Foreshock activity peaked just before the main event.

however, seismologists require information about seismic activity going back much farther than historical records. This information is provided by experts in **paleoseismology,** the study of prehistoric earthquakes. The primary goal of paleoseismology is to search the sedimentary rock record for evidence of major earthquakes and, if possible, to discern the intervals between them. The evidence may include vertical displacement of sedimentary layers, as well as indicators of liquefaction or lateral offsetting of sedimentary features. If the pattern of recurrence suggests an interval of, say, a century between major quakes, it may be possible to predict where and when a large quake may happen next. Certainly it is possible to monitor such areas closely when a big quake is thought to be imminent.

Studies of recurrence patterns have identified a number of **seismic gaps** around the Pacific rim (Figure 3.30). These are places along a fault where earthquakes have not occurred for a long time even though tectonic stresses are still active and elastic energy is steadily building up. Seismic gaps receive a lot of attention from seismologists because they are considered to be the places most likely to experience large earthquakes. A major seismic gap has been identified along the San Andreas Fault in the vicinity of Parkfield, California, about halfway between Los Angeles and San Francisco. Through paleoseismological study and close examination of historical records of earthquake activity, seismologists have determined that large earthquakes tend to occur along this stretch of the fault at intervals of about 20 to 27 years. The last large earthquake near Parkfield occurred in 1966, suggesting that the area may be overdue for another major quake. In 1985 seismologists at the U.S. Geological Survey predicted that a magnitude 6 earthquake would strike Parkfield in 1988 ± 5 years. The predicted time has come and gone with no large quakes along the

Parkfield portion of the fault. It may be that some of the stored energy was dissipated by large earthquakes in nearby areas, such as Landers and Loma Prieta. In the meantime, however, seismologists are keeping a close eye on Parkfield. If they cannot accurately predict the quake, at least they hope to be ready and waiting, with instruments in place, when it finally does occur.

Short-Term Prediction: Physical Precursors

Short-term prediction of earthquakes is based on observations of precursor phenomena: anomalous physical occurrences that may serve as early warning signs of earthquake activity. Most research on short-term earthquake prediction involves monitoring changes in the properties of elastically strained rocks—properties such as rock magnetism and electrical conductivity. Even simple observations, such as the level of water in wells or the amount of radon (an inert gas) in well water, might indicate changes in the properties of underlying rocks. Some researchers have also reported the detection of unusual radio waves near the epicenter of the Loma Prieta earthquake just before the quake.

Tilting or bulging of the ground and slow rises and falls in elevation are among the most reliable indications that strain is building up. Most significant are the small cracks and fractures that develop in severely strained rock. These can cause swarms of tiny earthquakes—**foreshocks**—that may be a clue that a big quake is coming. One of the most famous successful earthquake predictions, made by Chinese scientists in 1975, was based on slow tilting of the land surface, fluctuations in the magnetic field, and the numerous small foreshocks that preceded a magnitude 7.3 quake that struck the town of Haicheng (Fig. 3.31). Half the city was destroyed, but because authorities had evacuated more than a million people before the quake, only a few hundred were killed.

Because the People's Republic of China has suffered so many terrible earthquakes, Chinese scientists have used every technique they can think of in their efforts to predict quakes. They have even observed animal behavior, and on one occasion animal behavior apparently did foretell a quake. On July 18, 1969, zookeepers at the People's Park in Tianjin observed highly unusual behavior among animals in the park. Normally quiet pandas screamed, swans refused to go near water, yaks did not eat, and snakes would not go into their holes. The keepers reported their observations to the earthquake prediction office, and at about noon on the same day a magnitude 7.4 earthquake struck the region. Throughout the world there have been many reports, both documented and informal, of strange animal behavior before earthquakes (Fig. 3.32), and Japanese researchers have conducted extensive laboratory experiments to investigate the connections between animal behavior and earthquakes. However, the Tianjin earthquake remains the only well-documented case in which the behavior led to a successful prediction.

	Time before earthquake						
	1–2 minutes	10–30 minutes	1–4 hours	6–12 hours	1 day	Few days	Few weeks
Epicenter area							
20–50 km							
70–100 km							
150–200 km							
>250 km							

▲ F I G U R E 3.32

Spatial and temporal distribution of reported incidents of anomalous animal behavior before the main shocks of 36 major earthquakes in Europe, Asia, North America, and South America. The symbols indicate unusual behavior by catfish, eels, other fish, frogs, snakes, turtles, sea birds, chickens, other birds, dogs, cats, deer, horses, cows, rats, and mice; the circled numbers indicate the number of reported incidents.

The long-term prediction of earthquakes (i.e., predictions of quakes in particular areas within the next few years or decades) has met with reasonable success. Seismologists know where the most hazardous areas are; they can calculate the probability that a large earthquake will occur in a particular area within a given period, and they have a theory of earthquake generation that successfully unites their predictions and observations in a plate tectonic context. The short-term prediction of earthquakes (i.e., pinpointing the actual day or time of the event and issuing an early warning) has been rather less successful. The difficulties with short-term prediction stem partly from the fact that the mechanisms and processes associated with earthquakes are hidden under the ground, where they are not amenable to study and monitoring. But a more significant problem is that earthquakes are highly inconsistent in terms of their precursor phenomena. For example, after several notable successes in predicting earthquakes on the basis of foreshocks, radon anomalies, ground tilting, and anomalous animal behavior, Chinese scientists suffered a major setback when the T'ang Shan quake of 1976 struck with no discernible warning signs.

One of the most promising earthquake monitoring techniques involves the use of satellites to measure very small changes in the surface of the ground. Through the Global Positioning System (GPS), a network of ground-based receivers tracks distance ranging signals from satellites orbiting high above the Earth's surface. If there is a change in the distance from the ground station to the satellite, displacement or deformation of the ground surface has occurred. In an even more sophisticated application, satellite radar images taken immediately before and after the 1992 Landers, California, earthquake were superimposed on one another, yielding an "interferogram"—essentially a contour map of changes in the ground surface between the times when the two radar images were acquired (Fig. 3.33). This approach is particularly exciting because of its very high resolution and because it can be done entirely by satellite, without requiring the use of ground-based instrumentation. Such detailed measurements of changes in the ground surface allow seismologists to refine their understanding of the crustal deformation resulting from earthquakes and may one day allow them to issue early warnings based on observations of changes in the rate or extent of crustal deformation.

Controlling Earthquakes

If earthquakes are caused by the sudden, catastrophic release of huge amounts of stored elastic energy, might it not

▲ FIGURE 3.33
An "interferogram" of the 1992 Landers, California, earthquake. This image was created by D. Massonnet and colleagues by superimposing a satellite radar image of the region made immediately before the quake on an image made just after the quake. The result is essentially a contour map revealing the extent of changes in the position of ground reflectors between the time of acquisition of the two images. The fault itself can be seen as a jagged line running diagonally across the right-hand side of the image. Ground deformation was most severe in the areas nearest the fault, as evidenced by the close spacing of the interference contours in those regions. Each curved color fringe represents a vertical movement of 2.5 cm. The area shown is approximately 90 km by 90 km.

about 5 and two others had magnitudes of about 4. The cause of these quakes was thought to be the increased load of water in Lake Mead, which lubricated and reactivated some ancient underlying faults. Many other examples of *induced seismicity* associated with the filling of large reservoirs have since been documented, with magnitudes ranging as high as 6.5. It has been suggested that the filling of a nearby reservoir may have played a role in the Latur, India, earthquake, although this theory has yet to be substantiated.

In the late 1960s swarms of unexplained small earthquakes occurred in the area of Denver, Colorado, which normally is not seismically active. A correlation was soon made between the occurrence of the quakes and the pressure in a hazardous waste disposal well at the Rocky Mountain Arsenal, into which the United States Army was periodically injecting fluid wastes (a fully legal procedure). The epicenters of the earthquakes were located near the disposal well, and their focal depths ranged from 4 to 6 km, just below the depth of the 3.8 m well. In this case, too, the cause of the earthquakes was thought to be the reduction of frictional pressure and the resulting reactivation of old faults as a result of increased fluid pressure. Similar occurrences were later observed at the Rangely oil field in northwestern Colorado and near the Perry nuclear power plant in Cleveland, Ohio.

It has also been observed that earthquakes can be triggered by underground nuclear explosions. In the late 1960s the U.S. Geological Survey conducted seismologic monitoring of underground testing of explosives at the Nevada Test Site. These studies revealed the occurrence of thousands of small aftershocks (generally $M < 5$; the test itself yielded a M of 6.3) with focal depths as great as 7 km and epicenters as far as 13 km from ground zero. Most of the aftershocks occurred within the first week after the explosion, but a general increase in seismic activity was still evident 3 weeks later. The aftershocks appeared to have occurred along a system of fractures that was not evident at the surface.

These events and observations have not led to a mechanism for controlling or avoiding major earthquakes. What they do show is that it is possible to generate earthquakes by injecting fluids or setting off explosions underground. This, in turn, suggests that someday it may be possible to develop a way of artificially inducing slip along a locked section of a major fault in order to avoid the dangerous accumulation and sudden release of elastic energy.

be possible to control earthquakes by finding a way of releasing that energy more gradually? Scientific investigations into the possibility of control have centered on earthquakes that were unwittingly caused by human actions. The first such quakes were reported after the construction of the Hoover Dam in 1935. About 600 local earthquakes were reported in the 10 years following construction of the dam, during which time Lake Mead was being created. Most of the tremors were small, but one had a Richter magnitude of

SUMMARY

1. Earthquakes are caused by abrupt movements of the Earth's crust that release stored energy.

2. Fractures in rocks along which movement occurs are called faults. Normal faults are caused by tensional stresses that tend to pull the crust apart, whereas reverse faults arise from compressional stresses that squeeze, shorten, and thicken the crust. Strike-slip faults are caused by shear stresses; they are vertical fractures that have horizontal motion.

3. According to the elastic rebound theory, friction between two blocks on either side of a fault may inhibit smooth, continuous slippage along the fault. When this happens, the rocks in the blocks deform, changing their shape and/or volume and storing elastic energy in the process. When enough energy has built up to overcome the frictional lock, abrupt slippage occurs along the fault, energy is released suddenly in the form of an earthquake, and the rocks return to their original shapes.

4. The study of earthquakes is called seismology. Scientists who study them are seismologists, and earthquake vibrations are measured and recorded by seismographs.

5. Energy released at an earthquake's focus radiates outward as body waves, which are of two kinds—P waves (primary waves, which are compressional) and S waves (secondary waves, which are shear waves). Earthquake energy also causes the surface of the Earth to vibrate because of surface waves.

6. The focus and epicenter of an earthquake can be located by measuring the differences in travel time between P and S waves.

7. Earthquakes are measured on the Richter magnitude scale, which is based on the amount of energy released during an earthquake as determined from seismic wave amplitudes recorded on seismograms. Earthquake effects can also be quantified using the modified Mercalli intensity scale, which is based on reports of felt intensity of vibrations and damage to structures.

8. Earthquakes can be devastating in terms of loss of life and damage to structures. Factors such as population density, building codes and construction styles, and emergency preparedness are important in determining earthquake risk. Engineers are continually refining their knowledge about how structures respond to intense ground motion.

9. Seismic hazard mapping is based on known geologic structures, especially faults, as well as on records of previous earthquakes and calculations of the probability that ground motion will occur in a given location.

10. Primary hazards associated with earthquakes include ground motion, faulting, and ground rupture. The original shock may also trigger aftershocks. Secondary and tertiary hazards associated with earthquakes include fires, landslides, liquefaction, changes in ground level, tsunamis, and flooding.

11. Ninety-five percent of all earthquakes originate in the circum-Pacific seismic belt (80%) and the Mediterranean seismic belt (15%). The remaining 5 percent are widely distributed along the midocean ridges and elsewhere.

12. Zones of seismic activity define active plate boundaries. Each type of plate boundary is characterized by earthquakes that are distinctive in terms of fault motions and depth of foci. Very large earthquakes occasionally occur in the normally stable continental intraplate environment.

13. The association of earthquakes with the tectonic cycle and with specific geologic environments is the first step in the long-term prediction of earthquakes. Paleoseismology is used to determine the average interval between earthquakes in a given region. The most powerful tool in long-term prediction is the identification of seismic gaps—that is, regions along a fault where no earthquakes are occurring even though elastic energy is accumulating.

14. Short-term prediction of earthquakes is based on observations of precursor anomalies such as sudden changes in rock magnetic or electrical properties, levels of water or amounts of radon in wells, foreshocks, tilting or bulging of the ground surface, and even anomalous animal behavior. Short-term prediction is difficult because earthquake foci are deep underground and therefore are difficult to study and monitor, and because earthquakes are inconsistent in terms of precursor phenomena.

15. Someday it may be possible to control earthquakes by finding a technique through which built-up energy can be released slowly and safely. Possibilities include increasing fluid pressure along a fault and setting off controlled explosions.

IMPORTANT TERMS TO REMEMBER

aftershock (p. 88)
body wave (p. 75)
compressional wave (p. 75)
elastic rebound theory (p. 72)
epicenter (p. 76)
fault (p. 70)
focus (p. 76)
foreshock (p. 95)
liquefaction (p. 89)

modified Mercalli intensity scale (p. 79)
normal fault (p. 71)
P wave (p. 76)
paleoseismology (p. 95)
reverse fault (p. 71)
Richter magnitude scale (p. 78)
S wave (p. 76)
seismic belt (p. 92)
seismic gap (p. 95)

seismic wave (p. 75)
seismic sea wave (p. 90)
seismogram (p. 75)
seismograph (p. 73)
seismology (p. 73)
shear wave (p. 76)
strike-slip fault (p. 71)
surface wave (p. 75)

QUESTIONS AND ACTIVITIES

1. Do you live in a seismically active area? Are you sure? What is the nature of the geologic setting in which you live? Start your investigation by examining the seismic hazard maps in Fig. 3.18 and 3.19. You can find out more about the geology of your area by contacting the state, provincial, or national geologic survey or by visiting the library at your college or university and looking through the geology periodicals.

2. If there is a chance—even a remote one—of an earthquake occurring in your area, you might want to learn the answers to these questions: Do building codes and zoning regulations reflect the seismic hazard? Are possible secondary hazards, such as landslides, taken into account? Have emergency personnel rehearsed any response procedures? Would you know what to do if an earthquake occurred in your town? What steps might you take to prepare for such an event? Where would you go for emergency assistance?

3. Why are some earthquakes so devastating, whereas other, equally strong ones, are not? Write a brief account comparing the Latur, India earthquake of 1993 and the Northridge, California earthquake of 1994. What were the most significant differences between these two earthquakes in terms of the geologic setting and the impacts on people in the area?

4. Why do you think seismologists are so cautious about making predictions? Are their reasons valid, in your opinion? Write a fictional account of a predicted earthquake that never came to pass. What effects might a false prediction have? If you were a seismologist and thought you had evidence that an earthquake was imminent, would you issue a prediction? Consider this question from the points of view of science and society (including the public as well as leaders and decision makers) and from a personal perspective.

CHAPTER 4

VOLCANIC ERUPTIONS

A blast of burning sand pours out in whirling clouds.
Conspiring in their power, the rushing vapours
Carry up mountain blocks, black ash, and dazzling fire.

• Lucilius Junior (A.D. 50)

*I*n Indonesia during the summer of 1883 an apparently dormant volcano, Krakatau, began emitting steam and ash. On Sunday, August 26, the volcano's activity increased, and on the next day Krakatau literally blew up and disappeared. As a telegram of the time tersely reported, "Where once Mount Krakatau stood, the sea now plays." The explosion was heard on an island in the Indian Ocean, 4600 km away. Gigantic waves 40 m high spread out from the site of the explosion and crashed into the shores of Java, where more than 36,000 people lost their lives.

The effects of Krakatau were felt around the world. About 20 km³ of volcanic debris was ejected during the eruption, with some particles blasted as high as 50 km into the stratosphere. Within 13 days the stratospheric dust had encircled the globe, and for months afterward there were strangely colored sunsets—sometimes green or blue and other times scarlet or flaming orange. One November sunset over New York City looked so much like the glow from a massive fire that numerous calls were made to fire departments. The suspended dust made the atmosphere so opaque to the Sun's rays that the average temperature around the Earth dropped at least 0.5°C during 1884. Within 5 years all the dust had fallen to the ground and the climate returned to normal conditions.

Mt. Pinatubo, Philippines, during the eruption cycle of 1991. A vast cloud of hot, dust-laden gas has rolled down the flanks of the volcano and is spreading rapidly across the surrounding plains. The car and driver escaped harm, but the houses, trees, and fields were smothered with a blanket of hot, volcanic dust.

Every now and then a truly devastating eruption reminds us of the enormous magnitude of volcanic forces. The 1991 eruption of Mount Pinatubo in the Philippines stands out as one such event. Although the volcano had slumbered for more than 400 years, there were ominous signs of renewed activity in April 1991. By the end of the month the activity was increasing in intensity, and Filipino geologists requested the assistance of colleagues from the U.S. Geological Survey. From April to June the joint team of geologists monitored Pinatubo's activity and worked frantically to delineate the hazard and issue warnings to area residents. In the end, the top of the volcano was blasted off in an immense explosion of gas, ash, and rock fragments. The monitoring team had successfully predicted the date of the eruption. At least 58,000 people were evacuated and tens of thousands of lives saved.

VOLCANOES AND MAGMA

Magma is molten rock, sometimes containing suspended mineral grains and dissolved gases, that forms when temperatures rise sufficiently high for melting to occur in the Earth's crust or mantle. When magma reaches the Earth's surface, it does so through a **volcano,** a vent from which magma, solid rock debris, and gases are erupted. The term *volcano* comes from Vulcan, the Roman god of fire. For most people, the thought of a volcano conjures up visions of incandescent **lava**—magma that reaches the Earth's surface and pours out over the landscape.

Although lava often takes the form of hot, flowing streams, magma can also be erupted in other forms, such as clouds of tiny red-hot fragments, or large clots of molten material blasted high into the air. In fact, volcanoes and the magma that erupts from them are much more varied than is commonly realized. This chapter begins by giving some insight into the properties of magma and how they influ-

---

ence the style and explosiveness of volcanic activity. We then proceed to a discussion of volcanic eruptions and landforms; the regions where different types of volcanoes occur; the hazards and benefits associated with volcanoes; and finally how volcanologists monitor volcanoes and predict eruptions.

Characteristics of Magma

The characteristics and properties of magma have a significant influence on the style and, in particular, the explosiveness of volcanic eruptions. By observing and studying eruptions of lava, we can draw three important conclusions concerning magma:

1. Magma is characterized by a *range of compositions* in which silica (SiO_2) is predominant. Most magma contains some dissolved gases, as well as liquid (molten rock) and crystals.

2. Magma is characterized by *high temperatures.*

3. Magma has the properties of a liquid, including the *ability to flow.* This is true even though some magma is almost as stiff as window glass.

In the rest of this section we will examine each of these observations in greater detail.

Composition

The composition of magma is controlled by the most abundant elements in the Earth—Si, Al, Fe, Ca, Mg, Na, K, H, and O. (As we learned in Chapter 2, O^{2-} is the most abundant anion. Therefore it is usual to express the compositional variations of magmas in terms of oxides, such as SiO_2, Al_2O_3, CaO, and H_2O.) The most abundant component of magma is SiO_2.

The most common types of magma are *basaltic, andesitic,* and *rhyolitic* magmas. The first type contains about 50 percent SiO_2, the second about 60 percent, and the third about 70 percent. Three types of volcanic (extrusive) rocks—**basalt, andesite,** and **rhyolite** (Table 4.1)—are derived from these three types of magma.

For each of the common volcanic rocks there is a corresponding intrusive or plutonic rock. As discussed in Chap-

ter 1, plutonic rocks form from magmas that never make it to the surface of the Earth, instead cooling and crystallizing more slowly underground. The plutonic rock types corresponding to basalt, andesite, and rhyolite are *gabbro, diorite,* and *granite,* respectively.

The three common types of magmas are not formed in equal abundance. Approximately 80 percent of all magma erupted by volcanoes is basaltic, whereas andesitic and rhyolitic magmas each account for about 10 percent of the total. Hawaiian volcanoes such as Kilauea and Mauna Loa are basaltic; Mount St. Helens, Mount Pinatubo, and Krakatau are andesitic volcanoes; and the now dormant volcanoes at Yellowstone National Park are rhyolitic. As we will see later in the chapter, the origin and distribution of these different kinds of volcanoes are largely a function of plate tectonics.

Dissolved Gases

Small amounts of gas (0.2 to 3 percent by weight) are dissolved in all magmas. Even though they are present in very low abundance, these gases can strongly influence the properties of the magma, which in turn influence the style and explosiveness of the eruption.

As you can probably imagine, volcanic gases are particularly difficult to study. Aside from the dangers associated with eruptions and high temperatures, volcanic gases can be toxic or caustic. It is also exceedingly difficult to obtain a sample of volcanic gas without its becoming contaminated by the surrounding air (Fig. 4.1). According to popular accounts, two Russian volcanologists collected gas samples while riding down a river of lava on a "raft" of solidified volcanic rock during an eruption of Bilinkai, a volcano on the peninsula of Kamchatka. Other investigators have been less fortunate. In A.D. 79 the Roman statesman Pliny the Elder died, possibly from gas poisoning or asphyxiation, while trying to collect samples of volcanic gas during an eruption of Mount Vesuvius. Despite the obstacles, however, volcanologists have managed to gather large amounts of information about volcanic gas emissions.

The principal volcanic gas is water vapor, which, together with carbon dioxide, accounts for more than 98 percent of all gases emitted from volcanoes. Other volcanic

T A B L E 4.1 • Volcanic Rock Name, Plutonic Rock Name, Silica Content, Viscosity, and Melting Temperature for the Three Main Types Of Magma

Volcanic Rock Name and Lava Type	Corresponding Plutonic Rock Name	Silica Content (approx.)	Relative Viscosity	Melting Temperature (experimental) (°C)
Basalt	Gabbro	≈50%	Low	≈1400
Andesite	Diorite	≈60%	Intermediate	≈1100
Rhyolite	Granite	≈70%	High	≈ 800

◄ FIGURE 4.1 Geologist in a fire- and heat-resistant garment moving in to collect samples of volcanic gas during an eruption of Manua Loa volcano, Hawaii. The highest temperature recorded, 1,145°C, was measured at the time the gas samples were taken on the bright orange region in the center of the photo.

▲ FIGURE 4.2 Fountaining starts an eruption of Krafla, a basaltic volcano in Iceland. Use of a telephoto lens foreshortens the field of view. The geologist in a protective suit is making measurements several hundred meters away from the fountain.

gases include nitrogen, chlorine, sulfur, and argon, which are rarely present in amounts exceeding 1 percent. Throughout geologic history, volcanic gas emissions have been the primary mechanism whereby the Earth has released the volatile material from its interior. The chemical evolution of the atmosphere and the origin of the oceans are intimately linked to this outgassing process.

Temperature

It is difficult to measure the extrusion temperature of magma, but it can sometimes be done during volcanic eruptions. For obvious reasons, such measurements are usually made from a distance using optical devices (Fig. 4.2). Magma temperatures during eruptions of volcanoes such as Kilauea in Hawaii and Mount Vesuvius in Italy have been recorded as ranging from 800 to 1200°C. Experiments on synthetic magmas in the laboratory suggest that under some conditions magma temperatures might be as high as 1400°C (Table 4.1).

Viscosity

The degree to which a substance is resistant to flow is its **viscosity;** the more viscous a substance, the less fluid it is. The viscosity of a magma depends on its temperature and composition, especially its silica and dissolved gas content. Dramatic film footage of lava flowing rapidly down the sides of a volcano prove that some magmas are very fluid

▲ F I G U R E 4.3
This stream of low-viscosity basaltic lava moving smoothly away from an eruptive vent demonstrates how fluid and free flowing lava can be. The temperature of the lava is about 1100°C. The eruption occurred in Hawaii in 1983.

▲ F I G U R E 4.4
The flow of lava is controlled by viscosity. Two different flows are visible here. They have the same basaltic composition. The lower flow, on which the geologist is standing, is a pahoehoe flow formed from a low-viscosity lava such as that shown in Fig. 4.3. The upper flow (the one being sampled by the geologist), which is very viscous and slow moving, is an aa flow that erupted from Kilauea in 1989. The pahoehoe flow erupted in 1959.

(Fig. 4.3). Low-silica basaltic lava moving down a steep slope on Mauna Loa in Hawaii has been clocked at 64 km/h, indicating a very low viscosity. On the other hand, rhyolitic magma containing 70 percent or more SiO_2 flows so slowly that its movement can hardly be detected. The properties of such magma are more akin to those of solids than to those of liquids.

The SiO_4^{4-} anions that form the foundation for most common minerals are also present in magmas. Just as they do in minerals, these anions link together or *polymerize* by sharing oxygens. Unlike the anions in silicate minerals, however, those in magma form irregularly shaped groupings of chains, sheets, and networks. As the polymerized groups become larger, the magma becomes more viscous—that is, more resistant to flow—and behaves increasingly like a solid. The higher the silica content, the larger the polymerized groups. For this reason, high-silica rhyolitic magma is always more viscous than low-silica basaltic magma, whereas andesitic magma has a viscosity somewhere between those of rhyolitic and basaltic magma (Table 4.1).

Temperature and gas content also affect the viscosity of magmas. The higher the temperature and, in general, the higher the gas content, the lower the viscosity and the more readily the magma flows. A very hot magma erupted from a volcano may flow readily, but it soon begins to cool, becomes more viscous, and eventually comes to a halt. In Fig.

4.4 the smooth, ropy-surfaced lava, called *pahoehoe* (a Hawaiian word), formed from a hot, gas-charged, very fluid lava. The rubbly, rough-looking lava, by contrast, formed from a cooler, gas-poor lava with a high viscosity. Scientists call this rough lava *aa* (another Hawaiian word). Interestingly, these two types of lava are so distinctive that there are two equivalent words in the Icelandic language: *helluhraun* (for pahoehoe-type lavas) and *apalhraun* (for aa-type lavas).

VOLCANIC ERUPTIONS

It is impossible to produce a strict classification of volcanic eruptions. During most eruptive episodes the types of activity and the nature of the materials ejected from the volcano change, sometimes gradually (over weeks, months, or years) and sometimes from one day to the next or even from hour to hour. Nevertheless, it is possible to categorize volcanic eruptions on the basis of the volcano's eruptive style, the types of materials erupted, and the type of edifice or landform built up by the volcano. A general classification scheme is shown in Table 4.2. The table divides vol-

T A B L E 4.2 • Types of Eruptions

Eruptive Style	Characteristics	Type of Landform and Products	Examples
Principle Kinds of Central Eruptions			
Hawaiian	Discharge of fluid basaltic magma from vents within summit caldera and fissures on flanks	Shield volcano	Mauna Loa
		Lava flows	Kilauea
	Minor pyroclastics	Calderas	
	Fountaining	Lava tubes	
	Nonexplosive	Spatter cones	
Strombolian	Mild explosive eruptions of incandescent bombs, ash, and lapilli	Spatter cones	Stromboli
		Cinder cones	Parícutin
	Extrusion of lava from vents on flanks		
	Fumarolic activity		
Vulcanian	Viscous, silica-rich, gas-rich magmas	Pumiceous cones surrounded by wide sheets of fine pyroclastics and flows of glassy lava	Vulcano
	Dark eruption clouds composed of ash and lapilli fragments		Bárcena
	Pyroclastic flows		
Peléan	Viscous, silica-rich, gas-rich magmas	Steep-sided domes	Mont Pelée
	Violent, destructive eruptions	Short, thick flows	Santiaguito Domes
	Glowing avalanches	Hot flows of blocks and ash	
Plinian (or Kraka-toan)	Exceptionally powerful, continuous gas blast	Composite volcanoes (stratovolcanoes)	Vesuvius
	Gas-rich, siliceous magma		Krakatau
	Voluminous pyroclastics, glowing avalanches		Mount Mazama
	Summit caldera		
	Can be interlayered with flows		
Surtseyan (or phreatomagmatic)	Rising magma comes into contact with groundwater or seawater	Ash cones	Surtsey
		Cinder cones	Taal
	Violent explosive eruptions		
Steam-blast (or phreatic)	Immensely powerful blasts	Domes	Arenal
	No fresh magma is discharged, only fragments of solid rock mixed with steam	Collapse caldera	Bezymianny
		Glowing avalanches	Lassen Peak
	Result from groundwater coming into contact with hot rocks at depth	Hot lahars	Vesuvius
			Mount St. Helens
Other Kinds of Eruptions			
Fissures related to central vents	Caused by local or regional stresses on cones or shield volcanoes		Laki
			Hekla
	Nature of eruption and composition of magma can vary	Radial fractures and lines of vents, often related to central vents	
Fissures unrelated to central vents	Never witnessed in historic times	Linear or arcuate fissures	Valley of Ten Thousand Smokes
	Pyroclastic flows; pumice mixed with lapilli and ash	Form vast sheets of welded siliceous tuffs (ignimbrites)	Yellowstone Plateau
Plateau (flood) basalts	Most voluminous of all volcanic eruptions	Extensive sheets of basalt	Columbia Plateau
	Fluid basaltic lavas	Swarms of parallel fissures and multiple coalescing vents	Deccan Plateau
	All erupted during an interval of 2 to 3 million years about 15 million years ago	Fields of low, coalescing shield volcanoes	
Oceanic volcanism	Related to "hot spots" or sea-floor spreading along mid-ocean rifts	Pillow lavas	Mid-Atlantic Ridge
		Glassy flows	East Pacific Rise
	Fluid basaltic lavas	Extensive fissure systems related to seafloor spreading	Hawaiian–Emperor Seamount chain
	Essentially no pyroclastic debris		

[a] The central eruptions increase in explosiveness from top to bottom in the table. Note that a single volcano could combine several of these explosive styles, varying either over the long term or from day to day or hour to hour.

canic eruptions into those related to a central volcanic vent and those related to long, linear or arcuate fractures in the Earth's crust, called **fissures.**

Factors Influencing Eruptive Styles

Like most other liquids, magma is less dense than the solid rock from which it is formed. Once formed, therefore, the lower density magma exerts upward pressure on the enclosing higher density rock and slowly forces its way up. There is, of course, pressure on a rising mass of magma created by the weight of the overlying rock. This pressure is proportional to depth. As a magma rises, the pressure of the overlying rock decreases.

Pressure controls the amount of gas a magma can dissolve—more gas at high pressures, less at low pressures. Gas dissolved in a rising magma acts like gas dissolved in soda water. When a bottle of soda is opened, the pressure inside the bottle drops, gas comes out of solution, and bubbles form. Gas dissolved in an upward-moving magma also comes out of solution and forms bubbles. What happens to the bubbles is determined by the viscosity of the liquid. Thus, dissolved gases and viscosity are the two factors that are most influential in determining the eruptive style and explosiveness of a volcano.

Nonexplosive Eruptions

Understandably, people tend to regard any volcanic eruption as a hazardous event and to view active volcanoes as dangerous places that should be avoided. However, geologists have discovered that some volcanoes are comparatively safe and relatively easy to study. Nonexplosive eruptions, such as those we can witness in Hawaii, are relatively safe compared to violent, explosive events like the 1980 eruption of Mount St. Helens in Washington, the 1982 eruption of El Chichón in Mexico, and the 1991 eruption of Mount Pinatubo in the Philippines, each of which caused substantial destruction and loss of life. The differences between nonexplosive and explosive eruptions are largely a function of magma viscosity and dissolved gas content. Nonexplosive eruptions are associated with low-viscosity magmas and low dissolved-gas contents.

Even nonexplosive eruptions may appear violent during their initial stages. Gas bubbles in a low-viscosity basaltic magma rise rapidly upward like the gas bubbles in a glass of soda water. If a basaltic magma rises rapidly, resulting in a rapid decrease in pressure, gas can bubble out of solution so fast that spectacular fountaining occurs (Fig. 4.2).

When fountaining dies down, hot, fluid lava emerging from the vent flows rapidly downslope (Fig. 4.3). Because heat is lost quickly at the top of a flow, the surface forms a crust beneath which the liquid lava continues to flow downslope along well-defined channels. These enclosed lava tubes inhibit the upward loss of heat and enable a low-viscosity lava to move along just below the surface for great distances. As the lava cools and continues to lose dissolved gases, its viscosity increases and the character of the flow changes. The very fluid lava initially forms thin pahoehoe flows, but with increasing viscosity the rate of movement slows and the stickier lava may be transformed into a clinkery aa flow that moves very slowly. Thus, during a single, nonexplosive, Hawaiian-type eruption both pahoehoe and aa flows may be formed from the same batch of magma (Fig. 4.4).

As a basaltic lava cools and its viscosity rises, it becomes increasingly difficult for gas bubbles to escape. When the lava finally solidifies into rock, the last-formed bubbles become trapped, forming bubble holes called *vesicles.* The term *vesicular* is used to describe the texture produced by vesicles in an igneous rock (Fig. 4.5). In many vesicular basalts the vesicles are later filled with minerals deposited by heated groundwater.

Explosive Eruptions

In andesitic or rhyolitic magmas, gas bubbles can rise only very slowly because they are held back by the viscosity of the fluid. As these magmas move upward toward the Earth's surface, the decrease in pressure eventually allows the dissolved gas to expand and escape explosively. Bubbles that form quickly in a huge mass of sticky rhyolitic magma can shatter into a froth of innumerable tiny glass-walled bubbles, producing a rock called *pumice.* Some pumice has such a low density that it will float in water. Beaches on mid-Pacific islands are often littered with pumice that has floated in on currents from distant volcanic eruptions.

When little or no dissolved gas is present, a magma will be erupted as a lava flow regardless of its composition. If dissolved gas is present, however, it must escape somehow. The higher the viscosity, the more difficult it is for the gas to form bubbles and the greater the likelihood the escaping gas bubbles' causing an explosive eruption. We can extend

▲ F I G U R E 4.5
Vesicular basalt. The holes in the rock were left by small bubbles of gas in the magma. The specimen is 4.5 cm across.

T A B L E 4.3 • Names for Tephra and Pyroclastic Rock

Average Particle Diameter (mm)	Tephra (unconsolidated material)	Pyroclastic Rock (consolidated material)
>64	Bombs	Agglomerate
2–64	Lapilli	Lapilli tuff
<2	Ash	Ash tuff

the analogy of gas dissolved in soda by imagining that the soda bottle is full of a very thick, viscous liquid—such as wet cement—with gas held in solution under pressure. If we open the bottle and cause the pressure to drop very rapidly, gas bubbles will quickly form and escape—but with much more vigorous spattering of the liquid than would occur if the bottle had contained soda.

Tephra and Pyroclastic Rocks

A fragment of rock ejected during a volcanic eruption is called a **pyroclast** (from the Greek words *pyro*, meaning heat or fire, and *klastos*, meaning broken; hence, hot, broken fragments). Rocks formed from pyroclasts are **pyroclastic rocks.** Geologists also commonly refer to pyroclasts as **tephra,** a Greek word for ash. Tephra is employed as a collective term for all airborne pyroclasts, including fragments of newly solidified magma as well as fragments of older broken rock. It includes individual pyroclasts that fall directly to the ground and those that are transported over the ground as part of a hot, moving flow. Abundant tephra and pyroclastic materials are characteristic of violent, explosive eruptions.

The terms used to describe tephra of different size—*bombs, lapilli,* and *ash*—are listed in Table 4.3 and illustrated in Fig. 4.6. The term *volcanic ash,* referring to the fine pyroclastic material thrown out by volcanoes, is somewhat misleading. Strictly speaking, ash is the solid matter

▲ **F I G U R E 4.6**
Tephra. A. Large spindle-shaped bombs up to 50 cm long cover the surface of a tephra cone on Haleakala volcano in Maui. B. Intermediate-sized tephra called lapilli cover the Kau Desert, also located in Hawaii. The coin is about 1 cm in diameter. C. Volcanic ash, the smallest tephra, blankets a farm in Oregon after the eruption of Mount St. Helens in 1980.

that remains after something flammable, such as wood, has burned. However, fine tephra looks so much like true ash that it has become customary to use the word for this material also.

Volcanologists are fond of saying that tephra is igneous on the way up but sedimentary on the way down, meaning that the pyroclasts are ejected from the volcano in a molten state and solidify as igneous rocks in midair, but are deposited on the ground in the form of sedimentary fragments. As a result, pyroclastic rocks are a transitional form between igneous and sedimentary rocks. The names of pyroclastic rocks, like those of sedimentary rocks, are keyed to the size of the mineral grains of which they are composed. Pyroclastic rocks are called *agglomerates* when the tephra particles are large (i.e., bomb sized), and *tuffs* when they are smaller, either lapilli or ash sized (Table 4.3). Unconsolidated tephra can be converted into pyroclastic rock in two ways. The first, and most common, way is through the addition of a cementing agent such as quartz or calcite introduced by groundwater. The second way is through the welding of hot, glassy ash particles. When ash is very hot and plastic, the individual particles can fuse together to form a glassy pyroclastic rock.

Eruption Columns and Tephra Falls

The largest and most violent explosive eruptions are associated with silica-rich magmas that are high in dissolved gases. As the rising magma approaches the surface, rapid decompression causes the gases to expand and produces the violent upward thrust of a dense mixture of hot gas and tephra. This hot, turbulent mixture rises rapidly in the cooler air above the vent to form an **eruption column** (sometimes called a *Plinian column*) that may reach as high

as 45 km in the atmosphere. The photo at the beginning of the chapter shows the base of a huge eruption column rising from Mount Pinatubo in the Philippines.

Eruption clouds are typically dark, tempestuous, ominous-looking masses. Their internal dynamics are complex: the heat of the cloud and the force of the eruption create a tendency to rise, but the density of the ash produces a tendency toward collapse. At a height at which the density of the material in the column equals that of the surrounding atmosphere, the column will spread out laterally to form a mushroom-shaped cloud of the type seen in pictures of nuclear bomb explosions. As the cloud begins to drift with the upper atmospheric winds, particles of debris fall out and eventually accumulate on the ground as tephra deposits (Fig. 4.7). During exceptional explosive eruptions, tephra can be spread over distances of 1500 km or more.

Pyroclastic Flows

A hot, highly mobile flow of tephra that rushes down the flank of a volcano during a major eruption is called a **pyroclastic flow.** Often the material in the flow is dense, gas-charged, and so hot that it incandesces, in which case it may be referred to as a **glowing avalanche** (or *glowing ash cloud* or *nuée ardente*). These are among the most devastating and lethal forms of volcanic eruptions. The historical record of pyroclastic flows reveals that they can travel 100 km or more from source vents and reach velocities of more than 700 km/h. Pyroclastic flows can be caused by the gravitational or explosive collapse of a mass of hot lava near the top of a volcano, which produces a dense, downrushing mass of blocks, lapilli, ash, and hot gases. Geologists call the resulting poorly sorted deposit an *ignimbrite*. Pyroclastic flows can also be generated by the partial or continuous

A.

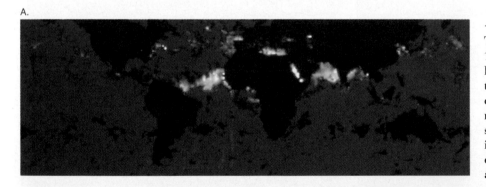

B.

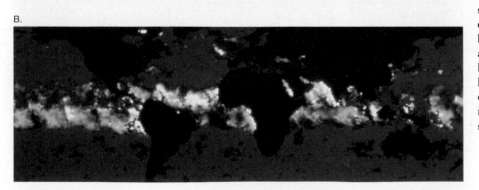

◄ F I G U R E 4.7
The ash-cloud from the June, 1991 eruption of Mt. Pinatubo, Philippines, spreading around the world. A. Image prior to the eruption. Optical density, a measure of particulate matter suspended in the atmosphere, is determined by satellite. High density areas (bright yellow) are regions of high population density or places where dust storms or some other local disturbance is producing a high level of suspended aerosols. B. After the eruption, between July 4 and 10, 1991, a bright band encircles the globe due to the spread of particulate matter erupted high into the stratosphere by Mt. Pinatubo.

collapse of an eruption column. During the 1980 eruption of Mount St. Helens, for example, a number of hot (850°C) pyroclastic flows, probably caused by column collapse, traveled up to 8 km down the north side of the mountain and covered an area of about 15 km².

Lateral Blasts

The 1980 eruption of Mount St. Helens displayed many of the features of a typical large explosive eruption. Nevertheless, the magnitude of the event caught geologists by surprise. The events leading to this eruption are shown diagrammatically in Fig. 4.8.

As magma moved upward under the volcano, the north flank of the mountain began to bulge upward and outward. Finally the slope became unstable, broke loose, and slid toward the valley as a gigantic landslide of rock and glacier ice. The landslide exposed the mass of hot magma in the core of the volcano. With the lid of rock removed, dissolved gases underwent such rapid decompression that a mighty blast resulted, blowing a mixture of pulverized rock and hot gases sideways as well as upward (Fig. 4.9). The sideways eruption, referred to as a **lateral blast,** initially traveled at the speed of sound. It roared across the landscape, killing every living thing in the blast zone. Within the devastated area, which extends as much as 30 km from the crater and covers some 600 km², trees in the formerly dense forest were blasted to the ground and covered with hot debris.

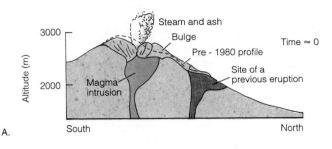

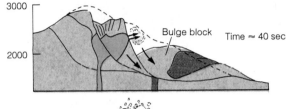

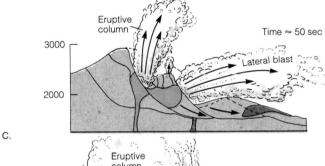

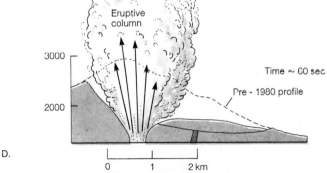

▲ F I G U R E 4.8
The sequence of events leading to the eruption of Mount St. Helens in May 1980. A. Earthquakes and then puffs of steam indicate that magma is rising; a small crater forms, and the north face of the mountain bulges alarmingly. B. On the morning of May 18, an earthquake shakes the mountain and the bulge breaks loose and slides downward. This reduces the pressure on the magma and initiates the lateral blast. C. The violence of the eruption causes a second block to slide downward, exposing more of the magma and initiating an eruption column. D. The eruption increases in intensity. The eruption column carries volcanic ash as high as 19 km into the atmosphere.

▲ F I G U R E 4.9
Plinian ash column from the May 1980 eruption of Mount St. Helens. At least 63 people died as a result of this eruption.

VOLCANIC LANDFORMS

During its lifetime a volcano builds up a volcanic landform or edifice whose shape is determined by the composition and characteristics of the volcanic materials and the types of eruptions that have occurred. These are summarized in Table 4.2. In this section we discuss each of the major types of volcanoes and the landforms it creates.

Shield Volcanoes

Perhaps the easiest kind of volcano to visualize is one built up of successive flows of fluid lava. Such lavas are capable of flowing great distances down gentle slopes, forming thin sheets of nearly uniform thickness. Eventually the pile of lava builds up a **shield volcano,** a broad formation, resembling a shield lying horizontally, with an average slope of only a few degrees (Fig. 4.10). Shield volcanoes are characteristically formed by the eruption of basaltic lava; the proportions of ash and other tephra are small. Hawaii, Tahiti, Samoa, the Galapagos, and many other oceanic islands are actually the upper portions of large shield volcanoes. The Hawaiian volcano Mauna Kea rises only about 4.2 km above sea level, but its true height when measured from the sea floor exceeds that of Mount Everest, making it the tallest mountain on Earth.

The slope of a shield volcano is slight near the summit because the magma is hot and very fluid; it will readily run down even a very slight slope. The farther the lava flows down the volcano's flanks, the cooler and more viscous it becomes and the steeper the slope must be in order for it to flow. Slopes on young, growing shield volcanoes, such as Kilauea in Hawaii, typically range from less than 5° near the summit to 10° on the flanks.

Stratovolcanoes

Stratovolcanoes, also called *composite volcanoes,* emit both tephra and viscous lava and build up steep, conical mounds of interlayered lava and pyroclastic deposits. They are most often composed of andesitic material. The volume of tephra in a stratovolcano may equal or exceed the volume of the lava.

The slopes of stratovolcanoes, which may be thousands of meters high, are steep. Near the summit the slope is typically about 30°, whereas toward the base the slope flattens to about 6° to 10°. The steep slope near the summit of a stratovolcano is due in part to the short, viscous lava flows that are erupted, and in part to tephra. The lava flows are a major factor in distinguishing between *tephra cones* (described below) and stratovolcanoes. As a stratovolcano develops, lava flows act as a cap to slow down erosion of the loose tephra, and as a result the volcano becomes much larger than a typical tephra cone.

Virtually all major continental volcanoes are stratovolcanoes. In general, however, stratovolcanoes are much smaller than the great oceanic shield volcanoes. For example, the total volume of Mauna Loa, the Earth's largest shield vol-

▼ F I G U R E 4.10

Mauna Kea, a 4200-m-high shield volcano in Hawaii, as seen from Mauna Loa. Note the gentle slopes formed by highly fluid basaltic lava. The view is almost directly north. A pahoehoe flow is in the foreground on the northeast flank of the volcano.

▲ F I G U R E 4.11
Mount Fuji, Japan, a snow-clad giant towering over
the surrounding countryside, displays the classic
profile of a stratovolcano.

A.

cano, is more than 300 times that of Mount Fuji in Japan,
one of the most voluminous of all stratovolcanoes.

The beautiful, steep-sided cones of stratovolcanoes are
among the most picturesque sights on Earth (Fig. 4.11).
The snow-capped peak of Mount Fuji has inspired poets
and writers for centuries. Mount Rainier and Mount Baker
in Washington and Mount Hood in Oregon are majestic
examples of such volcanoes in North America.

Other Features of Volcanoes

Tephra Cones

Rhyolitic and andesitic volcanoes tend to eject large quantities of tephra. As the debris showers down, a **tephra cone**
builds up around the vent (Fig. 4.12). The slope of the
cone is determined by the size of the tephra. Fine ash will
stand at a slope angle of 30° to 35°, whereas lapilli generally
stand at an angle of about 25°. The gradual decrease in the

B.

▶ F I G U R E 4.12
Tephra cones. A. Two small tephra cones forming as a
result of an eruption in Kivu, Zaire. Arcs of lights are
caused by the eruption of red-hot lapilli and bombs.
B. Tephra cone in Arizona built from lapilli-sized tephra.
Note the small basaltic lava flow coming from the base
of the cone.

◀ F I G U R E 4.13
Volcanic gas streams from a vent following a 1965 fissure eruption in Hawaii. Sulfur condensed from the acrid gases kills vegetation and covers the ground with a yellow coating of elemental sulfur.

volume of fallout material at greater distances from the vent leads to gentler slopes near the base of the cone.

Geysers, Fumaroles, and Thermal Springs

Gases that bubble up from magma far below the surface may emerge either from a central volcanic vent or from small satellite vents on the flanks of the volcano. A vent or fracture from which volcanic gases are emitted is referred to as a **fumarole.** The gases tend to be mostly water vapor, but other gases, such as evil-smelling sulfurous gas, may be present as well. Such gases can alter and discolor the rocks with which they come into contact (Fig. 4.13).

When active volcanism finally ceases, the rock in an old magma chamber may remain hot for hundreds of thousands of years. Descending groundwater that comes into contact with the hot rock is heated and tends to rise to the surface along rock fractures, forming a **thermal spring.** Thermal springs at volcanic sites in Italy, Iceland, Japan, and New Zealand, as well as many other locations, have become famous health spas.

A thermal spring equipped with a natural system of plumbing and heating that causes intermittent eruptions of water and steam is a **geyser.** The name comes from the Icelandic word *geysir,* meaning to gush, for Iceland is the home

◀ F I G U R E 4.14
The great Geysir, Iceland, from which all geysers take their name.

◄ F I G U R E 4.15
Crater Lake, Oregon, occupies
a caldera 8 km in diameter that
crowns the summit of a once
lofty stratovolcano called
Mount Mazama. Wizard Island,
a small tephra cone, was
formed after the collapse that
created the caldera.

of many geysers (Fig. 4.14). Most of the world's geysers outside Iceland are located in New Zealand and in Yellowstone National Park in the United States.

Craters and Calderas

Near the summit of most volcanoes is a **crater,** a funnel-shaped depression opening upward, from which gases, tephra, and lava are ejected. However, many volcanoes, both shield and stratovolcanoes, have a much larger depression known as a **caldera,** a roughly circular, steep-walled basin that may be several kilometers or more in diameter. Calderas are formed as a result of the partial emptying of a magma chamber. Rapid ejection of magma during a large

lava or tephra eruption can leave the magma chamber empty or partly empty. The unsupported roof of the chamber then collapses under its own weight, like a snow-laden roof on a shaky barn, dropping downward on a ring of steep vertical fractures.

Crater Lake in Oregon occupies a circular caldera 8 km in diameter that was formed after a great tephra eruption about 6600 years ago by a volcano posthumously called Mount Mazama (Fig. 4.15). The tephra deposits from that eruption can still be seen at many places in Crater Lake National Park and over a vast area of the northwestern United States and adjacent parts of Canada (Fig. 4.16). After the outpouring of about 75 km^3 of tephra, what remained of

◄ F I G U R E 4.16
The Pinnacles, Crater Lake
National Park. Striking
erosional forms developed
in the thick tephra blanket
left by the catastrophic
eruption of Mount Mazama
6600 years ago.

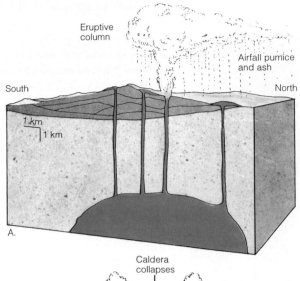

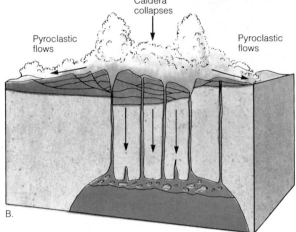

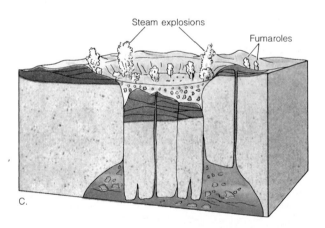

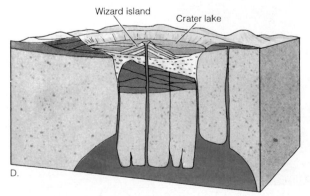

The sequence of events that formed Crater Lake following the eruption of Mount Mazama. A. An eruptive column of tephra rises from the flank of Mount Mazama. B. The eruption reaches a climax. Dense clouds of ash fill the air and hot pyroclastic flows sweep down the mountainside. It was at this stage that the deposits shown in Fig. 4.16 were formed. C. The top of Mount Mazama collapses into the partly empty magma chamber, forming a caldera 10 km in diameter. D. During a final phase of eruption, Wizard Island is formed. The water-filled caldera is Crater Lake, shown in Fig. 4.15.

the roof of the magma chamber collapsed into the partly empty chamber (Fig. 4.17).

Resurgent Domes

A volcano does not necessarily become inactive after the formation of a caldera. Magma starts reentering the chamber and in the process causes uplifting of the collapsed floor of a caldera to form a structural dome. Such a feature is called a **resurgent dome.** Subsequently, small tephra cones and lava flows build up in the interior of the caldera. Wizard Island in Crater Lake is a cone that was formed in this way.

When lava is extruded after a major volcanic eruption, it tends to have very little dissolved gas left in it. If the lava is sticky and viscous, it squeezes out to form a **lava dome.** A growing lava dome more than 200 m high can now be seen in the center of the crater of Mount St. Helens (Fig. 4. 18).

Fissure Eruptions

Some lava reaches the surface via elongate fissures through the crust. These events are called **fissure eruptions** (Fig. 4.19). Such eruptions, which can be very dramatic, are characteristically associated with basaltic magma. Successive lavas that emerge as fissure eruptions on land tend to spread widely and may create flat lava plains called *basalt plateaus.* In 1783 a fissure eruption at Laki in Iceland occurred along a fracture 32 km long. Lava flowed 64 km outward from one side of the fracture and nearly 48 km outward from the other side. Altogether the flow covered an area of 588 km². The volume of the lava extruded was 12 km³, the largest lava flow in historic times and also one of the most deadly. The flow destroyed homes and food supplies, killed livestock, and covered fields. In the ensuing famine 9336 people died. There is good evidence to prove that even larger eruptions occurred in prehistoric times. The Roza flow, a great sheet of basaltic lava in eastern Washington State, can be traced over an area of 22,000 km² and has been shown to have a volume of 650 km³.

◀ F I G U R E 4.18
A resurgent lava dome in the crater of
Mount St. Helens, Washington, in May
1982. The plume rising above the dome
is composed of steam.

VOLCANOES AND PLATE TECTONICS

We come now to some of the most interesting questions concerning magmas and volcanoes: How and where do magmas form, why do the different types of volcanoes occur in distinct tectonic environments, and why are there three major kinds of magmas—basaltic, andesitic, and rhy-

olitic? Many clues can be obtained from the distribution of different kinds of volcanoes. As noted earlier, there is a close relationship between plate tectonics and the locations of the different types of volcanoes. Understanding volcanism in the context of plate tectonics is important not only for improving our scientific knowledge of these processes but also for our ability to assess the hazards associated with volcanoes and to predict eruptions. A summary of present thinking about the distribution of the kinds of volcanoes in

◀ F I G U R E 4.19
Aerial view of a fissure
eruption, Mauna Loa,
Hawaii, in 1984. Basaltic
lava is erupting from a
series of parallel fissures.
Note the tephra cone
(upper left) formed during
an earlier eruption.

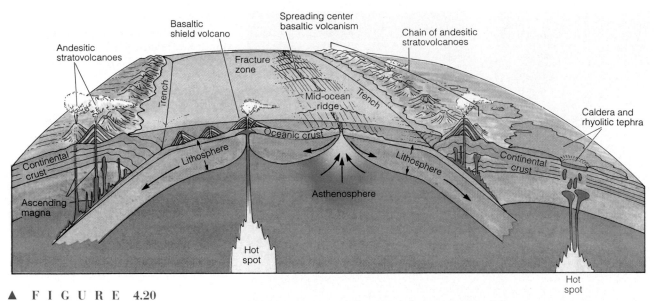

▲ F I G U R E 4.20
Diagram illustrating the locations of the major kinds of volcanoes in a plate-tectonic setting.

the context of plate tectonics is presented in Fig. 4.20 and in the following discussion.

Global Distribution of Volcanoes

Volcanoes that erupt rhyolitic magma are found primarily on the continental crust, suggesting that the processes that form rhyolitic magma must occur in continental crust but not in oceanic crust. Nor, presumably, do the processes that form rhyolitic magma occur in the mantle, because if they did, the magma would rise to the surface regardless of the kind of crust above it. Rhyolitic magmas are thought to form through partial melting of continental crust. Because the minerals in continental rocks are rich in water and carbon dioxide, as well as silica, the magmas that result from their melting also tend to have a high gas and silica content. Rhyolitic volcanoes therefore tend to erupt explosively. An example is the prehistoric eruption of Mount Mazama discussed earlier.

Volcanoes that erupt andesitic magma are found on both oceanic and continental crust. This suggests that andesitic magma forms in the mantle and rises, regardless of the nature of the overlying crust. An additional piece of information comes from the geographic distribution of andesitic volcanoes (Fig. 4.21). A ring of volcanoes surrounds the Pacific Ocean, forming the so-called *Ring of Fire*. The Ring of Fire, also called the *Andesite Line* by geologists, is home to many of the world's most active and explosive volcanoes, including Mount Pinatubo (Philippines), Mount Unzen and Mount Fuji (Japan), Krakatau and Tambora (Indonesia), and Mount Spurr (Alaska). The Ring of Fire is exactly parallel to the Pacific-rim plate subduction margins shown in Fig. 1.14. Andesitic magma is probably formed as

a result of the melting of old oceanic crust that has been subducted back into the mantle.

Volcanoes that erupt basaltic magma also occur on both oceanic and continental crust. Therefore, the mantle must be the source of basaltic as well as andesitic magma. The geographic distribution of basaltic volcanoes does not seem to be related to specific features of the crust, suggesting that basaltic magma must be formed by melting of the mantle itself, regardless of the kind of crust above it.

Two observations concerning basaltic volcanoes suggest something further about the origin of basaltic magma, however. First, everywhere along the midocean ridges volcanoes erupt basaltic magma. Midocean ridges coincide with plate spreading margins, so we must consider the possibility that plate motion and the generation of spreading-margin magma might somehow be connected. The second observation concerns large basaltic volcanoes that are not located along midocean ridges. An example is found in the volcanoes of Hawaii, which are located on oceanic crust in the middle of the Pacific Plate, far from any plate edges. The Hawaiian volcanoes that are active today—Mauna Loa, Kilauea, and Loihi (a submarine volcano)—are the youngest members of a chain of mostly extinct volcanoes. To the northwest, the volcanic rocks in the Hawaiian chain are progressively older (Fig. 4.22). The Hawaiian volcanic chain is believed to have formed as the Pacific Plate moved slowly northwest across a midocean "hot spot" above which frequent and voluminous eruptions built a succession of volcanoes. The hot spot is thought to have been building basaltic volcanoes on the moving plate for at least 70 million years. The exact causes of hot spot magmas are not yet known, but both spreading-margin magmas and hot-spot magmas are believed to be caused by convection in the mantle.

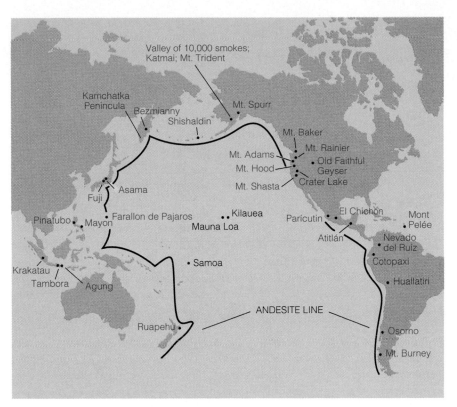

▲ FIGURE 4.21
The Ring of Fire around the Pacific Ocean basin is formed by andesitic volcanoes. The ring, called the Andesite Line by geologists, coincides with subduction zones where lithosphere capped with oceanic crust is being subducted into the asthenosphere. Volcanoes within the Pacific Ocean basin, such as Mauna Loa, erupt basaltic magma but not andesitic magma.

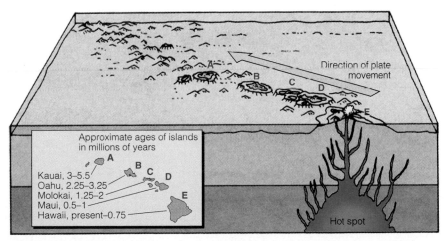

▲ FIGURE 4.22
The Hawaiian Island chain of volcanoes formed above a deep-seated source of basaltic magma within the mantle, called a hot spot. The magma source has remained stationary for at least 70 million years while the Pacific Plate has moved over it, carrying with it the old volcanic landforms built over the hot spot. The Hawaiian volcanoes are progressively older with increasing distance from the hot spot. The currently active volcanoes, Mauna Loa, Kilauea, and Loihi, are located on top of the hot spot. Eventually they, too, will be carried off with the moving plate, and a new volcanic edifice will be built over the hot spot.

VOLCANIC HAZARDS

Since A.D. 1800 there have been 18 volcanic eruptions in which a thousand or more people died (Table 4.4). Yet millions of people live in areas where there are active volcanoes and must come to terms with the associated hazards and risks. Like other natural phenomena, volcanic hazards have *primary*, *secondary*, and *tertiary* effects. In this section we discuss those effects in detail.

Primary Effects of Volcanism

Lava Flows

Most types of volcanoes produce at least some lava, but extensive lava flows are most characteristic of the quieter types of volcanoes such as those found in Hawaii. Because lava flows are closely controlled by topography, it is often possible to predict the general direction and course of a flow. Some lavas can travel downhill at remarkably high velocities. Basaltic lava, for example, can travel as fast as 64 km/h down a steep slope. Such fluidity is rare, however, and rates of flow are more commonly measured in meters per hour or even meters per day. As can be seen in Fig.

4.23, which shows basaltic lava destroying a house in Hawaii, lava flows are usually slow enough that people are not endangered. This means that in dealing with the hazardous effects of lava flows the main focus is on preventing property damage, not on saving lives.

Lava flows are one of the few aspects of volcanism that can be controlled, at least in part, by human intervention. Several methods of controlling lava flows have been tested. In 1935 bombing was tried, with limited success, during an eruption of Mauna Loa in order to spare the city of Hilo from excessive damage. It was attempted again at Mount Etna in Sicily in 1983 and 1992. The goal of bombing is to block or divert the advancing lava flow, either by altering the topography ahead of the flow or by creating a barrier by blocking the channel or causing a lava tube to collapse. The construction of artificial barriers is based on the same principle, that is, creating a blockage and diverting the flow from its natural course. There are numerous examples of preexisting walls that have withstood and even diverted oncoming lava flows. Walls and bulldozed rock barriers constructed for this purpose have been tested—again with somewhat limited success—in Hawaii, Iceland, and Japan.

Hydraulic chilling, first tested during an eruption of Kilauea in 1960, involves spraying water on an advancing lava

T A B L E   4.4 • **Volcanic Disasters Since A.D. 1800 in Which a Thousand or More People Lost Their Lives**

			Primary Cause of Fatalities			
Volcano	*Country*	*Year*	*Pyroclastic Eruption*	*Mudflow*	*Tsunami*	*Famine*
Mayon	Philippines	1814	1,200			
Tambora	Indonesia	1815	12,000			80,000
Galunggung	Indonesia	1822	1,500	4,000		
Mayon	Philippines	1825		1,500		
Awu	Indonesia	1826		3,000		
Cotopaxi	Ecuador	1877		1,000		
Krakatau	Indonesia	1883			36,417	
Awu	Indonesia	1892		1,532		
Soufrière	St. Vincent	1902	1,565			
Mt. Pelée	Martinique	1902	29,000			
Santa Maria	Guatemala	1902	6,000			
Taal	Philippines	1911	1,332			
Kelud	Indonesia	1919		5,110		
Merapi	Indonesia	1930	1,300			
Lamington	Paupa-New Guinea	1951	2,942			
Agung	Indonesia	1963	1,900			
El Chichón	Mexico	1982	1,700			
Nevado del Ruíz	Colombia	1985		23,000		

Source: From a Report by the Task Group for the International Decade of Natural Disaster Reduction, published in *Bull. Volcano. Soc. Japan*, Series 2, Vol. 35, No. 1 (1990): 80–95.

▲ F I G U R E 4.23
An advancing tongue of basaltic lava setting fire to a house in Kalapana, Hawaii, during an
eruption of Kilauea in June 1989. Flames at the edge of the flow are due to burning lawn grass.

flow so that the front of the flow solidifies. During a 1973
eruption of Heimay Island, off the coast of Iceland, fire
boats sprayed seawater on advancing lava flows. This action
is credited with having saved the harbor of the fishing village of Westmannaeyjar.

Violent Eruptions and Pyroclastic Activity

Many of the primary hazards of volcanism are directly related to the effects of violent eruptions, particularly pyroclastic activity. Unlike slowly moving lava flows, hot,
rapidly moving pyroclastic flows and laterally directed
blasts may overwhelm people before they can run away.
The most destructive pyroclastic flow this century (in terms
of loss of life) occurred on the Caribbean island of Mar-
tinique in 1902. In that eruption an avalanche of hot ash
rushed down the flanks of Mont Pelée at a speed of more
than 160 km/h, overwhelming the city of St. Pierre and instantly killing 29,000 people.

Pyroclastic debris sometimes contains blocks the size of
cars. In the 1968 eruption of Arenál Volcano in Costa Rica,
large falling blocks crashed through the roofs of houses 3
km away. However, much of the damage from pyroclastic
eruptions is caused by the widespread fall of ash. For example, in A.D. 79 many citizens in the nearby towns of Pompeii and Herculaneaum were killed in the eruption of
Mount Vesuvius. Most of the victims apparently were either buried and suffocated by falling ash or crushed by
buildings that collapsed under the weight of the ash (Fig.
4.24). The prehistoric eruption of Mount Mazama covered

▲ F I G U R E 4.24
Casts of the bodies of five citizens of Pompeii, Italy,
who were killed during the eruption of Mount Vesuvius in 79 A.D. Death was caused by poisonous gases;
then the bodies were buried by pyroclastic material.
Over the centuries the bodies decayed away, but their
shapes were imprinted in the tephra blanket. When
excavators discovered the imprints, they carefully
recorded them with plaster casts.

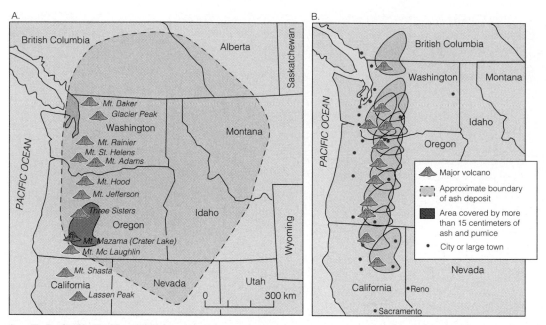

▲ F I G U R E 4.25
Area affected by ash fall from the prehistoric eruption of Mount Mazama. B. Area of Mount Mazama ash fall shown in A, transposed onto other dormant and active volcanoes in the region.

an area of 13,000 km² with a layer of ash more than 15 cm thick. The Cascade Range of volcanoes, which includes Mount Mazama and Mount St. Helens, stretches from California to southern British Columbia; imagine what the effect would be if another of the Cascade volcanoes were to undergo a similar eruption today (Fig. 4.25)! Even a relatively minor ash eruption can cause serious problems for residents of the area and wreak havoc on airplanes flying in the vicinity.

Poisonous Gas Emissions

Many volcanoes emit gases more or less continuously through fumaroles and geysers. Although water vapor is the main gas emitted by volcanoes, other kinds of gases are often present, many of them potentially harmful to people, animals, or vegetation. Some are toxic, such as carbon monoxide (CO). Some are acidic (e.g., hydrochloric acid, HCl, and hydrofluoric acid, HF); in other cases the emitted gases mix with water vapor to form acidic solutions (e.g., sulfuric acid, H_2SO_4).

Perhaps the best known examples of the destructive capabilities of volcanic gases occurred in Cameroon in West Africa. Lake Nyos and Lake Monoun are part of a series of summit crater lakes in small, young basaltic volcanoes in northwestern Cameroon. In August 1984 a release of car-

bon dioxide (CO_2) gas from Lake Monoun caused the deaths of 37 people. In 1986 a much more serious episode occurred at Lake Nyos. As one research team reported:

On 21 August at about 21:30 a series of rumbling sounds lasting perhaps 15 to 20 seconds caused people in the immediate area of the lake to come out of their homes. One observer reported hearing a bubbling sound, and after walking to a vantage point he saw a white cloud rise from the lake and a large water surge. Many people smelled the odor of rotten eggs or gunpowder, experienced a warm sensation, and rapidly lost consciousness. Survivors of the incident, who awakened from 6 to 36 hours later, felt weak and confused. Many found that their oil lamps had gone out, although they still contained oil, and that their animals and family members were dead. The bird, insect, and small mammal populations in the area were not seen for at least 48 hours after the event. The plant life was essentially unaffected.

At least 1700 people and 3000 cattle lost their lives in this episode. Scientists concluded that the cause of death was asphyxiation by CO_2 gas. Interestingly, the "rotten egg" smell reported by some of the survivors was later attributed to "olfactory hallucinations," a common side effect of exposure to nonlethal doses of CO_2. As discussed in Box 4.1, it is likely that the gas was slowly released from the underlying magma chamber, saturating both the lake water and bottom sediments with CO_2.

Secondary and Tertiary Effects of Volcanism

Mudflows and Debris Avalanches

Tephra can be dangerous long after an eruption has ceased. Rain or meltwater from summit snow can loosen tephra piled on a steep volcanic slope and start a deadly mudflow. Mudflows can also occur when a glowing avalanche enters a river channel. A volcanic mudflow is technically referred to as a **lahar.** A related phenomenon (actually a type of lahar) is the volcanic *debris avalanche,* in which many different materials—mud, blocks of pyroclastic material, trees, and so on—are mixed together.

Lahars and debris avalanches are common features of volcanic eruptions and can have devastating consequences. For example, in 1985, a small eruption of the volcano Nevado del Ruíz in Colombia melted part of the icecap on the mountain's summit. Mudflows were formed when the volcanic ash mixed with the meltwater. The massive lahars moved swiftly down river valleys on the flanks of the volcano, killing at least 23,000 people and causing more than US$212 million in property damage. It was the second worst volcanic disaster in the twentieth century.

Because they follow topography, lahars are relatively predictable. Sometimes they can be diverted by barriers or tunnels, and they are subject to the same types of control techniques as are lava flows. Fig. 4.26A shows a lahar hazard map that was hastily produced in the weeks preceding the eruption of Mount Pinatubo in 1991. A map of the actual lahars resulting from the eruption (Fig. 4.26B) illustrates the remarkable success of the team in predicting the paths taken by the hot volcanic mudflows. Even so, lahars were a major source of property damage and loss of life resulting from that eruption.

Flooding

It is common for volcanic eruptions to be accompanied or preceded by flooding, which may in turn cause mudflows. In some cases flooding may be caused by the rupture of a summit crater lake. In Iceland, for example, volcanoes buried under permanent ice caps cause subsurface accumulation of meltwater, which eventually escapes in a huge gush of water known as a *glacier burst.* Flooding can also result when rivers are blocked by lava flows or lahars. The resulting erosion and sedimentation may cause long-term disruption of downstream waterways; it can take years for a river system to clear itself of ash deposits from a major pyroclastic eruption. Some river courses in the vicinity of Mount Pinatubo were permanently altered after the 1991

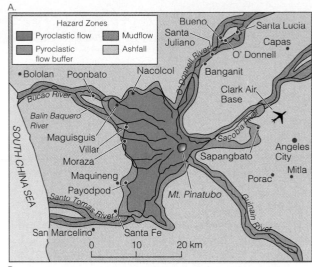

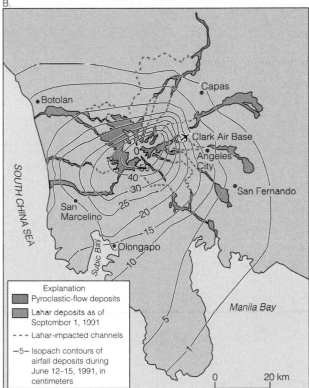

▲ F I G U R E 4.26
A. Hazard map produced by geologists immediately before the catastrophic eruption of Mount Pinatubo in 1991. B. Map of actual lahars resulting from the eruption shows the success of the monitoring team in predicting the volcanic hazards.

LAKES OF DEATH IN CAMEROON

*T*wo small lakes in a remote part of Cameroon, a small country in central Africa (Fig. B1.1), made international news in the mid-1980s when lethal clouds of carbon dioxide (CO_2) gas from deep beneath the surface of the lakes escaped into the surrounding atmosphere, killing animal and human populations far downwind. The first gas discharge, which occurred at Lake Monoun in 1984, asphyxiated 37 people. The second, which occurred at Lake Nyos in 1986, released a highly concentrated cloud of CO_2 that killed more than 1700 people. The two events have similarities other than location: both occurred at night during the rainy season; both involved volcanic crater lakes; and both are likely to recur without some type of technologic intervention.

Immediately after the disasters scientists began monitoring the lakes in an attempt to understand what had triggered these catastrophic incidents. They identified several factors that caused instability in the lakes and they produced models of the conditions that probably preceded the degassing. The researchers found that Lake Nyos had a huge reservoir of CO_2 stored deep beneath unstable, den-sity-layered or *stratified* lake waters. When it was disturbed by some event—wind at the surface, a landslide, an earthquake, or a minor eruption—the stratified column of water turned over, allowing approximately 100 million m^3 of CO_2 to bubble to the surface in just 2 hours (Figs. B1.2 and B1.3).

It soon became apparent that despite the release of huge amounts of CO_2 during the overturns, vast amounts of gas remain dissolved at great depths in the lakes, and more is being added each year from natural sources. On the basis of 6 years of records of dissolved gases, temperature, density, and recharge rates in each lake, scientists predicted that Lake Monoun was in danger of a violent degassing event within 10 years. Lake Nyos, which is much larger and deeper, would build up its CO_2 to dangerous concentrations within about 20 years.

Gas-rich volcanic crater lakes are known to occur in other localities as well, including Japan, Zaire, and Indonesia. In theory, the hazards associated with such lakes can be alleviated by controlled degassing. This technique involves installing a subsurface network of pipes to reduce the con-

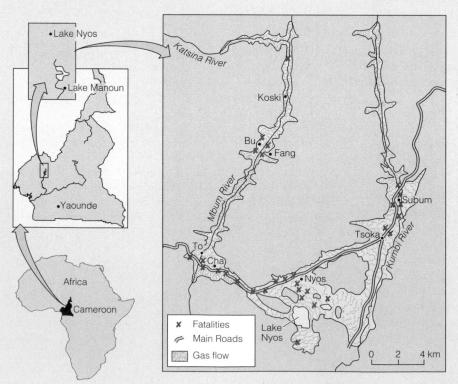

◀ F I G U R E B1.1
The path taken by the cloud of carbon dioxide emerging from Lake Nyos. Inset map shows the location of Cameroon and Lakes Monoun and Nyos.

Lake Nyos

Lake Manoun

Katsina River

Koski

Bu Fang

Mbum River

Subum

Tsoka

Kumbi River

To Cha

Nyos

Africa

Cameroon

Yaounde

✗ Fatalities
Main Roads
Gas flow

Lake Nyos

0   2   4 km

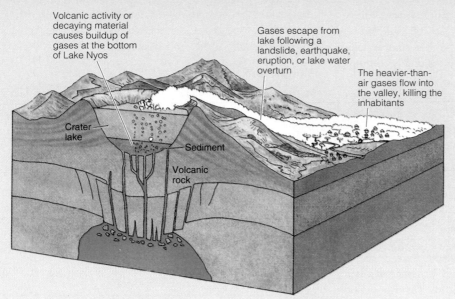

Volcanic activity or decaying material causes buildup of gases at the bottom of Lake Nyos

Gases escape from lake following a landslide, earthquake, eruption, or lake water overturn

The heavier-than-air gases flow into the valley, killing the inhabitants

Crater lake

Sediment

Volcanic rock

◄ F I G U R E   B1.2
A lethal cloud of CO_2 bubbled up from the bottom of Lake Nyos after the stratification of the lake water was disturbed. The gas, which is invisible and heavier than air, flowed over the natural dam surrounding the lake and down the adjacent valley, asphyxiating humans and animals in its path.

centrations of gas in deep water. Gas-charged water is pumped to the surface, where it releases gases to the air in nontoxic amounts. In Lake Monoun a prototype system of this type was tested successfully in 1992. It used three pipes, each 14 cm in diameter, placed deep beneath the surface of the water. Calculations showed that this procedure would restore CO_2 concentrations to stable levels in 2 to 3 years.

However, Lake Nyos presents a more complicated situation because of its greater size and depth. The challenge is

to control the flow of partially degassed water without upsetting the stratification of the lake water. One proposed solution is to divert water to a storage reservoir and then reinject it into the lake at the level of its natural buoyancy. The goal would be to establish a balance between the degassed water flowing into the lake and the bottom water removed from the lake. But unless the system is precisely calculated and perfectly designed, there is a danger that it will initiate the very situation the scientists are trying to prevent.

A.

B.

▲ F I G U R E   B1.3
Lake Nyos in Cameroon, Africa. A. The lake, muddy-brown in color, sits in a depression at the top of a volcano. One night in 1986 this lake discharged a mass of lethal carbon dioxide. B. Cattle killed by asphyxiation as a result of the 1986 gas discharge from Lake Nyos.

eruption, primarily by lahars that filled and blocked major drainage channels and diverted the rivers.

Tsunamis

Violent undersea eruptions can cause giant sea waves called *tsunamis*. Tsunamis set off by the eruption of Krakatau in 1883 killed more than 36,000 dwellers on the coast of Java and other Indonesian islands. Tsunamis are discussed more fully in Chapter 5.

Volcanic Tremors and Earthquakes

Seismicity is another common accompaniment of volcanic activity. Eruptions are commonly preceded by local earthquakes, which may be caused by the cracking and splitting open of fissures as the magma chamber inflates. At both Mount St. Helens (1980) and Mount Pinatubo (1991), hundreds of small earthquakes were recorded daily before the main eruption sequence, providing information that was used in predicting the onset of the eruption. Sometimes seismic activity comes in phases; sometimes it ceases when the eruption begins. The seismic activity may last only a few days or weeks or may continue for months or even years. The seismic prelude to the 79 A.D. eruption of Mount Vesuvius lasted 16 years!

Another type of seismicity that accompanies volcanic eruptions, *volcanic tremor* or *harmonic tremor*, consists of a more or less continuous, low-frequency, rhythmic ground motion. It may be associated with actual movement of the magma (e.g., boiling, convection, or drag of the magma against the chamber walls). Volcanic activity is associated with tectonic seismicity because the distribution of both volcanoes and earthquakes is controlled by the locations of active plate boundaries. It has even been suggested—though not confirmed—that large earthquakes may contribute to the onset of major volcanic eruptions.

Atmospheric Effects

Among the most significant long-term hazards associated with volcanism are atmospheric and climatic effects. Climatic effects result primarily from the ejection of ash and extremely fine particles or droplets called *aerosols* high into the stratosphere during major eruptions. Some eruption columns reach such great heights that high-level winds transport fine debris and sulfur-rich gas around the world. By blocking incoming solar energy, such atmospheric pollution can lower average temperatures at the land surface and cause spectacular sunsets as the Sun's rays are refracted by the airborne particles and aerosols. As mentioned earlier, the 1991 eruption of Mount Pinatubo blasted more than 8 km^3 of fine pyroclastic material and sulfur-rich gas high into the atmosphere, causing significant global cooling for as long as two years. And in 1815, three days of total darkness followed the major pyroclastic eruption of Tambora in Indonesia; the darkness extended as much as 500 km from the volcano. The following year was called "the year without a summer;" average global temperatures fell more than 1°C below normal, and crop failures were widespread.

Volcanic material ejected into the atmosphere can also cause toxic or acidic fallout. Hydrofluoric acid (HF) in volcanic fallout has been known to cause fluoride poisoning in livestock. Salty and acidic precipitation can damage crops, contaminate soil, and corrode materials.

Famine and Disease

As we will see in the next section, periodic light ash falls can contribute significantly to soil fertility. However, a major tephra eruption may wreak such havoc on agricultural land and livestock that famine results. The effects can be exacerbated by long-term climatic changes and by the dislocation of people and interruption of basic services associated with other aspects of the eruption. Most of the deaths that resulted from the 1991 eruption of Mount Pinatubo were caused not by the effects of the eruption itself but by disease, lack of water and sanitation, and related problems in temporary camps for the homeless.

BENEFICIAL ASPECTS OF VOLCANISM

Although we tend to focus on the hazards associated with volcanism, volcanoes have actually done much more good than harm to human beings. We mentioned earlier that the origin and evolution of the atmosphere and oceans were (and still are) directly dependent on the outgassing of volatile materials through volcanoes. The origin and evolution of life itself has been dependent on this process. Volcanoes have also created many thousands of square kilometers of new land, including some of the most beautiful and awe-inspiring settings in the natural environment (Fig. 4.27).

Aside from the beauty of the surroundings, it is no accident that people tend to live near volcanoes. These are areas with very fertile soils and high agricultural productivity. In a sense, volcanoes represent the only mechanism through which the Earth can directly revitalize soils that have been depleted through extended agricultural use. Periodic ash falls, especially when they are rich in potassium, phosphorus, and other essential elements, are effective natural fertilizers.

Volcanism is also linked with the formation of mineral deposits. Many types of ore deposits are associated with ancient volcanic systems, particularly collapsed calderas, ring fracture complexes, and subsurface fracture systems related to siliceous volcanism. Circulating groundwater heated by a subvolcanic magma chamber was a central feature in the formation of many major ore deposits. Recent explorations of oceanic rift zones have shown that the submarine rifting environment, in which seawater is heated by contact with hot basaltic magmas, is another important setting for the formation of major ore deposits.

▲ F I G U R E 4.27
Seemingly tranquil, lush fields grow on rich volcanic soils around an active volcano in the Philippines. Although an active volcano can be deadly dangerous during an eruption, the volcanic ash deposited fertilizes and so enriches the soil that people return quickly once an eruption ceases.

Volcanic heat also plays a role in geothermal energy. Using water warmed by hot rocks, people can heat or cool their houses, generate electricity, and swim in geothermally heated pools. The places on the Earth where geothermal energy is most easily exploitable are places of recent volcanic activity. Most volcanic and magmatic activity is close to plate margins, and it is here, in such places as New Zealand, the Philippines, Japan, Italy, Iceland, and the western United States, that geothermal power is being used.

PREDICTING ERUPTIONS

Volcanic eruptions are not rare events. Some 600 volcanoes are considered to be **active**—that is, they have erupted recently, or at least within recorded history—and every year about 50 to 60 volcanoes erupt. Volcanoes that have not erupted within recorded history, are deeply eroded, and show no signs of future activity are referred to as **extinct**. In between these two categories are volcanoes that are said to be **dormant**—that is, they have not erupted in recent memory and show no signs of current activity, but they are not deeply eroded. Dormant volcanoes can become active with unnerving ease—Mount St. Helens had been dormant for 123 years before its 1980 eruption; Mount Pinatubo had been dormant for more than 400 years before its 1991 eruption; and Mount Vesuvius was widely considered extinct before its eruption in A.D. 79.

To some extent, volcanic hazards can be anticipated if experts are able to gather data before, during, and after eruptions. With sufficient information, the experts can advise civil authorities on when to implement hazard warnings and when to move endangered populations to areas of lower risk. There is much to be gained through successful forecasting and prediction of eruptions. In the context of volcanism, a *prediction* offers a fairly specific date for an expected event, whereas a *forecast* is a more general statement of likelihood. For example, in 1975 D. R. Crandell and colleagues wrote that Mount St. Helens "will erupt again, perhaps before the end of this century." The volcano erupted 5 years later. This is a good example of a successful long-term forecast.

Recently scientists have begun to achieve some success in the short-term prediction of volcanic eruptions. Since the 1980 eruption of Mount St. Helens, for example, volcanologists monitoring the volcano have managed to maintain a near-perfect record in predicting minor dome-building eruptions. Successes have also been achieved at Mount Pinatubo and at Augustine Volcano in Alaska, among others. An important component of prediction is the identification and intensive study of high-risk volcanoes. Unfortunately, this process cannot yet guarantee that major disasters will be avoided altogether.

Aside from identifying a volcano as active, dormant, or extinct, the first step in predicting an eruption is to study the volcano's past behavior. This effort has two main goals. The first is to ascertain the historic style of eruption. This is important in predicting the type of activity to be expected and the area that might be affected by an eruption. The second goal—a crucial step in predicting the actual time of the eruption—is to determine the volcano's eruption interval. Some volcanoes, notably those of Hawaii, display reasonably regular eruptive sequences with definable intervals. Mauna Loa tends to alternate between flank and summit eruptions; a flank eruption can usually be expected to occur within three years after a summit eruption. Other volcanoes show little or no periodicity. In the case of Vesuvius, for example, no pattern or periodicity is discernible in 2000 years of recorded history.

Volcano Monitoring

The aspect of prediction in which the most progress has been made is in the monitoring of volcanoes. Such monitoring facilitates the prediction and forecasting of eruptions in two ways: (1) it allows scientists to track the distribution and movement of magma in the volcanic "plumbing" system; and (2) it makes possible the detection of anomalies and the identification of *precursor phenomena*—events or processes that signal the onset or progression of activity within the volcano.

Monitoring the Movement of Magma

There are also some techniques whereby volcanologists can actually attempt to monitor the distribution and movement of a magma body within a volcano. Changes in the magma body may indicate that an eruption is imminent. Several techniques are used in such studies.

Seismic studies Seismic waves are released both by earthquakes and by explosions. As discussed in Chapter 3, some types of seismic waves (S waves) cannot travel through liquids. Scientists can exploit this fact to study the distribution of magma underneath a volcano. The approach utilizes controlled explosives that are set off on one flank of the volcano, generating seismic waves. These are measured by seismographs set up on the other side of the mountain. If a body of liquid lies in the path of the seismic waves, the S waves will be blocked, creating a so-called *shadow zone* (Fig. 4.28). Through repetition of this technique, scientists can delineate the shape of the magma body and monitor any changes in its position.

Change in magnetic field Rocks contain minerals that are naturally magnetic; notable among these is the mineral magnetite (Fe_3O_4), which is relatively common. The magnetic field of a rock containing magnetite can be measured with a magnetometer. If magnetite is heated past its *Curie point,* 575°C, its magnetic field will exhibit a sudden decrease in strength. Recall that the temperature of molten rock may range anywhere from 800 to 1200°C. Thus, if a magma body were to move into close proximity to a rock, the rock would likely be heated past its Curie point and the strength of its magnetic field would decrease. This is another technique whereby scientists can sometimes track the movement of magma upward through a volcanic system.

Change in electrical resistivity The Earth has a natural (weak) electrical field. Changes in an underlying magma body should cause changes in the electrical resistivity of rocks. Such changes have been observed and measured and have been successfully correlated to magma movements that had not been detected by other methods.

Magma chamber modeling Detailed seismic studies and ground deformational studies, often derived from satellite imagery, can be combined with knowledge about the tectonic setting of a volcano to yield sophisticated three-di-

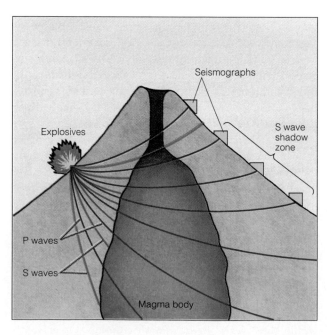

◀ F I G U R E 4.28
The use of controlled explosives to delineate an underlying magma body. Explosives set off on one flank of the volcano send seismic waves through the mountain. S waves, which cannot be transmitted through fluids, are blocked by any intervening body of fluid, creating an S-wave shadow zone on the other side of the mountain, where seismographs have been placed.

◀ F I G U R E 4.29
False color (infrared) image of Vesuvius and the Bay of Naples, Italy. Vesuvius, an active volcano, is the circular structure, center right. Lavas erupted from Vesuvius over the past 300 years are little weathered rocks and show up bright red. Older lavas and volcanic ash, such as those that buried the Roman towns of Pompeii and Herculaneum in 79 A.D., show in yellows and oranges. The dark blue and purple region at the head of the bay is the city of Naples. Left of Naples, near the center of the image, is a cluster of smaller volcanoes called the Flegreian Fields.

mensional models of the subvolcanic "plumbing" system. This is a new approach that is at the forefront of the science of volcanology.

Physical Anomalies and Precursor Phenomena

To a great extent, the prediction of volcanic eruptions (especially short-term predictions) is based on the detection of physical anomalies, and the recognition that these are precursors or warning signs pointing to the onset of a major event. A physical anomaly is any physical occurrence that is out of the ordinary and is linked in some way to activity within the volcano. Following are some of the most common anomalies that typically precede volcanic eruptions.

Ground deformation Major eruptions are sometimes preceded by a change in the shape or elevation of the ground, such as bulging, swelling, or doming related to the inflation of the underlying magma chamber. This type of change can be detected through the use of tiltmeters, geodolites, and electronic distance measurement instruments.

Change in the temperature of crater lakes, well water, fumaroles, or hot springs In some cases, the temperature

of water near a volcano rises, reaching a maximum at the onset of an eruption; in other cases, temperatures rise and then fall again prior to an eruption, or rise and fall with no eruption. Such physical anomalies are signs of activity in the volcano, even if there is no actual eruption.

Change in heat output at the surface Ground temperature in the vicinity of a volcano may also change before a major eruption. Recently, thermal (infrared) remote sensing techniques have been used to detect subtle changes in ground temperature near active volcanoes. Infrared satellite imagery can also be used to produce temperature maps of lava flows and to determine the total volume of a flow (Fig. 4.29).

Change in the composition of gases The composition of gases emitted from fumaroles has been observed to change before some eruptions. In general, the proportions of hydrochloric acid (HCl) and sulfur dioxide (SO_2) tend to increase relative to the proportion of water vapor, although other types of chemical changes have also been noted. Like changes in water temperature, changes in gas composition are not wholly reliable indicators, but they may contribute to a general scenario of an imminent event.

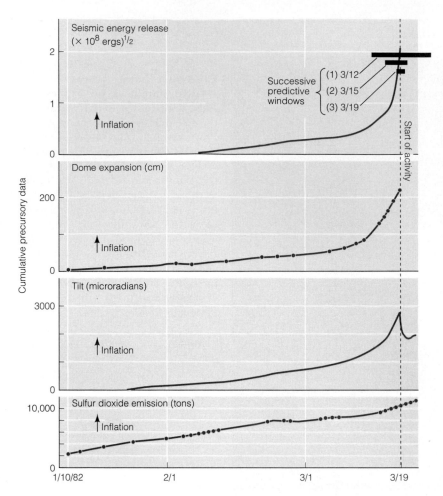

Seismic energy release
$(\times 10^8 \text{ ergs})^{1/2}$

Successive
predictive
windows
{ (1) 3/12
(2) 3/15
(3) 3/19

Start of activity

↑Inflation

Dome expansion (cm)

200

↑Inflation

0

Tilt (microradians)

3000

↑Inflation

0

Sulfur dioxide emission (tons)

10,000

↑Inflation

0

1/10/82   2/1   3/1   3/19

Cumulative precursory data

◄ F I G U R E 4.30
The acceleration of precursor activity before the March–April 1982 eruption of Mount St. Helens. The increasing accuracy of volcanologists' predictions, issued on March 12, 15, and 19, are shown by the sizes of the bars at top.

Local seismic activity   As discussed earlier in the chapter, several types of seismicity are associated with volcanic eruptions. Monitoring of precursory seismic activity has proved to be one of the most reliable tools for predicting eruptions. Seismic activity—especially local small earthquakes and harmonic tremor—often culminates in a major eruption. This occurred, for example, at Mount St. Helens and at Mount Pinatubo; in both of those cases monitoring of local seismic activity contributed to the issuance of successful predictions.

In most cases no single precursor indicator is sufficient to predict an eruption. Taken together, however, several physical or chemical anomalies may create an overall picture that provides a clear indication that an eruption is imminent (Fig. 4.30).

SUMMARY

1. Magma is molten rock, together with any suspended mineral grains and dissolved gases, that forms when temperatures rise sufficiently high for melting to happen in the Earth's crust or mantle. When magma reaches the surface, it does so through a volcano, a vent from which magma, solid rock debris, and gases are erupted. Magma that flows out and solidifies at the surface is called lava.

2. The most common types of magma are basaltic magma, which contains about 50 percent SiO_2; andesitic magma, containing about 60 percent SiO_2; and rhyolitic magma, containing about 70 percent SiO_2. The properties of magma—especially its composition (silica content and gas content) and viscosity—are important determinants of a volcano's eruptive style.

3. Nonexplosive eruptions are characterized by flows of fluid basaltic lava. Explosive eruptions are characterized by the eruption of large clouds of gas and pyroclastic material.

4. Some volcanoes erupt from fissures; others erupt from central vents. The most common landforms associated with central vent eruptions are shield volcanoes (gently

sloping volcanoes built up from successive lava flows) and stratovolcanoes (steep-sided composite volcanoes built up from alternating lava flows and deposits of pyroclastic material).

5. A close relationship exists between plate tectonics and the locations of different types of volcanoes. Rhyolitic volcanoes occur primarily on continental crust and are formed by the partial melting of continental rocks. Andesitic volcanoes form a Ring of Fire around the Pacific Ocean and are associated with the subduction and melting of crustal material. Basaltic magma, which is formed by the melting of mantle material, occurs at seafloor spreading centers and over deep-seated mantle hot spots.

6. Primary hazards associated with volcanism include lava flows, violent eruptions and pyroclastic activity, and emissions of poisonous gases. Secondary and tertiary hazards include mudflows and debris avalanches, flooding, tsunamis, local seismicity, atmospheric effects, and famine and disease.

7. Not all the effects of volcanism are negative. Volcanic eruptions have some beneficial aspects, including beautiful scenery, fertilization of agricultural soils, geothermal energy, and, in the long run, contributions to the chemical evolution of the atmosphere and the formation of mineral deposits.

8. Long-term forecasting of volcanic eruptions is based primarily on identification of a volcano's tectonic setting, past eruptive style, and eruptive interval.

9. Short-term prediction of volcanic eruptions and the issuance of warnings are based primarily on monitoring of volcanic activity. Monitoring has two main goals: to allow scientists to track the distribution and movement of magma in the volcanic "plumbing" system, and to permit the detection of physical anomalies and precursor phenomena—that is, events or processes that signal the onset or progression of activity within the volcano. Satellite imagery is an important aspect of modern volcano monitoring programs.

IMPORTANT TERMS TO REMEMBER

active volcano (p. 125)
andesite (p. 102)
basalt (p. 102)
caldera (p. 113)
crater (p. 113)
dormant volcano (p. 125)
eruption column (p. 108)
extinct volcano (p. 125)
fissure (p. 106)
fissure eruption (p. 114)

fumarole (p. 112)
geyser (p. 112)
glowing avalanche (p. 108)
lahar (p. 121)
lateral blast (p. 109)
lava (p. 101)
lava dome (p. 114)
magma (p. 101)
pyroclast (p. 107)
pyroclastic flow (p. 108)

pyroclastic rocks (p. 107)
resurgent dome (p. 114)
rhyolite (p. 102)
shield volcano (p. 110)
stratovolcano (p. 110)
tephra (p. 107)
tephra cone (p. 111)
thermal spring (p. 112)
viscosity (p. 103)
volcano (p. 101)

QUESTIONS AND ACTIVITIES

1. Imagine that you live in the vicinity of a shield volcano or a stratovolcano. What is the tectonic setting likely to be? What kinds of volcanic hazards do you think are most likely in each of these situations?

2. Are there any active volcanoes in your area? What about dormant or extinct volcanoes? You might wish to contact the local or regional Geological Survey for further information concerning ancient volcanic activity in your area.

3. Choose a volcanic event that interests you and investigate the impacts of the volcano on nearby residents. Some suggested events are the 1991 eruption of Mount Pinatubo or the 1980 eruption of Mount St. Helens.

4. Investigate a major eruption from long ago, such as the prehistoric eruption of Mount Mazama or the 1883 eruption of Krakatau. Try to project the events of the eruption into modern times. What hazards would be faced by residents if such an eruption were to occur today? On the basis of what you find out about the geology of the area, what is the likelihood that the event might recur in the near future?

5. Spend some time researching the role of volcanism and emissions of volcanic gas in the chemical evolution of the Earth's atmosphere. What kind of planet would we live on today if volcanism had been less active in this process?

TSUNAMIS

The bay was filling again; not slyly, as it had emptied, but in
a great rushing wave, climbing the shores . . . Right over the chariot-road
below me ran the salt sea, and climbed the plowland; spent itself,
and paused, and went sucking back from the scoured land.

• Mary Renault, The Bull From the Sea, 1962

According to Greek legend, Theseus, the king of Athens, had a handsome son named Hippolytos. Theseus' wife, Phaedra, fell in love with Hippolytos. When he refused her advances, she hanged herself and left a letter accusing Hippolytos of having raped her. Theseus, convinced of Hippolytos' guilt by the fact of Phaedra's death, invoked a curse entrusted to him by his father Poseidon, the god of the sea. As Hippolytos drove his chariot along the rocky coast, a huge wave swept over the land, bearing on its crest a sea-bull. After the wave receded, Hippolytos' battered corpse was found and carried to Theseus, who had learned the truth about Phaedra's deception too late.

Recent archeologic research has established a factual context for many of the events in Greek mythology. The wave referred to in the legend of Phaedra and Hippolytos may have been the *tsunami,* or *seismic sea wave,* generated in 1600 B.C. by an eruption of the volcano Santorin. This island volcano, now called Thíra, is located in the Mediterranean Sea to the northeast of Crete. The cataclysmic eruption of 1600 B.C.—the most powerful volcanic eruption in recorded history—has been blamed for the weakening and ultimate downfall of the Minoan civilization of Crete. Sea waves generated by the eruption would have inundated the nearby shores of Crete to depths of tens of

meters, totally destroying agricultural land. Coastal lowlands around much of the eastern Mediterranean also would have flooded. The legendary floods of Deucalion, reputed to have submerged the mythical lost city of Atlantis, have also been tentatively identified with these waves.

The eruption of Santorin in the seventeenth century B.C. was probably very similar to another, much more recent cataclysmic volcanic eruption. During the night of August 27, 1883, the Indonesian volcano Krakatau erupted, with catastrophic results. The eruption was significant for many reasons, but its most disastrous effect was the generation of a series of huge sea waves that inundated nearby coastal areas. The waves reached heights of at least 40 m above normal sea level. Blocks of coral and other objects weighing up to 600 tons were carried far inland by the force of the waves. At least 36,417 people were killed, most of them by drowning, and 165 coastal villages were destroyed.

WHAT IS A TSUNAMI?

A **tsunami** is a very long ocean wave that is generated by a sudden displacement of the sea floor. The term is derived from a Japanese word meaning "harbor wave." Tsunamis are sometimes referred to as **seismic sea waves** because submarine and near-coast earthquakes are their primary cause. They are also popularly called "tidal waves," but this is a misnomer; tsunamis have nothing to do with tides.

Tsunamis can occur with little or no warning, bringing death and massive destruction to coastal communities. In this chapter we examine how they are generated and what

An earthquake offshore of Japan on the 13th of July, 1993, caused a tsunami that destroyed the shorefront of Okushiri, Hokkaido, Japan.

kinds of damage they can cause. We also look at efforts to develop an integrated and effective early warning system for tsunami-prone areas.

Physical Characteristics of Tsunamis

Figure 5.1 illustrates the terminology used to describe tsunamis. These terms are essentially the same as those used for other types of waves, including light waves and seismic waves. For all types of waves, the term **wavelength** is used to refer to the distance from one crest to the next. Normal ocean waves average about 100 m in wavelength. A tsunami, in contrast, can exceed 200 km in wavelength.

Tsunamis also travel very quickly compared to normal ocean waves. This is particularly true in open water because wave velocities increase with water depth. In the open ocean, where the water is deepest, tsunami velocities can reach 950 km/h or more. (Normal ocean waves travel at velocities closer to 90 km/h.) Because of the relationship between velocity and water depth, when the waves reach shallower coastal waters they slow down abruptly—in a sense, they "pile up" on themselves. This causes the height of the wave from the still water line—its **amplitude**—to increase dramatically. Thus, in the open ocean tsunami amplitudes rarely exceed 1 m, little more than a broad, gentle bulge. But once they reach the shore it is not unusual for tsunamis to crest at 5 or 10 m. In the most dramatic documented cases these waves have risen to levels of 20, 30, and even 40 m above normal sea level.

The behavior of a tsunami after it reaches the shore is also different from that of a normal ocean wave. Sometimes the level of water at the coast recedes noticeably just before the onset of the tsunami. This phenomenon, called *drawdown,* can prove deadly to onlookers who may be tempted to walk down to the water's edge to investigate, only to be trapped when the wave finally hits. In other cases, the first observed movement of water may be a rise. Because of its long wavelength, it can take a long time for the tsunami to crest and recede. The water level may rise, reach a peak, and then remain high for several minutes. It can also take a long time, as much as an hour, for the successive crests in a series

of tsunamis to reach the shore. This is because the **frequency** of the wave—the time interval between crests—depends on both velocity and wavelength. Thus, although tsunamis have high velocities, their long wavelengths result in a long time interval between crests.

The Influence of the Shoreline

The water level achieved by a tsunami once it hits the shore is called its **run-up** (usually expressed as height in meters above normal high tide). Run-ups resulting from a particular tsunami vary from place to place along the coast because the height of a wave is strongly influenced by the depth of the water, the profile of the sea floor, and the shape of the coastline. In some cases the shoreline may bend or *refract* the wave, drawing it into otherwise protected harbors or bays. When the energy of a long section of the wave is concentrated on a particular stretch of coastline as a result of sea floor topography or shoreline configuration, it is called a **wave trap**. If the onrushing wave becomes concentrated in a long, narrow bay or river mouth, it can form a wall of water called a **bore**.

The run-up of a tsunami is usually determined by a combination of eyewitness accounts and physical indicators such as water marks on the sides of buildings, the locations of seaweed and other debris transported by the wave, the height to which vegetation has been damaged or killed by saltwater, the landward limits of sand deposition, and the height to which buildings or trees have been damaged by transported objects.

Measurement of Tsunami Magnitudes

Various scales are in use for the measurement of tsunami magnitudes. These scales define the magnitude of a tsunami in terms of the logarithm of the maximum wave amplitude observed locally. (In this respect, tsunami magnitude scales are analogous to the Richter earthquake magnitude scale.) One such scale is the *Imamura–Iida scale,* in which tsunami magnitude is calculated as a function of the maximum wave height along the coast. This type of formula leads to different estimates of magnitude for a given tsunami because of the variability of run-ups along a given

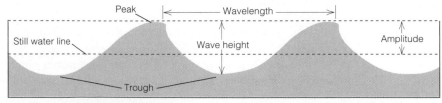

▲ F I G U R E 5.1
Terminology used in describing the characteristics of tsunamis. In most usages, *amplitude* **refers to half of the distance from peak to trough. In the context of ocean waves it is used differently, referring to the height of the wave from the still water line. The peak-to-trough distance is called the wave height.**

stretch of shoreline and the difficulties inherent in measuring the maximum run-up.

HOW TSUNAMIS ARE GENERATED

Any event that causes a significant displacement of the sea floor also causes the displacement of an equivalent volume of water. This, in turn, can generate tsunamis. Most tsunamis are generated by earthquakes, but they can also be caused by volcanic eruptions, submarine landslides, and human activity.

Earthquakes

Most tsunamis are produced by near-shore or offshore earthquakes. The primary causes of wave generation are the release of energy and the associated crustal deformation resulting from the earthquake. Any earthquake that generates a tsunami is called a **tsunamigenic earthquake**. Perhaps the most famous such earthquake occurred off the coast of Portugal in 1755. It produced a series of tsunamis with run-ups as high as 5 m above normal high tide that caused the deaths of 60,000 people in Lisbon (population 235,000). The waves were observed a few hours later as far away as the West Indies.

The magnitude of a tsunami is related to the magnitude of the earthquake that caused it; a large earthquake may be expected to generate a large tsunami. The correlation is not always that simple, however. For one thing, tsunamis are much more likely to result from vertical deformation of the crust than from horizontal deformation. No tsunami resulted from the great 1906 San Francisco earthquake, even though displacements of up to 6 m occurred along portions of the fault that are partly underwater. The San Andreas Fault, along which the quake occurred, is characterized by horizontal (strike-slip) motion, in which there is essentially no vertical displacement of the sea floor. By contrast, when the sea floor undergoes vertical deformation it acts like a huge paddle pushing the displaced volume of water outward from the zone of deformation. So earthquakes that occur along normal or reverse faults, in which the crustal displacement is primarily vertical, are much more efficient at displacing water and thus generating tsunamis.

Tsunami Earthquakes

Even along normal and reverse faults, very large earthquakes sometimes generate only moderate-size tsunamis or none at all. In other cases, earthquakes of small or moderate size generate unexpectedly large tsunamis. For this reason, Japanese tsunami experts have established a category of **tsunami earthquakes**, distinct from other tsunamigenic earthquakes. A tsunami earthquake generates a tsunami that is anomalously large with respect to the magnitude of the quake. Examples of tsunami earthquakes include the

▲ FIGURE 5.2
Tsunami striking Hilo, Hawaii, in 1946. Toward the lower left-hand corner of the photo you can see the figure of a man; he was swept away by the wave. The photograph was taken by a sailor on the *Brigham Victory.*

1986 Sanriku (Honshu, Japan) earthquake, which caused a tsunami 24 m high in which 26,000 people drowned, and the 1946 Unimak Island (Aleutian Islands, Alaska) earthquake, which caused a tsunami that traveled at a velocity of 800 km/h, inundating Hilo, Hawaii, 4.5 hours later with a run-up 18 m higher than normal high tide. The Unimak tsunami killed 150 people (90 of them in Hilo) and caused $25 million worth of property damage (Fig. 5.2). Both the Sanriku and Unimak earthquakes were moderate-sized quakes that were unexpectedly efficient at generating tsunamis.

Tsunami earthquakes occur along steeply dipping fault surfaces with vertical motion, the most efficient setting for generating tsunamis. The plate tectonic setting is typically, though not always, an active plate boundary characterized by deep oceanic trenches. There are two main reasons that a moderate-size earthquake in these settings may generate a very large tsunami: sediment slumping and surface rupture.

Along some active plate margins, called *accreting margins,* large amounts of sediment are accumulating in the deep trench along the fault zone (Fig. 5.3A). This mass of sediment is called an *accretionary prism.* When an earthquake strikes, large volumes of sediments in the accretionary prism may collapse and slump into the trench, causing unusually large tsunamis. This may have been the cause of the Sanriku and Unimak tsunamis; both followed quakes in areas characterized by large-scale slump features in underwater sediments near the trench.

The second category of tsunami earthquake is represented by the September 2, 1992, quake off the coast of Nicaragua, an earthquake of moderate size (Richter magnitude 7) that generated an unusually large tsunami. The related tsunami had a maximum run-up of about 10 m. About 170 people were killed by the wave and 500 were injured; 13,000 were left homeless and 1500 homes were destroyed. Unlike the trenches associated with Sanriku and the Aleutian Islands, the oceanic trench off the coast of Nicaragua does not contain a large volume of sediment; that is, it occurs along a *nonaccreting margin,* in which sediments are not accumulating but are being subducted into the trench (Fig. 5.3B). It has been proposed that because of the absence of sediments in the trench, the slip along the fault was able to propagate all the way from the earthquake's focus to the ocean floor. This means that the actual

sea floor ruptured, vertical displacement occurred, and a fault scarp was created, causing the "paddle" effect necessary to generate a sizable tsunami. (Along an accreting margin, by contrast, the presence of sediments would prevent the rupture from extending all the way to the ocean floor.)

For many years seismologists have used seismic wave forms to help them understand the complex tectonic movements of fault blocks that create earthquakes. This approach is called *seismic inversion.* Japanese scientists have applied this approach to the study of tsunami wave forms. To determine the type of crustal deformation that occurs during tsunami earthquakes, they look at historical records of tsunami arrival times and data from tidal gauges; then they work backward to recreate the wave form, its velocity and propagation pattern, and, ultimately, the characteristics of the crustal movement that have generated the wave. It is relatively easy to model the propagation of tsunamis because the factors that influence the velocities of water waves are better understood than those that influence the motion

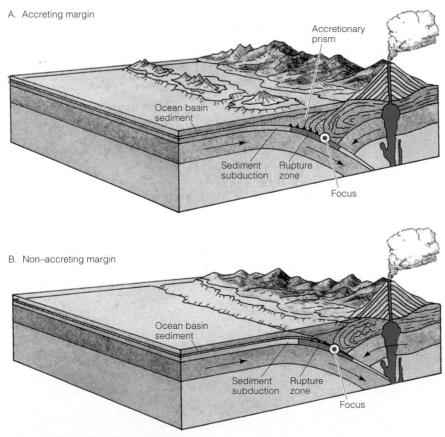

▲ F I G U R E 5.3
Diagram showing two types of active plate margins where tsunami earthquakes occur. A. An accreting margin, where slumping in the accretionary prism of sediments is the main tsunami-generating mechanism. B. A nonaccretionary margin, where surface rupture is the main tsunami-generating mechanism.

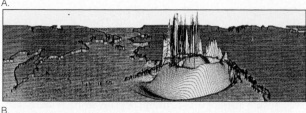

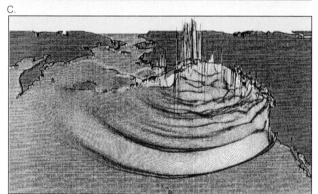

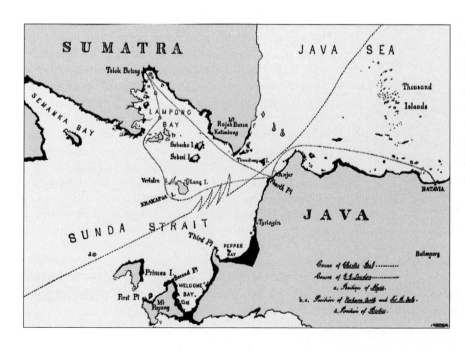

◀ F I G U R E 5.4
Computer simulation of the tsunamis resulting from the Alaska Good Friday (March 28, 1964) earthquake. The crustal deformation that accompanied the earthquake included about 3 m of uplift and 1 to 2 m of subsidence over an area of about 500×300 km². Tsunami propagation after (A) 1 hour, (B) 3 hours, and (C) 4 hours. Vertical exaggeration is about 2×10^6; the elevation of the continents above the ocean in the figure corresponds to a tsunami height of about 5 cm. The tsunami was 30 m high at Valdez Harbor in Alaska. When it reached the shore in Hawaii, the tsunami was 5 m high.

Volcanic Eruptions

Volcanic eruptions can also be efficient generators of tsunamis, as shown by the Santorin and Krakatau eruptions mentioned at the beginning of the chapter. An explosive eruption can displace enormous volumes of rock; if the volcano is partly or mostly submerged, a corresponding volume of water will also be displaced, causing a tsunami. The collapse of steep walls of a volcanic structure and associated underwater debris or ash flows can contribute to the formation of a tsunami.

In the case of Krakatau, an enormous Plinian eruption in 1883 generated a series of at least three great tsunamis and many smaller waves. Most of the 36,417 people who died as a result of the eruption were swept away by the tsunamis; hot ash caused a number of injuries but only a few fatalities. Underwater ash flows may have generated some of the tsunamis. Krakatau is located west of Java in the Strait of Sunda, which was an important shipping corridor at the time (Fig. 5.5). Shipboard reports of the erup-

of seismic waves. A computer simulation of tsunami propagation after the 1964 Good Friday earthquake in Prince William Sound, Alaska, is shown in Fig. 5.4.

◀ F I G U R E 5.5
Map of Krakatau Island and the Strait of Sunda. The areas shaded black are those that were submerged by the tsunamis following the eruption of Krakatau. The dotted and dashed lines and small circles (a, b, c, d) indicate the paths and positions of ships from which eyewitness accounts were made.

tion, the resulting tsunamis, and the destruction in its aftermath were among the most important eyewitness accounts of this event. After the eruption the sea was generally very rough, with unpredictable currents, and cluttered with blocks of pumice, fallen trees, and corpses. Crews reported that their ships had been tossed and tumbled by the waves, and several large vessels were run aground. The government steamer *Berouw* was carried 2.5 km inland and stranded 24 m above sea level; its crew of 28 drowned.

An elderly Dutch pilot who guided ships through the Strait gave this account of the first of the tsunamis caused by the eruption:

> *Looking out to sea I noticed a dark black object through the gloom, traveling towards the shore. At first sight it seemed like a low range of hills rising out of the water, but I knew there was nothing of the kind in that part of the Sunda Strait. A second glance—and a very hurried one it was—convinced me that it was a lofty ridge of water many feet high, and worse still, that it would soon break upon the coast near the town. There was no time to give any warning, and so I turned and ran for my life . . . In a few minutes I heard the water with a loud roar break upon the shore. Everything was engulfed. Another glance around showed the houses being swept away and the trees thrown down on every side . . . I gave up all for lost, as I saw with dismay how high the wave still was. I was soon taken off my feet and borne inland by the force of the resistless mass. I remember nothing more until a violent blow aroused me. Some hard firm substance seemed within my reach, and clutching it I found I had gained a place of safety. The waters swept past me, and I found myself clinging to a cocoanut palm-tree. Most of the trees near the town were uprooted and thrown down for miles, but this one fortunately had escaped and myself with it.*
>
> *The huge wave rolled on, gradually decreasing in height and strength until the mountain slopes at the back of [the town of] Anjer were reached, and then, its fury spent, the waters gradually receded and flowed back into the sea. The sight of those receding waters haunts me still. As I clung to the palm-tree, wet and exhausted, there floated past the dead bodies of many a friend and neighbour. Only a mere handful of the population escaped. Houses and streets were completely destroyed, and scarcely a trace remains of where the once busy, thriving town originally stood. Unless you go yourself to see the ruin you will never believe how completely the place has been swept away.*

This wave was followed by at least two other tsunamis of equal or greater magnitude. The wave run-ups varied, reaching as high as 41 m in some places. One observer reported that "up to a height of 30 to 40 m, the sea had erased or knocked down every object, and covered it with a gray layer of mud and pumice-like material." The waves destroyed buildings, uprooted trees, and transported some astonishingly heavy objects; for example, a block of coral weighing approximately 600 tons was ripped from an offshore coral reef and swept 100 m inland. Many smaller waves were also reported. In one report, from Telok Betong Harbor (where the *Berouw* ran aground), the water level was observed to be 1 m below the level of the pier at one moment and 1 m above the level of the pier a moment later.

Landslides

Tsunamis can also be caused by coastal and submarine landslides. In most such cases the landslides themselves have been generated by earthquakes or volcanoes. Sediments can build up along the walls of a submarine canyon and, when shaken by an earthquake, collapse or slough off into the bottom of the canyon. Many of the most damaging tsunamis that have occurred along the West Coast of North America have been generated in this manner. For example, the Good Friday earthquake triggered at least 20 separate landslide tsunamis. About 80 of the 119 deaths linked to the quake were caused by those tsunamis.

Underwater Explosions

Tsunamis are occasionally caused by human activity. For example, tsunamis were generated by submarine nuclear testing at Bikini Atoll in the Marshall Islands in the 1940s and 1950s.

Tsunamis in Lakes, Bays, and Reservoirs

Tsunami-like waves can occur in smaller, enclosed bodies of water such as lakes, bays, and reservoirs. In most cases, these waves are generated by falls or slides of rock or soil. The slides can occur spontaneously or be triggered by earthquakes, explosions, or human activity. A famous example of a spontaneous landslide leading to a tsunami-like wave is the Vaiont Dam disaster, which occurred in Italy in 1965. Hillslope failure resulted in an avalanche of soil and rock that fell into the reservoir, creating an enormous wave that surged over the dam and flooded the valley below. Another well-known episode occurred in Rissa, Norway. Excavations adjacent to the coast of a lake initiated a major "quick clay" landslide; the slide, in turn, generated a tsunami-like wave that swept across the lake, destroying a village on the opposite shore. Still another devastating landslide-induced tsunami occurred in Lituya Bay, Alaska, following an earthquake of Richter magnitude 7 that occurred on July 9, 1958. The earthquake triggered an avalanche that fell into one corner of the enclosed bay, producing a wave that stretched completely across the bay and ran up the opposite side to a height of 60 m, stripping vegetation from the shoreline (Fig. 5.6).

Seiches

A related phenomenon occurring in enclosed bodies of water is a **seiche**, a periodic standing-wave oscillation of the water surface. Once initiated, a seiche travels back and forth at regular intervals determined by the depth, size, and bottom and shoreline configurations of the body of water. In other words, each body of water will have a characteristic period of oscillation for such waves—or sometimes several characteristic periods, depending on the complexity of the water body, the shoreline, and so on. The size of the stand-ing waves can range from a few centimeters to several me-ters, and the interval of repetition can range anywhere from a few minutes to a few hours.

Seiches are usually caused by unusual winds, currents, or changes in atmospheric conditions. They can also be initi-ated by earthquakes. For example, the Loma Prieta (San Francisco) earthquake of October 17, 1989, caused seiche oscillations in Monterey Bay (Fig. 5.7). The waves oscil-lated at an interval of about 9 minutes and were about 0.4 m high from crest to trough.

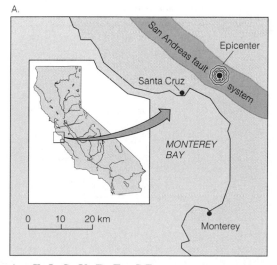

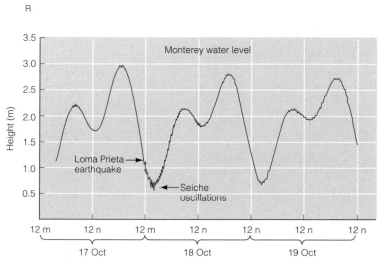

▲ F I G U R E 5.7
A. Location of Monterey Bay relative to the San Andreas fault and the epicenter of the Loma Prieta earthquake. B. Records from Monterey Bay showing water-level oscillations resulting from the Loma Prieta earthquake of 1989.

MITIGATION OF RISK AND HAZARDS

Tsunami Hazards

The main source of damage from a tsunami is the direct action of the wave on coastal structures. However, a variety of indirect mechanisms can also cause further damage. For example, flotation and drag can move houses, machinery, and railroad cars; debris such as trees, cars, and parts of destroyed structures can become projectiles; strong currents can erode foundations, leading to the collapse of buildings, bridges, and seawalls; and fires can result from the combustion of oil spilled by damaged ships and storage facilities.

The most serious hazard associated with tsunamis, of course, is loss of life. Over the past century, 94 destructive tsunamis have caused the deaths of a total of over 51,000 people. In 1992 alone, two tsunamis were responsible for the deaths of hundreds of people. In each of these cases (one caused by the earthquake in Nicaragua discussed earlier, the other in Riangkrok, Indonesia) the death toll was unusually high because the ferocity of the wave had not been adequately anticipated.

Prediction and Early Warning

The prediction of tsunamis centers on efforts to understand the mechanisms through which earthquakes generate tsunamis. Particular attention has been directed toward the so-called tsunami earthquakes, those that generate

tsunamis that are unexpectedly large in comparison with the magnitude of the quake. As mentioned earlier, the magnitude of an earthquake is not the sole determinant of the magnitude of a tsunami; the degree, direction, and disposition of crustal deformation are also important. For this reason, in predicting whether an earthquake will generate a sizable tsunami, the Richter magnitude scale may not be the most useful measure of earthquake size. Some scientists believe that an earthquake's *seismic moment* may be a more appropriate measure. Seismic moment is a different measure of the size of an earthquake and takes into account the elastic properties of the Earth material, the fault area, and the average area over which dislocations occurred during the earthquake—all factors that are important in the generation of tsunamis.

Tsunamis and tsunamigenic earthquakes are a particular hazard for Pacific Ocean islands and locations around the Pacific rim. Hawaii is especially vulnerable to dangerous tsunamis (Fig. 5.8 and Table 5.1) because of its location in the path of waves generated at many seismically active points around the Pacific rim. Sirens and radio newscasts alert Hawaii's population to arriving tsunamis, and maps printed in telephone books show the coastal zones at greatest risk (Fig. 5.9); both measures have helped reduce the loss of life from tsunamis.

Regional Warning Systems

In spite of the rapid onset and quick travel times of tsunamis, there is often ample opportunity to warn residents of coastal areas, especially if they are located more than 750 km (approximately 1 hour in tsunami travel time)

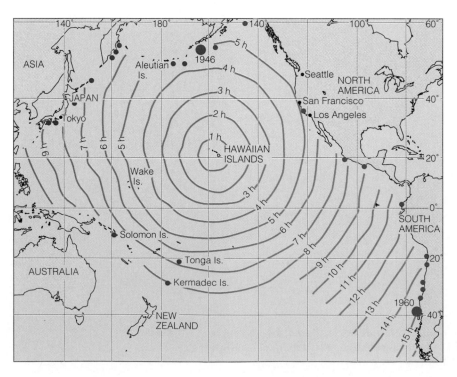

◄ F I G U R E 5.8
Map showing the time required for a tsunami to reach the island of Oahu, Hawaii. Small red dots mark the origins of tsunamis that have struck Hawaii. Large dots mark places where the disastrous tsunamis of 1946 and 1960 originated.

T A B L E 5.1 • Major Tsunamis in Recorded History.

Date	Source region	Visual run-up	Generated by	Comments
1600 B.C.	Santorin	?	Volcanic eruption	Devastation of Crete and Mediterranean coast
Nov. 1, 1755	Eastern Atlantic	5-10	Earthquake	Major damage in Lisbon, Portugal; tsunami reported from Europe to West Indies
Aug. 13, 1868	Peru–Chile	>10	Earthquake	Observed in New Zealand; damage in Hawaii
Aug. 27, 1883	Krakatau	40	Volcanic eruption	Over 30,000 drowned
June 15, 1896	Honshu	24	Earthquake	About 26,000 drowned
Mar. 2, 1933	Honshu	>20	Earthquake	3000 deaths from waves
April 1, 1946	Aleutian Islands	10	Earthquake	Over 150 drowned in Hilo, Hawaii; $25 million in property damage
May 23, 1960	Chile	>10	Earthquake	909 dead and 834 missing along Chilean coast; 120 dead in Japan
Mar. 28, 1964	Alaska	6	Earthquake	In California, 119 deaths and $104 million damage
Dec. 2, 1992	Indonesia	26	Earthquake	137 people killed, village destroyed
Sept. 2, 1992	Nicaragua	10	Earthquake	170 deaths, 500 injured, 13,000 homeless

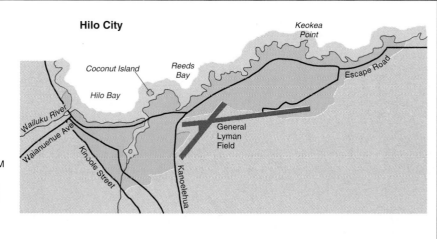

CIVIL DEFENSE
Tsunami Inundation Maps

Tsunamis

In Hawaii, tsunamis (tidal waves) have been a recurring menace and are given special emphasis. Scientific studies and historical records indicate that anticipated flooding generally will be limited to the shaded areas shown on the sectional maps on the following pages. When the

ATTENTION/ALERT SIGNAL is sounded, the EMERGENCY BROADCAST SYSTEM will direct evacuation of threatened areas. If a tsunami is generated by a local earthquake, limits of flooding may exceed shaded areas and there may be no time for an official warning. *Any violent earthquake* – one that causes you to fall or hold onto something to keep from falling – *is a natural tsunami warning.* Immediately evacuate beaches and low-lying coastal areas as soon as the shaking stops. Go to an area that is safe from flooding.

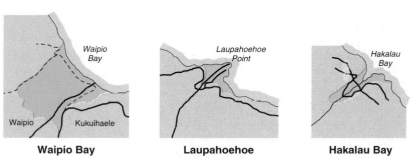

▲ F I G U R E 5.9
Page from a telephone book showing areas (in green) along the coast of the island of Hawaii that are susceptible to inundation by tsunamis.

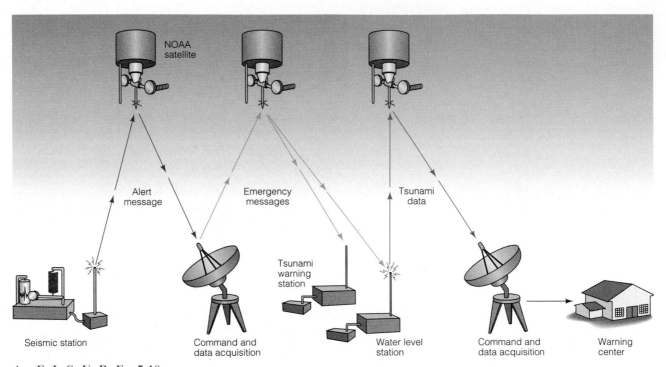

▲ F I G U R E 5.10
Early warning systems in the Pacific Ocean basin make use of information relayed from seismic stations to tsunami warning stations via NOAA (National Oceanographic and Atmospheric Administration) satellites.

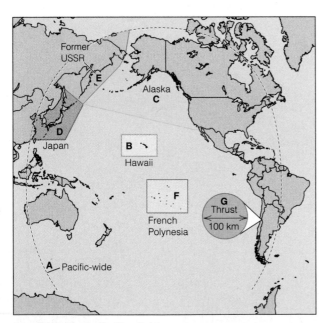

▲ F I G U R E 5.11
Map of the Pacific Ocean basin showing: A. the Pacific-wide early-warning system, which warns populations about 1 hour or more than 750 km from the source of the tsunami; B. through F. regional warning systems, which warn residents about 10 minutes or 100 to 750 km from the source; G. THRUST, a local warning system designed to warn populations within 100 km of the source.

from the source of the wave. Several different types of early warning systems now exist in the Pacific basin. One such system is operated by the National Oceanic and Atmospheric Administration (NOAA) Pacific Tsunami Warning Center (PTWC) near Honolulu. It is an international monitoring network consisting of about 30 seismic stations and 78 tide stations located around the Pacific basin. When an earthquake occurs, it is detected by the seismic stations. If the quake meets certain criteria with respect to location and magnitude, local tidal gauges are monitored for signs of a tsunami. If a tsunami is detected, and if it is large enough to be potentially hazardous, a Pacific-wide warning is issued. Information is relayed from the seismic station to the tsunami warning station via NOAA satellites (Fig. 5.10). Arrival times for the tsunami can be calculated for different localities around the Pacific rim. The minimum time required to gather this information and issue a warning is 1 hour; since tsunamis typically travel at 750 km/h, this means that the Pacific-wide early-warning system is effective only for communities located more than 750 km from the epicenter.

Regional warning systems are designed to provide warnings to areas in a 100- to 750-km radius (i.e., between about 10 minutes and 1 hour) from the epicenter (Fig. 5.11). Regional systems have been established in areas that are known to be particularly susceptible to earthquake-generated tsunamis, including Japan, Alaska, and French Poly-

BOX 5.1

•

THE HUMAN PERSPECTIVE

A TALE OF TWO TSUNAMIS

*T*wice in 15 months (July 1993 and October 1994) major earthquakes occurred near the coasts of Japan. Although the two quakes were of comparable magnitude (7.8 and 8.2, respectively), the first created a damaging tsunami of catastrophic proportions, whereas the second caused only minor tsunamis and a 6-hour media event. The differing impacts of the quakes may be attributed to differences in the topography of the ocean floor on Japan's east and west coasts as well as to the early-warning system installed by the government in 1994.

Tsunamis generated by earthquakes sometimes cause more deaths and damage in coastal areas than the earthquakes themselves. Japan is especially vulnerable to these devastating events because of its location in a tectonically active region and the high-density population centers located on its coasts (Fig. B1.1). The July 1993 quake, for example, triggered a 33-m seismic sea wave that struck the west coast of Okushiri Island within seconds of the first tremors, killing 239 people and destroying 558 houses.

Although the west coast of Japan historically has experienced fewer tsunamis than the east coast, current plate tectonic theory suggests that the west has a greater potential for quakes and tsunamis than was previously recognized. During the past 50 years a series of earthquakes has occurred in the Sea of Japan near Honshu and Hokkaido islands. The locations of these quakes is consistent with the theory that the North American and Eurasian plates meet just west of Honshu Island, whereas the Eurasian Plate is being subducted beneath western Japan. In addition, the islands in the area have coastlines with numerous bays, reflecting the shallow topography of the ocean bottom. Scientists predict that these characteristics of the seabed topography and the tectonic setting create conditions that may produce more damaging tsunamis in the future.

In contrast to the devastation on Okushiri Island caused by the 1993 tsunami, the tsunamis produced by an undersea earthquake east of Hokkaido in 1994 caused no casualties. The epicenter of the *M* 8.2 quake was located about 160 km off the northern coast of Japan. Fatalities were averted through a new system that links the seismographs of the Japanese Meteorological Agency with the national media. Eight seconds after the first tremors were recorded at the Kushiro Observatory, warnings were automatically sent to radio and television stations. Those warnings were available to a nationwide network in screen flashes and voiceovers 2 minutes later, even before the jolt of the earthquake was felt in Tokyo. The local early-warning system had been operational for just 3 months, 1 year after the fatal Okushiri tsunami.

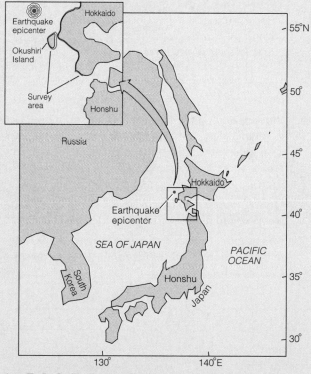

▲ F I G U R E   B1.1
Map of Japan, showing locations mentioned in the text.

nesia. The regional centers issue warnings on the basis of earthquake magnitude and location alone; warnings typically can be issued within 10 or 12 minutes of the earthquake's occurrence. In addition, local early-warning systems can be designed to provide warnings for populations less than 10 minutes (less than 100 km) from the source. An example of such a system is THRUST (Tsunami Hazards Reduction Utilizing Systems Technology), a pilot project in Valparaiso, Chile (Fig. 5.11).

The effectiveness of tsunami early-warning systems is impressive. For example, before the Japanese system was established, a total of over 6000 people had been killed by 14 tsunamis; since then, 20 tsunamis have killed a total of 215 people in Japan. In addition to the regional early-warning systems, some localities have established systems to issue warnings to the immediate coastal area within a few minutes of a large tsunami-generating event. The success of local early-warning systems depends heavily on emergency operations planning, as well as on access to timely information about earthquake occurrences and water levels. Other important factors are the ability of local authorities to assess the danger; the ability to disseminate information very quickly; and education of the public to respond appropriately in the event of a tsunami emergency.

SUMMARY

1. A tsunami is a very long ocean wave that is generated by a sudden displacement of the sea floor. Although they are sometimes mislabeled "tidal waves," they have nothing to do with tides.

2. The wavelengths of tsunamis typically exceed 200 km. In the open ocean, their velocities can reach 950 km/h or more. When the waves reach shallower coastal waters they slow down very abruptly and "pile up" on themselves. In open water, the amplitude of a tsunami rarely exceeds 1 m, but on shore it is not unusual for tsunamis to crest at 5 or 10 m above normal sea level.

3. Run-ups resulting from a tsunami vary from one place to another along the coast because the height of a wave is strongly influenced by the depth of the water, the profile of the sea floor, and the shape of the coastline.

4. Most tsunamis are generated by earthquakes, but volcanic eruptions, submarine landslides, underwater explosions, and human activity can also cause them.

5. The magnitude of an earthquake is not the only factor that determines whether the earthquake will generate a tsunami. Some earthquakes are unexpectedly efficient at generating tsunamis.

6. Tsunami-like waves sometimes occur in enclosed bodies of water such as lakes, bays, and reservoirs. Seiches, or oscillating standing waves, are a related phenomenon.

7. The direct action of waves is the main source of damage and loss of life caused by tsunamis. Other sources of damage include strong currents and debris acting as projectiles.

8. The prediction of tsunamis focuses on identifying earthquakes that are likely to generate tsunamis and on estimating the travel times of tsunamis across ocean basins.

9. Regional warning systems around the Pacific rim have been quite effective at minimizing loss of life from tsunamis.

IMPORTANT TERMS TO REMEMBER

amplitude (p. 132)
bore (p. 132)
frequency (p. 132)
run-up (p. 132)

seiche (p. 137)
seismic sea wave (p. 131)
tsunami (p. 131)
tsunami earthquake (p. 133)

tsunamigenic earthquake (p. 133)
wave trap (p. 132)
wavelength (p. 132)

QUESTIONS AND ACTIVITIES

1. Not all tsunamis originate on active plate margins. Investigate the Newfoundland earthquake of 1929, which occurred on a *passive,* or trailing, continental margin yet caused significant tsunamis. What was the tsunami-generating mechanism in that case?

2. Imagine that you live in a coastal community in a region that is susceptible to tsunamis (maybe you really do). What kinds of action do you think your community should carry out to minimize the potential damage and loss of life from tsunamis? What kinds of planning exercises do you think your local or regional government should undertake to determine the risk to your community?

3. If you do live in a tsunami-prone area, how well prepared is your community? Is there an early-warning system? If so, find out how it works. What kinds of action could you take to minimize risk if you own property near the coast? What would you do if you learned that a tsunami was heading toward your coast?

4. As discussed in this chapter, Japan is particularly susceptible to destructive tsunamis. Assess the tsunami risk to the Japanese public by investigating (a) earthquake activity in Japan, (b) the distribution of population centers along Japan's coastlines, (c) the history of damaging tsunamis in Japan, and (d) the status of Japan's participation in local, regional, and Pacific-wide early-warning systems. Write a summary of your findings.

LANDSLIDES AND MASS-WASTING

The heights of our land are thus leveled with the shores; our fertile plains are formed from the ruins of mountains.

• James Hutton

*I*n the high Andes of South America, steep, unstable slopes rise above densely populated valleys. Here many active volcanoes with rugged peaks have been built up along converging lithospheric plates. These conditions hold the potential for disaster in a landscape where major earthquakes and volcanic eruptions can cause the steep slopes to collapse.

In Colombia, a group of active volcanoes lies west of Bogotá. One of them, Nevado del Ruíz, has a history of eruptive activity extending back to at least 1595. In late 1984 the dormant volcano awakened and began belching clouds of steam and ash; this activity continued through the autumn of 1985. People in the city of Armero, located far downvalley from the volcano, grew alarmed. Local authorities reassured them, even though recent geologic studies of the volcano had disclosed a history of repeated large volcanic mudflows. In early November, when the volcano showed signs of increasing activity, geologists warned that in the event of an eruption the resulting mudflows could pose a danger for Armero. At 3 p.m. on November 13 a technical emergency committee urged that Armero be evacuated, but the warning went unheeded.

That night, as the local radio station played cheerful music and urged people to be calm, the volcano erupted. Torrents of water released from rapidly melting ice and snow near the summit sent huge waves of muddy debris

◄ _____

Mudflow of water-saturated volcanic ash in the town of Armero, Colombia, 1985. The thick, muddy slurry was released by an eruption of an ice-clad Andean volcano. More than 23,000 people were killed and very few buildings were left standing.

surging down the volcano's slopes into the surrounding valleys. The largest of several mudflows moved rapidly toward Armero. Just after 11 p.m., as most of the citizens of Armero were sleeping soundly, a turbulent wall of mud came rushing out of a canyon and inundated the city. At least 23,000 people were buried in sulfurous volcanic mud. Had the geologists' warning been heeded in time, the tragedy might have been avoided.

MASS-WASTING AND ITS HUMAN IMPACTS

The landscapes we see about us may appear fixed and unchanging, but if we were to make a time-lapse motion picture of almost any slope it would be clear that the slope is constantly changing. Much of the recorded motion would be a result of **mass-wasting**, the movement of Earth materials downslope as a result of the pull of gravity. Any perceptible downslope movement of bedrock, regolith, or a mixture of the two is commonly referred to as a **landslide**, but, as we shall see, many different types of movement, materials, and triggering events may be involved in downslope mass movements of Earth material.

As the human population increases and cities and roads expand across the landscape, mass-wasting processes become increasingly likely to affect people. Landslides occur throughout the world; virtually no area is immune. The impacts of mass-wasting, in terms of loss of life and damage to property, can be devastating (Table 6.1). In the United States alone, landslides cause about $1.5 billion in economic losses and 25 to 50 deaths in a typical year. In less developed countries, the toll in damage and loss of life can be considerably more significant because of population density, lack of stringent zoning laws, scarcity of information about landslide hazards, and inadequate preparedness.

Although it may not always be possible to predict or prevent the occurrence of mass-wasting events, a knowl-

T A B L E 6.1 • **Fatalities Resulting from Some Major Landslides During This Century**[a]

Year	Location[b]	Fatalities
1916	Italy, Austria	10,000
1920	China[e]	200,000
1945	Japan[f]	1,200
1949	USSR[e]	12,000–20,000
1954	Austria	200
1962	Peru	4,000–5,000
1963	Italy	2,000
1970	Peru[e]	70,000
1985	Columbia[v]	23,000
1987	Ecuador[e]	1,000

[a]Source: National Research Council (1987).

[b]Landslides related to earthquakes ([e]), floods ([f]), and volcanic eruptions ([v]).

edge of the processes and their relationship to local geology can lead to intelligent planning that will help reduce losses of life and property. In this chapter we look at the factors that control slope stability, the different types of mass movements, and the impacts of mass-wasting on humans.

TYPES OF MASS-WASTING PROCESSES

All mass-wasting processes take place on slopes. There are many kinds of slope movements, but there is no simple way to classify such movements. The composition and texture of the material involved, the amount of water or air in the mixture, and the steepness of the slope all influence the type and velocity of movement. In effect, there is a progression from the flow of clear stream water to sediment-laden stream water to an array of mass-wasting processes ranging from those in which water promotes downslope movement to those in which water plays no direct or significant role.

For the purposes of this discussion, we will divide mass-wasting processes into two basic categories: (1) those involving the sudden failure of a slope, which results in the downslope transfer of relatively coherent masses of rock or rock debris by slumping, falling, or sliding; and (2) those involving the downslope flow of mixtures of sediment, water, and air. In the latter category, which involves internal motion of flowing masses of debris, processes are distinguished on the basis of their velocity and the amount of water in the flowing mixture. We also briefly examine some processes and deposits that are representative of mass-wasting in cold regions and on the ocean floor. This approach to the classification of mass movements is outlined in Table 6.2.

Slope Failures

The constant pull of gravity makes all hillslopes and mountain cliffs susceptible to failure. When failure occurs, material is transferred downslope until a stable slope condition is reestablished. Some of the most common types of slope failure are illustrated in Fig. 6.1.

Slumps

A **slump** is a type of slope failure involving *rotational* movement of rock or regolith, that is, downward and outward movement along a curved, concave-up surface (Fig. 6.1). Slumps can range from small displacements covering only one or two square meters to large complexes that cover hundreds or even thousands of square meters.

Slumps frequently result from artificial modification of the landscape. They are common along roads and highways

T A B L E 6.2 • **Outline for Classification of Types of Mass-Wasting Processes**

Slope failures	Sediment flows	Mass-wasting in cold climates	Subaqueous mass-wasting
Slumps	Slurry flows	Frost heaving	Slumps
Falls	Solifluction	Gelifluction	Slides
Rockfall	Debris flows		Flows
Debris fall	Mudflows		
Slides	Granular flows		
Block glide	Creep		
Rockslide	Earthflows		
Debris slide	Grain flows		
	Debris avalanches		

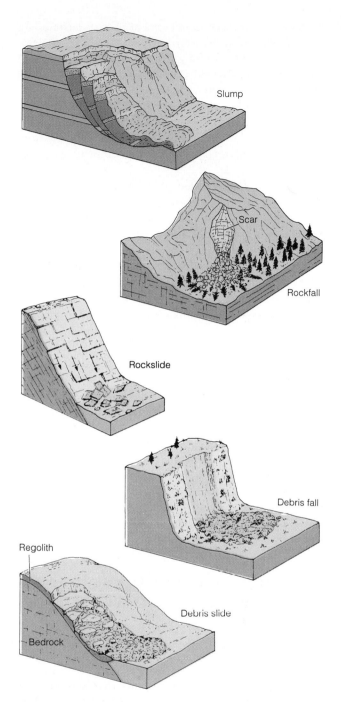

▲ F I G U R E 6.1
Examples of slope failures giving rise to slumps, falls, and slides.

▲ F I G U R E 6.2
Slump in agricultural land, Guatemala. Curvature of the slip surface has caused the toe of the slump block to rotate outward. The slump is believed to have been caused by ground saturation as a result of crop irrigation.

where bordering slopes have been oversteepened by construction activity. We can also see them along river banks or seacoasts where currents or waves have undercut the base of a slope. In some places slumping recurs seasonally and is associated with seepage of water into the ground during the rainy season (Fig. 6.2). Some slumps may be related to changing climatic conditions.

Falls

If you ask mountain climbers about the greatest dangers associated with their sport, they are likely to place falling rock near the top of the list. A **fall** is a sudden, vertical movement of Earth material, for example, from an overhanging cliff. *Rockfall,* the free fall of detached bodies of bedrock from a cliff or steep slope, is common in precipitous mountainous terrain, where rockfall debris forms conspicuous deposits at the base of steep slopes (Fig. 6.1).

As a rock falls, its speed increases. If we know the distance of the fall *(h)*, we can calculate the velocity *(v)* on impact as

$$v = \sqrt{2gh}$$

where *g* is the acceleration due to gravity. What this formula tells us is that a rock of a given size will be traveling at a much higher velocity if it falls from a point high on a steep mountain face than if it falls from a low cliff.

A rockfall may involve the dislodgement and fall of a single fragment or the sudden collapse of a huge mass of rock that plunges hundreds of meters, gathering speed until it breaks into smaller pieces on impact. The pieces continue to bounce, roll, and slide downslope before friction and decreasing slope angle bring them to a halt. Sometimes not only rock but overlying sediment and plants are dislodged. The resulting *debris fall* is similar to a rockfall, but it consists of a mixture of rock and weathered regolith as well as vegetation (Fig. 6.1).

Slides

Slides, like slumps and falls, involve the rapid displacement of masses of rock or sediment. In slides the movement is *translational,* that is, uniform movement in one direction with no rotation. *Translational slides* are also called *block glides* because they involve the movement of relatively coherent blocks of material along well-defined, inclined surfaces such as faults, foliation planes (in metamorphic rocks), or layering (in sedimentary rocks or alternating sequences of rock types). A *rockslide* is the sudden downslope movement of detached masses of bedrock (or of debris, in the case of a *debris slide*) (Fig. 6.1). Like falls, rockslides and debris slides are common in high mountains where steep slopes abound.

Flows

When a sufficiently large force is applied, any deformable material will begin to flow. In mass-wasting, the force is gravity and the material consists of dense mixtures of sediment and water (or sediment, water, and air). Mass-wasting

processes that involve the movement of such mixtures are called **flows**. The way a sediment flows depends on the relative proportions of solids, water, and air in the mixture and on the physical and chemical properties of the sediment. All streams carry at least some sediment, but if the sediment becomes so concentrated that the water can no longer transport it, a sediment-laden stream becomes a very fluid sediment flow. In such a case the water helps promote flow, but the pull of gravity remains the primary reason for movement.

In Fig. 6.3 flows are subdivided into two classes—*slurry flows* and *granular flows*—based on water content (i.e., concentration of sediment). Slurry flows are water saturated mixtures, whereas granular flows are not water saturated. Each of these two classes is further subdivided on the basis of the velocity of the flow, which can range from very slow (millimeters or centimeters per year) to very fast (kilometers per hour). In this classification of flows, the boundaries between processes are approximate and depend on the size distribution of the grains, the concentration of the sediment, and other factors.

Slurry Flows

A *slurry flow* is a moving mass of water-saturated sediment. In slurry flows, the sediment mixture is often so dense that large boulders can be suspended in it. Boulders that are too large to remain in suspension may be rolled along by the flow. When the flow ceases, fine and coarse particles remain mixed, resulting in an unsorted sediment.

Very slow downslope movement of water-saturated soil and regolith is known as *solifluction.* As can be seen in Fig. 6.3, solifluction lies at the lower end of the velocity scale

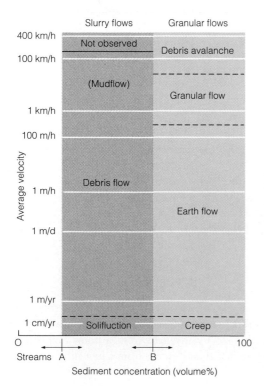

◄ F I G U R E 6.3

Classification of sediment flows on the basis of their average velocity and sediment concentration. The transition from a sediment-laden stream to a slurry flow occurs when the concentration of sediment becomes so high that the stream no longer acts as a transporting agent; instead, gravity becomes the primary force causing the saturated sediment to flow. As the percentage of water decreases further, a transition from slurry flow to granular flow takes place. Now the sediment may contain water and/or air. The boundaries between muddy streams and slurry flows (A) and between slurry and granular flows (B) are not assigned sediment-concentration percentages because they can shift to the left or right depending on the physical and compositional characteristics of the mixture. Different types of slurry and granular flows are recognized on the basis of their mean velocity.

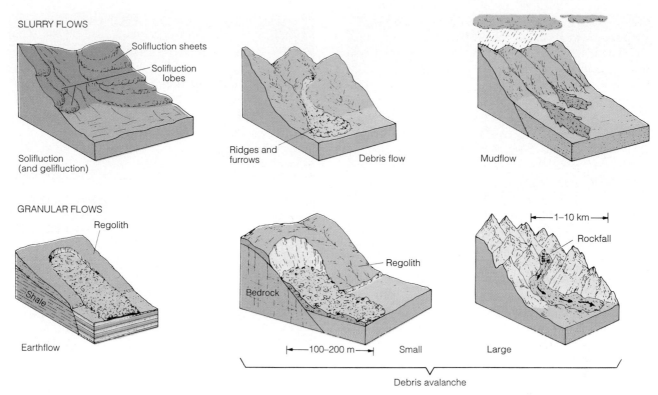

SLURRY FLOWS

Solifluction sheets
Solifluction lobes
Solifluction (and gelifluction)

Ridges and furrows
Debris flow

Mudflow

GRANULAR FLOWS

Regolith
Shale
Earthflow

Regolith
Bedrock
100–200 m
Small

1–10 km
Rockfall
Large

Debris avalanche

▲ F I G U R E 6.4
Examples of slurry flows and granular flows.

for flowing sediment–water mixtures. The rates of movement of such mixtures are generally so slow as to be detectable only by measurements made over several seasons. Solifluction occurs on hill slopes where sediment remains saturated with water for long intervals. It results in distinctive surface features, including lobes and sheets of debris that sometimes override one another (Figs. 6.4 and 6.5).

A *debris flow* involves the downslope movement of regolith whose consistency is coarser than that of sand, at rates ranging from 1 m/year to as much as 100 m/h (Fig. 6.3). In some cases a debris flow begins with a slump or debris slide, whose lower part then continues to flow downslope (Figs. 6.4 and 6.6). Once mobilized, a typical debris flow moves along a stream channel and may then spread

◄ F I G U R E 6.5
A meter-thick solifluction lobe has slowly moved downslope and covers glacial deposits on the floor of the Orgière Valley in the Italian Alps.

▲ F I G U R E 6.6
Debris slides that turned into debris flows on a steep mountainside in southern Puerto Rico have stripped away vegetation and inundated two houses at the base of the slope.

A.

B.

C.

▲ F I G U R E 6.7
Passage of a muddy debris flow along a canyon near Farmington, Utah, in June 1983. A. The boulder-laden front of a muddy debris flow advances from left to right along a stream channel in the wake of an earlier surge of muddy debris. B. The steep front, about 2 m high and advancing at 1.3 m/s, acts as a moving dam, holding back the flow of muddy sediment upstream. C. The main slurry, having a sediment concentration of about 80 percent and now moving at about 3 m/s, is thick enough to carry cobbles and boulders in suspension.

out to form a poorly sorted deposit. Debris flow deposits commonly have a tongue-like front and a very irregular surface, often with concentric ridges and depressions. They are frequently associated with intervals of extremely heavy rainfall that lead to oversaturation of the ground.

A debris flow that has a water content sufficient to make it highly fluid is commonly called a *mudflow*. In Fig. 6.3, the velocity of mudflows lies at the upper range of the velocity scale for debris flows (more than about 1 km/h). Most mudflows are highly mobile and tend to travel rapidly along valley floors (Fig. 6.4). The consistency of mudflow sediment can range from freshly poured concrete to a souplike mixture only slightly denser than very muddy water. After a heavy rain in a mountain canyon, a mudflow can start as a muddy stream that continues to pick up loose sediment until its front becomes a moving dam of mud and rubble, extending to each wall of the canyon and urged along by the force of the flowing water behind it (Fig. 6.7). When it reaches open country the moving dam collapses, floodwater pours around and over it, and mud mixed with boulders is spread out in a wide, thin sheet.

On active volcanoes in wet climates, layers of tephra and volcanic debris commonly cover the surface and are easily mobilized as mudflows called *lahars*. When closely associated with an actual eruption, lahars can be very hot. A particularly large mudflow that originated on the slopes of Mount Rainier about 5700 years ago traveled at least 72 km. The sediment spread out as a broad lobe as much as 25 m thick; its volume is estimated to be well over a billion cubic meters. Throughout much of its history Mount St.

Helens has produced mudflows, the most recent of which occurred during the huge eruption of May 1980 (Fig. 6.8). And as discussed in Chapter 4, lahars that resulted from the mixing of ash with typhoon rains caused extensive damage during the 1991 eruption of Mount Pinatubo in the Philippines.

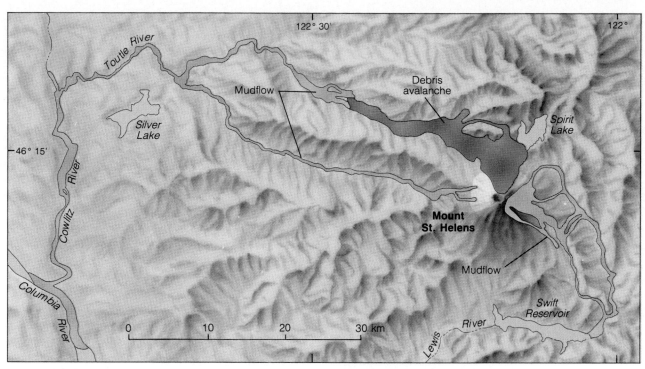

▲ F I G U R E 6.8
During the 1980 eruption of Mount St. Helens in Washington, volcanic mudflows were channeled down valleys west and east of the mountain. Some mudflows reached the Columbia River after traveling more than 90 km. Flow velocities were as high as 40 m/s and averaged 7 m/s.

Granular Flows

A *granular flow* is a mixture of sediment, air, and water but, unlike a slurry flow, it is not saturated with water; instead, the weight of the flowing sediment is supported by contact or collision between grains. The sediment of granular flows may be largely dry, with air filling the pores, or it may con-tain water but include a range of grain sizes and shapes that allows the water to escape easily.

Creep is an imperceptibly slow granular flow. Most of us have seen evidence of creep in curved tree trunks or old fences, telephone poles, or gravestones leaning at an angle on hill slopes (Fig. 6.9). Steeply inclined rock layers may be

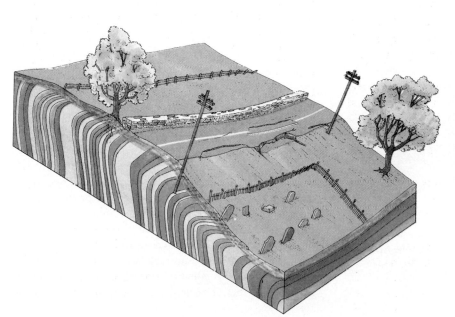

◄ F I G U R E 6.9
Effects of creep on surface features and bedrock. Steeply inclined rock layers have been dragged over near the surface by creep so they appear folded. Telephone poles and fence posts affected by creep are tilted, stone fences are deformed, roadbeds are locally displaced, and gravestones are tilted or have fallen.

bent over in the downslope direction just below the surface of the ground, another sign of creep. A number of factors contribute to creep, including the growth and decay of plants (which can bind sediment particles together or wedge them apart); the activities of animals (such as bur-

▲ F I G U R E 6.10
Colored targets placed in a straight line across a hill-slope in Greenland (A) had been moved differentially by creep when photographed a year later (B). The maximum recorded movement along the slope averaged 12 cm/year.

rowing or trampling); and heating, cooling, wetting, and drying (all of which cause changes in the volume of mineral particles). However, as with all types of mass-wasting, gravity is the main downslope force.

Although creep occurs too slowly to be seen, careful measurement of the downslope displacement of objects at the surface enables scientists to record the rates involved (Fig. 6.10). As might be expected, rates of creep tend to be higher on steep slopes than on gentle slopes. Measurements in Colorado, for example, document a creep rate of 9.5 mm/year on a slope of 39° but a rate of only 1.5 mm/year on a 19° slope. Creep rates also tend to increase as the amount of moisture in the soil increases. However, in wet climates the density of vegetation also increases, and roots, which bind the soil together, tend to inhibit creep. Despite the slow rates of movement involved, creep affects all hillslopes covered with regolith, and its cumulative effect is therefore very great.

Earthflows are among the more common types of mass-wasting. An *earthflow* is a downslope granular flow that is more rapid than creep (Fig. 6.3). Earthflows may continue for several days, months, or even years. Even after their initial motion ceases, they may be highly susceptible to renewed movement. Earthflows occur where the ground is saturated intermittently, and they are frequently associated with intervals of excessive rainfall. An earthflow typically heads in a steep **scarp**, a cliff that is formed where displaced material has moved away from undisturbed ground upslope (Fig. 6.11).

If you have ever walked along the crest of a sand dune and stepped too close to the steep slope that faces away from the wind, your footsteps likely started a cascade of sand flowing down the dune face. This is an example of still another type of mass-wasting, called *grain flow,* which involves the movement of a dry or nearly dry granular sediment with air filling the pore spaces.

A *debris avalanche* is a type of granular flow that travels at high velocity (tens to hundreds of kilometers per hour) and can be extremely destructive (Fig. 6.4 and Table 6.3). Large debris avalanches are rare but spectacular events involving huge masses of falling rock and debris that break up, pulverize on impact, and then continue to travel downslope, often for great distances. Because they are infrequent and extremely difficult to study while they are occurring, there are few observational data about the processes involved. It has been suggested that the debris actually rides on a layer of compressed air. If this is true, debris avalanches behave somewhat like a commercial hovercraft that travels across land or water on air compressed by a large propeller. Alternatively, air trapped and compressed within the moving debris may reduce friction between particles and cause the mass to behave in a highly fluid manner.

The slopes of steep, unstable stratovolcanoes are especially susceptible to collapse, leading to the production of debris avalanches. The deposits of such avalanches can be

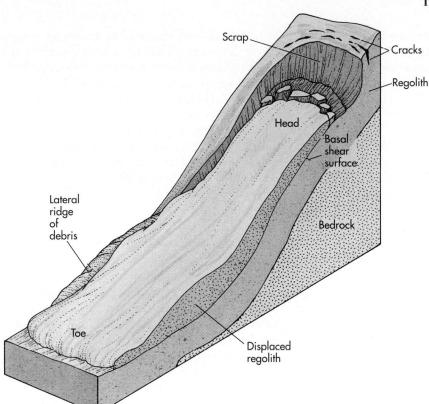

▲ F I G U R E 6.11
In this section through an idealized earthflow, regolith moves downslope across a basal shear surface, leaving a scarp at the head, where the sediment separated from the slope above. A bulging toe protrudes beyond ridges of sediment that have piled up along the lower margins of the earthflow.

difficult to recognize because of their huge dimensions. For example, a broad valley that extends some 40 km north of Mount Shasta in northern California contains a complex of hills and mounds of volcanic rock that resulted from the collapse of a flank of the volcano about 300,000 years ago. The volume of rock involved was at least 26 km³—almost 10 times the volume of the huge debris avalanche associated with the 1980 eruption of Mount St. Helens. Whereas the St. Helens deposits cover an area of 60 km², the Shasta debris avalanche overwhelmed at least 450 km².

Mass-Wasting in Cold Climates

Mass-wasting is especially prevalent at high latitudes and high altitudes, where average temperatures are very low. In such regions much of the landscape is underlain by perennially frozen ground, and frost action is an important geologic process. When water freezes, its volume increases. Ice forming in saturated regolith therefore pushes up the ground surface in a process called **frost heaving**. Frost heaving strongly influences the downslope creep of sedi-

T A B L E 6.3 • **Characteristics of Some Large Debris Avalanches**

Locality	Date	Volume (million m³)	Vertical Movement (m)	Horizontal Movement (km)	Calculated Velocity (km/h)
Huascarán, Peru	1971	10	4000	14.5	400
Sherman Glacier, Alaska	1964	30	600	5.0	185
Mount Rainier, Washington	1963	11	1890	6.9	150
Madison, Wyoming	1959	30	400	1.6	175
Elm, Switzerland	1881	10	560	2.0	160
Triolet Glacier, Italy	1717	20	1860	7.2	≥125

ment in cold climates. When freezing occurs, the ground surface is lifted essentially at right angles to the slope. As the ground thaws, each particle tends to drop vertically, pulled downward by gravity. The net result of repeated episodes of freezing and thawing, during which a particle experiences a succession of upward and downward movements, is slow but progressive downslope creep.

In cold regions that are underlain by frozen ground year round, a thin surface layer thaws in summer and refreezes in winter. During the summer the thawed layer becomes saturated with meltwater and is very unstable, especially on hillsides. As gravity pulls the thawed sediment slowly downslope, distinctive lobes and sheets of debris are produced. This process, which is similar to solifluction in temperate and tropical climates, is known as *gelifluction.* Although measured rates of movement are generally less than 10 cm/year, gelifluction is so widespread on high-latitude landscapes that it constitutes a highly important agent of mass transport. Hill slopes in arctic Alaska and Canada, for example, are often mantled by sheets or lobes of gelifluction regolith.

▲ F I G U R E 6.12
In the Wrangell Mountains of southern Alaska, a jumbled mass of angular rock debris supplied from a steep cliff moves slowly downslope as a rock glacier.

A *rock glacier,* another characteristic feature of many cold, relatively dry mountain regions, is a tongue or lobe of ice-cemented rock debris that moves slowly downslope in a manner similar to the movement of a glacier (Fig. 6.12). Rock glaciers generally originate below steep cliffs, which provide a source of rock debris. Active rock glaciers may reach a thickness of 50 m or more and advance at rates of up to about 5 m/year. They are especially common in high interior mountain ranges such as the Swiss Alps, the Argentine Andes, and the Rocky Mountains.

Subaqueous Mass-Wasting

As geologists have extended the search for petroleum to offshore regions, their explorations have shown that mass-wasting is an extremely common and widespread means of sediment transport on the sea floor. Mass-wasting also has been documented in lakes. As on land, the potential for gravity-induced movement of rock and sediment exists wherever there are *subaqueous* (underwater) slopes.

Extensive studies of the offshore slopes of eastern North America using a variety of modern techniques (deep ocean drilling, sonar, echograms, and submersible vessels, among others) have shown that vast areas of the sea floor are disrupted by submarine slumps, slides, and flows. Some large slide complexes cover areas of more than 40,000 km^2 and reach depths as great as 5400 m. A subaqueous slope failure can give rise to a *turbidity current,* a type of sediment flow that travels down submarine canyons and deposits sediments off the continental shelf. Such flows have been known to cause extensive damage to underwater communications cables.

Major marine deltas also frequently display surface features and sediments that can be attributed to slope failures. In such subaqueous environments, failure can occur even on slopes as low as 1°. The slope failures generally display three distinct zones: a source region where subsidence (sinking or downward settling) and slumping take place; a central channel where sediment is transported; and a zone where sediment is deposited, often in the form of overlapping lobes of debris. These and other features are common on the submarine slopes of the Mississippi Delta, among the best studied deltas in the world (Fig. 6.13). In places, more than 30 m of sediment have been deposited by sediment flows in the last 100 years.

FACTORS THAT INFLUENCE SLOPE STABILITY

Under natural conditions, a slope evolves toward an angle that allows the quantity of regolith reaching any point from upslope to be balanced by the quantity that is moving downslope from that point. Such a slope is said to be in a balanced, or *steady-state,* condition. The slope may appear

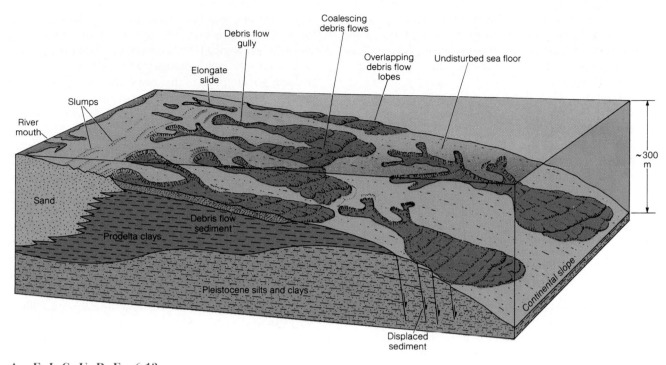

▲ F I G U R E 6.13
**Block diagram showing various mass-wasting features on the submarine surface of
the Mississippi Delta.**

stable and show little evidence of geologic activity. Yet if we examine the regolith beneath the surface we will probably find some rock particles derived from bedrock located farther upslope. We can deduce, therefore, that the particles have moved downslope. As you have just learned, the particle's downward journey can be very slow or very fast, but in either case the movement is controlled primarily by gravity.

Many factors affect slope stability. A change in any one or a combination of these factors can alter the steady-state condition of the slope, decreasing its stability and sometimes leading to slope failure. In some cases, the change may take the form of a relatively sudden triggering event, whether natural (such as an earthquake) or human-generated (such as an explosion). In other cases, the slope may have been slowly changing over time; again, the cause can be natural (such as a long period of intense precipitation) or a result of human activities (such as the construction of a dam).

The main factors that influence slope stability are (1) the force of gravity, and therefore the gradient of the slope; (2) water, and therefore the hydrologic characteristics of the slope; (3) the presence of troublesome Earth materials; and (4) the occurrence of a triggering event. In this section we examine each of these factors in detail.

Gravity and Slope Gradient

Two opposing forces determine whether a body of rock or debris located on a slope will move or remain stationary.

These forces are *shear stress* and *shear strength*. The first of these forces, **shear stress**, causes movement of the body parallel to the slope. The primary factor influencing shear stress is the pull of gravity, which is related to the slope's **gradient**, or steepness. On a horizontal surface, gravity holds objects in place by pulling on them in a direction perpendicular to the surface (Fig. 6.14). On any slope, however, gravity consists of two component forces. The *perpendicular* component (g_p in Fig. 6.14) acts at right angles to the slope and tends to hold objects in place. The

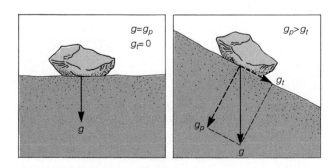

▲ F I G U R E 6.14
**Effects of gravity on a rock lying on a hillslope.
Gravity acts vertically and can be resolved into two
components, one perpendicular (g_p) and the other
parallel (g_t) to the surface.**

tangential component (g_t in Fig. 6.14) acts along and down the slope and causes objects to move downhill. As a slope becomes steeper, the tangential component increases relative to the perpendicular component and the shear stress becomes larger.

The second force, **shear strength**, is the internal resistance of the body to movement. Shear strength is governed by factors inherent in the body of rock or regolith, such as friction and cohesion between particles and the binding action of plant roots. As long as shear strength exceeds shear stress, the rock or debris will not move. However, as these two forces approach a balance, the likelihood of movement increases. This relationship is expressed in a ratio known as the **safety factor** (denoted *Fs*):

$$Fs = \frac{shear\ strength}{shear\ stress}$$

When the safety factor is less than 1 (i.e., shear strength is less than shear stress), slope failure is imminent.

The relationship between shear stress and slope gradient (the steeper the slope, the greater the shear stress) means that conditions favoring mass movement tend to increase as slope angle increases. Steep slopes, of course, are most common in mountainous areas, so it is not surprising that mass-wasting is most frequent in high mountains.

Water

Water is almost always present within rocks and regolith near the Earth's surface, and it plays a variety of important roles in mass-wasting of both solid rock and regolith. Unconsolidated (loose, uncemented) sediments behave in different ways depending on whether they are dry or wet, as anyone knows who has constructed a sand castle at the beach. Dry sand is unstable and difficult or impossible to mold. When poured from a bucket, dry sand (or any other dry, unconsolidated sediment) will form a cone-shaped mound. The steepness of the cone's sides, called the *angle of repose,* is determined by the characteristics of the material, primarily the size and angularity of the particles. Sand, for example, will always pile up with slopes of about 32° to 34° (Fig. 6.15). When a little water is added, the sand gains strength; its angle of repose is greater, so it can be shaped into vertical walls. The water and sand grains are drawn together by *surface tension,* a property of liquids that causes the exposed surface to contract to the smallest possible area. This force tends to hold the wet sand together as a cohesive mass. However, the addition of *too* much water saturates the sand; the spaces fill with water, and the sand grains lose contact with one another. The mixture turns into a slurry that easily flows away, as the sand castle builder sees with dismay when the rising tide destroys the elaborate work of an afternoon.

Moist or weakly cemented fine-grained sediments, such as fine silt and clay, may be so cohesive that they can stand in near-vertical cliffs. But if the silt or clay becomes saturated with water and the internal fluid pressure rises, the fine-grained sediment may also become unstable and begin to flow like the water-saturated sand castle.

The movement of some large masses of rock has been attributed to water pressure in voids in the rock. If the voids along a surface separating two rock masses are filled with water, and the water is under pressure, a buoying effect may result. In other words, the water pressure may be high enough to support the weight of the overlying rock mass, thereby reducing friction along the points of contact. The result can be a sudden failure. An analogous situation can make driving in a heavy rainstorm extremely dangerous.

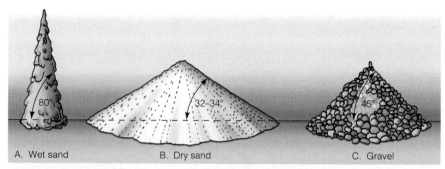

▲ F I G U R E 6.15
Angle of repose in unconsolidated materials. A. Wet sand can be piled up steeply, but (B) dry sand will always come to rest in a cone-shaped mound with slopes of about 32° to 34°. If more sand is poured onto the pile, it will simply roll down the slopes. C. Coarser material, such as gravel, will have a steeper angle of repose. Moisture content and the angularity of particles can also affect a material's angle of repose.

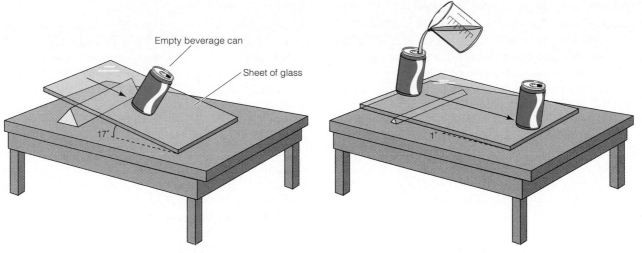

▲ F I G U R E 6.16
An experiment illustrating how water can reduce friction at the base of a mass resting on a slope. A. An empty beverage can placed on a wet sheet of glass will begin to slide down the surface when the angle reaches about 17°. B. If a small hole is made in the bottom of the can and water is poured in through the top, the can will begin to slide down the slope at a much lower angle. In an analogous way, high water pressure at the base of a large rock mass may promote downslope movement of the rock.

When water is compressed beneath the wheels of a moving car, the increasing fluid pressure can cause the tires to "float" off the roadway, a condition known as hydroplaning.

You can demonstrate this principle by conducting a simple experiment (a variation on a classic experiment first described by geologists M. King Hubbert and William Rubey in 1959). An empty beverage can is placed in an upright position on the wetted surface of a sheet of glass (Fig. 6.16A). If the glass is slowly tilted, the can will not begin to slide until a certain critical angle is reached. For the particular substances used in this demonstration (metal and wet glass), sliding begins at an angle of approximately 17°. Next, punch a small hole in the bottom of the can. Place it on the glass sheet and slowly pour some water into the open top (Fig. 6.16B). The can will begin to slide at a much gentler angle because the conditions at the base of the can have changed. In an analogous way, high water pressure at the base of a large mass of rock may promote downslope movement of the rock.

These examples show that water can be instrumental in reducing shear strength and thereby promoting the movement of rock and sediment downslope under the pull of gravity. It does so by reducing the natural cohesiveness between grains or by reducing friction at the base of a mass of rock through increased water pressure. As we will see shortly, water can act in other ways that contribute to slope instability, such as undercutting the base of the slope or altering the chemical composition of the sediment.

Troublesome Earth Materials

Some Earth materials are particularly susceptible to the types of changes and disturbances that can lead to slope failure. Such materials, sometimes referred to as *problem soils,* are often involved in mass-wasting.

Liquefaction

As discussed earlier, when water is added gradually to a dry soil, the material first becomes *plastic,* or moldable. If enough water is added, the particles lose contact with one another and the material turns into a loose slurry, losing its shear strength in the process. The transformation of a soil from a solid to a liquid state, usually (but not always) as a result of increased water content, is called **liquefaction**. The point at which this transition occurs, called the *liquid limit,* varies from one soil to another. Some materials, particularly some clay-bearing soils, have very high liquid limits and may remain plastic over a broad range of water contents. These soils can be particularly troublesome because by the time the liquid limit is exceeded, the moisture content of the soil is so high that the material behaves in an extremely fluid manner.

Expansive and Hydrocompacting Soils

Expansive soils, also referred to as *shrink–swell soils,* expand greatly when they are saturated with water and shrink when they dry out. Much of the increase in volume is caused by the chemical attraction of water molecules between the submicroscopic layers of clay minerals called *smectites* (Fig. 6.17A). Expansion resulting from increased water content can drastically reduce the shear strength of an Earth material, often contributing to downslope movement. It can also cause extensive damage to structures built on that material (Fig. 6.17B).

When expansive clays dry out, they undergo a decrease in volume. The process of shrinkage and/or collapse resulting from water loss is referred to as *compaction* (or, more precisely, *hydrocompaction,* since soil compaction can be caused by other processes besides water loss). In addition to expansive clays, there are other types of soils that can exhibit substantial decreases in volume when they dry out. Extreme compaction associated with drying is a particular problem in water-saturated, organic-rich soils such as peat. Soil compaction often results in *subsidence,* a type of mass movement involving the lowering or collapse of the ground surface. (Subsidence is discussed in greater detail in Chapter 7.)

Sensitive Soils

In some clay-rich soils, the particles are arranged in an open, porous structure like a house of cards (Fig. 6.18A). Such an arrangement can occur in very fine marine clays, in which salt acts to stabilize the "house of cards" by "gluing" the particles together end to end. Eventually, fresh groundwater may move through the area, changing the chemical composition of the clay and washing away the salt. Without the stabilizing effect of the salt, the clay particles collapse and take on a new, more compact arrangement, as shown in Fig. 6.18B. The transition from open to compact arrangement causes a sudden and dramatic loss of shear strength—in other words, liquefaction—which can propagate with astonishing speed throughout the entire mass of clay. Liquefaction or compaction that results from a disturbance of the internal structure of a soil is referred to as *remolding.* Materials that lose shear strength as a result of remolding are called **sensitive soils**. Those that are the most susceptible to remolding and liquefaction are called *quick clays.* Some other types of soils lose their shear strength suddenly when disturbed but gradually strengthen and resume their original properties when left undisturbed; these are referred to as *thixotropic clays.*

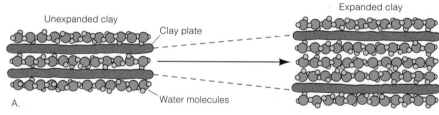

Unexpanded clay

Clay plate

Expanded clay

Water molecules

A.

◄ F I G U R E 6.17
A. Expansion of a single smectite grain as a result of the addition of water between clay layers. B. The type of structural damage that can result from the expansion of soil beneath a building.

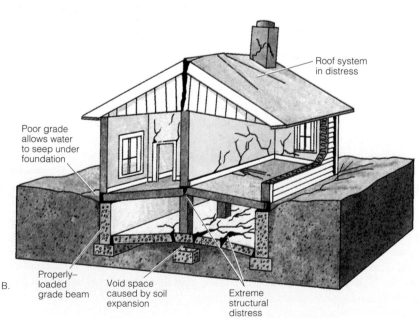

Roof system in distress

Poor grade allows water to seep under foundation

Properly-loaded grade beam

Void space caused by soil expansion

Extreme structural distress

B.

A. B.

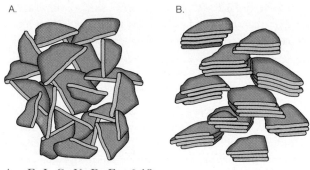

▲ F I G U R E 6.18
**Sensitive clays. A. The "house-of-cards" structure
of microscopic plates in a sensitive clay. B. Structural
change in a sensitive clay as a result of compaction,
remolding, or thixotropism.**

The most extensive deposits of quick clay are found in
Canada (in the St. Lawrence River Valley) and Scandinavia
(in Norway and Sweden). In both cases the clays were deposited in shallow, quiet marine environments at the edges
of glaciers. Both areas have been subjected to damaging
mass-wasting events resulting from the remolding and liquefaction of the quick clays. A well-documented example is
the quick clay slide that occurred in Rissa, Norway, in
1978. This event began when a farmer made a small excavation for a new barn, piling the excavated soil along the
shore of a nearby lake. The extra load on the quick clay at
the edge of the lake initiated a small slide, which quickly
propagated more than a kilometer up the valley behind the
farmhouse. What had been solid ground completely liquefied. Entire buildings were carried intact toward the lake by
the fluidized mass, traveling at rates of 20 km/h or more.
After the slide, the Norwegian Geotechnical Institute carried out extensive studies and testing in the area. Some remaining quick clay was removed, and the rest of the slopes
in the area were regraded and stabilized.

Triggering Events

As in the example of the Rissa quick clay slide, slope failures are often triggered by some extraordinary activity or
occurrence. It is also very common for a combination of
conditions to lead to slope failure: a moderate-size earthquake might not generate landslides under normal conditions, but it could do so in an area underlain by sensitive
soils; a slope steepened by construction might not fail
under normal conditions, but it might during a period of
exceptionally heavy precipitation. Among the most common types of triggering events are earthquakes, volcanic
eruptions, slope modifications, and changes in the hydrologic characteristics of an area (including the effects of prolonged or exceptionally intense rainfall).

Earthquakes and Other Shocks

An abrupt shock, such as an explosion, an earthquake, an
electrical storm, or even a truck passing by, can increase
shear stress and contribute to slope failure. Intense shaking
can cause a buildup of water pressure in the pore spaces of a
sediment, leading to liquefaction. In other words, liquefaction is not always related to an increase in water content;
sometimes shaking causes the pore water *already* present in
the sediment to coalesce so that the sediment grains lose
contact with one another. The result is fluidization of the
sediment and abrupt failure. Any structure built on such
sediments or in their path may be demolished (Fig. 6.19).

▲ F I G U R E 6.19
**Chaotically tilted trees and houses in suburban
Anchorage, Alaska show how violent shaking of the
ground during the great 1964 earthquake caused
sudden liquefaction of underlying clays and
widespread slumping.**

Earthquakes frequently generate landslides. In 1970 a large earthquake in Peru triggered a debris avalanche that roared more than 3.5 km down the steep, rocky slopes of Mount Huascarán, reaching speeds of 400 km/hr. The villages of Yungay and Ranrahirca were destroyed and as many as 20,000 people killed (Fig. 6.20). An eyewitness to the event (a geophysicist, by chance) made the following report:

I heard a great roar coming from Huascarán. Looking up, I saw what appeared to be a great cloud of dust and it looked as though a large mass of rock and ice was breaking loose from the north peak. . . The crest of the wave [of rock and ice] had a curl, like a huge breaker coming in from the ocean. I estimated the wave to be at least 80 m high. I observed hundreds of people in Yungay running in all directions and many of them towards

Cemetery Hill. All the while, there was a continuous loud roar and rumble. I reached the upper level of the cemetery near the top just as the debris flow struck the base of the hill and I was probably only 10 seconds ahead of it . . . It was the most horrible thing I have ever experienced and I will never forget it.

Some earthquakes release so much energy that many landslides of different types and sizes are triggered simultaneously. For example, in 1929 a major earthquake in northwestern South Island, New Zealand, triggered at least 1850 landslides larger than 2500 m² within a 1200 km² area near the quake's center. As mentioned in Chapter 3, the great 1964 earthquake in Alaska also generated many landslides, including a spectacular one in Turnagain Heights, a residential section of Anchorage. In that slide, a section of coastline 2.5 km long broke off up to 0.5 km in-

▲ F I G U R E 6.20
View of Mt. Huascarán, Peru and the debris avalanche that destroyed the villages of Yungay (remains are lower right) and Ranrahirca, in the May 1970 earthquake.

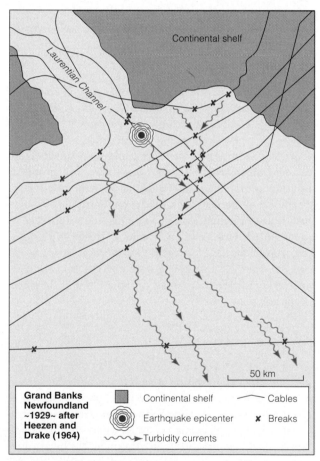

▲ F I G U R E 6.21
Submarine slope failures off Grand Banks, Newfoundland, 1929. Turbidity currents generated by an earthquake traveled at least 470 km, breaking telephone cables on the ocean floor.

BOX 6.1

•

THE HUMAN PERSPECTIVE

VALLEY FEVER LINKED TO LANDSLIDES

After the initial devastation of the 1994 Northridge (Los Angeles) earthquake, residents in a widespread area surrounding Los Angeles continued to experience repercussions for several months. In Ventura County, for example, 166 cases of acute coccidioidomycosis, commonly known as valley fever, were reported from January through March—a dramatic increase over the 53 cases reported in the entire previous year.

Valley fever is contracted by inhaling airborne spores *(Coccidioides immitis),* a fungus that lives in the top few inches of surface soils. The illness is not passed from person to person. In January 1994, the Northridge earthquake triggered thousands of landslides in the Santa Susana Mountains, which form the northern edge of the San

Fernando and Simi Valleys. The landslides generated huge clouds of dust that were blown by prevailing northeasterly winds across Simi Valley and eastern Ventura County.

The Santa Susana Mountains consist of very young, weak sedimentary rock that is extremely susceptible to sliding during earthquakes. The highest concentration of valley fever cases were found directly downwind from the area of the most concentrated landslides. Little change in the number of valley fever cases was reported in the more highly populated San Fernando Valley, however, where the earthquake was centered. Therefore, researchers from the U.S. Geological Survey and the Centers for Disease Control deduced that landslides triggered the Simi Valley/Ventura County valley fever outbreak.

land, and slid toward the ocean in great blocks. The movement of the blocks was facilitated by liquefaction, in which sandy layers underlying the blocks lost their shear strength and were reduced to a slurry as a result of the shaking. The landslide began approximately 2 minutes after the earthquake and lasted for 5 minutes, during which time the sliding blocks moved as far as 300 m and created an entirely new coastline and local landscape.

Major submarine turbidity currents and slumps off eastern North America are also known to have been caused by strong earthquakes. One such quake occurred in Charleston, South Carolina, in 1886, and another in Cape Ann, Massachusetts, in 1755. In 1929 an earthquake off Grand Banks, Newfoundland, was followed by a succession of breaks in underwater telephone cables (Fig. 6.21). In piecing together the evidence from the sequence and locations of the cable breaks, it was determined that the quake had generated a turbidity current at least 150 km wide in which an enormous volume of sediment was moved. The distance from the source area to the farthest cable break was 470 km, and the maximum velocity of the submarine slide was 93 km/h.

Volcanic Eruptions

Volcanic eruptions are another mechanism for triggering mass-wasting events. Large stratovolcanoes consist of inher-

ently unstable accumulations of interlayered lava flows, rubble, and pyroclastic material that form steep slopes. The slopes of high, ice-clad volcanoes may be further steepened by glacial erosion. Large volumes of water, released when summit glaciers and snowfields melt during an eruption of hot lavas or pyroclastic debris, can combine with unconsolidated deposits to form rapidly moving lahars. As in the case of Armero, Colombia, these highly fluid mudflows can travel great distances and at such high velocities that they constitute one of the major hazards associated with volcanic eruptions.

Slope Modifications and Undercutting

Landslides often result when natural slopes are modified, either by natural processes or by human activities. Translational slides can occur, for example, where roads have been cut into regolith or unstable rock, creating an artificial slope that exceeds the angle of repose or exposes natural planes of weakness (Fig. 6.22). Such landslides are especially common along mountainous and coastal cliffs where roads have been carved into deformed sedimentary or metamorphic rocks.

Overloading—placing a building or a mass of excavated material at the top of a slope, for example—can also contribute to slope failure because of the added weight as well as the steepening effect of the load. In 1966, for example, ex-

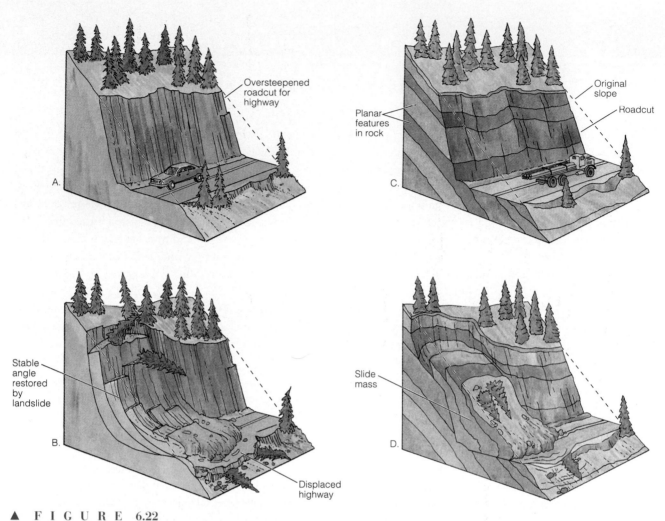

▲ F I G U R E 6.22
Modification of a slope during road construction can lead to slope failure. A. The natural angle of repose of the material in the slope is exceeded. B. The oversteepened slope fails, and a landslide buries the road. In the process, the natural angle of repose is reestablished. C. Roadcuts can expose natural planes of weakness, such as bedding planes in sedimentary rock or foliation in metamorphic rock, that can contribute to slope failure (D).

cessive rain in Rio de Janeiro caused an oversteepened slope in a road cut to fail. The mass of material involved in this small slide overloaded the top of the underlying slope, triggering a large landslide that destroyed several houses and two apartment buildings and killed 132 people (Fig. 6.23).

Oversteepened artificial slopes also fail occasionally. Perhaps the most famous example occurred in 1966 in the coal-mining town of Aberfan in Wales. The debris left over from the mining process—mostly very fine clay, in this case—was routinely piled up in large artificial hills called *tips.* One morning in October, a major slope failure occurred on one of the tips (Fig. 6.24). It was later determined that drainage within the tip had been inadequate, resulting in saturation and eventually liquefaction of the lower portion. The upper part of the tip moved as a coher-

ent mass, carried along on the liquefied material below. When the mass of fluidized material came to rest, it had lost most of its water content and returned to its original solid state. Part of the town was engulfed by the landslide, including the elementary school where 116 children and 5 teachers were killed; 144 people were killed in all.

Slumps and other types of landslides can also be triggered by the natural steepening of slopes as a result of the undercutting action of a stream along its bank or waves along a coast. This is a contributing factor in the continuing movement of a large slide block at Portugese Bend, California (Box 6.2). Coastal landslides are often associated with major storms that direct their energy against rocky headlands or the bases of cliffs composed of unconsolidated sediments (Fig. 6.25).

▲ F I G U R E 6.23

An oversteepened roadcut caused an initial failure of this slope in Rio de Janeiro, Brazil. The weight of the initial slide overloaded the slope and caused a major landslide. One hundred and thirty-two people are known to have been killed; others probably remain buried.

▲ F I G U R E 6.24

The failure of coal tip No. 7 at Aberfan, Wales, in October 1966.

Changes in Hydrologic Characteristics

Changes in the characteristics of subsurface water or drainage in an area often contribute to landslides. For example, heavy or persistent rains may saturate the ground and make it unstable. Such was the case in 1925 when prolonged rains, coupled with melting snow, started a large debris flow in the Gros Ventre River basin of western Wyoming. The water saturated a porous sandstone overlying an impermeable rock unit that sloped toward the valley floor. This water-saturated condition was an ideal trigger for slope failure. An estimated 37 million m³ of rock, regolith, and organic debris moved rapidly downslope and created a natural dam that ponded the river. Two years later the natural dam failed, causing a flood that resulted in several deaths. Today, more than 65 years after the debris flow

▲ F I G U R E 6.25

Steep cliffs along the coast of Hawaii are undercut by pounding surf. When a cliff collapses, the resulting landslide debris is rapidly reworked by waves and currents, and the process begins anew.

BOX 6.2

•

THE HUMAN PERSPECTIVE

THE PORTUGESE BEND LANDSLIDE

*R*esidents of Los Angeles face a variety of geologic hazards, including steep, unstable slopes; frequent earthquakes; intense storms separated by extensive periods of drought; and coastal flooding. Major slope failures have been particularly damaging to urban developments on the slopes surrounding Los Angeles. An example is the large and costly landslide at Portugese Bend on the Palos Verdes Peninsula, which has become a classic example for hazard mitigation studies.

The topography of the Palos Verdes Peninsula is cliffed and hummocky, with poorly drained depressions. Slopes are steepened and made unstable by marine erosion at the base of the cliffs, which plunge steeply to the Pacific Ocean. Portugese Bend itself consists of an enormous block of sedimentary rock interlayered with volcanic ash deposits. The volcanic ash has weathered to a clay mineral, bentonite, which readily absorbs water and thus becomes susceptible to downslope movement. The bentonite layers provide a convenient slide surface for the block of sedimentary rock.

The region had been landslide prone for a long time—probably hundreds or thousands of years—before its development as real estate. Evidence of a prehistoric landslide mass at Portugese Bend was documented on geologic maps published in the 1940s, suggesting that the area might best have been left undeveloped. The ancient slide block was dormant until 1956, when unusually heavy precipitation following years of urban development initiated intermit-

tent movement of the mass. The situation was aggravated by earthquakes and by high tides that contributed to erosion. Septic tank drainage, which passed directly into the hillside, and construction of new roads, which added weight to the top of the slide block, compounded the problem.

By the 1980s, 150 homes in the Portugese Bend area had been damaged or destroyed by the continuing movement of the slide block, and $10 million in losses were assessed (Fig. B2.1). The County of Los Angeles was held accountable in court for at least some of the private property losses and had to pay the land owners more than $1 million.

During the 1970s a smaller slide-prone area to the north of Portugese Bend, called Abalone Cove, became active. A management plan was devised to stabilize its slopes by installing dewatering wells. The plan was so successful that in 1985 a similar plan was created for Portugese Bend. In the latter case, engineers recommended surface grading to inhibit infiltration, as well as a network of drainage wells to pump out excess groundwater. New development in the Portugese Bend area has stopped, and stricter building codes have been adopted. These efforts have alleviated the slide hazard in Portugese Bend. The people who have chosen to remain in the active slide area have adjusted to the constant slow movement of their homes; for example, they periodically employ contractors to level buildings, and all utilities are placed at the surface of the property.

▲ FIGURE B2.1
Damage to homes and road as a result of landslides near Los Angeles, California. A. Houses on Point Fermun sliding into the sea, 1969. B. Fractures resulting from landslides offset lines of a tennis court, Abalone Cove, 1983.

◄ F I G U R E 6.26
Debris flow in the Gros Ventre River basin, Wyoming. The flow, which occurred in 1925, left massive scars on the landscape. The source of the flow was the bare area, upper center; the toe of the flow is in the foreground, so the direction of the flow is toward the viewer.

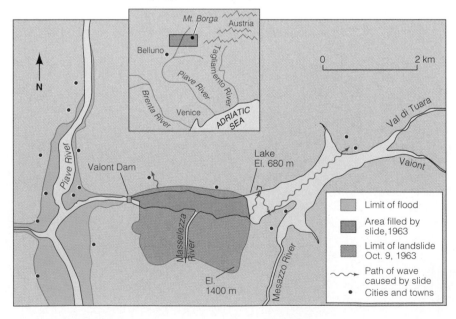

◄ F I G U R E 6.27
Sketch map of the Vaiont Dam and Reservoir showing the extent of the 1963 landslide. When the slope failed, a large mass slid into the reservoir, displacing water and causing a giant wave to overflow the dam. Both downstream and upstream areas suffered flooding and damage as a result.

▲ F I G U R E 6.28
Photo of Scarborough Bluffs, Toronto, along the shore of Lake Ontario.

began, the scar at the head of the slide is still quite visible, as is the distinctive chaotic topography downslope (Fig. 6.26).

The filling of a large reservoir can also cause changes in subsurface water conditions. Sometimes the increased water pressure in the pores of the underlying rock combines with other destabilizing factors to produce mass-wasting. Such factors caused the world's worst dam disaster, which occurred in Italy in 1963. A huge mass—almost 250,000,000 m^3—of rock and debris slid into the reservoir behind the Vaiont Dam (Fig. 6.27). The material filled the reservoir and created a wave 100 m high, which overflowed the dam and swept both up and down the valley, killing almost 3000 people. There were several driving forces in this event: the slopes were composed of inherently unstable blocks of limestone, with open fractures dipping toward the reservoir; the water impounded in the reservoir had increased pore water pressure in the rock walls of the valley; and a long period of above-average precipitation had also contributed to high water pressures in the valley walls. The landslide itself followed a period of very slow creep that extended over 3 years. The rate of creep had increased dramatically in the days prior to the event, but engineers were caught off guard when they discovered, just 1 day before the landslide occurred, that a huge portion of the valley wall had been moving downslope as a single mass.

Urban and suburban development on hillslopes can lead to hydrologic changes that ultimately contribute to slope failure. For example, in Toronto, Ontario (Fig. 6.28), houses have been constructed along the edge of the Scarborough Bluffs, a series of dramatic cliffs along the shore of Lake Ontario. The cliffs, which are made of fine glacial silts and clays, are continuously steepened by wave action at their base. The sediments are fractured, and subsurface drainage tends to be focused along zones of weakness. Periodic slope failures in these zones have created a series of deeply incised ravines. The problems inherent in the natural drainage system have been amplified by clifftop deforestation, an increased proportion of impermeable (paved) surfaces on the clifftops, and focused drainage from septic tank systems, swimming pools, and other household sources. Along some portions of the shoreline, the cliffs are retreating at a rate of 5 m/year as a result of slope failures.

ASSESSING AND MITIGATING MASS-WASTING HAZARDS

Landslides and other forms of mass-wasting are ubiquitous, and they cause extensive damage and loss of life each year.

◄ F I G U R E 6.29
A new apartment building at the base of a steep mountain slope in the Italian Alps was struck by a large boulder falling from the cliffs above. This relatively small rockfall, which occurred just one day before the new owners were to move in, demolished the bedroom and most of the living room.

With careful analysis and planning, together with appropriate stabilization techniques, the impacts of mass-wasting processes on humans can often be reduced or eliminated.

Prediction and Hazard Assessment

Assessments of the hazards posed by potential mass-wasting events are based on reconstruction of similar past events in order to evaluate their magnitude and frequency; mapping and testing of soil and rock properties; and analysis of slopes to determine their susceptibility to destabilizing processes. Such information can be used in determining how often an event of a certain magnitude is likely to recur in a given locality.

Maps showing areas that could be affected by mass-wasting events are important tools for land-use planners. For example, large debris avalanches and small rockfalls are ever-present hazards in the northern Italian Alps (Fig. 6.29). Field studies have shown that large debris avalanches have repeatedly blanketed valley floors with rocky debris during the last 3000 years. From this evidence, a map has been constructed showing areas that could be affected by future rockfalls with various trajectories and distribution patterns. A number of small communities are at risk from small or medium-size rock avalanches (traveling 3 to 5 km), and several large communities, including one village with a population of several thousand, could be affected by large debris avalanches (traveling up to 7 km) such as those recorded in the deposits on the valley floors.

Valleys in the Cascade Range of Washington and Oregon contain deposits created by large mudflows that repeatedly spread from high volcanoes during the last 10,000 years. Lahar hazard maps have been prepared from information about the number and extent of such deposits, the current status of volcanic activity, and the topography of the slopes. Hazard maps prepared before the 1980 eruption of Mount St. Helens and the 1991 eruption of Mount Pinatubo (Fig. 4.26) proved prophetic, for the mudflows generated by those eruptions had distributions very similar to those predicted on the basis of the geologic studies.

Eliminating or restricting human activities in areas where slides are likely to occur may be the best way to mitigate such hazards. For example, land that is susceptible to mild failures might be suitable for some types of development (e.g., recreation or parkland) but not others (e.g., intensive agriculture or housing). In Fig. II.1, landslide hazard mapping in California provided a basis for land-use recommendations. Scientific understanding of the geology of an area and its potential hazards is combined with building codes and zoning laws in setting limits on the types of activities permitted in the area.

Early-warning systems can also help reduce loss of life and, in some cases, property damage caused by landslides. The U.S. Geological Survey has developed a system for forecasting landslides in coordination with National Weather Service forecasts in some regions. The system combines analyses of rainfall data and forecasts with the delineation of areas known to be susceptible to landslides.

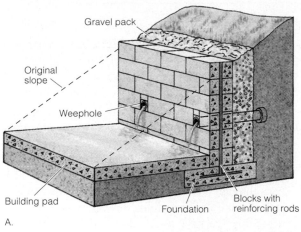

Gravel pack

Original slope

Weephole

Building pad

Foundation

Blocks with reinforcing rods

A.

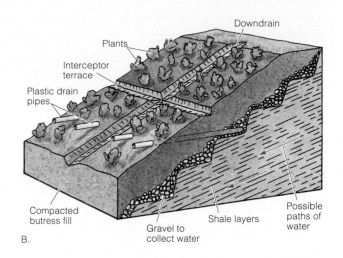

Downdrain

Plants

Interceptor terrace

Plastic drain pipes

Compacted butress fill

Gravel to collect water

Shale layers

Possible paths of water

B.

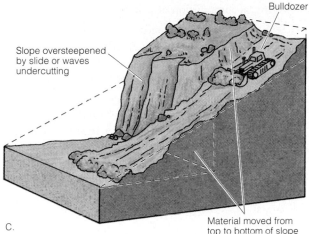

Bulldozer

Slope oversteepened by slide or waves undercutting

Material moved from top to bottom of slope

C.

▲ F I G U R E 6.30
Engineering techniques to stabilize slopes and prevent failure. A. Rock bolts and retaining wall. B. Drainage pipes. C. Diversion walls.

Prevention and Mitigation

In addition to assessment, prediction, and early warning, some engineering techniques can be used to mitigate or even prevent landslides. These include retaining devices; drainage pipes; grading; and diversion walls (Fig. 6.30). One of the most common approaches is the use of concrete block walls, poured or sprayed concrete, rock bolts, or gabions (rocks contained in wire mesh cages) to strengthen slopes (Fig. 6.30A). Slopes that are subject to creep can be stabilized by draining or pumping water from saturated sediment (Fig. 6.30B); this is accomplished by the insertion of permanent drainage pipes, often in combination with a wall. Oversteepened hill slopes can be prevented from slumping if they are regraded to angles equal to or less than the natural angle of repose. Sometimes the slope itself cannot be stabilized but downslope structures can be protected by the construction of diversion walls (Fig. 6.30C). In some mountain valleys subject to mudflows from active volcanoes, reservoirs can be quickly emptied so that dams will halt mudflows before they reach population centers.

SUMMARY

1. Mass-wasting (the movement of Earth materials downslope as a result of the pull of gravity) has devastating impacts in terms of both property damage and loss of life.

2. Two basic types of movement may be involved in mass-wasting: the sudden failure of a slope, resulting in the downslope transfer of relatively coherent masses of rock or rock debris by slumping, falling, or sliding; and the downslope flow of mixtures of sediment, water, and air.

3. Slumps involve rotational movement of blocks of material; falls involve vertical free-fall of material; and slides involve translational movement along a well-defined plane.

4. Flows can be divided into slurry (water-saturated) and granular (unsaturated) flows. The water content of a flow may range from heavily sediment-laden stream flow to dry grain flow in a sand dune. Flows also vary in velocity, from imperceptibly slow (creep, solifluction) to extremely rapid (mudflows, debris avalanches).

5. Mass-wasting is especially prevalent at high latitudes and high altitudes, where average temperatures are very low and frost-heaving is an important force. Subaqueous (underwater) mass-wasting is also very common, especially on continental slopes and in major marine deltas.

6. The main factors that influence slope stability are (a) the force of gravity, and therefore the gradient of the slope; (b) water, and therefore the hydrologic characteristics of the slope; (c) the presence of troublesome Earth materials; and (d) the occurrence of a triggering event.

7. In any body of rock or rock debris located on a slope, two opposing forces—shear stress and shear strength—determine whether the body will move or remain stationary. As long as shear strength exceeds shear stress, the rock or debris will not move. However, as these two forces approach a balance, the likelihood of movement increases.

8. Water plays a variety of important roles in mass-wasting of both solid rock and regolith. Water can decrease slope stability by reducing the natural cohesiveness between grains or reducing friction at the base of a rock mass through increased water pressure. Water can also contribute to slope instability through wave action undercutting the base of the slope, or by altering the chemical composition of sediment.

9. Some types of Earth materials are considered "troublesome" because they are particularly susceptible to failure under certain conditions. These include sensitive soils, expansive clays, and hydrocompacting clays.

10. Among the most common triggering events are earthquakes, volcanic eruptions, slope modifications, and changes in the hydrologic characteristics of an area (including the effects of prolonged or exceptionally intense rainfall).

11. Scientific understanding of the types of rock and characteristics of slopes in a given area can be combined with weather analysis to delineate landslide hazards and, sometimes, to issue predictions and early warnings of major landslides.

12. With careful planning, building regulations, and zoning laws, and the use of appropriate stabilization techniques, the impacts of mass-wasting processes on humans can often be reduced. Commonly used techniques include retaining devices (concrete, rock bolts, gabions), drainage pipes, grading, and diversion walls.

IMPORTANT TERMS TO REMEMBER

expansive soils (p. 158)
fall (p. 147)
flow (p. 148)
frost heaving (p. 153)
gradient (p. 155)

landslide (p. 145)
liquefaction (p. 157)
mass-wasting (p. 145)
safety factor (p. 156)
scarp (p. 152)

sensitive soils (p. 158)
shear strength (p. 156)
shear stress (p. 155)
slide (p. 148)
slump (p. 146)

QUESTIONS AND ACTIVITIES

1. Is the region where you live particularly susceptible to landslides? Have there been any major mass-wasting events there in recent history? If so, were the causes primarily natural (e.g., undercutting of a coastal cliff by waves, or slope failure in a mountainous region)? Or were human activities involved (e.g., alterations in natural drainage patterns or slope steepening by roadcuts)?

2. Has systematic landslide hazard mapping been carried out in your area? If so, which organization(s) was responsible? Has the mapping been coordinated with planning for new building regulations and zoning laws? Is an emergency response plan or early-warning system in effect for areas at greatest risk?

3. Investigate a major mass-wasting event. Some of the most interesting are the Gros Ventre, Vaiont Dam, and Turnagain Heights events mentioned in this chapter, but you may also want to find out about a major historic event that occurred in an area near where you live.

4. Make a table that classifies major mass-wasting events according to the conditions or triggering event primar-ily responsible for slope failure. Start with the examples given in this chapter and see how many you can add using other sources of information. For how many of the events in your table was more than one condition or triggering event responsible for slope failure?

5. Almost all parts of the world—even relatively flat areas—are susceptible to mass-wasting by creep. Go on a walking or driving tour of your neighborhood and see how many signs of creep you can discover on hillslopes. Look for bent tree trunks, curved fences, lobes of soil on grassy slopes, tilted gravestones, and other types of evidence.

6. As you walk or drive around your town, keep an eye out for the structures used to stabilize slopes or protect property from mass-wasting. Are the slopes in your area heavily engineered, or have they been left more or less in their natural state? Where you find such structures as retaining walls or drainage pipes, do they appear to have stabilized the slope as intended?

CHAPTER 7

SUBSIDENCE

In nature things move violently to their place,
and calmly in their place.

• **Sir Francis Bacon**

*A*t 7:00 P.M. on May 8, 1981, a small tree in a vacant lot in Winter Park, Florida, suddenly disappeared. Within 10 hours a steep-sided crater 20 m in diameter had developed around the central depression where the tree originally stood. The crater continued to grow and deepen throughout the night, and by noon on the following day it was 100 m wide and about 30 m deep and was beginning to fill with water. By the time it stabilized, the depression had swallowed up parts of a house, six commercial buildings, and the municipal swimming pool, as well as several automobiles. The total cost of the damages was estimated at over $2 million.

The Winter Park event was the largest in a series of collapses that occurred in the Orlando area in May 1981. Much of Florida is underlain by carbonate rocks, which can be dissolved by slightly acidic groundwater. Dissolution of this rock has produced extensive cave systems in Florida and other areas in the southern United States. The caves are enlarged whenever groundwater levels are high, but the pressure of the water in the caverns helps support the weight of the overlying rocks. The collapses in the Orlando area occurred during a time of drought, when groundwater levels were particularly low. This left underground spaces and passageways unsupported, facilitating the collapse of overlying rocks into the spaces below.

Sudden, dramatic collapses like the Winter Park occurrence are unusual. However, a wide variety of factors can contribute to changes in the surface of the land, whether abrupt and dramatic or gradual and almost imperceptible. The process of sinking or collapse of the land surface, *subsidence,* is the subject of this chapter.

◄ _____

Sinkhole in Winter Park, near Orlando, Florida. The crater appeared at 7 pm, May 8, 1981 and grew to be 100 m wide within about 20 hours.

SURFACE SUBSIDENCE AND COLLAPSE

Subsidence is the sinking or collapse of a portion of the land surface. The movement involved in subsidence is essentially vertical; little or no horizontal motion is involved. It may take the form of a sudden, dramatic collapse or a slow, almost imperceptible lowering. In some cases subsidence is a localized phenomenon; in others it is regional in extent.

Like mass-wasting, subsidence differs from other types of erosion, transportation, and deposition of Earth materials in that it does not require a transporting medium such as water or ice. However, water can facilitate ground collapse in a variety of ways, and it usually plays an important role in the process of subsidence. Also like mass-wasting, subsidence is driven by gravity. Subsidence differs from mass-wasting, however, in that it is controlled by the physical properties of the rocks and sediments underlying the collapse area and is not a function primarily of slope stability.

The mechanisms of collapse, and sometimes the conditions existing before the collapse, result from natural physical processes. In many cases, however, the conditions leading up to a subsidence event are exacerbated or even created by human actions. In this chapter we will examine some of the conditions that lead to subsidence, how those conditions can be caused or aggravated by human activities, what types of damage may result, and how subsidence-related hazards can be addressed.

CARBONATE DISSOLUTION AND KARST TOPOGRAPHY

In regions underlain by rocks that are highly susceptible to chemical weathering, groundwater creates underground caverns and distinctive landscapes that are among the most interesting and picturesque on our planet. These extensive cave systems, however, can set the stage for catastrophic events such as the one in Winter Park. To appreciate the mechanisms through which underground caverns can lead to surface subsidence, we must first consider the processes of *carbonate dissolution* and *cave formation*.

Dissolution

As soon as rainwater infiltrates the ground, it begins to react with minerals in the regolith and bedrock, causing chemical weathering. An important part of that process is **dissolution,** in which minerals and rock materials pass directly into solution.

Among the rocks of the Earth's crust, the *carbonate* rocks (those consisting primarily of minerals based on the CO_3^{2-} anion, such as calcite and dolomite) are most readily attacked by dissolution. Limestone, dolostone, and marble are the most common carbonate rocks; they underlie millions of square kilometers of the Earth's surface. Carbonate minerals are nearly insoluble in pure water but are readily dissolved by carbonic acid (H_2CO_3), a common constituent of rainwater. The weathering attack occurs mainly along fractures and other partings and openings in the carbonate bedrock, often with impressive results. When granite is weathered chemically, quartz and other resistant minerals are little affected and remain as part of the weathered regolith. However, when limestone weathers, nearly all of its volume may be dissolved away in slowly moving groundwater.

In carbonate terrains, the rate of dissolution can exceed the average rate of erosion of the surface by streams, masswasting, and other processes. Through periodic measurement of the amount of dissolution observed on small, precisely weighed limestone tablets placed at open sites in various areas, geologists have obtained estimates of the average rate at which limestone landscapes are being lowered by dissolution. In temperate regions with high rainfall, a high water table, and nearly continuous vegetation cover, carbonate landscapes are being lowered at average rates of up to 10 mm/1000 year. In dry regions with scanty rainfall, low water tables, and discontinuous vegetation, dissolution rates are lower.

Caves

Caves are large underground open spaces. The earliest evidence we have of human dwellings comes from limestone caves in Europe and Asia that provided shelter for paleolithic peoples during the Pleistocene glacial ages. The walls of those caves served as rocky canvases for prehistoric artists, whose polychrome paintings provide us with superb renditions of the prey of ice-age big-game hunters.

Caves come in many shapes and sizes. Although most caves are small, some are very large. A large cave or system of interconnected cave chambers is often called a **cavern.** The Carlsbad Caverns in southeastern New Mexico (Fig. 7.1) include a chamber that is 1200 m long, 190 m wide,

▲ FIGURE 7.1
Caves and caverns are underground openings that are created when carbonate rocks undergo dissolution. This spectacular room is part of the Carlsbad Caverns region of New Mexico.

and 100 m high. Mammoth Cave in Kentucky consists of interconnected caverns with an aggregate length of at least 48 km. The recently discovered Good Luck Cave on the island of Borneo includes a chamber so large that into it could be fitted not only the world's largest previously known chamber (in Carlsbad Caverns) but also the largest chamber in Europe (in Gouffre St. Pierre Martin, France) and the largest chamber in Britain (Gaping Ghyll).

Cave Formation

Caves are formed through a chemical process in which carbonate rock is dissolved by circulating groundwater. The process begins with dissolution by percolating groundwater along a system of interconnected open fractures and bedding planes. A cave passage then develops along the most favorable flow route. Carbonate formations (called *speleothems*) are deposited like icicles on the cave walls while a stream occupies the floor. After the stream has stopped flowing, similar formations are deposited on the floor. Although geologists disagree as to the exact conditions under which cave formation occurs, available evidence favors the idea that most caves are excavated near the top of a seasonally fluctuating water table.

The rate of cave formation is related to the rate of dissolution. In areas where the water is acidic, the rate of disso-

▲ F I G U R E 7.2
The Karst region in the southeastern corner of the former Yugoslavia, adjacent to the border with Albania. The Adriatic Sea is in the lower left-hand corner. The curiously patterned landscape is produced largely by solution of underlying limestones.

lution increases with increasing velocity of flow. As a passage grows and the flow becomes more rapid and turbulent, the rate of dissolution increases. The development of a continuous passage by slowly percolating waters has been estimated to take up to 10,000 years, and the further enlarge-

ment of the passage by more rapidly flowing water to create a fully developed cave system may take an additional 10,000 to 1 million years.

Karst Topography

In regions where rock is exceptionally soluble, the topography is characterized by many small, closed basins and a disrupted drainage pattern. Streams disappear into the ground and eventually reappear elsewhere as large springs. Such terrain is called **karst topography** after the Karst region of the former Yugoslavia (Fig. 7.2). Karst is most typical of limestone landscapes but can also develop in areas underlain by dolomite, gypsum, and salt.

Several factors control the development of karst landscapes: The topography must permit the flow of groundwater through soluble rock under the pull of gravity. Precipitation must be adequate to supply the groundwater system. Soil and plant cover must supply an adequate amount of carbon dioxide (to make carbonic acid from rainwater), and temperatures must be high enough to promote dissolution. Although karst terrain occurs in a wide range of latitudes and at varied altitudes, it is most fully developed in moist temperate or tropical regions underlain by thick and widespread soluble rocks.

The most common type of karst terrain is **sinkhole karst,** a landscape dotted with closely spaced circular collapse basins of various sizes and shapes (Fig. 7.3). Such landscapes are seen in southern Indiana, south-central Tennessee, and Jamaica, among other places.

▼ F I G U R E 7.3
Steep limestone pinnacles up to 200 m high, surrounded by flat expanses of alluvium form a spectacular karst landscape around the Li River near Guilin, China.

Sinkholes

In contrast to a cave, a **sinkhole** is a large dissolution cavity that is open to the sky. Some sinkholes are caves whose roofs have collapsed; these are sometimes called *collapse sinkholes* or *solution sinkholes*. Others are formed at the surface in places where rainwater is freshly charged with carbon dioxide and hence is most effective as a solvent.

Some sinkholes are funnel shaped; others have high, vertical sides. Sinkholes on the Yucatan Peninsula in Mexico, locally called *cenotes* (a word of Mayan origin), are steep sided and contain water because their floors lie below the water table (Fig. 7.4). The cenotes were the primary source of water for the ancient Maya and formerly supported a considerable population. A large cenote at the ruined city of Chichen Itza was dedicated to the rain god. Remains of more than 40 human sacrifices, mostly young children, have been recovered from the cenote, together with jade, gold, and copper offerings.

Sinkhole Formation

In many carbonate landscapes, such as those in Florida, new sinkholes are forming constantly. In one small area of about 25 km² more than 1000 collapses have occurred in recent years. In this case, lowering of the water table as a result of drought and excessive pumping of water wells has led to extensive collapse of cave roofs.

Some sinkholes are formed catastrophically. An account of one such event, which took place in rural Alabama, describes how a resident was startled by a rumble that shook his house. He then distinctly heard the sound of trees snapping and breaking. A short time later, hunters walking through nearby woods discovered a 50-m-deep sinkhole that was 140 m long and 115 m wide.

The sinkhole in Winter Park, Florida, described at the beginning of the chapter, was formed by the collapse of surface materials into a preexisting cavity in underlying carbonate rocks. In this case, the cavern was connected to near-surface rocks and unconsolidated overburden by an open, chimneylike passageway called an **aven** (Fig. 7.5A). Sinkhole formation can be initiated either by the sudden wholesale collapse of the bedrock "roof" into the underground void, a process called **stoping,** or by much more gradual downward movement of unconsolidated material into the aven. The latter process, called **raveling,** eventually leaves the roof materials unsupported; surface fractures begin to develop and the roof eventually collapses (Fig. 7.5B), filling the aven with sediment and debris. This type of process is thought to have led to the formation of the Winter Park sinkhole. The lowering of the water table in the region as a result of drought probably contributed to the sediment raveling. Groundwater percolating downward through the passageway provided by the aven carried sediment with it, facilitating the downward movement of material.

REMOVAL OF SOLIDS AND MINE-RELATED COLLAPSE

Sinkholes in karst terrains develop naturally in response to the dissolution of underlying carbonate rocks. However, as noted earlier, other types of rock are also susceptible to dissolution, and there are other ways in which the removal of solid materials can create spaces underground that may result in subsidence.

Removal of Salt

Because rock salt (composed mainly of the mineral halite, NaCl) can be dissolved by groundwater, karst terrains and sinkholes can develop in areas underlain by salt as well as in carbonate terrains. In 1980 a sinkhole 110 m wide and 34 m deep developed in this manner near Kermit, Texas. This depression, called the Wink Sink (Fig. 7.6), appeared without warning and grew to its full size within 48 hours. Groundwater had dissolved caverns in underlying salt beds in much the same manner as that described earlier for carbonate terrains.

One technique used for mining salt is to inject fluids and induce dissolution of the salt underground so that it can be withdrawn in a liquid state. This is called *solution mining*. When the salt-saturated solution is pumped out it leaves cavities in the rock, thereby weakening the support for overlying material. In 1974 solution mining of salt was directly responsible for the rapid development of a water-filled sinkhole 300 m in diameter in Hutchinson, Kansas (Fig. 7.7).

▲ **FIGURE 7.4**
The sacred well at Chichen Itza, a ruined Mayan city on the Yucatan Peninsula of Mexico. This cenote, formed in flat-lying limestone rocks, contained a rich store of archaeological treasures that had been cast into the water along with human sacrifices centuries ago.

A. One quiet day in Winterpark. . .

B. Roof has collapsed, aven is filled with sediment

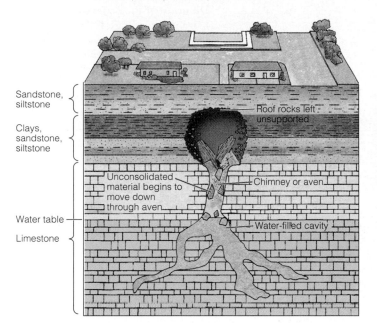

▲ F I G U R E 7.5

**Formation of the sinkhole in Winter Park, Florida,
as a result of the collapse of surface materials into a
preexisting cavity in underlying carbonate rocks. A.
The cavern was connected to near-surface rocks and
unconsolidated overburden by an open, chimneylike
passageway called an aven. B. Formation of the sinkhole was initiated by raveling, or gradual downward
movement of unconsolidated material into the aven.
This eventually left the roof materials unsupported;
surface fractures began to develop and the roof eventually collapsed, filling the aven with sediment and
debris.**

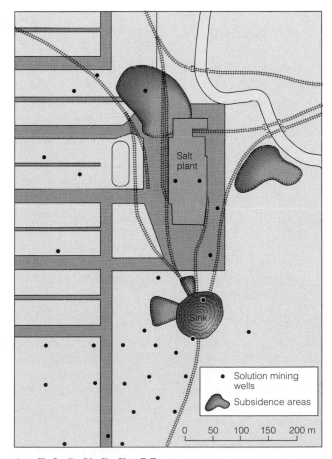

▲ F I G U R E 7.7

**Map of the sinkhole area near Hutchinson, Kansas,
where solution mining of salt led to the development
of collapse sinkholes.**

▲ F I G U R E 7.6

**The Wink Sink, Kermit, Texas. The road and cars
provide scale.**

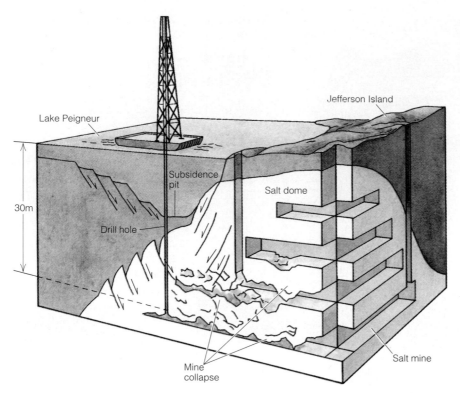

◀ F I G U R E 7.8
Diagram of the Jefferson Salt Dome under Lake Peigneur, Louisiana. In 1980 an oil-drilling rig accidentally punched a small hole through the lake bottom into a mining shaft. The hole grew rapidly as lake water drained into it, dissolving the salt and eventually causing the mine roof to collapse.

Perhaps the most astonishing collapse associated with a salt mine occurred at Lake Peigneur, a shallow lake in southern Louisiana. Under the lake was a still-active, multi-million-dollar salt mine in the Jefferson Island Salt Dome. On November 21, 1980, an oil-drilling rig accidentally punched a small hole through the lake bottom into a mining shaft (Fig. 7.8). The hole grew rapidly as lake water drained into it, dissolving the salt and eventually causing the roof of the mine to collapse. The entire lake drained into the mine in a whirlpool, carrying with it ten barges, a tugboat, and an oil-drilling rig. More than 25 hectares of land were lost, including a private home and substantial portions of a botanical garden. Incredibly, nine of the barges popped to the surface 2 days later as the depression began to stabilize and refill with water!

Salt and Oil

Salt is often associated with oil because of the shallow marine (i.e., saltwater) setting in which many oil deposits were formed. The Wink Sink, for example, occurred in an oil field, although the collapse itself was caused by the natural dissolution of salt layers. In the Lake Peigneur event, an oil-drilling operation was responsible for the hole that triggered the collapse, although it was the dissolution of salt that ultimately led to the collapse of the mine roof. The oil company was held legally responsible for the event and was required to pay damages to the salt mining company as well as to affected landowners.

Because of the frequent association of salt and oil, oil drilling commonly brings large quantities of salty fluid, or *brine*, to the surface along with the oil. The brine is often reinjected into waste disposal wells, which are drilled to depths of several hundred meters and may pass through thick salt layers. If the disposal well casing leaks, the brine may escape and dissolve large underground cavities in the salt layers (Fig. 7.9). Such an occurrence was responsible for the formation of an 80-m-wide sinkhole in Macksville, Kansas, in the summer of 1988. The underground cavern had probably existed for some time prior to the collapse; the disposal well had been plugged since 1984. The cavern that ultimately caused the Macksville sinkhole was probably the size of the Houston Astrodome! Other underground voids—some with no evidence of surface subsidence—have been detected in nearby areas, leading to concerns that a similar event could occur in the future.

Coal Mining

Removal of solid material from underground can lead to subsidence even when the rock is not soluble in water. Areas in which coal is mined are particularly susceptible to such events. Coal is commonly mined underground using the *"room-and-pillar"* method, in which large masses of coal are removed, creating the "rooms," and columns of coal are left behind to form the supporting "pillars." In the past, many mining companies removed as much coal as possible,

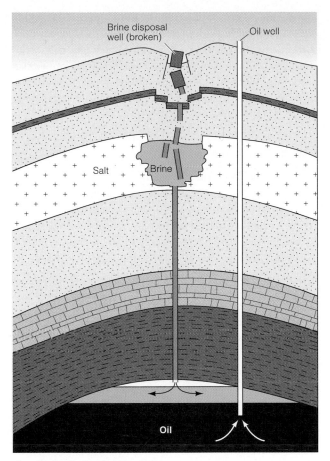

▲ F I G U R E 7.9
Brine (salty water) is a waste product generated during oil drilling. The brine is often reinjected into disposal wells. If the well casings break, the brine can leak out and dissolve large openings in salt layers. This can ultimately lead to subsidence or collapse, as occurred in Macksville, Kansas.

even from seams relatively near the surface. In some cases, subsidence occurred because too much coal was removed, and the pillars left behind could not support the weight of the overburden. In other cases, the mine was excavated too close to the surface and the roof material between the pillars collapsed.

Some relatively recent episodes of subsidence can be attributed to mining done 50 years ago or more. Grasslands near the town of Sheridan, Wyoming, are scarred and pitted with features resulting from subsidence of the ground surface into underground rooms from which coal was mined. Most mining activities associated with the subsidence ended decades ago. The pattern of the subsidence features reveals the layout of the underground rooms and pillars of the former mines (Fig. 7.10). Some of the features are broad, shallow depressions; others are deep, steep-sided pits.

Another hazard commonly associated with coal mining is underground fires, which are caused by spontaneous ignition of coal dust or methane gas. When such fires burn out of control they remove even more coal, increasing the likelihood of surface collapse. Extensive coal fires have burned for more than a quarter of a century in old mines underneath the towns of Laurel Run and Centralia, Pennsylvania, with subsequent ground fracturing, subsidence, and release of toxic gases. The fires, which have migrated several kilometers since they were ignited, are rekindled by oxygen passing through fractures and cracks in the ground surface.

A study investigating the land-use impacts of coal mining estimated that of the 8 million acres that have been undermined for coal in the United States, about 1.85 million—nearly 25 percent—have subsided. Loss of productive farmland is the main problem associated with subsidence in rural areas. In some cases groundwater supplies

◄ F I G U R E 7.10
Subsidence above abandoned coal mine workings, 15 km north of Sheridan, Wyoming. The Dietz Mine was active from the 1890s to the 1920s. The subsidence pattern reflects the design of support pillars and mine openings.

and underground drainage systems may also be adversely affected. Urbanization in previously mined areas can cause further problems, not only because of the potential damage to structures but also because the associated surface loading can increase the possibility of subsidence.

SUBSIDENCE CAUSED BY FLUID WITHDRAWAL

Withdrawal of fluids from underground can lead to surface subsidence, but the mechanisms involved generally differ from those described earlier. Sometimes fluid withdrawal and subsidence occur naturally. During a drought, for example, rows of trees or plants with very deep roots (such as alfalfa) may deplete soil moisture to such an extent that fractures as wide as 1 m are formed. Surface soils may bridge the subsurface fractures, hiding them from view; when it rains, the soil bridge may collapse, producing a gaping hole.

Subsidence resulting from the withdrawal of fluids can also be caused or facilitated by human activities. The fluids most often involved are groundwater, oil, natural gas, and associated brines, as well as mixtures of steam and water used for geothermal energy. In contrast to the sudden, localized events associated with mine-related collapses and sinkhole karsts, subsidence associated with fluid withdrawal is often gradual but may be regional in extent.

The subsidence that occurs when fluids are withdrawn is caused primarily by **compaction,** that is, a decrease in the thickness of a layer of sediment or rock. The weight of overlying materials causes the mineral particles to pack to-gether more tightly when the supporting pressure of their pore water is withdrawn. The amount of subsidence is proportional to the volume of fluid withdrawn and the thickness of the compressible layers. Crystalline (i.e., igneous or metamorphic) rock formations subside least; highly compressible materials, such as clays, or organic-rich sediments, such as peat, subside most.

Water

Subsidence is often caused by removal of subsurface water, especially if the rate of removal is faster than the rate of replenishment. *Hydrocompaction*—compaction caused by the drying or dewatering of sediments—and associated subsidence can occur whenever water is removed from layers of compressible materials, as in the draining of a marsh or wetland. (Hydrocompaction in clays and organic-rich sediments and its relationship to slope failure is discussed in Chapter 6.)

When groundwater is pumped out of an **aquifer** (a water-bearing rock or sediment unit), the pumping creates a depression in the surface of the water table in the shape of an inverted cone (Fig. 7.11). When many pumped wells are close together, as in some urban areas or irrigated farmlands, these features—called **cones of depression**—may begin to overlap. This can cause a regional depression in the level of the water table. In addition to creating water supply problems, regional lowering of the water table commonly leads to ground subsidence in the area overlying the aquifer.

Sometimes it is possible (although costly) to detect and reverse the process of subsidence by pumping water into the aquifer to raise the fluid pressure. However, the real

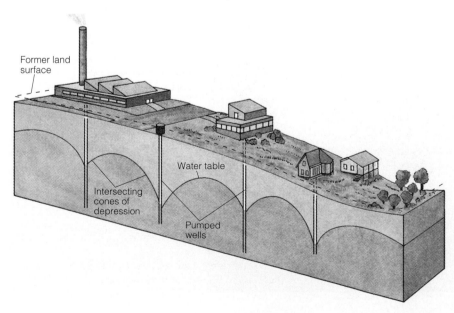

◄ F I G U R E 7.11
Overlapping cones of depression in an area of closely spaced pumped wells, leading to regional subsidence and depression of the surface of the aquifer (the water table).

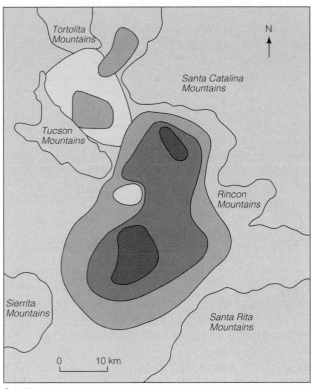

Subsidence rates

☐ This area "rose"	▨ 1.0–2.0 cm/year
▨ 0–1.0 cm/year	■ > 2.0 cm/year

▲ F I G U R E 7.12
Subsidence rates in the Tucson Basin. Studies show that the weight of the basin, combined with over-pumping of the aquifer, is damaging its water storage capacity.

face dropped 3 mm for every meter of decline in the water table. More recent measurements show that the ground surface is now dropping 24 mm for every meter drop in water level—eight times the previous rate (Fig. 7.12). Researchers have suggested that the dramatic increase in rate of compaction may signal a transition from elastic to inelastic subsidence, with potentially serious implications for the aquifer's water-storage capacity.

Subsidence caused by excessive withdrawal of groundwater is of particular concern in the southern and southwestern United States, where many thousands of square kilometers of land are affected. In the Los Banos–Kettleman City area of California, 6200 km² of land have subsided more than 0.3 m, with a maximum subsidence of 8.5 m. In areas where subsidence is especially pronounced, cracks or fissures as much as 1 m wide and 1 km long have opened around the edges of the subsiding basins.

Oil and Gas

The withdrawal of oil, natural gas, and associated briny fluids can lead to subsidence through the same mechanisms as the withdrawal of water. A map of the Inglewood oil field in California (Fig. 7.13) shows the correlation between the area from which oil was withdrawn, beginning in 1924,

problems begin when an aquifer becomes so compacted that it passes from *elastic* to *inelastic* subsidence. As noted in Chapter 3, inelastic refers to an irreversible change. In other words, the inelastic compaction of sediment or clay particles is permanent (not reversible, as would be an elastic compaction). Even if water were pumped back into the aquifer, it would not regain its original thickness. In such cases, the decrease in *porosity* (the amount of open or *pore* space in the sediment or rock) caused by compaction means that the water-bearing capacity of the sediment or rock has become permanently impaired.

The rate of ground subsidence relative to the decline in the level of the water table can be used to determine whether compaction is elastic or inelastic: Large water loss with only slight subsidence indicates elastic subsidence, whereas small water loss with significant subsidence suggests inelastic subsidence. Records from the Tucson Basin in southern Arizona show that before 1981 the ground sur-

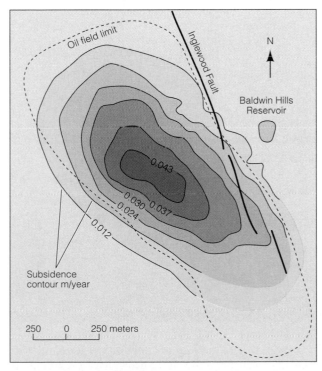

▲ F I G U R E 7.13
Map of the Inglewood oil field in Los Angeles showing the area from which oil was withdrawn and the area of surface subsidence.

and the resulting subsidence. As in the cases of groundwater withdrawal just described, the subsidence proceeded gradually and continued over an extended period. Decades after the subsidence began, fissures appeared around the edges of the subsidence basin. This would ultimately prove disastrous for the Baldwin Hills Reservoir, located on the northeast edge of the basin. In 1963 the Baldwin Hills Dam failed, releasing all of the water in the reservoir. The resulting flood killed five people and caused $12 million in damage. The conditions that led to the failure of the dam were caused primarily by the subsidence, which was associated with the reactivation of a preexisting fault underlying the dam. The injection of water (used, in this case, to flood the oil field, not to control the subsidence) may have played a minor role in reactivating the fault.

▲ FIGURE 7.14
The friendly fire hydrant is high and dry. Ground subsidence at Long Beach, California as a result of oil removal. The ground sank but the pipe structures remained at their former height. Repressuring of the former oil reservoirs has restored the land surface closer to, but not equal to, its former level.

Subsidence associated with the Wilmington oil field near Long Beach, California, has cost over $100 million since it was first recognized in 1940. The area of the subsidence basin reached 50 km^2 at its maximum; near the center of the basin the vertical subsidence was as great as 9 m, causing localized flooding. In 1958 an aggressive program was undertaken with the goal of repressurizing the rocks from which the oil had been withdrawn. Large quantities of water were injected into the rocks to raise the fluid pressure. By 1962 the sinking was stopped and the area of the subsidence basin had been reduced to 8 km^2. A small amount of vertical rebound was observed at some sites (Fig. 7.14).

OTHER CAUSES OF SUBSIDENCE

In certain circumstances mechanisms other than those discussed so far can lead to subsidence. For example, in areas of former glaciation, depressions known as *kettles* are often observed. These were created where masses of debris-covered ice melted away, causing collapse of the overlying sediment. The resulting bowl-shaped depressions often contain a lake or swamp. A famous example of a kettle lake is Walden Pond in Massachusetts. Kettles can range up to tens of meters in depth and a kilometer or more in diameter.

Another special circumstance arises when flowing lava forms *tubes* or tunnels. As the lava cools and solidifies, it forms a lid of solid rock under which molten lava can continue to flow. If the source of molten lava is cut off, the flow of lava ceases, leaving a hollow tube or "skin" of solidified material. If the roof of the tube is thin enough, it will collapse; this produces a long, sinuous depression where the lava river once flowed.

Endogenous Subsidence

The types of subsidence discussed so far originate near the surface of the Earth, either as a result of natural weathering and groundwater processes or as a consequence of human activities such as mining or fluid extraction. Another broad category of subsidence mechanisms is associated with internal or *endogenous* Earth processes. These include such tectonic processes as folding, faulting, and plate motion, as well as sedimentation in certain tectonic environments. Except in the case of catastrophic episodes related to earthquakes or volcanism, the rate of subsidence associated with such processes rarely exceeds 10 mm/year.

Coastal areas, especially deltas, are particularly susceptible to the gradual subsidence that results from endogenous processes. Downwarping of the crust in such areas is normally offset by the influx of sediment carried by rivers. But the amount of sedimentation can decrease dramatically if a river is dammed, channeled, or diverted, resulting in higher rates of erosion. Natural isostatic subsidence combined with decreased sedimentation, increased erosion, and, in some cases, accelerated subsidence induced by the withdrawal of water or oil may create problems for some low-lying coastal areas in the future. For example, researchers at Woods Hole Oceanographic Institution have speculated that by the year 2100 local sea level in the Nile Delta could be as much as 3.3 m higher than it is now, meaning that Egypt could lose up to 26 percent of its habitable land as well as invaluable coastal assets such as fisheries and mangrove forests. Potential increases in sea level because of global warming over the next few decades could further exacerbate the situation.

SINKING CITIES

In urban settings, factors conducive to subsidence often occur together. Given the high density of people, buildings, and utilities, this can create a significant potential for urban subsidence, which may cause extensive and costly damage (Table 7.1).

All of the cities that have experienced major problems with subsidence are located on unconsolidated sediments consisting of silt, clay, peat, and sand associated with river floodplains (e.g., New Orleans, London, Bangkok), coastal marshes and/or delta complexes (e.g., Venice, Houston/Galveston, Tokyo, Shanghai), or lake beds (e.g., Mexico

▲ F I G U R E 7.15
Land subsidence and the relative rise of the water table forces the citizens of New Orleans, Louisiana, to bury their dead in above-ground cemeteries.

T A B L E 7.1 • Subsiding Lands[a]

	Maximum subsidence (m)	Area affected (km²)
Coastal		
London	0.30	295
Venice	0.22	150
Po Delta	1.40	700
Taipei	1.90	130
Shanghai	2.63	121
Bangkok	1.00	800
Tokyo	4.50	3,000
Osaka	3.00	500
Niigata	2.50	8,300
Nagoya	2.37	1,300
San Jose	3.90	800
Houston	2.70	12,100
New Orleans	2.00	175
Long Beach/Los Angeles	9.00	50
Savannah	0.20	35
Inland		
Mexico City	8.50	225
Denver	0.30	320
San Joaquin Valley	8.80	13,500
Baton Rouge	0.30	650

[a]From Dolan and Goodell, "Sinking Cities," *American Scientist,* Vol. 74, Jan.–Feb. 1986.

City). Subsidence initially occurs in response to the sheer weight of urban development. Unconsolidated sediments settle, and fluids are squeezed out of pore spaces. As discussed earlier, in such highly compressible sediments as clay and peat the particles may rearrange themselves permanently, flattening and compacting in response to loading, with the result that the subsidence becomes inelastic.

In addition, many cities in which subsidence is a problem (with the notable exception of Mexico City) are located in coastal areas that are prone to natural subsidence. In New Orleans, regional long-term subsidence from the compaction of delta sediments and isostatic downwarping ranges between 12 and 24 cm per 100 years, with local subsidence caused by groundwater withdrawal as high as 3 m in 50 years. The situation has been aggravated by the construction of levees and channels that direct the sediment-laden waters of the Mississippi River into deep water offshore. As a result of this combination of natural and human activities, the Mississippi Delta is sediment-starved, land is being lost rather than created, and 45 percent of the urbanized area of New Orleans is at or below sea level (Fig. 7.15).

Many large cities also contain concentrations of buildings that were constructed before the emergence of zoning laws based on an understanding of subsidence and soil mechanics. Capuchinas Church in Mexico City, constructed

BOX 7.1
•
THE HUMAN PERSPECTIVE

SUBSIDENCE IN MEXICO CITY

*N*atural subsidence has occurred in the Mexico Basin since before the Aztec period. In 1325 A.D. the Aztecs established Mexico City (originally named Tenochtitlan). At that time the Mexico Basin was a network of lakes and islands surrounded by volcanic mountains, and the city was established on several of the islands. Large structures built by the Aztecs, and by the Spanish several hundred years later, placed excessive loads on the fine-grained volcanic ash and lake sediments of the Mexico Basin aquifers. Also during this period the lakes were drained and canals were built. These actions initiated the first subsidence episodes directly related to human activity in the Mexico Basin.

Water confined within the layers of the aquifer provides pressure that supports the overburden of sediment by pushing apart the individual grains of subsurface material. This pressure can also support buildings and other structures at the surface. However, in this century a critical subsidence problem has developed as a result of the exploitation of groundwater reserves to meet the needs of Mexico City's rapidly growing population. When groundwater is removed at rates that exceed the natural rate of replenishment, subsurface material is compacted vertically and the land subsides. The subsidence creates problems such as uneven settling of large buildings; stranding of sidewalks, well casings, and other structures above the land surface; and weakening of foundations (a particular concern in this earthquake-prone region). Even after the withdrawal of groundwater is stopped, the subsidence may continue because of the weight of the surface material and structures. Furthermore, the compacting of sediments caused by the subsidence may be irreversible.

Most of the critical problems of subsidence in Mexico City originated in the early 1900s, when the population began placing tremendous demands on freshwater resources. In the 1850s there were 140 free-flowing artesian wells in the Mexico Basin aquifers. By 1922, the water table in the regional aquifers had declined more than 35 meters and land subsidence resulting from groundwater extraction in Mexico City totaled 1.25 m. In 1940, more than 100 deep-water production wells were dug to reach additional water supplies, greatly accelerating the rate and severity of subsidence. Subsidence reached depths of 8 m in some places and occurred at rates of between 15 and 46 cm/year. By 1954 most of the wells in the metropolitan area were closed to alleviate subsidence. However, as the population continued to grow, new large-capacity wells were installed at the periphery of the Mexico Basin aquifer system. During this time the rate of subsidence was accelerated in outlying areas even as it was reduced in the central city.

Chalco Plain, an area south of Mexico City, has been used since 1960 to provide additional water for the central metropolitan area. As a consequence of the pumping of water from wells, this region has subsided 6 m in places. If extraction of groundwater continues, the rate of subsidence will exceed that of the Valley of Mexico. Subsidence in Chalco Plain has already caused contamination of groundwater and flooding of agricultural and urban land. It remains to be seen whether land use planners for the world's largest city can manage their limited water resources without perpetuating the problems associated with subsidence.

in the eighteenth century, was seriously out of plumb as a result of differential settling until extensive restorations were undertaken. The Leaning Tower of Pisa (Fig. 7.16) is probably the best known example of a sinking building. Immediately after its construction was begun in 1174, the tower began to sink as the foundations settled into the soft sediment of the Arno floodplain. In the 1960s the rate of tilting increased as a result of the withdrawal of groundwater from aquifers underlying the city. Recent efforts to secure the tower have centered on the injection of concrete to stabilize the foundations.

The most severe problem facing urban areas, however, is the accelerated subsidence caused by intensive withdrawal of oil and/or water from underlying reservoirs. In the past 50 years, groundwater pumping in response to the pressures of industrialization and rapid population growth has

◄ F I G U R E 7.16
The Leaning Tower of Pisa is a campanile adjacent to the Cathedral. It leans because it settled unevenly into the soft sediment of the River Arno floodplains. Ground water withdrawal in modern times has increased the angle of tilt.

greatly accelerated rates of urban subsidence. In Venice, Italy, for example, subsidence is a critical problem (Fig. 7.17). Venice is built on a series of interconnected marshy islands in a saltwater lagoon on the coast of the Adriatic Sea. Subsidence has made the city and its historic buildings extremely vulnerable to storm surges and flooding associated with high tides. In the 1950s the withdrawal of large quantities of groundwater to meet industrial needs led to a 10-fold increase in the rate of subsidence, accompanied by a sharp decrease in well water pressures. When groundwater pumping was stopped in 1969, subsidence stopped and water pressures increased; there was actually a slight (2 cm) rebound in elevation in some parts of the city, although most of the subsidence became permanent because of inelastic compaction in underlying clay layers.

London is another city where a combination of factors has led to significant and damaging subsidence. The gradual regional tilting of southeastern England, which has occurred at a rate of about 30 cm per century, combined with

◄ F I G U R E 7.17
Venice, one of the world's most beautiful and fascinating cities, is slowly sinking. Periodic flooding is the result. Here, people are crossing St. Mark's Square during a time of flooding.

▲ F I G U R E 7.18
Flood barriers on the River Thames, London, looking upstream. The barriers are designed to prevent high tides and storm surges from flooding central London. The dome-shaped pedestals are fixed. The cylindrical structures between the domes are barriers that can be rotated up, to prevent flow, or down, to allow flow.

the dewatering and compaction of clay layers and the pumping of groundwater from an underlying chalk aquifer, has resulted in an increase in tide height (measured at London Bridge) of 60 cm over the past 100 years. This has greatly increased the potential for urban flooding. It is estimated that 40 stations and 127 km of subway track in the London underground rail system are in flood-prone areas. In response to the growing danger, the equivalent of $1 billion has been spent to construct the Thames Barrier, a series of 10 floodgates designed to prevent floods and storm surges in the lower Thames from reaching central London (Fig. 7.18).

Urban flooding is a common side effect of subsidence in cities located on river floodplains, coastal areas, or deltas. Bangkok, for example, is located on the floodplain of the Chao Phraya River, 25 km north of the Gulf of Thailand. Intensive withdrawal of groundwater to support the rapidly growing urban population has resulted in subsidence of up to 1 m, which has occurred at rates as high as 14 cm/year in the southeastern part of the city. Subsidence and urban development have interfered with the drainage of water into the Chao Phraya via a series of canals constructed for this purpose. As a result, the city is now regularly inundated with sewage-choked floodwaters after rainstorms. Differential subsidence in Bangkok has also damaged buildings, cracked roads and sidewalks, and broken utility pipes.

PREDICTING AND MITIGATING SUBSIDENCE HAZARDS

When removal of solid material is the primary cause of subsidence, prediction of subsidence and assessment of the related hazards are generally based on mapping of susceptible areas. In former coal-mining areas, for example, maps showing the distribution of previously mined areas are combined with information about subsidence in similar areas, the depth of the mine workings, and the locations where mine fires have occurred or are in progress. This results in a map showing the areas where the risk of mine-related subsidence is greatest. Prediction of subsidence in carbonate terrains in Florida has used aerial photography to identify areas of disrupted drainage. This information can be combined with studies of historic and prehistoric sinkhole occurrences to produce a map of subsidence potential (Fig. 7.19).

Where fluid withdrawal is the main cause of subsidence, it is important to know the rate of fluid withdrawal and the characteristics of the underlying rock layers, particularly with respect to compressibility. For example, in a study of the Houston area, researchers compared historical records of the drawdown of water table surfaces with records of land surface subsidence. They found that subsidence was greatest in areas of maximum drawdown. Specific predic-

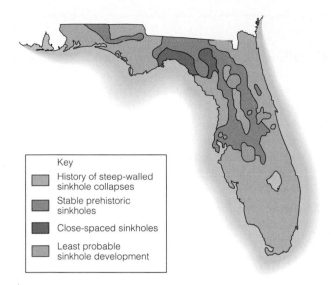

◀ F I G U R E 7.19
Map of Florida showing areas of differing potential for subsidence due to dissolution.

Key
History of steep-walled sinkhole collapses
Stable prehistoric sinkholes
Close-spaced sinkholes
Least probable sinkhole development

tions of future subsidence are based on an understanding of historical relationships between subsidence and water withdrawal, combined with models of the maximum compressibility of underlying materials (Fig. 7.20). Predictions of subsidence can be calculated for different water withdrawal scenarios.

So far, the most viable approach to the prevention, control, and mitigation of hazards associated with subsidence is to modify the human activities that contribute to its occurrence. In part, this can be accomplished through zoning reg-

ulations based on an understanding of existing geologic hazards. A complicating factor in the United States is that laws concerning the right to sue for damages from subsidence conflict with laws concerning the right to withdraw an underground resource (such as groundwater), even when such withdrawal may lead to subsidence in a neighboring property. Subsidence insurance and other commercial and government programs are increasingly available to property owners in areas where the risk of subsidence is high.

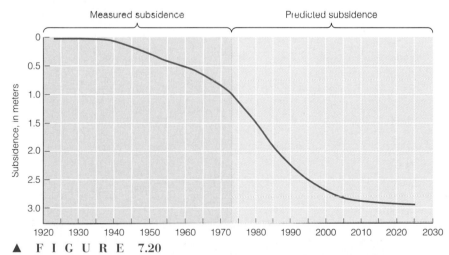

▲ F I G U R E 7.20
Measured rates of surface subsidence and predicted rates of subsidence. The predictions are based on measured land surface subsidence, measured drop in level of water table, and rates of fluid withdrawal.

SUMMARY

1. Subsidence—the vertical sinking or collapse of a portion of surface land—can be slow or sudden, localized or regional in extent. It can result from human actions as well as from natural processes.

2. In regions underlain by rocks, such as carbonates, that are highly susceptible to chemical weathering, extensive systems of caves and caverns may form. Cave formation is mainly a chemical process involving dissolution of carbonate rock by circulating groundwater; the rate of cave formation is related to the rate of dissolution.

3. In regions where rocks are exceptionally soluble, karst topography may form. This peculiar terrain is characterized by many small, closed basins and a disrupted drainage pattern.

4. Sinkholes—large dissolution cavities open to the sky—are a common feature of karst terrains. Some sinkholes are formed when the ground surface collapses into existing underground cavities, whereas others are formed directly at the surface as a result of interaction with acidic rainwater. Sinkholes can result from sudden collapses or from the gradual raveling of sediment through a chimneylike passageway called an aven.

5. The use of solution mining to mine salt sometimes causes subsidence. Near-surface coal mining also is commonly associated with subsidence.

6. The withdrawal of fluids—especially water, oil, or gas—from underground rock or sediment units can also lead to subsidence. This type of subsidence is usually caused by the compaction of a rock or sediment when fluid no longer fills the spaces between the particles. Sometimes the compaction and associated subsidence are inelastic and, hence, irreversible.

7. Internal tectonic processes such as folding, faulting, and plate motion, as well as sedimentation in certain tectonic environments, can lead to subsidence. Coastal deltas are particularly susceptible to isostatic downwarping and subsidence.

8. In urban settings a variety of factors conducive to subsidence (such as excessive groundwater withdrawal, mining activity, and the weight of urban development) commonly occur together. Given the density of people, buildings, and utilities, this can create a high potential for extensive and costly damage. Cities located on unconsolidated silt, clay, peat, and sand associated with river floodplains, coastal marshes and/or delta complexes, or lake beds are most likely to experience problems of subsidence.

9. When the removal of solid material is the primary cause of subsidence, prediction of subsidence and assessment of related hazards are generally based on mapping of susceptible areas. Knowledge of the characteristics of underlying rock and rates of fluid withdrawal are needed when fluid withdrawal and/or compaction of compressible layers is the primary cause of subsidence.

IMPORTANT TERMS TO REMEMBER

aquifer (p. 180)
aven (p. 176)
cave (p. 174)
cavern (p. 174)
compaction (p. 180)

cone of depression (p. 180)
dissolution (p. 174)
karst topography (p. 175)
raveling (p. 176)
sinkhole (p. 176)

sinkhole karst (p. 175)
stoping (p. 176)
subsidence (p. 173)

QUESTIONS AND ACTIVITIES

1. If you live (or vacation) in a region underlain by carbonate rocks, make a point of visiting caves or caverns in the area. Many spectacular caverns in North America and elsewhere in the world are open to visitors and offer guided tours. Try to observe the processes of cave formation in action. Where can you see carbonate being dissolved? Where is it being precipitated to form speleothems? A word of warning: don't go caving ("spelunking") without an experienced guide; it can be very dangerous.

2. If you visit a cavern or live in an area underlain by caverns, see if you can spot the signs of karst topography on the surface. Can you see sinkholes or small, pothole-shaped lakes? Get a map of the area—a geologic map, if possible—and try to determine the drainage patterns of streams and rivers.

3. What do you think would be the implications of karst topography and underground cave and drainage systems in the event of a chemical spill? How would the presence of karst topography affect your assessment of the suitability of an area for a hazardous waste disposal site or a landfill? (You may wish to revisit this question after reading Chapters 16 and 17.)

4. Select an urban subsidence problem to investigate in detail. There are lots of interesting examples to choose from—Bangkok, Venice, New Orleans, and Mexico City, among many others. Try to find out the history of subsidence in the city you have chosen. To what extent was the area naturally predisposed to subsidence, and to what extent have human actions caused or contributed to the problem? What is the estimated cost (both economic and environmental) to the city? What solutions to the problem are being considered or undertaken?

FLOODS

Rain added to a river that is rank
Perforce will force it overflow the bank.

• **William Shakespeare**

The Mississippi River has a habit of shifting its channel from time to time in order to avoid flowing around large bends. In seeking a new and shorter route, the river cuts across a bend at its narrowest point, thereby abandoning a piece of the old channel. In the nineteenth century riverboat pilots spoke of such an event as a "cutoff." Former steamboat pilot Mark Twain, an observant amateur geologist, speculated about the future history of the Mississippi River in *Life on the Mississippi* (1883):

> The Mississippi between Cairo and New Orleans was 1215 miles long 176 years ago. It was 1180 after the cutoff of 1722. It was 1040 after the American Bend cutoff. It has lost 67 miles since. Consequently, its length is only 973 miles at present.
>
> Now, if I wanted to be one of those ponderous scientific people, and "let on" to prove what had occurred in the remote past by what had occurred in a given time in the recent past, or what will occur in the far future by what has occurred in late years, what an opportunity here! Geology never had such a chance, nor such exact data to argue from! Please observe:
>
> In the space of 176 years the lower Mississippi has shortened itself 242 miles. That is an average of a trifle over one mile and a third per year. Therefore, any calm person, who is not blind or idiotic, can see that in the Old Oolitic Silurian Period, just over a million years ago next November, the Lower Mississippi River was upwards of 1,300,000 miles long, and stuck out over the Gulf of

Mexico like a fishing rod. And by the same token any person can see that 742 years from now the Lower Mississippi will be only a mile and three-quarters long, and Cairo and New Orleans will have joined their streets together, and be plodding comfortably along under a single mayor and a mutual board of aldermen. There is something fascinating about science. One gets such wholesale returns of conjecture out of such a trifling investment of fact.

Twain wrote of the river shortening itself through natural processes, but the Mississippi has also been shortened by human intervention. During the 1930s, for example, a massive flood protection and channel modification program resulted in the shortening of the river's course by 185 km between Cairo and New Orleans. In spite of those efforts, the Mississippi River asserted itself on a grand scale in the summer of 1993. The main portion of the "Great Flood" of 1993 lasted for 79 continuous days, surpassing the previous record of 77 days set by the flood of 1973. Fifty people were killed and 55,000 homes destroyed or damaged. At the height of the flooding, more than 40,000 km² of land bordering the Missouri and Mississippi rivers were under water. The impact of this record-shattering flood on the lives of people in the midwestern United States will not soon be forgotten.

Fleeing the flood. An estimated 55,000 homes were flooded when the Mississippi River overtopped its levees during the great flood of 1993. This farm family's home was in St. Charles County, Missouri.

THE WATER'S EDGE

From the beginnings of civilization, people have tended to live by the water's edge—beside streams, rivers, or lakes; in delta regions; or along the sea coast. Land adjacent to water has traditionally offered many advantages to settlers, advantages that at first were necessary for survival and later were conducive to development and industrialization (Fig. 8.1).

◄ F I G U R E 8.1
Like many of the world's great cities, Paris is situated along the banks of a major river. The Seine provides water for human and industrial use, serves as an avenue of transportation, and has great aesthetic and recreational value. However, people and buildings in the urbanized floodplain must be protected from the constant threat of flooding. This has been accomplished mainly through extensive structural modifications of the river's channel.

Those advantages include fertile soil; access to water, food, energy, and other resources; ready transportation routes; and the capacity for absorption and dispersal of wastes. River valleys, in particular, are favored localities for human settlement because valley floors are flat and easy to build on and the soils typically are deep and arable. But living by the water's edge has some disadvantages, of which the most obvious is the repeated threat of flooding.

A **flood** occurs when the level of a body of water rises until it overflows its natural or artificial confines and submerges land in the surrounding area. Because of the human tendency to develop and inhabit such land, floods can take a high toll in loss of life and damage to structures. This is especially true in less developed countries, where factors such as population density, absence of strict zoning regulations, and lack of effective response, control, and early-warning systems combine to increase the risks associated with flooding.

Bangladesh is more susceptible to flood disasters than any other nation in the world (Fig. 8.2). Half of that nation's land is less than 8 m above sea level, and as much as 30 percent of its total area has been covered during a single flood! High population density in an active delta with constantly shifting rivers combines with heavy monsoon rainfalls and cyclones to increase the hazard. In the spring of 1991 more than 200,000 people were killed in delta flooding associated with cyclone activity.

In industrialized countries loss of life because of flooding is typically much lower than in developing countries, but the amount of damage and disruption can be staggering. Governments and communities spend countless dollars and much effort perfecting and installing flood-control devices. Yet, as demonstrated by the flooding of the Mississippi River in the summer of 1993, such measures sometimes create a false sense of security.

Understanding the causes of flooding and the dynamics of the natural systems in which floods occur is the first step toward predicting and—to a certain extent—controlling floods. There are many different types of floods and many different causes of flooding. In this chapter we concentrate primarily on *river flooding* because of its impact on human activities and interests.

▲ FIGURE 8.2
Cutting of forests in the foothills of the Himalaya, combined with high rainfall, has led to repeated instances of massive downstream flooding of the Brahmaputra River in Bangladesh.

CAUSES OF FLOODING

People who experience floods are frequently surprised and even outraged at what a rampaging torrent of water can do. Geologists, however, tend to view floods as normal and expectable events, for the geologic record shows that floods have been occurring as long as rain has been falling on the Earth.

The Role of Precipitation

In general, natural systems of water flow, cycling, and containment are *mass-balanced;* that is, water flowing into each part of the system is balanced by water flowing out of that part. In the hydrologic cycle as a whole, evaporation is bal-

anced by precipitation. But precipitation is not uniformly distributed over the surface of the Earth. The balance of inflow and outflow in one part of a surface water system may be temporarily disturbed by a period of exceptional precipitation. *Drought* can result when precipitation is exceptionally low, and flooding may occur when precipitation is unusually intense or prolonged. Regardless of their impact on humans, these processes help maintain the long-term balance of the hydrologic cycle as a whole. Flooding therefore is a necessary consequence of the uneven distribution of precipitation and a normal part of the hydrologic cycle. Most floods are caused or exacerbated by intense precipitation, usually in combination with other factors. When a period of exceptional precipitation is combined with such factors as melting snow, inadequate drainage, water-saturated ground, or unusually high tides, the potential for flooding may increase dramatically.

Coastal Flooding

A variety of processes can lead to flooding in coastal areas. Some are natural processes, such as tsunamis (Chapter 5), hurricanes, cyclones, and unusually high tides (Chapter 9). Other floods in coastal areas are caused by human activities. For example, urban development on unconsolidated, compressible sediments in deltas and coastal regions can lead to subsidence of the ground surface, with associated flooding. As noted in Chapter 7, urban flooding in Bangkok is an oft-cited example of this phenomenon.

The failure of protective seawalls during storms can also lead to flooding in low-lying coastal lands reclaimed from the sea. In January 1953 a combination of raging storms and exceptionally high tides led to failure of dike systems in The Netherlands, with tragic consequences. When the flooding was over, 100,000 hectares (6 percent of the country's cultivated land) had been inundated by salt water, affecting well over half of the population. About 5000 houses were destroyed, 60,000 animals drowned, and 1490 people killed. The total cost of the damages was estimated to be at least $250 million.

About 27 percent of the current land area of The Netherlands has been reclaimed from below sea level as a result of efforts going back a thousand years or more. But the costs—both economic and environmental—of reclaiming this land have been extremely high. The new land must be protected by a vast network of dikes, dams, canals, and drainage stations; the Dutch government spends more than $400 million a year on drainage alone. Other environmental problems, primarily sediment compaction and water pollution, have been linked to these structural interventions. These problems, along with concerns about a possible rise in sea level because of global warming, have led the Dutch government to undertake limited flooding of some of the reclaimed land, with the goal of eventually reverting it to

◄ F I G U R E 8.3
The flooding of low-lying land
in the Netherlands, January
1995.

marshland. The plan may have been hastened in the spring of 1995 by extensive coastal flooding (Fig. 8.3) associated with a long period of above-average rainfall in the region.

Dam Failures

Dam failures are another cause of flooding related to human activities. At midnight on March 12, 1928, the St. Francis Dam in Saugus, California, failed, releasing 46,500,000 m³ of water in 1 hour; 450 people lost their lives in the flood. Another spectacular dam failure occurred in Johnstown, Pennsylvania on May 31, 1889, when an old, poorly maintained earthen dam collapsed. The dam contained a reservoir with a maximum capacity of 13,568,060 m³. A period of intense rainfall in April and May had caused the dam to deteriorate. When the structure collapsed, a wave estimated to be 12 m high roared toward Johnstown, killing 2200 people. In the Vaiont, Italy, dam disaster of 1963 (discussed in Chapter 6), the dam itself did not fail but landsliding in the walls of the reservoir caused a tsunami-like wave, with associated flooding both upstream and downstream from the dam.

Natural dams may also fail. It is common, especially in mountainous terrain, for debris from large landslides to block the narrow part of a valley, thereby creating a natural reservoir. Natural dams made of weathered, unconsolidated material are likely to collapse eventually, causing flooding in downstream areas. Lakes in the summit craters of volcanoes are highly susceptible to breaching. Valleys at high elevations can be temporarily blocked by ice; when spring melting occurs, the blockage may be breached and lake waters released.

A curious landscape in eastern Washington, called the Channeled Scablands (Fig. 8.4), provides evidence that a prehistoric flood of catastrophic proportions occurred when a large glacially impounded lake was suddenly released. During the last glaciation, part of the ice sheet covering western Canada dammed a huge lake in the vicinity of Missoula, Montana. The lake, which contained between 2000 and 2500 km³ of water when it was filled, remained in existence only as long as the ice dam was stable. When the glac-

▲ F I G U R E 8.4
The Channeled Scablands of the Columbia Plateau.
Giant ripples formed by raging floodwaters as they
swept around a bend of the Columbia River. Composed of coarse gravel, the ripples are up to several
meters high and their crests as much as 100 m apart.

ier retreated or began to float in the rising lake water, the dam failed and the water was released from the basin, flowing across the Channeled Scablands region and down the Columbia River to the sea. This catastrophic flood created such landforms as huge current ripples, dry canyons with abrupt cliffs marking the sites of former waterfalls, and massive piles of gravel containing huge boulders.

RIVER SYSTEMS

A body of water that flows downslope along a clearly defined natural passageway, transporting particles and dissolved substances, is called a **stream.** A **river** is a stream of considerable volume with a well-defined passageway. This discussion is concerned primarily with river flooding. However, the processes described are generally applicable to stream systems of all sizes. By studying the dynamics of streams and rivers we can begin to understand some of the factors that contribute to flooding.

In addition to their immediate practical and aesthetic importance, rivers are vital geologic agents; they carry most of the water that goes from land to sea and hence are an essential part of the hydrologic cycle. Rivers and streams of all sizes transport billions of tons of sediment to the oceans each year. They also carry soluble salts released by weathering; these play an essential role in maintaining the saltiness of seawater. Rivers shape the surface of the Earth; most landscapes consist of river valleys separated by higher ground and have been produced by weathering, mass-wasting, and river erosion working in combination.

Channels

The passageway of a river is its **channel,** and the particles of sediment and dissolved matter it carries with it constitute the bulk of its **load.** The quantity of water passing by a given point on the bank within a given interval of time is the river's **discharge.** Discharge varies both along the channel and over time, mainly because of changes in precipitation.

A river's channel is an efficient conduit for running water. In response to variations in discharge and load, the channel continuously adjusts its shape and orientation. Therefore, a river and its channel are dynamic elements of the landscape. The dimensions and characteristics of a river's channel determine its ability to handle the flow of water. Developing an understanding of channel dynamics is therefore an important step toward understanding and predicting river flooding.

The size and shape of a channel's cross-section reflect the river conditions prevailing at that point. Very small streams may be as deep as they are wide, whereas very large rivers are usually many times wider than they are deep. Because the volume of water moving through a channel generally increases with increasing distance from the source, the ratio of channel width to depth is also likely to increase.

If we measure the vertical distance that a channel falls between two points along its course, we obtain a measure of a stream's **gradient.** The average gradient of a steep mountain stream, such as the Sacramento River in California, may reach 60 m/km or even more, whereas near the mouth of a very large river, such as the Missouri, the gradient may be as low as 0.1 m/km or even less (Fig. 8.5).

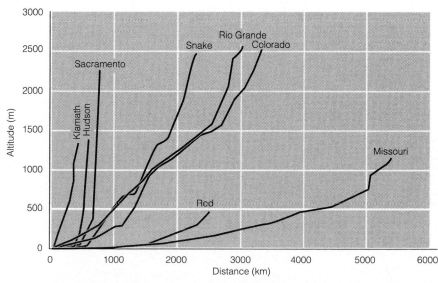

▲ F I G U R E 8.5
Long profiles of some rivers in the United States. The Klamath, Hudson, and Sacramento are relatively short, steep-gradient streams, whereas the Missouri is a long river with a low average gradient.

A river's *long profile* (a line drawn along the river's surface from its source to its mouth) is a curve that decreases in gradient with distance from the source (Fig. 8.5). However, the curve is not perfectly smooth because of irregularities in gradient along the channel. For example, a local change in gradient may occur at the point where a channel passes from a bed of resistant rock into one that is more erodible, or when a landslide or lava flow forms a temporary dam across the channel. A hydroelectric dam also introduces an irregularity in the long profile of a river channel and may create an extensive reservoir upstream.

Channel Patterns

From an airplane, it is easy to see that no two rivers are alike. The variety of channel patterns can be explained if we understand the relationships among river gradient, discharge, and sediment load. In this section we explore some of the most common types of river channels.

Straight Channels

Straight channel segments are rare. Generally they occur for only brief stretches before the channel begins to curve. Close examination of a straight segment of natural channel shows that it has some of the features of *sinuous* (curving) channels. For example, a line connecting the deepest parts of a straight channel typically wanders back and forth across the channel (Fig. 8.6A). This may initially be due to random variations in channel depth. However, when the deepest water lies at one side of a channel, a deposit of sediment (a *bar*) tends to accumulate on the opposite side, where the velocity of flow is lower. The sinuous flow of water within a channel thus causes a succession of bars to form on alternating sides of the channel.

Meandering Channels

In many rivers the channel follows a series of smooth bends like the switchbacks of a mountain road (Fig. 8.6B). Such a bend is called a *meander,* after the Menderes River (Latin *Meander*) in southwestern Turkey, which is noted for its winding course. Meanders occur most frequently in channels that lie in fine-grained river sediments and have gentle gradients. The meandering pattern reflects the way the river minimizes resistance to flow and dissipates energy as uniformly as possible along its course.

If you try to wade or swim across a meandering river, you will soon realize that the velocity of the flowing water is not uniform in all parts of the river. Velocity is lowest along the bottom and sides of the channel because that is where the flow encounters the greatest frictional resistance. Along a straight segment of the channel, the highest velocity is usually found near the surface in midchannel. Wherever the water rounds a bend, however, the zone of highest velocity swings toward the outside of the channel. This type of variation makes it very difficult to obtain accurate measurements of the average velocity of flow in a large river like the Mississippi.

Over time, meanders migrate slowly down a valley. As water sweeps around a bend, strong turbulence causes undercutting and slumping of sediment where the fast-moving water meets the steep bank. Meanwhile, along the inner side of the bend, where the water is shallow and the flow is less rapid, coarse sediment accumulates to form a *point bar* (Fig. 8.6B). As a result of these processes, meanders slowly change shape and shift position as sediment is subtracted from and added to their banks. Sometimes the water in the channel manages to find a shorter route downstream, cutting off a meander. As sediment is deposited along the margin of the new channel, the bypassed meander is transformed into a curved *oxbow lake* (Fig. 8.7).

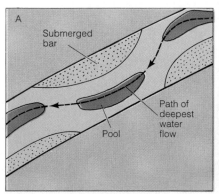
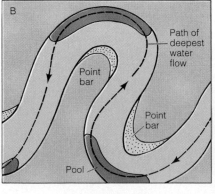
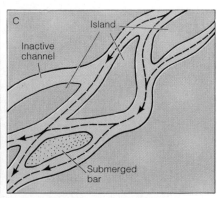

Straight channel   Meandering channel   Braided channel

▲ **F I G U R E 8.6**
Features associated with (A) straight, (B) meandering, and (C) braided channels. Pools are places along a channel where the water is deepest. Arrows indicate the direction of flow and trace the path of the deepest water.

▲ **FIGURE 8.7**
A meandering river near Phnom Penh, Cambodia, flows past agricultural fields that cover the river's floodplain. Light-colored sandy bars lie opposite cutbanks on the outsides of the meander bends. Two oxbow lakes, the products of past meander cutoffs, lie adjacent to the present channel.

Nearly 600 km of the Mississippi River's channel have been abandoned through cutoffs since 1776. However, contrary to the analysis offered by Mark Twain at the beginning of the chapter, the river has not been shortened appreciably, because the loss of channel resulting from cutoffs has been balanced by lengthening of the channel as other meanders have grown.

Braided Channels

The intricate geometry of a *braided stream* resembles the pattern of braided hair, for the water repeatedly divides and reunites as it flows through two or more interconnected channels separated by bars or islands (Fig. 8.6C). Braiding is related to a river's ability to transport sediment. If a river is unable to move all the available load, it tends to deposit the coarsest sediment in the form of a bar that divides the flow and concentrates it in the deeper segments of the channel on either side. As the bar builds up, it may emerge above the surface as an island and become stabilized by vegetation that anchors the sediment and inhibits erosion.

A braided pattern tends to form in rivers with a highly variable discharge and easily erodible banks that can supply abundant sediment to the channel system. It seems to represent an adjustment by which a river increases its efficiency in transporting sediment. Large braided rivers typically have numerous constantly shifting shallow channels (Fig. 8.8). Although at any given time the active channels may cover no more than 10 percent of the width of the entire channel system, within a single season all or most of the surface sediment may be reworked by the laterally shifting

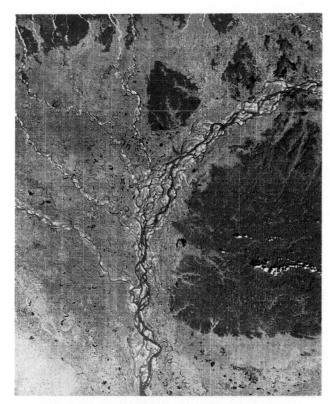

▲ **FIGURE 8.8**
The intricate braided pattern of the Brahmaputra River where it flows out of the Himalaya en route to the Ganges Delta. Noted for its huge sediment load, the river may be 8 km wide during the rainy monsoon season.

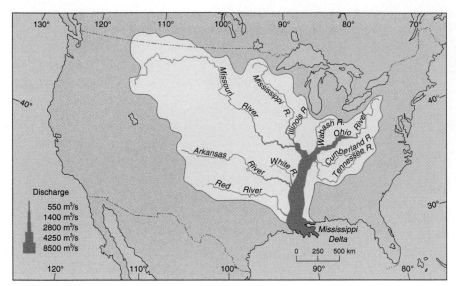

◄ F I G U R E 8.9
The drainage basin of the Mississippi River encompasses a major portion of the central United States and extends into southern Canada. In this diagram the width of the river and its major tributaries reflect discharge values, that is, the average amount of water carried by each portion of the river per unit of time (given in m³/s).

channels. The continuously shifting braided channels of the Brahmaputra River in the Ganges Delta represent a major flood hazard for residents of that densely populated region.

Drainage Basins and Divides

Every river is surrounded by its **drainage basin,** the total area that contributes water to the river. The line that separates adjacent drainage basins is a **divide.** Drainage basins range from very small (less than a square kilometer) to vast areas of subcontinental dimension. The drainage basin of the Mississippi River encompasses more than 40 percent of the area of the contiguous United States (Fig. 8.9). Not surprisingly, the area of a drainage basin is proportional to both the length and the mean annual discharge of the river into which its waters drain.

As a drainage system develops, details of its pattern change. Rivers acquire new *tributaries*—smaller streams that flow into it. Some old tributaries are lost as a result of *capture,* the interception and diversion of one stream by another. When capture occurs, some river segments are lengthened and others are shortened. Just as the river channel and its discharge and load are constantly changing, so too is the drainage system. Like a river channel, a drainage system is a dynamic system tending toward a condition of equilibrium.

Not surprisingly, there is a close relationship between drainage patterns and the nature of the underlying rocks. The ease with which a particular rock is eroded depends on its structure and composition. The course a river takes across the land is therefore strongly influenced by these factors. Structures such as faults and fractures in the underlying rocks can influence the direction of flow and the type of

drainage pattern that develops. Fig. 8.10 shows some of the most common drainage patterns and the geologic factors that control them.

Dendritic

Irregular branching of channels ("treelike") in many directions. Common in massive rock and in flat-lying strata. In such situations, differences in rock resistance are so slight that their control of the directions in which valleys grow headward is negligible.

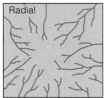

Radial

Channels radiate out, like the spokes of a wheel, from a topographically high area, such as a dome or a volcanic cone.

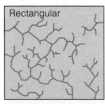

Rectangular

Channel system marked by right-angle bends. Generally results from the presence of joints and fractures in massive rocks of foliation in metamorphic rocks. Such structures, with their cross-cutting patterns, have guided the directions of valleys.

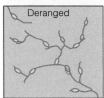

Deranged

Streams show complete lack of adjustment to underlying structural or lithologic control. Characteristic of recently deglaciated terrain whose preglacial features have been remodeled by glacial processes.

▲ F I G U R E 8.10
Some common drainage patterns and their relationship to rock type and structure.

Dynamics of Streamflow

If you stand outside during a heavy rain, you can see that water initially tends to move down slopes in broad, thin sheets, a process called *overland flow*. After traveling a short distance, overland flow begins to concentrate into well-defined channels, thereby becoming **streamflow.** Streamflow and overland flow together constitute **runoff,** the portion of precipitation that flows over the surface of the land. The factors that control runoff determine how much of the water that falls as precipitation eventually reaches a given channel; the dynamics of streamflow determine how well that channel can handle the water.

Several basic factors control the way a river behaves. The most important are gradient, the channel's cross-sectional area (width × average depth), the average velocity of water flow, discharge, and load. Unlike the sediment that forms the bulk of a river's load, dissolved matter generally does not affect the behavior of a river. The relationship among discharge, velocity, and channel shape is expressed by the following equation:

$$\underset{\substack{\text{Discharge} \\ (m^3/s)}}{Q} = \underset{\substack{\text{Cross-sectional area} \\ (m^2)}}{a} \times \underset{\substack{\text{Average} \\ \text{velocity} \\ (m/s)}}{v}$$

This equation is critical to understanding streamflow and, therefore, flooding. It tells us that when discharge changes, one or more of the factors on the right side of the equation must also change if equilibrium is to be maintained. For example, if discharge increases, both velocity and channel cross-sectional area are likely to increase so that the channel can accommodate the added flow. Conversely, a decrease in discharge can lead to a corresponding decrease in channel dimensions and velocity.

Flooding can produce dramatic changes in these factors. In one 6-month period the channel of the Colorado River at Lees Ferry, Arizona, underwent major changes in dimensions as discharge increased and then fell (Fig. 8.11). Before this flood, the channel averaged about 2 m in depth and 100 m in width. As discharge increased in late spring, the water rose in the channel and erosion scoured the bed until the channel was about 7 m deep and 125 m wide. With an associated increase in velocity, the enlarged channel was now able to accommodate the increased flood discharge and carry a greater load. As discharge fell, the river was unable to transport as much sediment and the excess load was dropped in the channel, causing its floor to rise. At the same time, decreasing discharge caused the water level to fall, thereby returning the cross-sectional area to its pre-flood dimensions.

RIVER FLOODING

Now that a basic understanding of river systems and streamflow has been established, let's look more closely at the processes involved in river flooding.

Flood Stage

Hydrologists use the term **stage** to refer to the height of a river (or any body of water) above a locally defined reference surface. When a river's discharge increases as a result of prolonged or intense precipitation or meltwater from a spring thaw, its channel may fill completely, so that any further increase in discharge results in water overflowing the banks of the river. In this condition the river is said to have reached **bankfull stage,** or **flood stage.**

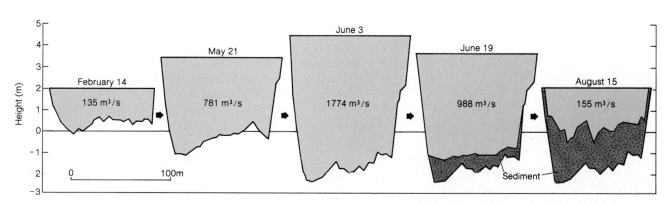

▲ F I G U R E 8.11
Change in the cross-sectional area of the Colorado River at Lees Ferry, Arizona, during a 6-month period in 1956. As discharge increased from February to June, the channel floor was scoured and deepened and the water level rose higher against the banks. During the falling-water phase, from June to August, the river level fell and sediment was deposited in the channel, decreasing its depth.

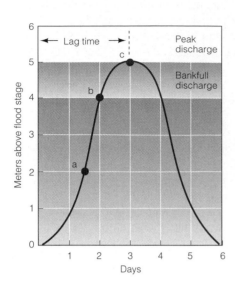

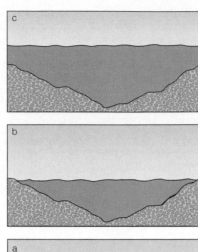

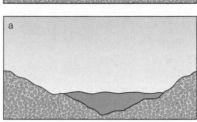

◀ F I G U R E 8.12
A hypothetical example of a flood. The discharge hydrograph (discharge plotted against time after the beginning of the storm) shows the discharges measured over several days at a single hydrologic gauging station. Point *a* shows preflood discharge, *b* shows the discharge at bankfull stage, and *c* shows peak discharge. The cross-sections and water levels in the channels at each of these times are shown for comparison. The peak or crest of the flood occurred about 3 days after the storm began (lag time), and the flood crested at 1 m above the local flood stage.

The discharge of a river is typically displayed on a **hydrograph,** a graph in which river discharge (Q) is plotted against time. Figure 8.12 shows a plot of discharges measured at one gauging station over a period of 6 days. The three channel cross-sections show what the river's cross-section and stage would have looked like on the 3 days corresponding to the points marked on the hydrograph. In this hypothetical example, the river rose from its normal level (at point *a*) to flood stage (*b*). It then exceeded flood stage (*c*), flowing over the banks and onto adjacent land and reaching its maximum discharge about 3 days after the beginning of the storm. The time that elapses between the

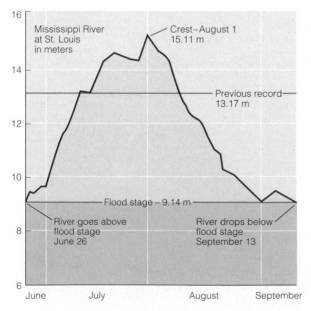

◀ F I G U R E 8.13
A discharge hydrograph for the Mississippi River flooding of 1993, taken from the *St. Louis Post-Dispatch.* The hydrograph shows the river cresting on August 1 at 49.58 feet (more than 15 m), 19.58 feet (about 6 m) above the local flood stage of 30 feet (just over 9 m). The river did not drop below flood stage until September 13, and local flooding continued even after that date.

onset of precipitation and the peak flood stage (3 days in Fig. 8.12) is called **lag time.** The lag time makes it possible to track the progress of the storm and prepare for the flood or evacuate the area if necessary.

The discharge measured when the river is at bankfull stage (point *b* on the hydrograph in Fig. 8.12) is referred to as **bankfull discharge.** Point *c*, at which the maximum discharge for this particular event was reached, is called the **peak discharge.** If you were listening to a report of this flood on the news, you might hear that "the river is expected to peak (or *crest*) tomorrow at 5 meters." This means that the peak discharge is projected to be 5 m vertically above the local reference surface, or 1 m above flood stage as shown on the hydrograph. On August 1, 1993, the Mississippi River crested at more than 15 m at St. Louis, or 6 m above the local flood stage of just over 9 m (Fig. 8.13).

Floodplains and Levees

During a major flood, the water overflowing a river's banks inundates the adjacent **floodplain** (Fig. 8.14). The fine silts and muds that settle out of floodwaters to form the broad, flat floodplain produce the deep, arable soils that make floodplain development so attractive. The size of a river's floodplain is a function of the size of the river itself. For example, the floodplain of the Mississippi River encompasses some 80,000 km^2; a smaller stream develops a much smaller floodplain.

Often the boundary between a channel and its floodplain is the site of a *natural levee*—a broad, low ridge of fine sediment built along the side of the channel by debris-laden floodwater. Natural levees are created only during floods that are high enough to submerge the river's banks completely and flow over them onto the floodplain. As sediment-laden water flows over the banks, its depth, velocity, and turbulence decrease abruptly. This results in sudden, rapid deposition of the coarser part of the river's load along the margins of the channel, creating the natural levee. Farther away, finer silt and clay settle out of the quieter water covering the floodplain.

Precipitation and Infiltration

As mentioned earlier, most flooding involves a period of exceptionally long or intense precipitation. Of particular significance is the balance between precipitation and infiltration. When it rains, some of the water is intercepted either by impermeable surfaces or by vegetation. Much of the intercepted water is returned to the atmosphere directly through evaporation and transpiration. Some of the rainfall enters the ground through the process of infiltration. The remainder flows over the surface as runoff, that is, overland flow and streamflow. Runoff thus can be expressed as follows:

Runoff = Precipitation − Infiltration − Interception − Evaporation

During a storm, problems arise when the ground's *infiltration capacity*—that is, the maximum rate at which water can be absorbed—is exceeded. When that happens river channels and other surface depressions fill to overflowing and flooding occurs. The nature, magnitude, and likelihood of flooding therefore depend not only on the amount and distribution of rainfall but also on factors affecting the infiltration capacity of the ground.

Rainfall Distribution: Upstream and Downstream Flooding

The distribution of rainfall over the area affected by a storm is sometimes depicted in the form of an **isohyetal map** (an *isohyet* is a line connecting points of equal precipitation). The intensity of rainfall in a particular area is plotted using contours of equal rainfall depth, either for a specified time interval or for the entire duration of the storm. An isohyetal map of the upper part of the Mississippi River Basin is shown in Fig. 8.15. In this example, rainfall during the first half of 1993 is expressed in terms of the percentage by which it exceeded the 30-year average rainfall for the same period (January to July). The isohyetal map is superimposed on a map showing the general area where flooding occurred during that period.

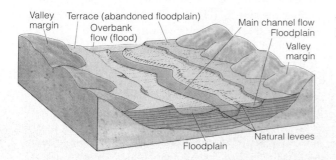

Valley margin | Terrace (abandoned floodplain) | Overbank flow (flood) | Main channel flow | Floodplain | Valley margin | Floodplain | Natural levees

◀ **F I G U R E   8.14**
Main features of a river valley.

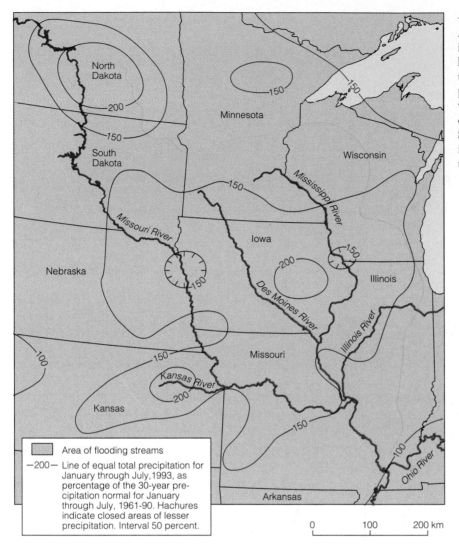

◄ F I G U R E 8.15
An isohyetal map showing rainfall in the upper Mississippi River basin during the period from January to July 1993. Rainfall is expressed in terms of the percent by which it exceeded the 30-year average rainfall for the same period. Superimposed is the general area in which flooding occurred during the summer of 1993.

Passing storms often generate brief intervals of intense rainfall. As the runoff from such a storm moves into the river channel, the river's discharge quickly rises. The crest of the resulting flood, the point at which peak discharge is reached, passes the gauging station about 2 hours after the storm has passed. It takes an additional 8 hours for the flood runoff to pass through the channel and for discharge to return to the normal level. This type of storm—intense, but infrequent and of short duration—often causes flooding that is severe but local in extent. This is called **upstream flooding,** because the effects of the storm runoff usually do not extend to the larger streams joining the system farther downstream (Fig. 8.16A).

Upstream flooding can be very hazardous. Such floods typically have a short lag time (the interval between maximum rainfall intensity and the peak of flooding). Floods in which the lag time is exceptionally short—hours or minutes—are called **flash floods.** A devastating flash flood occurred in the Black Hills of South Dakota in 1972, when 40 cm of rain, almost the total yearly average, fell in a period of under 6 hours. In this brief but intense event, 2932 people were injured and 237 killed; 750 homes were destroyed and as many as 2000 cars were damaged by floodwaters.

Storms that continue for a long time and extend over an entire region are responsible for **downstream flooding.** In this type of flooding, the total discharge increases downstream as a result of increased flows in the many tributaries joining the system (Fig. 8.16B). Downstream flooding is often intensified by spring meltwater. The 1993 Mississippi River flood was a case of downstream flooding. Persistent heavy rains began in the spring and continued on an almost daily basis throughout much of the summer. Precipitation in upstream areas, including Iowa, Nebraska, and the Dakotas, greatly exacerbated flooding downstream in Missouri and Illinois. The total area inundated by the flooding of the Mississippi and Missouri rivers and their tributaries has been estimated at more than 40,000 km².

A.

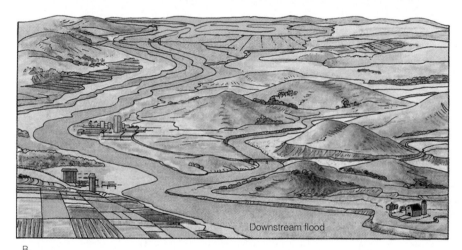

B.

◄ F I G U R E 8.16
A. Upstream floods are generally local in extent, usually with short lag times. They typically result from intense storms of short duration. B. Downstream floods are regional in extent, with longer lag times and higher peak discharges. They result from regional storms of long duration or extended periods of above-normal precipitation.

Factors Affecting Infiltration

The capacity of the ground to absorb rainfall is influenced by a variety of factors. Of particular importance are the physical characteristics of underlying soil and rock formations. These include particle size and type (which in turn influence the porosity and permeability of the material), amount and type of vegetation, the amount of moisture and organic material in the soil, and the degree to which the material has been fragmented or disrupted by roots or burrowing animals. Climatic factors can also influence infiltration capacity. For example, soils in arid and semiarid regions sometimes develop a thick, highly impermeable crust composed primarily of calcium carbonate and variously referred to as *duricrust, hardpan,* or *caliche.* Permanently or seasonally frozen ground, even in the subsurface, also reduces the rate of infiltration.

Of all these factors, soil moisture is arguably the most important. Infiltration capacity may be high at the beginning of a storm, but in periods of intense or extended rainfall the ground can become completely saturated. A wet fall and big winter snows set the stage for the 1993 Mississippi River flood; by the time the heavy spring and summer rains began, large areas of ground were already well saturated, effectively preventing infiltration.

T A B L E 8.1 • **Fatalities from Some Disastrous Floods**[a]

River	Type of Flood	Date	Fatalities	Remarks
Huang He, China	River flooding	1887	ca. 900,000	Flood inundated 130,000 km^2 and swept many villages away
Johnstown, Pennsylvania	Dam failure	1889	2200	Dam failed. Wave 10–12 m high rushed down valley
Yangtze, China	River flooding	1911	ca. 100,000	Formed lake 130 km long and 50 km wide
Yangtze, China	River flooding	1931	ca. 200,000	Flood extended from Hankow to Shanghai (>800 km), leaving 10s of millions homeless
Vaiont, Italy	Dam failure	1963	2000	Landslide into lake caused wave that overtopped dam and inundated villages below

[a]Source: *Encyclopedia Americana* (1983).

Factors related to human use are critical in determining the absorptive characteristics of the ground. The primary effect of development is to increase runoff by increasing the amount of impervious cover, that is, roads, buildings, parking lots, and other structures that are impermeable to water.

HAZARDS ASSOCIATED WITH FLOODING

We can divide the hazards associated with flooding into primary, secondary, and tertiary effects. The primary impacts of flooding are those caused by actual contact with flowing water; they include death by drowning, structural damage to buildings, crop loss, and erosion of flooded areas. The secondary and tertiary effects are longer term impacts that are indirectly related to the flooding. As shown in Tables 8.1 and 8.2, river flooding can be disastrous to human interests, causing both loss of life and extensive property damage.

Primary Effects

Major floods can have a truly devastating impact on humans. The Huang He in China (Fig. 8.17), sometimes called the Yellow River because of the yellowish-brown color produced by its heavy load of silt, has a long history of disastrous floods. In 1887, the river inundated 130,000 km^2 and swept away many villages in the heavily populated floodplain. In August 1931, another flood killed a staggering 3,700,000 people.

The strength of flood currents can be surprising. People affected by flooding often express astonishment at the incredible force exerted by the flowing water, which may tear out trees by their roots or knock buildings off their foundations. Somehow one expects floodwaters to spread peacefully across the floodplain. Unfortunately, the twisted

T A B L E 8.2 • **Some Primary, Secondary, and Tertiary Impacts of the 1993 Flooding of the Mississippi**[a]

Primary impacts
Water damage to household items
Structural damage to buildings
Destruction of:
  Roads
  Rail lines
  Bridges
  Engineered structures (levees, etc.)
  Boats, barges, moorings
Historical sites destroyed
Crop loss
Cemeteries flooded, graves disrupted
Loss of life

Secondary and tertiary impacts
Destruction of farmlands
Destruction of parklands and wildlife habitat
Health impacts:
  Disease related to pollution
  Injuries (back, electric shock, etc.)
  Fatigue
  Stress, depression
Disruption of transportation services
Gas leaks
Disruption of electrical services
Lack of clean water
Impacts on crop prices; food shortages
Job loss and worker displacement
Economic impacts on industries:
  Construction (beneficial impact)
  Insurance (negative impact)
  Legal (beneficial impact)
  Farming (negative impact)
Misuse of government relief funds
Changes in river channels
Collapse of whole community structures

[a]This is not a comprehensive list but rather a sampling taken from newspaper reports during the event.

◀ F I G U R E 8.17
A large suspended load of sediment gives the Huang He a very muddy appearance and its English name (Yellow River). The Huang He has a long history of disastrous floods. The Chinese government plans to construct a gigantic dam system to control the flooding and to generate electricity.

wreckage left behind in flooded towns tells a very different story.

As discharge increases during a flood, so does velocity. The increased velocity enables the river to carry not only a greater load but also larger particles. Damage caused by moving debris thus is one of the main hazards associated with flooding. One of the most frightening episodes in the 1993 Mississippi River flooding involved a group of large propane tanks that were torn from their moorings by the force of rising water. The tanks' steel cable moorings snapped, so they were linked to their supports only by connecting pipes. The tanks, which had been filled almost to capacity in order to make them heavier, bobbed and strained at the pipes and began to develop leaks. On August 2 the entire surrounding area was evacuated; residents were given only 3 minutes to leave their homes. The hazard was extremely serious: if a large cloud of propane vapor had ignited, much of south St. Louis could have been obliterated in a fiery explosion. However, when the floodwaters began to recede, emergency crews managed to retether the tanks and repair the leaks.

The collapse of the St. Francis Dam in 1928 provides another example of the exceptional force of floodwaters. When the dam gave way, the water behind it rushed down the valley, moving blocks of concrete weighing as much as 9000 metric tons over distances of more than 750 m.

The massive load of finer particles carried by a river during flooding represents a different type of hazard. After the Mississippi and Missouri river floodwaters had receded, many people in Missouri, Illinois, and Iowa found their homes almost knee-deep in fine river mud. Volunteer crews from all over North America helped homeowners dig out their belongings (Fig. 8.18). Floodwaters can also concentrate garbage, debris, and pollutants, as described in this

◀ F I G U R E 8.18
Cleaning up after the great Mississippi River flood of 1993. When flood waters finally receded they left behind a gooey blanket of muddy sediment.

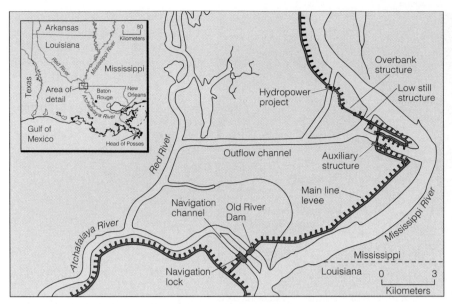

▲ F I G U R E 8.19

The Old River flood-control and channel-control structures at the convergence of the Mississippi and Atchafalaya rivers. The Atchafalaya takes water and suspended sediment from the Mississippi, its "parent," but leaves behind the river's bottom load of sediment. Meanwhile, the Atchafalaya is scouring its bottom, looking for sediment to match its flow. The scouring causes the Atchafalaya's channel to deepen, allowing it to draw even more water from the Mississippi. Eventually, when the flows in the two rivers are approximately equal, a great flood will cause the main channel to switch; the parent channel will quickly silt up and the new channel will enlarge rapidly and irretrievably. The engineering structures at this location are partly intended to avoid this type of occurrence.

eyewitness account of the aftermath of a flood caused by a dam failure in Buffalo Creek, West Virginia, in 1972:

Well, there was mud all over the house, no place to sleep. There was mud in the beds, about a foot and a half of it on the floors. And I had all that garbage. It was up over the windows. In the yard I had poles and trees and those big railroad ties from where they had washed out at Becco—furniture, garbage cans, anything that would float in the water.

Secondary and Tertiary Effects

Disruption of services such as electricity, water, and gas is a common secondary hazard associated with flooding. One of the main challenges during any flood is maintaining the delivery of drinkable water to areas where water supplies have been contaminated. Road, bridge, and rail closures can contribute to the shortage of basic supplies in flooded communities. In regions where emergency response procedures are not well orchestrated, problems such as disease, hunger, and homelessness may result from flooding and the associated disruption of services.

Flooding can also have longer term effects, such as loss of wildlife habitat or permanent changes in local ecosystems. River channels can be permanently changed by flooding, sometimes quite dramatically, as illustrated by the Colorado River example discussed earlier (Fig. 8.11). During the 1993 Mississippi River flooding, a major concern was that the river might actually abandon its main channel in southern Louisiana and seek a shortcut to the Gulf of Mexico via the Atchafalaya River (Fig. 8.19). So far, engineering structures at the juncture of the two channels have prevented this from occurring.

Muddy deposits left by receding floodwaters can clog channels, interfering with transportation or recreational use. Some agricultural fields in the Midwest are now covered by thick deposits of white sand, the remnants of sand bars that built up behind artificial levees. When the levees were breached, the floodwaters scoured deep holes and carried immense loads of sand out onto the fields; some of the fields were irretrievably damaged.

In the long run, however, the silt deposited by floodwaters is extremely beneficial. It replenishes the rich, arable soil that is typical of floodplains. Before the construction of the Aswân High Dam and other dams on the Nile River, annual flooding deposited sediment over the river's flood-plain and delta, adding to the soil at a rate of 6 to 15 cm per century. The construction of the dams eliminated seasonal flooding, but it also cut off the natural source of enrichment and replenishment of the floodplain soil; Egyptian farmers must now use artificial fertilizers and soil additives to keep their land productive.

PREDICTING RIVER FLOODING

There are three main approaches to flood prediction. They involve (1) using statistical techniques to predict the frequency of floods of a given magnitude; (2) using models and mapping to determine the areal extent of hazards associated with floods of a given magnitude; and (3) monitoring the progress of a storm in order to provide a forecast or early warning to those who may be affected by a flood.

Frequency of Flooding

Predictions of the frequency of flooding are based on statistical analyses of records of hydrologic events at a specific locality. For one particular gauging station, a range of discharge values will have been measured over the space of a year; among these, one value will be the *annual maximum discharge*—the highest value measured at that station in that year. A record of annual maxima over a period of years is an annual maximum discharge *series* or *array*.

Recurrence Interval

The first step in the analysis of a maximum discharge series is to rank the annual maxima. Rank is given the notation *m;* the highest annual maximum discharge in the series is given the rank $m = 1$, the next largest $m = 2$, and so on. The smallest annual event thus will have a rank that is equivalent to the number of years (n) represented in the data series, or $m = n$. These values are used to calculate the **recurrence interval (R),** which is the average interval between occurrences of two hydrologic events of equal magnitude. Recurrence interval is determined from the following relationship, called the *Weibull equation:*

$$R = \frac{n + 1}{m}$$

It is important to remember that R represents the *average* interval between two floods of equal magnitude, not the actual *recorded* interval. In other words, calculations based on the existing data series may indicate that a flood with a discharge of 2500 m³/s should recur approximately once every 50 years, but this is just a statistical average; in reality, two floods of that magnitude could occur in successive years, or 10 years apart, or 100 years apart, or even twice in the same year. For example, in 1986–1987 two floods with calculated recurrence intervals of 100 years occurred within one 10-month period in the Chicago area.

A different form of the Weibull equation can be used to determine the probability that a discharge value of given magnitude will be exceeded in any given year:

$$P_e = \frac{m}{n + 1}$$

This is simply the reciprocal of R (i.e., $1/R$). The term P_e is referred to as the *annual exceedance probability*. A discharge with a recurrence interval of, say, 5 years would be expected to occur approximately 10 times over a 50-year time span; in any single year, the probability that a discharge of equal or greater magnitude will occur is given by $1/R$, or 0.20 (20 percent). A discharge with a recurrence interval of 100 years would have an annual exceedance probability of 0.1, so there is a 1 percent probability that a flood of that magnitude will occur or be exceeded in any given year.

Flood Frequency Curve

Flood magnitudes (i.e., discharges) are often plotted with respect to the recurrence interval calculated for a flood of that magnitude at the given locality. This type of graph produces a **flood frequency curve.** Figure 8.20, for example, shows the flood frequency curve determined from the annual maximum discharge series for the Skykomish River at

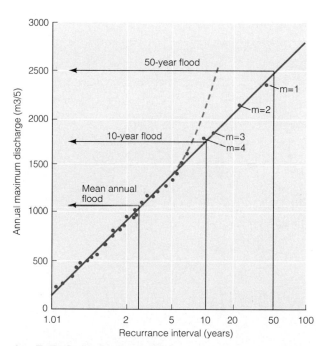

▲ F I G U R E 8.20
Curve of flood frequency for the Skykomish River at Gold Bar, Washington, plotted on a probability graph. A flood with discharge of 1750 m³/s has a 1 in 10 probability of occurring in any year (a 10-year flood), whereas a flood of 2500 m³/s has a 1 in 50 probability (a 50-year flood). The dashed line shows a possible extrapolation of the curve if the annual maximum discharge array from which the graph has been constructed did not include the four largest events. This extrapolation would lead to a misinterpretation of the expected size and recurrence interval for a flood of large magnitude.

Gold Bar, Washington. The curve shows that a flood with a discharge of at least 1750 m³/s has a 1 in 10 (10 percent) probability of occurring in any given year, whereas a flood with a discharge of at least 2500 m³/s has a 1 in 50 (2 percent) probability. We refer to a flood that has a recurrence interval of 10 years as a 10-year flood; if the interval is 50 years, it is a 50-year flood; and so on.

Flood frequency curves can also be used to predict the recurrence interval for events that are of greater magnitude than any historically recorded event. This is done by extending the flood frequency curve. For example, the curve shown in Fig. 8.20 suggests that a flood with a discharge of 2700 m³/s would have a recurrence interval of about 100 years and, therefore, an annual exceedance probability of 1 percent. The longer the available data set, the more valid are the predictions based on the data. A long data set is critical in determining the recurrence interval for very large floods. For example, if the series represented in Fig. 8.20 did not include the four largest events (ranks m = 1, 2, 3, and 4), a projection of the flood frequency curve might suggest a very different recurrence interval and probability for a major flood.

Flood Hazard Mapping

Another important aspect of flood prediction is *flood hazard mapping*, the use of mapping to determine the impact and areal extent of a potential flood. Hazard mapping is concerned with the following types of questions: Given the occurrence of a flood of a specific magnitude in this locality, what will the event be like? How high will the water levels rise? What areas will be under water, and how deep will the water be in those areas? Flood hazard mapping can be based on known, recorded events or on prehistoric events.

Hazard mapping based on known events relies on recorded observations. Probably the most important sources of observational data for recent floods are aerial photographs and satellite images (Fig. 8.21). Remote sensing can provide information not only about the areal extent of floodwaters but also about the amounts of debris and silty deposits left behind after the floodwaters recede. On-site observations are also valuable sources of information concerning both recent floods and earlier ones, for which satellite images may not be available. This type of observation includes, for example, the elevation of high-water

A.

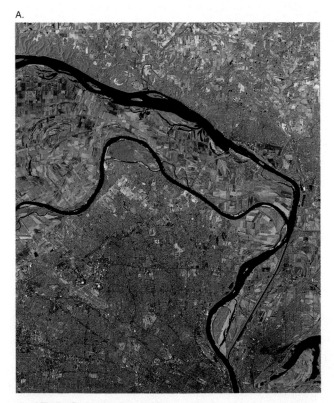

B.

▲ F I G U R E 8.21

A pair of satellite images shows the region where the Missouri River joins the Mississippi River at St. Louis, Missouri, (A) in a typical summer (July 1988) and (B) during the disastrous flood of July 1993, when weeks of torrential rains caused the rivers to overflow protective levees and inundate numerous towns and vast areas of farmland.

marks on the sides of buildings, or the location of a deposit left by floodwaters.

Experts also use physical models of river systems in their efforts to predict floods and the associated hazards. The U.S. Army Corps of Engineers formerly used a 13-km-long, 200-acre scale model of the Mississippi River and its tributaries to predict the system's flooding behavior, but the model became outmoded because of the extent of engineering interventions and modifications in the river system. Computer programs now enable experts to account for many variables in an integrated, coordinated manner; to constantly update their information; to predict the downstream impacts of events such as levee breaks; and to determine the potential effects of various flood management scenarios. Both the Army Corps of Engineers and the National Weather Service were criticized for failing to upgrade their computer modeling capabilities before the 1993 flooding.

Hazard mapping can be extended to encompass floods of greater magnitude than any in recorded history for the area under study. One way to accomplish this is by extrapolating from a known event. For example, if water depths and the area of inundation have been measured for a particular discharge value, the areal extent and depth of a flood of greater magnitude can be extrapolated by calculating the additional volume of water involved, taking into account the local topography. Another approach to flood hazard mapping is **paleoflood hydrology,** the study of ancient floods. Paleoflood hydrology relies on the study of landforms and sediment deposits to determine the areal extent, volume, depth, direction, and rate of flow of ancient floodwaters. The enormous prehistoric flooding in the Channeled Scablands of eastern Washington State was studied in this manner.

Monitoring the Progress of a Flood

So far we have looked at the procedures used to determine the frequency, probability, extent, and characteristics of floods of given magnitudes. Another important aspect of flood prediction is real-time monitoring of storms, which sometimes leads to the issuance of a *forecast* or *early warning* of flooding. Hydrologists use the term *forecast* to refer to a short-term prediction specifying the actual time when the peak or crest of a flood will pass a particular location, as well as the magnitude of the peak. An *early warning* is a statement issued by authorities to the general public. It is based on the forecast as well as on other types of flood hazard information, and it includes recommendations for actions such as evacuation of specific areas.

Forecasting depends on accurate estimates of the expected magnitude of peak discharge. These, in turn, depend on constant monitoring of the extent and distribution of rainfall. Information about the characteristics of the storm must be combined with knowledge of the area, including such features as topography, vegetation, and impermeable cover. This information can be used in predicting the amount of storm runoff, its velocity, and its probable course. The rate at which the peak is moving downstream can be determined by taking successive readings from a series of hydrologic gauging stations. The issuance of an early warning to downstream areas thus takes advantage of the lag time between the beginning of the storm and the arrival of the flood crest. Forecasting is particularly difficult in floods with short time lags, that is, flash floods.

HUMAN INTERVENTION

Flooding is highly susceptible to the effects of human intervention. Urban development and other human activities can worsen or even cause flooding in a variety of ways. People also intervene in natural systems for the purpose of controlling or mitigating flood hazards, and these interventions also have consequences. Therefore, approaches to the prediction and control of flooding must take human factors into account.

Channel Modifications

River channels are modified or "engineered" for many different reasons. Some common goals of channel modification include flood control, enhancing drainage in wetlands, increasing access to floodplain lands for development, facilitating transportation, and controlling erosion. Occasionally the goal is to improve the appearance of the river channel. The modifications usually consist of some combination of straightening, deepening, widening, clearing, or lining of the natural channel. These processes are collectively referred to as **channelization** (or sometimes *channel enhancement*).

In the context of flood control, channelization is generally undertaken with the aim of increasing the channel's cross-sectional area. Given the relationships among discharge, velocity of flow, and cross-sectional area discussed earlier, an increase in the width and depth of the channel should enable it to handle a greater discharge at a higher velocity. Straightening of the channel can also lead to higher flow velocities, thereby also facilitating the passage of more water per unit of time.

The Channelization Controversy

Channelization is a controversial process. Many opponents believe that such modifications interfere with natural habi-

VALMEYER: TRADING FLOODS FOR SINKHOLES

*T*he people of Valmeyer, Illinois, had had enough. After Mississippi River floodwaters inundated their town not once but twice last summer, destroying 90 percent of the homes, offices, and public buildings, the 900 citizens decided to move to higher ground. In October 1993 they asked the Illinois State Geological Survey (ISGS) to examine a new site for geologic hazards.

The new site is a 500-acre tract about 3 km inland from the present town on a limestone bluff 90 m above the Mississippi floodplain. After studying well logs, borehole data, and maps, eight ISGS scientists walked the area, drilled for subsurface samples, ran tests to determine the depth to bedrock and nature of the bedrock surface, and determined the likelihood of collapse at a nearby limestone mine. Four potential types of geologic hazards were identified:

- About 9 m of highly erodible loess on the surface; therefore, the town will have to be careful about water disposal, including storm water and runoff from lawns.

- Karst features, such as sinkholes and other collapsible structures that must be avoided during construction, including 14 possible sinkholes that were mapped, as well as others that probably exist.

- Some radon (radioactive soil gas), but not enough to be a health hazard.

- The remote possibility that the limestone mine will someday extend under the town.

At a town meeting in December 1993 the citizens voted to apply for relocation. Approval came in May 1994; 95 percent of Valmeyer's buildings and homes will be rebuilt at the new site (a few families chose to remain on the floodplain). The site on the floodplain will be graded and seeded with prairie grasses and other native vegetation. The Federal Emergency Management Agency (FEMA) is moving or reconstructing about 6600 structures damaged in the floods, but only a few towns have asked for relocation.

Valmeyer's relocation costs are estimated at $9.6 million; FEMA will provide $7.2 million. As a condition of the grant, the town must adopt drainage, sediment, and erosion controls; a storm-water management plan; construction guidelines; and other mitigation ordinances to guard against damage to the loess, construction around sinkholes, and other actions that might result in activation of known geologic hazards.

---

tats and ecosystems (Fig. 8.22). The aesthetic value of the river can be degraded, the groundwater regime disrupted, and water pollution aggravated by channelization. Paradoxically, it has also been demonstrated that, although channelization may control flooding in the immediate area, it may actually contribute to more intense flooding further downstream.

An interesting example of channelization gone awry is the Kissimmee River in Florida. In 1962 a channelization project was undertaken by the Army Corps of Engineers, primarily for the purpose of flood control. Over a period of 9 years and at a cost of $24 million, the river was straightened and parts of it were lined with concrete. Thirty years later, the river had been degraded aesthetically; drainage problems and water pollution had worsened, with ensuing public health problems; 40,000 acres of wetland habitat had been lost; and the original goal of flood control had not been fully achieved. Now the South Florida Water

Management District has received approval from the Florida government to restore parts of the Kissimmee River to its original condition. Work has already begun on a test segment of the restoration. If the restoration is completed, the final cost will far exceed that of the original channelization project.

Controversy aside, perhaps the most important thing to understand regarding channelization is that *any* modification of a channel's course or cross-section renders invalid any series of hydrologic data collected there in the past. During the Mississippi River floods of 1973 and 1993, the actual water levels were much higher than had been predicted for the discharges recorded during those events. For example, the 1973 flood was referred to as a "200-year flood" on the basis of its extent and water levels, even though it occurred during a period when measured discharges were only at the 30-year level. Most analysts agree that extensive modifications in the river's channel over

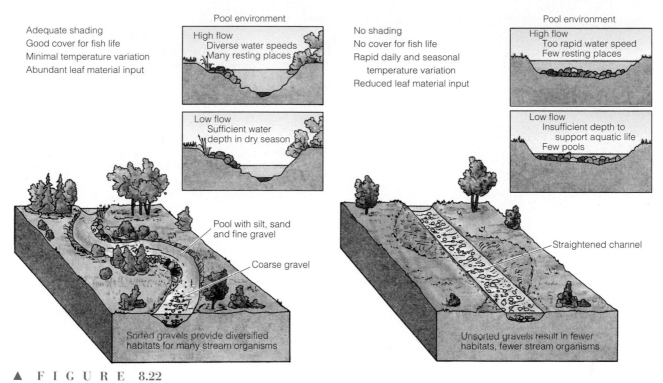

▲ F I G U R E 8.22
Characteristics of a natural river, compared with those of a channelized river.

many years contributed significantly to the failure to predict the extent of damage associated with these floods.

Effects of Development on Flood Hazards

Development can contribute to flooding in a variety of ways. As mentioned earlier, urban construction on unconsolidated, compressible sediments, often accompanied by withdrawal of groundwater, can lead to subsidence and urban flooding. Floodplain development that decreases the cross-sectional area of the river channel can also increase

flood hazards because it reduces the amount of discharge the river can handle (Fig. 8.23).

Although the effects of development, particularly urbanization, can be complex, flood hazards are associated mainly with the problems created by impermeable cover and storm sewer drainage. An increase in the amount of impermeable (or impervious) cover reduces infiltration capacity. In urban areas, impermeable ground cover—buildings, roads, parking lots, and the like—is ubiquitous, and runoff from paved areas can add substantially to total runoff. Storm sewers also can be important contributors to

◄ F I G U R E 8.23
Development on a floodplain can alter the cross-section of the channel, which, in turn, may increase water levels in subsequent floods.

After development   Before development   Increase in flood height

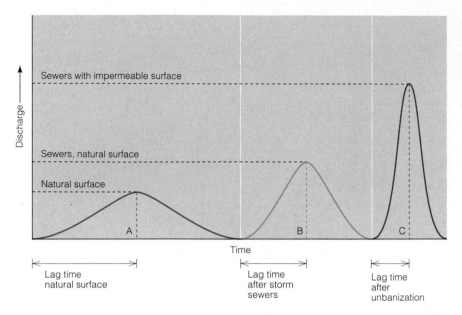

◄ F I G U R E 8.24
Discharge hydrographs for a given hydrologic event on (A) a natural surface, (B) a surface with storm sewers installed, and (C) an impermeable surface with sewers. The increase in impermeable surface and enhanced runoff through storm sewers is associated with a decrease in lag time and an increase in peak discharge and total discharge for the event.

flooding because they allow the runoff from paved areas (sometimes called *urban runoff*) to reach the river channel more quickly.

Because of the effects of development, floods in urban areas differ from floods in natural river valleys in several important respects. Urbanization typically leads to a decrease in lag time, an increase in the peak discharge, and an increase in the total discharge for a particular flood. These relationships are illustrated in Fig. 8.24.

Organized Response to Flood Hazards

Organized response to flood hazards takes two main forms: (1) structural and engineering approaches aimed at controlling flooding and (2) regulations, emergency response programs, and compensation packages aimed at decreasing vulnerability and adjusting to the hazards.

Structural Approaches

A wide range of structural approaches can be used to control flooding or protect human interests from its impacts. In addition to the channel modifications discussed earlier, structural approaches include dams and barrages; retention ponds and reservoirs designed to release drainage from paved areas in a controlled manner; and levees, dikes, and flood walls designed to keep out floodwaters. Sometimes these approaches backfire. In the 1973 Mississippi River flooding, for example, levees preserved the downtown area of St. Louis from damage, while surrounding areas were inundated by floodwaters. The flood walls managed to protect St. Louis again in the summer of 1993, but some experts feel that the flood control system may have worsened flooding both upstream and downstream, essentially squeezing the river into higher crests. This raises several obvious questions: To what extent can or should structural approaches to flood control be carried out? Do such approaches create a false sense of security? Can they worsen problems elsewhere along the river's path? And when does the cost—economic or environmental—of engineering the river channel become prohibitive?

Adjustment and Reduction of Vulnerability

Since the passage in the United States of the National Flood Insurance Act in the late 1960s, with resulting legislation at the state level, significant efforts have been made to implement more nonstructural means of reducing damage from flooding. Nonstructural approaches have gained renewed support since the damaging floods in the Midwest in 1993 and the southeastern United States in 1994, which led many people—experts as well as the general public—to question the effectiveness of structural interventions for flood control. Among the nonstructural approaches now in use are floodplain zoning, specialized building codes, open-space planning in flood zones, floodplain buyout programs, mortgage limitations, and development rights transfers. Federal Emergency Management Agency (FEMA) regulations stipulate that if a home or business owner is compensated for more than 50 percent of the value of a building, the structure must be demolished and the site rezoned to prohibit future development. The goal of all these approaches is to limit floodplain development to applications that are appropriate or adaptable to recurrent inundations, such as recreational uses or certain types of farming. Most experts agree that engineering interventions are still needed, at least for the protection of urban areas already existing in the floodplains. However, calls for the buyout and

HOW FEMA WORKS

*T*he U.S. Federal Emergency Management Agency (FEMA) has two broad mandates: to ensure preparedness for a nuclear attack and to respond to natural disasters such as hurricanes and floods. Until the end of the Cold War in the late 1980s, FEMA's primary focus was on preparedness for nuclear war. Since 1989, however, the emphasis has shifted to responding to natural disasters, largely because of several closely timed catastrophes: the Loma Prieta earthquake, the Midwest floods, and Hurricanes Hugo, Andrew, and Iniki.

A military mission has some things in common with a natural disaster mission, such as dealing with the details of evacuation, medical attention, and shelter for victims. However, the National Performance Review (1993) concluded that FEMA's original dual mission policy had diluted the effectiveness of the agency staff and created a barrier to agency communications during recent emergencies. The review recommended that FEMA shift the efforts of its employees and resources from the national security component of their mission to preparation for and response to the consequences of *all* disasters, natural and human generated.

The primary responsibility for disaster response rests with state and local governments. The federal government, through FEMA, offers only supplemental support during an emergency. When a disaster overwhelms the resources of a local or state jurisdiction, FEMA steps in to coordinate the efforts of 26 federal agencies and the Red Cross. As it now exists, however, the system encourages state and local elected officials to apply for maximum federal disaster assistance.

Annual budget appropriations for FEMA total nearly $1 billion ($827 million in 1993), of which one-third is designated for disaster relief and two-thirds for military preparedness. The U.S. Congress has the authority to supplement these funds with special appropriations. For example, Congress appropriated nearly $3 billion in additional funding for FEMA operations in response to Hurricanes Andrew and Iniki in 1992 and $1.7 billion for assistance to areas affected by the Midwest floods in 1993. Requests to FEMA have increased by approximately 50 percent in the past 10 years, and even minor emergencies have been awarded full compensation. For example, 17 states and the District of Columbia were awarded a total of $126 million to cover the costs of snow removal following the storms of March 1993.

The disasters that struck the United States during the early 1990s highlighted the need for communities to increase their general level of preparedness. According to FEMA reports, fewer than 15 percent of the people affected by the Midwest floods had flood insurance, even though most lived on a recognized high-risk floodplain. Better building construction and zoning regulations can also reduce losses during disasters; noncompliance to building codes accounted for $5 billion of the $20 billion insurance costs resulting from Hurricane Andrew.

In response to criticism of its handling of recent national emergencies, FEMA has adopted a broad plan for reorganization. The new organizational design will consolidate the military and natural disaster divisions into an all-hazards approach to emergency management. In addition, to limit federal assistance to situations of real need, FEMA is developing objective criteria for declaring emergencies and "major" disasters. The resulting "disaster hierarchy" will assist in the allocation of federal aid. FEMA also intends to implement a competitive program of state-level grants for emergency preparedness and to enforce existing building codes and insurance requirements for high-risk areas.

restoration of bottom lands to their original wetland state are becoming increasingly common.

Following a particularly devastating flood in Brisbane in 1974, the Australian Minister for Environment and Conservation stated that

Floodplains are for floods, although this is not to say that there is nothing we can do to minimize the damage which flooding can cause. [We] should remember that floods and droughts are in many areas the norm; it is good years which are the exception. . . . There are often perfectly sound reasons for using land subject to flooding for agricultural and other activities. However, it is essential in planning such enterprises that full provision is made for foreseeable losses.

His statement summarizes the philosophy behind the regulation of floodplain development, that is, to obtain full and reasonable use of floodplain resources while maintaining a sensible attitude concerning the hazards associated with flooding. An integrated approach is probably the only realistic way to reduce vulnerability to the risks of living on a floodplain. Such an approach combines floodplain zoning regulations, scientific information and public education about flood hazards, engineered flood-control methods, efficient early-warning and emergency response mechanisms, and compensation for loss caused by flooding.

SUMMARY

1. The advantages of living near water include fertile soil; access to water, food, energy, and other resources; waste absorption and dispersal; and transportation routes. However, flooding can take a high toll in loss of life and damage to structures.

2. The majority of floods are caused or exacerbated by intense precipitation, usually in combination with other geologic factors. Processes that lead to flooding in coastal zones include tsunamis, hurricanes, cyclones, and unusually high tides. Urban development on compressible sediments can lead to subsidence, with associated flooding. Failure of dams and protective structures can also cause flooding.

3. A river's channel is an efficient conduit for the movement of water and sediment. Channels continuously change shape and orientation in response to variations in discharge and load. The dimensions and characteristics of a river's channel determine its ability to handle the flow of water, influencing the nature and extent of flooding.

4. No two rivers are alike. The development of different types of channels—straight, meandering, or braided—depends on variations in gradient, sediment load, and discharge.

5. The area of land drained by a given river—its drainage basin—depends on the river's size and discharge. The type of drainage pattern that develops depends on the composition and structure of underlying rocks.

6. The main factors that control streamflow are gradient, channel cross-sectional area, average velocity of water flow, discharge, and load. Discharge, velocity, and cross-sectional area are interrelated such that when discharge changes, the product of the other two factors also changes.

7. During floods, streams overflow their banks and construct natural levees, which grade laterally into the silt and clay that make up the adjacent floodplain.

8. Typically there is a lag between the onset of precipitation and the peak of flooding downstream. This makes it possible to track the progress of the storm and make preparations for flooding. The lag time is typically shorter in upstream flooding than in downstream flooding.

9. When the rate or amount of precipitation exceeds the ground's infiltration capacity, runoff is increased and flooding may occur. The nature, magnitude, and likelihood of flooding depend on the amount and distribution of rainfall and factors that affect infiltration capacity.

10. The primary impacts of flooding are those caused by actual contact with flowing water, including death by drowning, structural damage to buildings, crop loss, and erosion of flooded areas. The secondary and tertiary effects are longer term impacts indirectly caused by the flooding, such as interruption of services, permanent changes in the river channel, or loss of wildlife habitat.

11. Prediction of flood frequency is based on statistical analysis of hydrologic records. Such records permit the calculation of recurrence intervals and annual exceedance probabilities for a discharge of given magnitude. These are plotted on flood frequency curves, which allow for extrapolation to events of greater magnitude than those represented in the historical records.

12. Flood hazard mapping, which is based on paleoflood hydrology and observations of known events, is used to predict the extent of impacts for discharges of a given magnitude. Real-time monitoring of floods allows ex-

perts to forecast the arrival time and magnitude of the flood peak and, if necessary, issue early warnings.

13. Human activities influence flooding in many ways. Development can change runoff patterns and affect infiltration capacity and other characteristics of the ground. Channel modifications can alter the relationship between a channel's cross-section and its discharge, thereby changing flood patterns and invalidating historical records.

14. Structural approaches to flood control include channelization; dams and barrages; retention ponds and reservoirs; and levees, dikes, and flood walls designed to keep out floodwaters. Recently efforts have been made to implement more nonstructural means of flood damage reduction, including floodplain zoning, specialized codes, floodplain buyout programs, mortgage limitations, and development rights transfers.

TERMS TO REMEMBER

bankfull discharge (p. 201)
bankfull stage (p. 199)
channel (p. 195)
channelization (p. 209)
discharge (p. 195)
divide (p. 198)
downstream flooding (p. 202)
drainage basin (p. 198)
flash flood (p. 202)

flood (p. 192)
flood frequency curve (p. 207)
floodplain (p. 201)
flood stage (p. 199)
gradient (p. 195)
hydrograph (p. 200)
isohyetal map (p. 201)
lag time (p. 201)
load (p. 195)

paleoflood hydrology (p. 209)
peak discharge (p. 201)
recurrence interval (p. 207)
river (p. 195)
runoff (p. 199)
stage (p. 199)
stream (p. 195)
streamflow (p. 199)
upstream flooding (p. 202)

QUESTIONS AND ACTIVITIES

1. Has your family ever been affected by flooding? Was it coastal or river flooding? Upstream or downstream? Did human activities have anything to do with causing the flood (as in a dam failure) or worsening it?

2. If you have lived through a devastating flood, write an account of the event. Even if your own family was not affected, it is worthwhile to record your impressions of events that affected your neighbors or nearby communities.

3. Become aware of your own environment. To what extent are the rivers, streams, or coastlines in your area artificially controlled? You don't have to live near a major river or seacoast to answer this question; even small streams are often heavily engineered. Try to discern the motivation for the intervention: Is it primarily for flood control? ease of transportation? erosion control? How successful do you think the intervention has been?

4. What do you think should happen to a family whose home or business is destroyed in a major flood? What if they have no flood insurance? Who should be responsible for the cost of rebuilding or relocating? Should they be allowed to rebuild in the same location?

5. Imagine the discussion that went on in Valmeyer, Illinois, concerning whether or not the town should relocate. What do you think were the main arguments for and against the plan? How might you have voted?

6. Investigate the Federal Emergency Management Agency. What kinds of emergencies besides floods does FEMA respond to? How good is its track record? Do you think past criticisms of FEMA have been valid? Do you think the proposed changes in FEMA's structure are appropriate?

7. Table 8.2 was constructed from newspaper clippings during and after the 1993 Mississippi River flood. Try the same thing yourself the next time a major natural disaster occurs (it doesn't necessarily have to be a flood). Keep track of newspaper, television, and radio reports and make a list of the reported impacts of the event. Then divide your list into primary, secondary, and tertiary hazards.

8. What do you think of the plan to restore parts of the Kissimmee River to their natural state? Will the restoration justify its high cost? The restoration will be an integrated program involving the buyout of privately owned lands along the channel. In what ways might the program's coordinators convince private landowners to cooperate with the restoration project? These would be good topics for a group discussion or report. Start with some individual research into the original goals of the channelization, the negative impacts of the original project, the reasons for undertaking the restoration, and the geologic setting of the south Florida region.

CHAPTER 9

HAZARDS OF OCEAN AND WEATHER

For Hot, Cold, Moist, and Dry, four champions fierce,
Strive here for mastery . . .
• **John Milton (1608–1674)**

*A*round Christmastime along the Pacific coast of South America, the temperature of the ocean and air rises, heavy rainfall begins, and changes in marine life are observed. This weather pattern and its ecological effects have been recognized by fishers since ancient times. They call it *El Niño*—"the child"—after the Christ child, because of the season when the changes begin.

El Niño begins when warmer, less saline tropical water starts moving toward the south along the coast of South America, deflecting the cooler, more nutrient-rich waters of the Humboldt Current that normally flow northward along the coast. In most years, the trend reverses itself by March or April, but once every 3 to 7 years El Niño is more intense, extensive, and prolonged. Ocean waters in the eastern equatorial Pacific become anomalously warm, the trade winds weaken, and the normally westward-flowing South Pacific Equatorial Current weakens or even reverses its flow. These exceptional El Niños can have devastating effects on marine life and on coastal communities that depend on the sea. They also bring torrential rains, drought, and other disruptions of normal weather patterns, whose effects are by no means restricted to the South American coastal region.

The most severe El Niño of this century occurred in 1982–1983. It caused heavy rains in Colombia, Ecuador, and northern Peru, drought in southern Peru and Bolivia, and storms that lashed the western coast of the continent. Its effects were also felt around the Pacific. There were

◀

Fishers on the Pacific coast of South America have poor catches during an El Niño event. These fishers in northern Chile are sorting the day's meager catch.

droughts in Australia, New Guinea, and Indonesia and increased storm activity over the Pacific. Unusual weather patterns in other parts of the world—intensified drought in northeast Brazil and the Sahel region of Africa, warm winters in North America and Europe—may have been caused by El Niño as well.

There are two important lessons to be learned from El Niño. First, complex interactions and feedback loops between the oceans and the atmosphere play a critical role in determining climate and weather; in effect, the oceans and the atmosphere are interdependent elements of the global climate system. Second, El Niño serves as a reminder of human vulnerability to environmental change. When weather patterns fluctuate as dramatically as they do during an exceptional El Niño, the ecologic and economic impacts can be devastating.

THE OCEAN–ATMOSPHERE SYSTEM

In many respects, the oceans and the atmosphere are inextricably linked; they are complementary parts of one huge, complex, dynamic system. They are the two great water reservoirs fundamental to the functioning of the global hydrologic cycle as well as that of other global biogeochemical cycles. The oceans play a critical role in climate and weather systems, particularly in regulating the temperature and humidity of the lower part of the atmosphere. Atmospheric circulation, in turn, drives ocean waves and currents.

In this chapter we describe the atmosphere and the oceans, consider their interrelationships, and show how oceanic and atmospheric processes function together in the formation of weather and climatic patterns. The goal of the chapter is to examine some of the many ways in which weather, waves, currents, and tides affect the lives of people around the world. We focus in particular on geologic processes in coastal zones, which are especially vulnerable to the hazards of ocean and weather.

In this chapter we are concerned with natural oceanic and atmospheric processes that are sometimes damaging to human interests. As with other types of natural hazards, the impacts of some ocean and weather processes—notably desertification and coastal erosion—can be worsened by human activities. Air pollution, ozone depletion, and global warming, which are primarily technological or anthropogenic hazards, are addressed in Chapter 18. Similarly, offshore dumping of wastes and other types of pollution in the marine environment—which are wholly anthropogenic—are discussed in Chapter 17.

Weather and Climate

Weather is an extremely complicated phenomenon. It is much studied but not very well understood and even less successfully predicted. The term **weather** refers to conditions in the atmosphere at any given time, particularly temperature, barometric pressure, humidity, clouds, precipitation, and wind velocity. The Sun's rays do not strike the surface of the Earth with equal force everywhere. Weather fluctuations and the restlessness of the atmosphere simply represent the natural, but never completed, process by which the system attempts to smooth out these differences in surface temperature.

Weather phenomena are local and transient. When the characteristic weather patterns for a given region are averaged over a significant period the term **climate** is used. Thus, the patterns and processes that we think of as weather—wind, rain, snow, sunshine, storms, and even floods and droughts—are temporary, local aberrations when viewed against the more stable, longer term background of global climate.

The "Butterfly Effect"

The interactions among air, land, and water that create weather are not only complex but also highly sensitive to changing conditions. Meteorologists sometimes characterize this sensitivity by saying that a change as small as the draft created by a butterfly's wing can become magnified through a network of feedback systems, ultimately developing into a wind or even a cyclone. They even have a name for this phenomenon: the *"butterfly effect."* The variability, complexity, and extreme sensitivity of atmospheric processes make weather prediction a very tricky business.

THE ATMOSPHERE

The Earth's atmosphere consists of a mixture of gases dominated by nitrogen (N_2) and oxygen (O_2), with small amounts of other gases, water vapor, and clouds of water droplets. Scientists describe the atmosphere as composed of

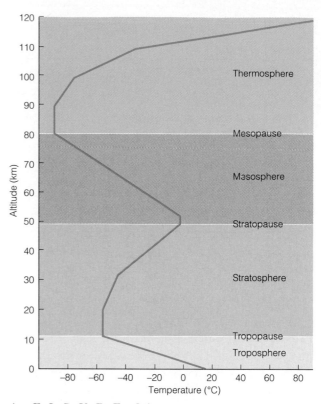

▲ F I G U R E 9.1
The atmosphere is divided into four zones on the basis of variations of temperature with altitude. The outermost zone, the thermosphere, continues to an altitude of about 700 km.

several layers, each with its own physical, chemical, and temperature characteristics (Fig. 9.1). From the bottom up, these layers are: (1) the *troposphere,* from the ground surface to approximately 15 km; (2) the *stratosphere,* up to 50 km; (3) the *mesosphere,* up to 90 km; and (4) the *thermosphere,* up to 700 km. Most weather-related phenomena originate in the lowest layer, the **troposphere**. The troposphere also contains about 90 percent of the actual mass of the atmosphere, including virtually all the water vapor and clouds. Although little mixing occurs between the troposphere and the layers of the atmosphere above it, the troposphere itself is extremely dynamic and thoroughly mixed.

Circulation in the Atmosphere

The reason the troposphere is so dynamic is that more of the Sun's heat is received per unit of land surface near the equator than near the poles. This unequal heating gives rise to *convection currents.* The heated air near the equator expands, becomes lighter, and rises. Near the top of the troposphere it spreads outward in the direction of both poles. As the upper air travels northward and southward, it gradually cools, becomes heavier, and sinks. On reaching the Earth's surface, this cool, descending air flows back toward the equator, warms up, and rises, thereby completing a convective cycle (Fig. 9.2A). In **meteorology** (the study of the Earth's atmosphere and weather processes), the term

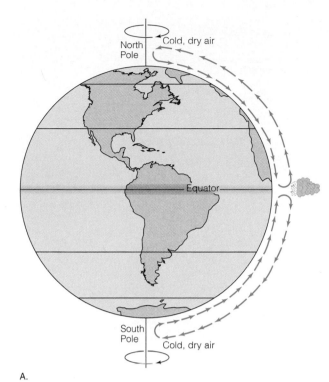

A.

◀ F I G U R E 9.2
A. Global circulation as it would happen on a nonrotating Earth. Huge convection cells would transfer heat from equatorial regions, where the input of solar energy is greatest, to the poles, where the solar input is least. B. The Earth's global wind systems, shown schematically. Because the Earth is not stationary, the flow of air toward the poles and the return flow toward the equator are influenced by the Coriolis effect. Convection does operate, but the flow is not as simple as the case shown for a nonrotating Earth. Moist air, heated in the warm equatorial zone, rises convectively and forms clouds that produce abundant rain. Cool, dry air descending at latitudes 20°–30° produces a belt of subtropical high pressure in which many of the world's great deserts lie.

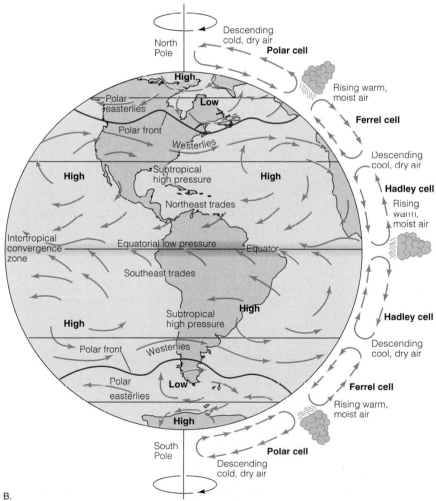

B.

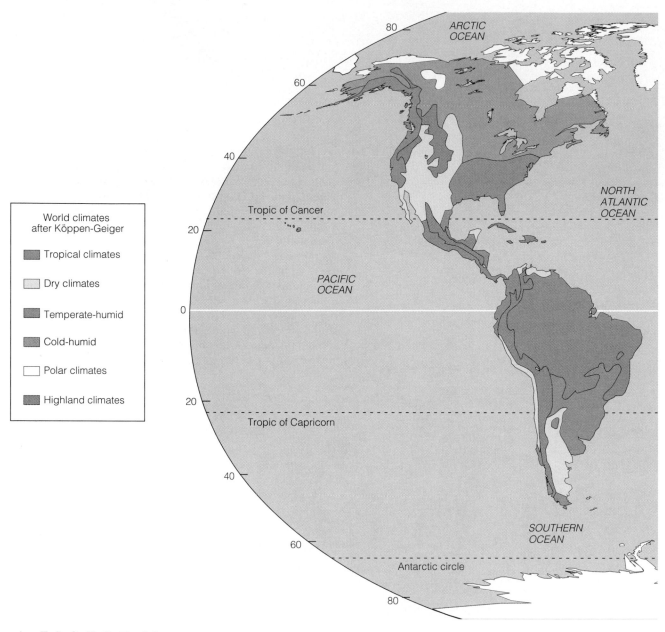

▲ F I G U R E 9.3
**The Köppen climatic classification system. There are five basic types of climates: (A)
tropical climates, (B) dry climates, (C) temperate humid climates, (D) cold humid climates,
and (E) polar climates.**

convection is used specifically to refer to the vertical component of an atmospheric convective cycle. The horizontal
motion of air in the cycle is referred to as *advection*.

The Coriolis Effect

The Earth's rotation modifies what would otherwise be a
simple convective cycle (Fig. 9.2). The **Coriolis effect**,
named after the nineteenth-century French mathematician
who first analyzed it, causes any body that moves freely
with respect to the rotating Earth to veer to the right in the
Northern Hemisphere and to the left in the Southern

Hemisphere. This is true regardless of the direction in
which the body may be moving. Both flowing water and
flowing air respond to the Earth's rotation; the global patterns of ocean currents and wind systems therefore are influenced by the Coriolis effect.

Wind Systems

In the atmosphere, the Coriolis effect breaks up the flow of
air between the equator and the poles into belts (Fig. 9.2B).
The result is a large belt or cell of circulating air lying between the equator and about 30° latitude in both the

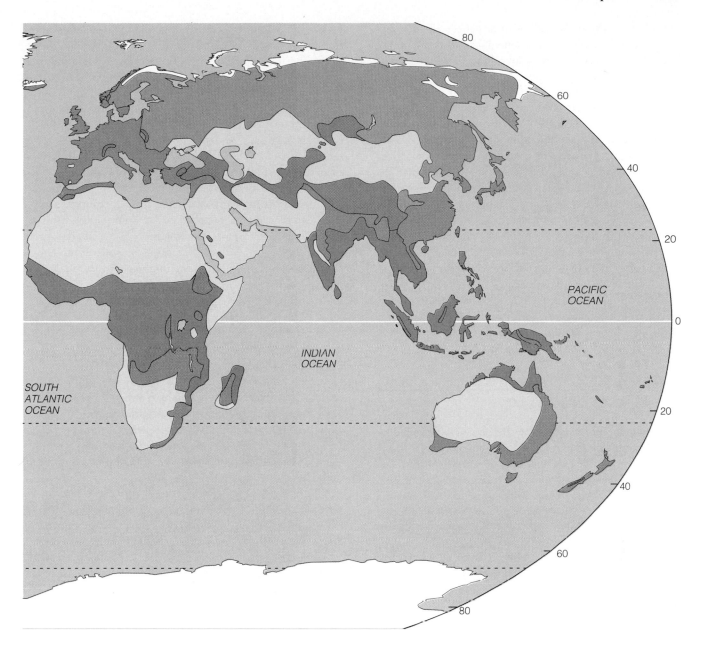

Northern and Southern hemispheres. In these low-latitude cells (called *Hadley cells*), the prevailing winds are northeasterly in the Northern Hemisphere (i.e., they flow *from* the northeast toward the southwest), whereas in the Southern Hemisphere they are southeasterly. These winds are called **trade winds** because their consistent direction and flow carried trade ships across the tropical oceans at a time when winds were the chief source of power.

In each hemisphere, a second cell of circulating air lies poleward of the low-latitude cell. In these middle-latitude cells (called *Ferrel cells*), westerly winds prevail (i.e., blowing *from* the west). Cold upper air flowing toward the equator descends near 20°–30° latitude in both hemispheres, and northward-moving surface air rises in higher latitudes at

the point where it meets dense, cold air flowing from the polar regions.

A third set of circulating air cells (called *polar cells*) lies over the polar regions. In each polar cell, cold, dry, upper air descends near the pole and moves toward the equator in a wind system called the polar easterlies. As this air slowly warms and encounters the belt of westerlies, it rises along the *polar front* (the zone where the polar cells and the Ferrel cells meet) and returns toward the pole.

Effects of Air Circulation on Climate

The global patterns of air flow ultimately control the variety and distribution of the Earth's climatic zones. Those patterns, in turn, are determined by the nonuniform heating of the Earth's surface, the Coriolis effect, the distribution of land and sea, and the topography of the land. For example, the cold upper air converging where the low- and mid-latitude cells meet cannot hold as much moisture as warm air, so at the point where this cold air descends, dry conditions are created at the land surface. As a result, much of the arid land in both the Northern and Southern hemispheres is centered between latitudes 20° and 30°. By contrast, abundant moisture in the warm air surrounding the equator condenses as the air rises and becomes cooler and denser, creating clouds that release their moisture as tropical rains.

If the Earth had no mountains and no oceans to affect the moving atmosphere, the major climatic zones would all lie parallel to the equator. However, the pattern of climatic zones is distorted by the distribution of oceans, continents, high mountains, and plateaus (Fig. 9.3). As a result, average temperature, precipitation, cloudiness, and windiness vary greatly from one place to another and give rise to an array of distinct climatic regions.

OCEANS AND COASTAL ZONES

By now it should be clear that the processes that produce weather and climate are extremely complex. There is a delicately balanced, dynamic interplay between the oceans and the atmosphere that involves a constant exchange of heat, moisture, and gases. Oceanic processes and their influence on precipitation and wind systems have a major influence on local weather fluctuations and regional climatic patterns. In this section we examine the ocean ecosystem and oceanic processes in detail.

The Ocean Ecosystem

Close to 97 percent of the world's surface water is salt water, which circulates throughout the oceans in currents, wave systems, and tides. The Southern Hemisphere is dominated by oceans, and the Northern Hemisphere by land masses. Because of the role of the oceans in regulating atmospheric temperature and moisture, this difference between the hemispheres is very important in determining the distribution of weather and climatic patterns around the world.

The oceans are all interconnected, and water is able to flow from one into the other; in a sense, they can be considered a single huge ecosystem. However, the oceans are far from a uniform environment; it is more precise to view them as a set of different but interconnected ecosystems.

The study of the physical, chemical, biologic, and geologic aspects of the Earth's oceans is called **oceanography.** Although there are many bodies of salt water on the Earth, oceanographers recognize four main ocean basins: the Atlantic, the Pacific, the Arctic, and the Indian oceans. The Pacific is the largest and deepest; it is greater in area than all of the land masses on Earth put together and contains more than half of the world's surface water.

Oceanic Environments

The ocean ecosystem comprises a number of separate but interrelated environments, each with its own physical and biologic characteristics. Offshore from the continental coasts are the shallow, barely sloping *continental shelves* (Fig. 9.4). These typically are fertile environments that support abundant fisheries because of the presence of nutrient-rich sediments that have been washed off the land into coastal waters. The steeper *continental slopes* lead downward to the ocean floor.

The ocean floor itself is remarkably variable in topography. Ocean floor terrains range from the extremely flat *abyssal plains* to rugged underwater volcanic mountains and long, narrow *oceanic trenches* (Fig. 9.4). The deepest spots on the Earth's surface, almost 12 kilometers below sea level, occur within these trenches. The ocean floor is a virtually unexplored source of mineral resources. The depths of the oceans also support unique and exotic life forms, some of which have only recently been observed for the first time. The variability of oceanic environments reflects the fact that the ocean floors are among the most tectonically active regions in the planet's geologic cycle.

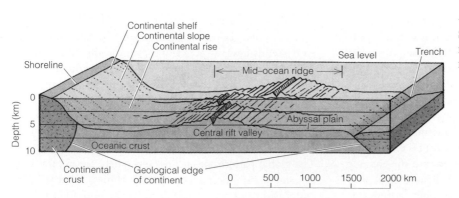

◄ F I G U R E 9.4
Schematic drawing of the ocean floor showing major topographic features.

The Nature of the Coastal Zone

A coastal zone represents the interface between land and sea and typically includes communities that span the biotic spectrum from terrestrial to marine. Physically, the coast is a constantly changing environment. Coastal zones encompass a wide variety of ecosystems and landforms, such as mangrove swamps, estuaries, salt marshes, beaches, barrier islands, rocky cliffs, and coral reefs. Each type of coastal ecosystem exhibits its own physical and biologic dynamics. Each contributes its own unique resources, and each performs environmental services, such as buffering the shoreline against the erosive force of storm-generated waves.

A majority of the world's population lives within 100 km of the ocean. This reflects our dependence on the oceans and the economic benefits afforded by easy access to ocean resources and services, as well as the particular richness of resources in coastal zones. However, the concentration of such large numbers of people in coastal areas means that the coastal environment must absorb the impacts of a wide range of human activities. It also means that human vulnerability to hazards can be particularly high in coastal zones; infrequent events such as large storms can cause major loss of life and damage to property. For these reasons, it is important to understand the special types of geologic processes that characterize coastal zones as well as the hazards of ocean and atmosphere to which the inhabitants of these areas are particularly susceptible.

TIDES, CURRENTS, AND WAVES

If you visit almost any coastal zone on two occasions a year apart, you will see changes. Sometimes the changes are small, but often they are substantial. Large dunes may have shifted; sand may have built up behind barriers, or been eroded away; steep sections of coastline may have collapsed;

channels may have broken through from the sea to lagoons on the landward side, where no channels were before. The energy driving these changes comes from ocean currents and waves, which in turn derive their energy from winds and tides.

Tides and Water Levels

Tides are the cycles of regular rise and fall of the level of water in oceans and other large bodies of water. They result from the gravitational attraction of the Moon, and to a lesser degree the Sun, acting on the Earth. Gravitational attraction causes ocean water to bulge upward on the side of the Earth nearest to the Moon. On the opposite side of the Earth, inertia (the force that tends to maintain a body in uniform linear motion) created by the Earth's rotation also raises a bulge in the ocean, but in the opposite direction. The result is two *tidal bulges,* one on either side of the Earth.

To visualize how tides work, consider the tidal bulges oriented with their maximum amplitude lying along a line extending through the center of the Earth and the center of the Moon, as shown in Fig. 9.5. While the Earth rotates, the tidal bulges remain essentially stationary beneath the Moon. Thus, during a full rotation of the Earth, any given point of land will move westward through both tidal bulges each day. Every time a land mass encounters a tidal bulge, the water level along the coast rises. As the Earth continues to rotate, the coast passes through the highest point of the tidal bulge (high tide) and then the water level begins to fall.

At most places on the margins of the oceans, two high tides and two low tides are observed each day. There are two tidal cycles because any point on a coast passes across both tidal bulges during every complete rotation of the Earth. The Sun's gravitational force also affects the tides, sometimes opposing the Moon by pulling at a right angle and sometimes aiding it by pulling in the same direction. Because the distance from the Earth to the Sun is so much

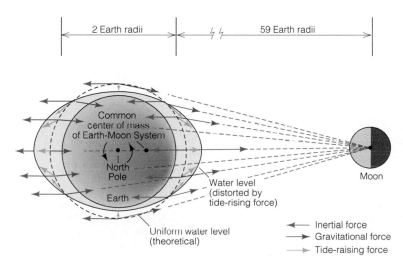

◀ FIGURE 9.5
Tidal forces. Tides are raised by the Moon's gravitational attraction and by inertial force. On the side toward the Moon, both forces combine to distort the water level, creating a tidal bulge. On the opposite side of the Earth, where inertial forces are greater than the Moon's gravitational force, a tidal bulge forms in the opposite direction. As the Earth rotates, the tidal bulges remain essentially stationary and land masses essentially "run into" them twice each day, creating two high tides and two low tides for one full rotation of the Earth.

A.

B.

▲ F I G U R E 9.6
The tidal range in the Bay of Fundy in eastern Canada is one of the largest in the world.
A. Coastal harbor of Alma, New Brunswick, at high tide. B. The same view at low tide.

greater than that from the Earth to the Moon, the Sun is only about half as effective as the Moon in producing tides. Therefore, the opposing tidal effects never entirely cancel each other out.

Tidal Bores

In the open sea the effect of tides is small. However, the configuration of a coastline can greatly influence tidal *run-up* height, the highest elevation reached by the incoming water. Narrow openings into bays, rivers, estuaries, and straits can amplify normal tidal fluctuations. At the Bay of Fundy in Nova Scotia, tidal ranges (i.e., the difference between high and low tide) of up to 16 m are reported (Fig. 9.6). The bay has a very long, narrow configuration that causes the incoming tide to rush in, forming a steep-fronted, rapidly moving wall of water called a **tidal bore.** The extreme tidal range at the Bay of Fundy makes it one of the few localities in the world that may be suited for the exploitation of tidal energy. However, a wall of water moving as fast as 25 km/h can easily move large quantities of sediment. Minimizing the impacts of water-borne sand is a major engineering challenge in the development of tidal energy technologies.

Fluctuations in Water Level

Tides are not the only cause of fluctuations in water level in seas and large lakes. Global sea levels change in response to a wide variety of factors, including long-term climatic fluc-

tuations, wind-driven atmospheric forces, seismic and submarine landslide activity, and even changes in the size and shape of the ocean basins. Of these, long-term climatic variations—especially glaciation—are the most important. At the height of the most recent glacial period, about 18,000 years ago, sea levels were about 120 m lower than they are today, reflecting the fact that a great quantity of water was tied up in ice sheets at the time.

Changes in water level that are global in extent are referred to as *eustatic* changes. As we will see in Chapter 18, eustatic changes in sea level may prove to be one of the most problematic effects of accelerated global warming caused by human activities.

Water levels in other large bodies of water, such as the Great Lakes, also change in response to factors such as barometric pressure, seasonal changes, amount of precipitation, and storm waves. Wind-driven processes, such as the piling up of water in the downwind part of a lake, can also be important in causing water-level fluctuations. In Lake Ontario, for example, this effect can raise water levels by as much as 60 cm. When combined with other seasonal or meteorological effects that cause high water levels, the result can be significant flooding in coastal areas.

Ocean Currents

Surface ocean **currents** are broad, slow drifts of water in a particular direction. They are set in motion by the prevailing surface winds. Air that flows across a water surface drags the water slowly forward, creating a current of water as broad as the current of air, but rarely more than 50 to 100 m deep.

In low latitudes surface seawater moves westward with the trade winds (Fig. 9.7). The general westerly direction of the North and South equatorial currents, which are driven by these winds, is reinforced by the Earth's rotation. The moving currents are influenced by the Coriolis effect and are deflected wherever they encounter a coast. On reaching middle latitudes they travel eastward, moved by the prevailing westerly winds. The result is a circular motion of water in each major ocean basin, both north and south of the equator.

Along most continental margins the predominant flow of water roughly parallels the coast. Warm surface water originating in the equatorial region moves north or south along the eastern margins of continents to about latitude 45°, whereas cold, east-flowing water at higher latitudes encounters the western margins of continents and is deflected south (see Fig. 9.7). At still higher latitudes, cold currents generally prevail.

Ocean Waves

Like surface currents, ocean waves receive their energy from winds. The size of a wave depends on how fast, how far, and how long the wind blows across the water surface. A gentle breeze blowing across a bay may ripple the water or form low waves less than a meter high. By contrast, storm waves produced by intense winds blowing for days across hundreds or thousands of kilometers of open water may become so high that they tower over ships unfortunate enough to be caught in them.

The Dynamics of Wave Motion

In deep water, each small parcel of water in a wave moves in a loop, returning very nearly to its former position as the wave passes (Fig. 9.8). The distance between successive wave crests or troughs is the *wavelength,* usually denoted λ. Downward from the surface of the water a progressive loss of energy occurs; this is expressed as a decrease in the diameter of the looplike motion of the water parcels. Eventually, at a depth of about $\lambda/2$, the motion of the water becomes negligible. This effective lower limit of wave movement

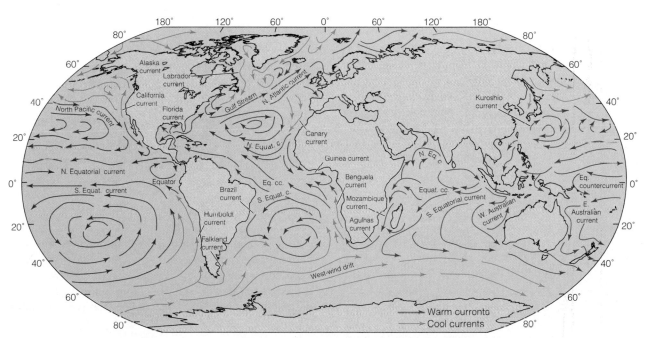

▲ **F I G U R E   9.7**
Surface ocean currents form a distinctive pattern, curving to the right (clockwise) in the Northern Hemisphere and to the left (counterclockwise) in the Southern Hemisphere because of the Coriolis effect. The westward flow of tropical Atlantic and Pacific waters is interrupted by continents, which deflect the water toward the poles. The flow then turns away from the poles and returns as eastward-moving currents in the middle latitudes.

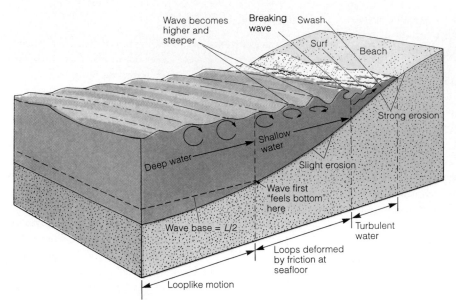

◄ F I G U R E 9.8
Waves change as they travel from deep water through shallow water to the shore. In the process the circular motion of water parcels, found in deep water, changes to elliptical motion as the water becomes shallower and the wave encounters frictional resistance to forward movement. In this figure, vertical scale is exaggerated, as is the size of the loops relative to the scale of the waves.

(which, by extension, is also the lower limit of erosion by the bottoms of waves) is called the **wave base**. Wavelengths as long as 600 m have been measured in the Pacific Ocean. For such large waves the lower limit of wave action (λ/2) is about 300 m, one and a half times as deep as the average depth of the outer edges of continental shelves (200 m). Thus, it is possible for large waves to affect even the outer parts of the continental shelves.

Wave Action Along Coastlines

Waves change as they travel from deep water through shallower water toward the shore. As shown in Fig. 9.8, the circular loops that characterize wave motion in deep water become flatter as the water becomes shallower and frictional resistance against the bottom restricts forward movement. Landward of depth λ/2, the circular wave motion is influenced by the increasingly shallow seafloor, which restricts vertical movement. As depth decreases, the orbits of the water parcels become progressively more elliptical until the movement of water is limited to a back-and-forth motion.

As a wave approaches the shore it undergoes a rapid transformation. The frictional resistance of the shallow seafloor interferes with wave motion and distorts the wave's shape, causing the height to increase and the wavelength to decrease. Now the front of the wave is in shallower water than the rear and is also steeper than the rear. Eventually the front becomes too steep to support the advancing wave, and as the rear part continues to move forward, the wave collapses, or *breaks* (see Fig. 9.8).

When a wave breaks, the motion of the water instantly becomes turbulent, like that of a swift river. Such "broken water," called **surf**, is found between the line of breakers and the shore, an area known as the *surf zone*. Each wave finally dashes against rock or rushes up a sloping beach until its en-

ergy is expended; then the water flows back toward the open sea. Water that has piled up against the shore returns seaward in an irregular and complex way, partly as a broad sheet along the bottom and partly in localized narrow channels as *rip currents*, which are responsible for dangerous undertows that can sweep unwary swimmers out to sea.

The geologic work of waves is accomplished primarily by the direct action of surf. Surf is a powerful erosional agent because it possesses most of the original energy of each wave that created it. This energy is eventually consumed in turbulence, in friction at the bottom, and in movement of sediment that is thrown violently into suspension from the bottom.

COASTAL EROSION AND SEDIMENT TRANSPORT

The world's coastlines are dynamic zones of conflict where land and water meet. Through erosion and the creation, transport, and deposition of sediment, the form of a coastline changes, often slowly but sometimes very rapidly. Erosional forces tear away at the land while other forces move and deposit sediment, thereby adding to the land. Because the conflict is unending, few coastlines achieve a condition of complete equilibrium.

Erosion by Waves

Most erosion along a seacoast is accomplished by waves moving onshore. Wave erosion takes place not only at sea level, in the surf zone, but also below and—especially during storms—above sea level. Most coastal erosion is con-

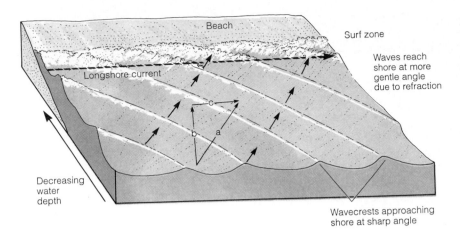

◄ F I G U R E 9.9
A longshore current develops parallel to the shore as waves approach a beach at an oblique angle. A line drawn perpendicular to the front of each approaching wave (a) can be resolved into two components: the component oriented perpendicular to the shore (b) produces surf, whereas that oriented parallel to the shore (c) is responsible for the longshore current. Such a current can transport considerable amounts of sediment along a coast.

fined to a zone that lies within 10 m above and 10 m below mean sea level. Ocean waves typically break at depths that range between wave height and 1.5 times wave height. Because waves are seldom more than 6 m high, the depth of vigorous erosion by surf should be limited to 6 m times 1.5, or 9 m below sea level. This theoretical limit is confirmed by observation of coastal structures such as seawalls, which are only rarely affected by surf to depths of more than 7 m.

An important kind of erosion in the surf zone is the wearing down of rock by particles transported by waves. Through continuous rubbing and grinding with these particles, the surf wears down and deepens the bottom and eats into the land. At the same time, the particles themselves become smoother, rounder, and smaller. Because surf is the active agent, this activity is limited to a depth of only a few meters below sea level. In effect, the surf is like an erosional saw cutting horizontally into the land.

Transport of Sediment by Waves and Currents

Sediment, either produced by waves pounding against a coast or brought to the sea by rivers, is redistributed by currents that build distinctive shoreline deposits or move sediment offshore onto the continental shelves. The transport of sediment is accomplished by several processes, including longshore transport and beach drift.

Longshore Transport

Most waves reach the shore at an oblique angle (Fig. 9.9). The path of an incoming wave can be resolved into two directional components, one oriented perpendicular to the shore and the other parallel to the shore. The perpendicular component produces the crashing surf. The parallel component sets up a **longshore current** within the surf zone that flows parallel to the shore. While surf erodes sediment at the shore, the longshore current moves the sediment along the beach. The direction of longshore currents may change sea-

sonally if the prevailing wind directions change, thereby causing changes in the direction of the arriving waves.

Beach Drift

Meanwhile, on the exposed beach, incoming waves produce an irregular pattern of water movement along the shore. Because waves generally strike the beach at an angle, the *swash* (uprushing water) of each wave travels obliquely up the beach before gravity pulls the water down the slope of the beach (Fig. 9.10). This zigzag movement of water carries sand and pebbles first up, then down the beach

▼ F I G U R E 9.10
Surf swashes obliquely onto a Brazilian beach and forms a series of arc-shaped cusps as the water loses momentum and flows back down the sandy slope.

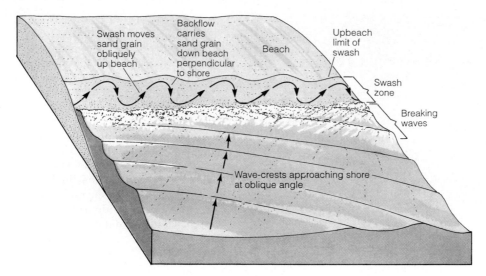

▲ F I G U R E 9.11

As surf rushes up a beach with each incoming wave, sand grains are picked up and carried toward the shore. Surf approaching the shore at an angle travels obliquely up the beach. The return flow, pulled by gravity, flows back nearly perpendicular to the shoreline. A grain of sand therefore moves along a zigzag path as successive waves reach the shore. Net motion down the beach is called beach drift.

slope. The net effect of successive movements of this type is the progressive transport of sediment along the shore, a process known as **beach drift** (Fig. 9.11). The greater the angle of waves to shore, the greater the rate of drift. Marked pebbles have been observed to drift along a beach at a rate of more than 800 m/day. When the volume of sand moved by beach drift is added to that moved by longshore currents, the total can be very large.

Coastal Dynamics During Storms

The approximate equilibrium among the erosional and depositional forces operating along coasts is occasionally interrupted by exceptional storms that erode cliffs and beaches at rates far greater than the long-term average. After a single storm, cliffs of compact sediment on Cape Cod were observed to have retreated up to 5 m—more than 50 times the normal annual rate of retreat. Infrequent bursts of rapid erosion of this type can have a significant impact on the inhabitants of coastal areas (Fig. 9.12).

The west coasts of Ireland and Britain are exposed to the full force of Atlantic storm waves. During one great storm in Scotland a solid mass of stone, iron, and concrete weighing 1200 metric tons was ripped from the end of a breakwater and moved inshore. The damage was repaired with a block weighing more than 2300 metric tons, but 5 years

later storm waves broke off that block and moved it too. These incidents involved pressures of about 27 metric tons/m². Even waves with much less force can break bedrock from sea cliffs. Waves pounding against a cliff compress the air trapped in fissures, and the force of the compressed air can be great enough to dislodge blocks of rock.

In calm weather an exposed beach is likely to receive more sediment than it loses and therefore to become wider. But during storms the increased energy in the surf erodes the exposed part of a beach and makes it narrower. Because storminess is seasonal, changes in beach profiles are more likely to occur at some times of the year than at others. Along parts of the Pacific coast of North America, winter storm surf tends to carry away fine sediment, and the remaining coarse material assumes a steep profile. In calm summer weather, fine sediment drifts in and the beach assumes a gentler profile.

A Variety of Coastlines

The end result of the constant interplay between erosive and depositional forces operating along coastlines is a wide variety of shorelines and coastal landforms. The world's coasts do not fall easily into identifiable classes. Their configurations depend on the active geologic processes at work; on the structure and erodability of coastal rocks; and on the

◀ F I G U R E 9.12
Storm waves damage the pier, Seal Beach, California.

length of time these processes have operated. Eustatic sea level changes can also influence the development of coastal features; many coastal and offshore landforms are relics of times when sea level was either higher or lower than it is now (Fig. 9.13). Repeated emergence and submergence of coastlines over many glacial–interglacial cycles, each accompanied by erosion and redeposition of shoreline deposits, has resulted in complex assemblages of coastal landforms. The types of embayments that occur along a coastline also contribute to the variety of coastal types.

Despite the variability of coasts and shorelines, three basic types are most common. They are the rocky (cliffed) coast, the lowland beach and barrier island coast, and the

coral reef. Each of these is characterized by a particular set of erosional and depositional landforms.

Rocky (Cliffed) Coasts

The most common type of coast, comprising about 80 percent of ocean coasts worldwide, is a rocky or cliffed coast (Fig. 9.14). Seen in profile, the usual elements of a cliffed coast are a wave-cut cliff and wave-cut bench or terrace, both the work of erosion. A **wave-cut cliff** is a coastal cliff cut by wave action at the base of a rocky coast (Fig. 9.15). As the upper part of the cliff is undermined, it collapses and the resulting debris is redistributed by waves. An undercut cliff that has not yet collapsed may have a well-de-

◀ F I G U R E 9.13
An uplifted wave-cut bench at Tongue Point, New Zealand. Crustal uplift along this coast has raised the former seafloor to expose a broad bench. Light-colored beach sediment overlies the darker rocks of the wave-cut cliff at the seaward edge of the bench. Below the uplifted bench, a younger one is forming.

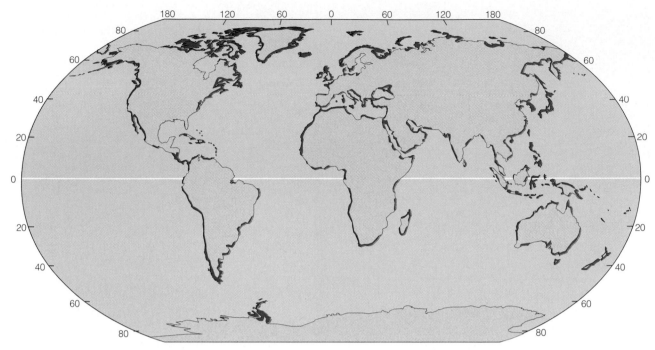

▲ F I G U R E 9.14
World map showing location of sea-cliff coasts (dark lines).

veloped notch at its base. Below a wave-cut cliff you can often find a *wave-cut bench* or *terrace*, a platform that has been cut across bedrock by the surf. The shoreward parts of some benches are exposed at low tide. If the coast has been uplifted, a wave-cut bench and its sediment cover can be completely exposed (see Fig. 9.13).

ABC=Material removed by mass wasting (rockfalls and slides)

ACD=Material eroded by surf

Wave base

Wave-cut cliff

Beach

Wave-cut notch

Wave-cut bench

▲ F I G U R E 9.15
Principal features of a shore profile along a cliffed coast. Notching of the cliff by surf action undermines the rock, which collapses and is reworked by surf. Note the large proportion of material removed by mass-wasting (ABC) relative to that eroded by surf (ACD).

The rocky character of cliffed coasts may be misleading to those in search of a stable, unchanging platform on which to build a home. Cliffed shorelines are susceptible to frequent landslides and rock falls as erosion eats away at the base of the cliff. Roads, buildings, and other structures built too close to such cliffs can become casualties when sliding occurs.

Beaches and Barrier Islands

Beaches are characteristic features of many coasts, even those with steep, rocky cliffs. Along sandy coasts that lack cliffs, the beach constitutes the primary shore environment. A beach is regarded by most people as the sandy surface above the water along a shore. Actually, a **beach** is defined as wave-washed sediment along a coast. A beach thus includes sediment in the surf zone, which is continually in motion.

On low, open shores an exposed beach typically has several distinct elements (Fig. 9.16). The first is a rather gently sloping *foreshore*, a zone extending from the level of lowest tide to the average high-tide level. Here is found a *berm*, a nearly horizontal or landward-sloping bench formed from sediment deposited by waves. Beyond the berm lies the *backshore*, a zone extending inland from the berm to the farthest point reached by surf. On some coasts, beach sand is blown inland by onshore winds to form belts of coastal dunes.

Sediment that has been transported to the beach and offshore areas by rivers or by longshore currents is reworked

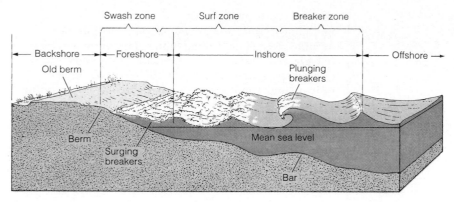

▲ F I G U R E 9.16
**Section across a beach showing the principal elements of the shore profile.
The section is about 75 m long. Vertical scale is exaggerated in this figure.**

by wave action, forming a variety of depositional land-forms. A landform commonly associated with beaches is the **barrier island**, a long, narrow, sandy island lying offshore and parallel to the coast (Fig. 9.17). A barrier island generally consists of one or more ridges of sand dunes associated with successive shorelines. Barrier islands are found along most of the world's lowland coasts. For example, the Atlantic coast of the United States consists mainly of a series of barrier beaches ranging from 15 to 30 km in length and 1.5 to 5 km in width, located 3 to 30 km offshore. Sand dunes are typically the highest topographic features along this coastline, which includes Coney Island, New York City's coastal playground, and the long chain of islands cen-

tered at Cape Hatteras on the North Carolina coast. As seen in Figure 9.18, land along this coastline is eroding in some places and *accreting* (building up) in others.

During major storms, surf washes across low places on barrier islands and erodes them, cutting inlets that may remain open permanently (Fig. 9.19; also see Box 9.1). At such times fine sediment is washed into the lagoon between the barrier island and the mainland; this is called *overwash*. In this way the length and shape of barrier islands is always changing. Studies of the response of barrier islands to changing environmental conditions suggest that island development is closely related to the amount of sediment in the system, the direction and intensity of waves and

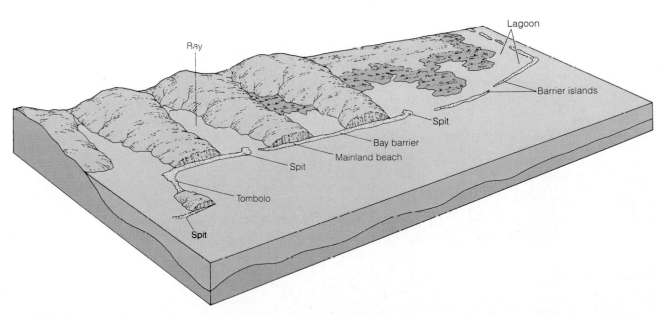

▲ F I G U R E 9.17
**Some depositional features along a stretch of coast. The local direction of beach drift is
toward the free end of the spits.**

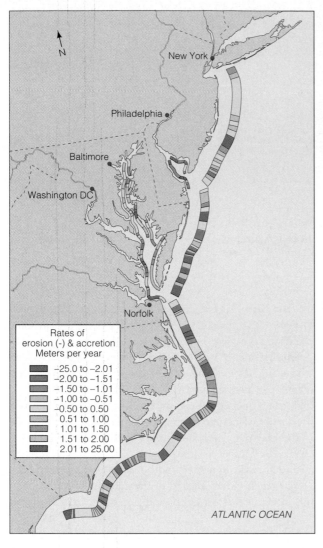

◄ **F I G U R E 9.18**
Shoreline erosion and accretion along the middle
Atlantic coast of the United States. The warmer colors
(purple, red, orange) show segments of the coast where
erosion is occurring, while the cooler colors (green,
brown, blue) show segments where the land is accreting
through deposition of sediment. Rates of erosion and
deposition are given in the legend. Overall, 44 percent
of the shoreline on this map is eroding at rates greater
than 0.6 m/year and 17 percent is accreting.

◄ **F I G U R E 9.19**
Storm driven erosion on Fire
Island, a barrier beach on the
south side of Long Island,
New York.

BOX 9.1
•
THE HUMAN PERSPECTIVE

PEA ISLAND AND THE BEACH STABILIZATION CONTROVERSY

*I*f engineers, politicians, scientists, environmentalists, and beachfront property owners were to meet on the beach at the Pea Island National Wildlife Area, they would soon be embroiled in intense discussions. This tiny barrier island south of Oregon Inlet, North Carolina, provides a setting for the most difficult issues related to beach stabilization: private versus public rights, dune grasses versus dune buggies, ecology versus economy, and beach stabilization versus nonintervention.

Oregon Inlet (Fig. B1.1) was cut during a storm in 1846 and has migrated south at a rate of almost 30 m/year to its current position, which is stabilized through the ongoing efforts of the Army Corps of Engineers. These efforts include constant dredging to maintain a navigation channel to the mainland, construction of a rock groin on the north shore of the island, and plans to construct two jetties—each 2.4 km long—on either side of the inlet. The projected cost of the latter project is more than $100 million.

Barrier islands like Pea Island are dynamic places. The islands are washed over, dunes shift, inlets migrate, and structures are perpetually pounded by ocean waves. Rising sea level, even at gradual rates, is pushing many barrier islands toward the mainland. Without new sources of sediment, anchored barrier islands will be submerged. Beachfront property owners see this as a problem; however, the movement of the islands is a natural process and becomes a problem only when static structures are placed on unstable surfaces.

The north end of Pea Island has been designated as a national wildlife refuge and hence cannot be used for commercial or residential development. But many questions remain about the stabilization of other parts of the island: What environmental impacts will it have? How long should it be continued? Will it be successful? Are the costs, both economic and environmental, justified? And, most importantly, will the long-term costs of stabilization eventually exceed the value of the property that is being protected?

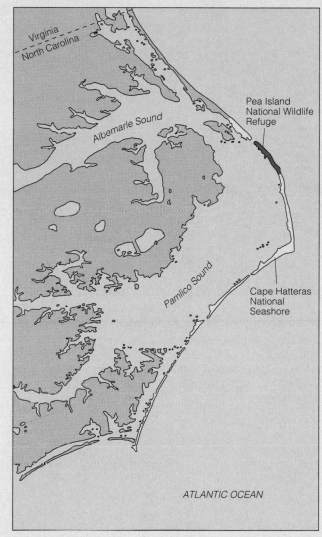

▲ FIGURE B1.1
The distribution of federally owned and managed lands along the Outer Banks, showing locations mentioned in the text.

▲ F I G U R E 9.20
The complex spit of Cape Cod. Waves and currents rework sediment eroded from the peninsula, forming the south side of Cape Cod Bay, and transport the sediment northward and southward. An eddy carries sediment around the north point of the spit and into the bay.

nearshore currents, the shape of the seabed, and the stability of sea level.

Other common depositional landforms associated with beaches include the *spit* (an elongated ridge of sand or gravel that projects from land and ends in open water) and the *tombolo* (a spitlike ridge of sand or gravel that connects an island to the mainland); these are also illustrated in Figure 9.17. A well-known example of a large, complex spit is Cape Cod, Massachusetts (Fig. 9.20).

The elongate bay lying inshore from a barrier island or other low, enclosing strip of land (such as a coral reef) is called a *lagoon.* Lagoons are commonly fed by *estuaries,* the wide, funnel-shaped mouths of rivers in the tidal zone where fresh and salt water meet. Lagoons and estuaries are important habitats for a wide variety of plants, birds, and animals. They also play an important role in the protection of mainland shorelines because they serve as buffers against the erosive impact of storm waves. Unfortunately, these sensitive environments are particularly susceptible to the impacts of human activities.

Coral Reefs

Many of the world's tropical coastlines consist of limestone **reefs** built by vast colonies of tiny organisms, principally corals, that secrete calcium carbonate. Reefs are built up very slowly over thousands of years. Each of the tiny coral animals, called *polyps,* deposits a protective layer of calcareous material; the layers eventually build to form a complex reef structure. *Fringing reefs* form coastlines that closely border the adjacent land, while *barrier reefs* are separated from the land by a lagoon, as in the case of the Great Barrier Reef off Queensland, Australia (Fig. 9.21).

Reefs are highly productive ecosystems inhabited by a diversity of marine life forms. They also perform an important role in the recycling of nutrients in shallow coastal environments. They provide physical barriers that dissipate the force of high-energy waves, protecting the ports, lagoons, and beaches that lie behind them, and they are an important aesthetic and economic resource.

Corals require shallow, clear water in which the temperature remains above 18°C. Reefs therefore are built only at or close to sea level and are characteristic of low latitudes. Because of their very specific water temperature and light level requirements, coral reefs are particularly susceptible to damage from human activities as well as from natural causes such as tropical storms.

Adapting to Coastal Erosion

Shoreline homes command premium prices in the real estate market. If they are built on solid rock, they can be a lasting investment; however, coastlines are among the most dynamic places on the Earth's surface, and a house built on a coast composed of erodible sediment can prove to be a poor bargain. There are several ways of stabilizing the shoreline and protecting coastal property; all of them have some drawbacks.

Protection of the Shoreline

Measures for protecting the shoreline fall into two general categories: *hard stabilization,* which relies on engineered structures; and *soft stabilization,* or nonstructural approaches to shoreline stabilization.

Hard Stabilization The structures involved in hard stabilization are of two main types: those that interrupt the force of the waves and those that interrupt the flow of sand along the shore. Examples of the first type are *seawalls* and *breakwaters* (Fig. 9.22A). Since these are built parallel to the shoreline, they deflect wave energy away from the protected property (Fig. 9.22B). However, these structures can actually accelerate loss of beach sand on the ocean side, making the beach steeper and narrower until it is finally destroyed.

The second type of hard stabilization structure includes *groins* and *jetties* (Fig. 9.23A). These structures are built

◄ F I G U R E 9.21
The Great Barrier Reef, a barrier coral reef on the continental shelf of northeastern Australia, is one of the world's most diverse marine ecosystems.

A.

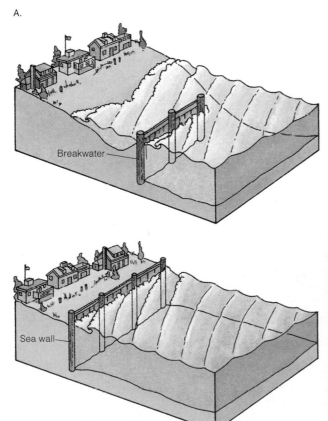

▼ F I G U R E 9.22
Hard stabilization of a shoreline: seawalls and breakwaters. A. Structures such as seawalls and breakwaters are built parallel to the shoreline to interrupt the force of the waves. B. Breakwaters constructed offshore from Tel Aviv, Israel, protect the beach zone from incoming waves. Sediment builds up behind each breakwater, creating a scalloped coastline. Meanwhile wave action at the base of such structures can hasten erosion and steepen the beach on the ocean side.

B.

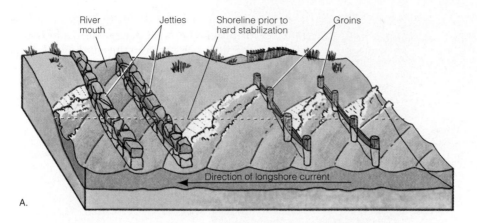

A.

B.

▲ F I G U R E 9.23
Hard stabilization of a shoreline: groins
and jetties. A. Structures such as groins
and jetties are built perpendicular to the
shoreline to interrupt the flow of sediment
with the longshore current. B. Groins have
been built perpendicular to the shoreline
of Miami Beach, Florida, to prevent
excessive loss of sand by longshore drift
at this popular resort area.

perpendicular to the shoreline and widen protected beaches by trapping sand updrift of the groin or jetty. At the same time, however, they degrade the adjacent beaches downdrift (Fig. 9.23B). Once a property owner has installed a groin, neighboring downdrift property owners are cut off from their sand supply and may be forced to add a groin field to their beach fronts.

Soft Stabilization Soft stabilization refers primarily to the process of beach "nourishment," in which sand is dredged offshore and deposited onto the beach as a slurry via pipelines. The newest techniques use offshore hopper dredges instead of onshore pipeline systems to transfer sand to a point close to the beach from which natural wave dynamics distribute the sand onto the beach itself.

Beach nourishment and other soft stabilization approaches, such as the stabilization of dunes by planting certain types of grasses, have generally been considered more desirable than hard, structural approaches. However, beach nourishment raises several concerns, not the least of which is high cost; once it is begun, beach nourishment must be repeated periodically or it will not succeed. The dredging of millions of cubic meters of sand can also cause extensive disruption to shoreline ecosystems. Some species, such as sea turtles, are very susceptible to changes in the characteristics of the sand: if the sand is too coarse, they may have trouble nesting; if it is too dark, their eggs may overheat and die in the nest. Also, it is unclear how resistant beach nourishment is to the effects of storms. For example, in 1993 the Army Corps of Engineers was in the process of replenishing Folly Beach, North Carolina, when Hurricane Andrew hit. In a single day most of the piped-in sand was washed into shallow offshore areas.

The Controversy over Coastal Erosion

The long-term effectiveness of hard stabilization techniques for stabilizing shorelines is a subject of intense controversy. Many experts argue that, in the long run, extensive interference in the natural processes affecting shorelines can lead to expensive and possibly irreversible damage. Because of their high recreational value, beaches in densely populated regions may justify the costs of stabilization and protection against erosion. A beach, however, presents a special sort of problem. As a result of longshore currents and beach drift, what happens on one part of a beach affects all the parts lying in the downdrift direction. The net result of intervention therefore may be to protect one part of the beach at the expense of another part; small beaches have been completely destroyed in this way in only a few years.

Beaches around the world are deteriorating because of human interference. Along North Carolina's Outer Banks, for example, there is a noticeable difference between the barrier islands that have been built on and those that have not: the undeveloped islands tend to have beaches 100–200 m wide, whereas developed beaches, such as those on Hat-

teras Island, have been reduced to widths of 30 m or less by accelerated erosion. In southern California, most of the sand on beaches is not supplied directly by the erosion of wave-cut cliffs but is carried to the sea by streamflow during floods. However, because buildings and other structures in stream valleys are vulnerable, dams have been built across the streams to control flooding. The dams trap sand and gravel carried by the streams, preventing it from reaching the sea. As a consequence, the natural balance among the factors involved in the longshore transport of sediment has been upset, resulting in significant erosion of some beaches (Fig. 9.24).

A further dramatic example of human interference can be seen along the coast of the Black Sea. Of the sand and pebbles that form the natural beaches there, 90 percent used to be supplied by rivers as they entered the sea. During the 1940s and 1950s three things occurred: large resort developments were built at the beaches; large breakwaters were constructed so that two major harbors could be extended into the sea; and dams were built across some rivers inland from the coast. All this construction upset the equilibrium among the supply of sediment to the coast, longshore currents and beach drift, and deposition of sediment on beaches. By 1960 it was estimated that the combined area of all the beaches along the coast had decreased by 50 percent. Then beachfront buildings began to sag or collapse as the surf ate away at their foundations. An ironic twist to this chain of events lies in the fact that large volumes of sand and gravel were removed from beaches and used as concrete aggregate, not only to construct the resort buildings but also to build the dams that eventually cut off the supply of sediment to the coast.

▲ FIGURE 9.24
Damage to homes on Fire Island, New York caused by coastal erosion. The storm that caused the damage occurred in December 1992.

There is, of course, another way to adjust to coastal erosion and protect shorelines: leave them alone and move away. This may seem like an extreme measure, but more and more communities are adopting this stance. Shoreline protection measures sometimes succeed in protecting homes and other structures along the beach, but they are rarely able to protect the beach system itself in the long term. Zoning laws and policies for some coastal areas specify that undeveloped beachfront sections on barrier islands must be left to natural processes, rather than built upon and "stabilized." For beachfront areas where construction is permitted, building codes can dictate appropriate site-specific structural designs. If the buildings predate the codes, insurance programs will sometimes pay a property owner to move the structure or to not rebuild a house in a location threatened by beach erosion.

EXCEPTIONAL WEATHER

Many of the exceptional weather-related occurrences that can be disastrous for humans are actually normal, natural aspects of weather formation—explicable and, in fact, inevitable in the broader context of global oceanic, atmospheric, and climatic processes. In this last section of the chapter we shall look at the natural causes and human impacts of weather-related occurrences of exceptional intensity. These occurrences include a variety of intense storms; drought and desertification; and El Niño, a complex weather pattern with far-reaching consequences.

Cyclonic Storms

A **cyclone** is an atmospheric low-pressure system that gives rise to roughly circular, inward-spiraling wind motion, called *vorticity*. Although we tend to think of cyclones as being destructive tropical storms, the scientific connotation of the term is more general and includes a variety of broader, weaker low-pressure systems. Because of the Coriolis effect, such systems rotate in a counterclockwise direction in the Northern Hemisphere and clockwise in the Southern Hemisphere (Fig. 9.25).

Tornadoes

A **tornado** is a cyclonic storm with a very intense low-pressure center. Tornadoes are short-lived and local in extent, but they can be extremely violent (Fig. 9.26). They typically follow a very narrow, sharply defined path, usually in the range of 300–400 m wide. U.S. National Weather Service records show that tornadoes have the strength to drive 2 x 4 wooden boards through brick walls, lift an 83-ton railroad car, and carry a home freezer over a distance of 2 km. Their rate of forward motion is highly variable; some

◄ F I G U R E 9.25
A low-pressure center (cyclone) centered over Ireland and moving eastward over Europe. The counterclockwise winds of a Northern Hemisphere low are clearly shown by the spiral cloud pattern.

◀ F I G U R E 9.26
A tornado crossing the plains of North Dakota.

are almost stationary, whereas others race along at over 100 km/h. The strength of a tornado can be estimated from the damage it causes by referring to the *F-scale,* which was devised by Professor T. Theodore Fujita of the University of Chicago (Table 9.1).

Although they can occur anywhere in the world, tornadoes are particularly common in the central and southeastern United States. Tornadoes typically form along the *cold front* of a fast-moving mid-latitude cyclonic storm system. When a moving mass of cold air overtakes and traps an underlying layer of warm, moist air, the warm air is drawn up into the core of the storm in a spiraling upward motion. At the same time the cold air spirals downward, creating a *vortex,* or twisting funnel cloud.

The destructive capabilities of tornadoes come partially from their extremely high wind velocities, which have been clocked at 450 km/h, and partially from the near-vacuum that exists within the vortex. The air pressure inside the vortex may be as low as 60 percent of normal atmospheric pressure; buildings can actually explode and disintegrate as a result of the pressure differential between the inside and outside air. The partial vacuum also causes the tornado to suck up soil and debris, which give the funnel cloud its typical dark, ominous appearance.

T A B L E 9.1 • The F-Scale for Tornado Intensity

F-Scale	Category	Estimated Wind Speed, km/h	Damage
0	weak	65–118	Minor. Twigs broken.
1		119–181	Trees down, mobile homes moved off foundation.
2	strong	182–253	Demolish mobile home; roof off frame houses.
3		254–332	Lift motor vehicles. Destroy well-constructed building.
4	violent	333–419	Level buildings, toss automobiles around.
5		420–513	Lift and toss around houses.

HURRICANE ANDREW
August 25, 1992 20:20 GMT

▲ F I G U R E 9.27
Hurricane Andrew, one of the largest and strongest hurricanes of modern times, formed over the Atlantic Ocean and slammed into Florida in August 1992. The hurricane, here photographed from above, packed winds in excess of 200 km/h.

Typhoons and Hurricanes

Tropical cyclones that form over the ocean are also characterized by low-pressure centers and cyclonic wind circulation. In contrast to tornadoes, however, tropical cyclones are longer lived and much more regional in extent. Their low-pressure centers are less intense than those of tornadoes, but the total power of such a storm can be awesome—the energy flow in a tropical cyclone in one day can be equivalent to the energy released by 400 20-megaton hydrogen bombs.

Tropical cyclones are called **hurricanes** in the Caribbean and North America, **typhoons** in the western Pacific, and tropical cyclones in the Indian Ocean. When fully developed, a tropical cyclone is a circular storm resembling a huge whirlpool up to 600 km in diameter (Fig. 9.27). Such storms travel erratically and can last several weeks. The intense low-pressure center creates winds with velocities up to 300 km/h. At the center itself, called the *eye* of the storm, the inward-rushing winds are drawn upward; they never actually reach the center of the storm. This phenomenon is responsible for the eerie calm found in the eye of a storm.

Damage from hurricanes and typhoons comes from their intense winds and torrential rains as well as from the battering and erosional effects of ocean waves (Fig. 9.28). Flooding—both river flooding caused by intense rainfall in a short period and coastal flooding caused by *storm surges* (to be discussed shortly)—is another hazard commonly associated with hurricanes.

▲ F I G U R E 9.28
All that remained of a Florida town after Hurricane Andrew passed through. All the damage seen here was caused by high-speed winds.

In countries with adequate early-warning systems, the main hazard associated with hurricanes and typhoons is property damage. When Hurricane Andrew roared through Florida, Louisiana, and the Bahamas in 1992, it caused an estimated $20 billion in property damage, but only 25 people were killed. In contrast, a hurricane that hit Galveston, Texas, in August 1900 reportedly killed 8000 people, mainly because there was no early warning. In 1991 cyclones devastated the coast of Bangladesh and killed 200,000 people; perhaps 500,000 others died there in a cyclone in 1970. Bangladesh is particularly vulnerable to cyclone hazards because of its high population density in low-lying coastal areas, combined with the lack of an effective early warning system or emergency response mechanism.

Nor'easters

Another devastating type of cyclonic storm is the **nor'easter.** Nor'easters are *extratropical*—that is, they form outside of tropical regions. They originate in regions of atmospheric instability in middle latitudes over land or coastal regions and are named for their winds, which blow from the northeast. The low-pressure system of a nor'easter is typically weaker and more diffuse than that of a tropical cyclone, and the associated winds are less intense. However, they can cover very large areas, mostly along the eastern coast of North America, and are associated with very high waves that often cause significant damage. Nor'easters often generate storm surges as high as 5 m above normal tidal levels. These can be particularly damaging because the storms themselves often last several days (i.e., through several tidal cycles).

A particularly damaging nor'easter, known as the Ash Wednesday storm, hit the Atlantic coast of the United States on March 7, 1962. The storm's 10-m-high waves, which battered parts of the coast for days, caused damage estimated at over $300 million. Like tornadoes and other types of storms, nor'easters are classified by their intensity, from I (weak) through V (extreme). The Ash Wednesday nor'easter was a class V storm; it created many new tidal in-

lets and caused extensive erosion of dunes and beaches. Table 9.2 shows the Dolan/Davis scale, which relates storm classes to the types of coastal damage observed along the Outer Banks of North Carolina.

Storm Surges

A phenomenon often associated with intense wind systems is the **storm surge**, an abnormal, temporary rise in water levels in oceans or lakes. As mentioned earlier, friction between wind and the surface of the water can cause water to "pile up" in the downwind portion of an enclosed body like a lake or bay. A sharp drop in atmospheric pressure, such as that associated with an intense storm, can also create a storm surge. The very low pressures in the eye of a cyclone cause the surface of the water to bulge upward as much as 1 m.

Storm surges can cause extensive damage and coastal flooding, particularly if they coincide with high tides or standing waves *(seiches)* in an enclosed body of water. The configuration of the coastline and the angle at which the storm is approaching can also influence the height of the surge. Much of the loss of life associated with storms is actually caused by coastal flooding resulting from storm surges. The devastating 1970 cyclone in Bangladesh created a storm surge with water levels officially reported at 4.5 m above normal; survivors reported water levels up to 9 m above normal. The extensive flooding of lowland coastal areas in the Netherlands in 1953 (discussed in Chapter 8) was associated with a storm surge in the North Sea, which is essentially a closed body of water. The surge was generated by a mid-latitude cyclone off the southern coast of Iceland.

Droughts, Dust Storms, and Desertification

In contrast to the exceptional weather occurrences we have considered thus far, a **drought** is a period of weather *inactivity*—specifically, an extended period of exceptionally low precipitation. The effects of drought can be devastating to local agriculture and livestock and the people who depend on them. Severe droughts that occurred in China from

T A B L E 9.2 • The Dolan/Davis Scale: The Relationship Between Storm Class and Coastal Damage, Inferred from Observations Along the Outer Banks of North Carolina[a]

Storm Class	Beach Erosion	Dune Erosion	Overwash	Property Damage
I (weak)	Minor	None	None	None
II (moderate)	Moderate	Minor	None	None
III (significant)	Extending across beach	Significant	None	Moderate
IV (severe)	Severe with recession	Severe or localized destruction	On low-profile beaches	Loss of structures at community scale
V (extreme)	Extreme	Dunes destroyed over extensive areas	Massive, in sheets and channels	Extensive regional-scale losses in millions of dollars

[a] From Davis and Dolan, Nor'easters, *American Scientist,* September–October 1993, pp. 428–439.

1876 to 1879 caused an estimated 10 million deaths from famine and malnutrition.

The Nature of Drought

Three categories of drought are recognized by the U.S. National Weather Service: (1) a *dry spell,* a minimum of 15 consecutive days during which less than 0.8 mm of rain falls; (2) a *partial drought,* or 29 consecutive days in which the mean daily rainfall does not exceed 0.2 mm; and (3) an *absolute drought,* a period of at least 15 days without any measurable rainfall. Obviously, classifications of this type depend on the ability to define the "normal" amount of precipitation for a particular locality. Often this is easier said than done; some regions, particularly semiarid regions adjacent to large deserts, are characterized by extreme variability in precipitation.

Parts of the world where there are distinct wet and dry seasons may experience *seasonal droughts.* Seasonal droughts are more predictable than other droughts but may be equally stressful for residents of the area if the dry season is unusually long or if precipitation has been below average during the preceding wet season.

Although lack of moisture is associated with both droughts and deserts, it is important to distinguish between the two. The word "desert" literally means a deserted (relatively uninhabited) region that is nearly devoid of vegetation. However, the development of artificial water supplies has changed the meaning of the word by making many desert regions suitable for agriculture. As a result, the term **desert** is now generally used as a synonym for land where annual rainfall is less than 250 mm or in which the rate of evaporation exceeds the rate of precipitation, regardless of whether the land is "deserted" (Fig. 9.29). Thus the term "drought," which refers to an exceptional weather occurrence, may not be strictly applicable to true desert areas where low precipitation and moisture deficiency are everyday phenomena. Drought conditions, during which precipitation is *abnormally* low for a period ranging anywhere from weeks to decades, may occur at any time in any part of the world.

Dust Storms

One of the most striking features of deserts is dust storms, in which visibility at eye level is reduced to 1000 m or less by dust raised from the ground by blowing winds. Such storms are most frequent in the vast arid and semiarid regions of central Australia, western China, Central Asia, the Middle East, and North Africa (see Fig. 9.29). In North America, blowing dust is especially common in the Great Plains and in the desert regions of the southwestern United States.

In the "dust-bowl" years of the mid-1930s, many farm families in the southern Great Plains region of the United States abandoned their homes and lands and trekked west,

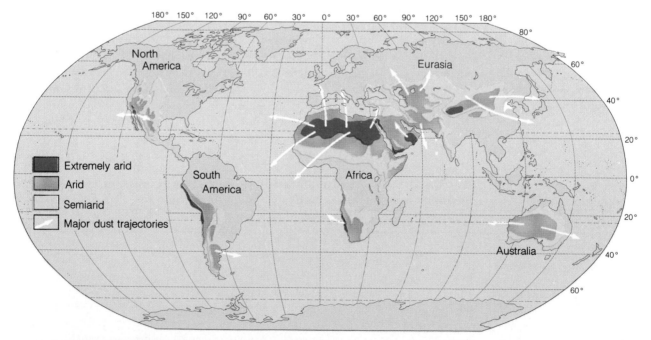

▲ **FIGURE 9.29**
Major dust storms are most frequent in arid and semiarid regions that are concentrated in the areas of subtropical high-pressure belts north and south of the equatorial zone. Arrows show the most common trajectories of dust transported during major storms.

part of a great migration described by John Steinbeck in his award-winning novel *The Grapes of Wrath*. A primary factor behind the exodus was a severe drought, which led to major dust storms that destroyed crops and buried formerly productive fields with drifting sand and dust. In one storm on March 20, 1935, a cloud of suspended sediment extended 3.6 km above Wichita, Kansas. The dust load in the lowermost 1.6 km of this cloud was estimated at 35 million kg per cubic kilometer. Samples of sediment collected from flat roofs showed that on the day of the storm about 280,000 kg of rock particles, or about 5 percent of the load suspended in the lowermost layer of air, was deposited on each square kilometer of land. Enough sediment was carried eastward to cause a temporary twilight at midday over New York and New England.

The frequency of dust storms is related to cycles of drought, with a marked rise in atmospheric dust concentrations coinciding with severe droughts. The frequency has also risen with increasing agricultural activity, especially in semiarid lands. The effectiveness of the wind in creating the 1930s dust bowl was aided by decades of poor land-use practices. Grasses growing on the prairies when the original settlers arrived protected the rich topsoil from wind erosion. However, over time the grasses were replaced by plowed fields and seasonal grain crops that left the ground bare and vulnerable part of the year. Although today the land is still potentially vulnerable in drought years, improved farming practices should reduce the likelihood of similar catastrophes in the future.

Desertification and Land Degradation

Desertification, the spread of desert into nondesert areas, can result from natural environmental and climatic changes as well as from human activities. The major symptoms are declining water tables, increasing saltiness of water and topsoil, decreasing surface water supplies, unnaturally high rates of erosion, and destruction of vegetation. Although we can find abundant evidence of natural desertification in the geologic record, there is increasing concern that human activities can promote desertification regardless of natural climatic trends.

In the region south of the Sahara lies a belt of dry grassland known as the Sahel (Arabic for *border* or *shore*). There the annual rainfall is normally only 100 to 300 mm, most of it falling during a single brief rainy season. In the early 1970s the drought-prone Sahel experienced the worst drought of this century (Fig. 9.30). For several years the annual rains failed to appear, causing the adjacent desert to spread southward as much as 150 km. The drought zone extended from the Atlantic to the Indian Ocean, a distance of 6000 km, and affected at least 20 million people, many of them seminomadic herders. The remarkable persistence of the Sahelian drought was among the most outstanding climate anomalies of the second half of the twentieth century.

The results of the drought and associated desertification were intensified by the fact that between about 1935 and

▲ F I G U R E 9.30
Overgrazing during years of drought killed most of the vegetation around wells in the Azaouak Valley of Mali. Without vegetation, soil blows away and the desert advances.

1970 the human population of the Sahel region had doubled, and the number of livestock had also increased dramatically. This increase led to severe overgrazing, so that with the coming of the drought the grass cover almost completely failed. About 40 percent of the cattle died. Millions of people suffered from thirst and starvation, and many died as vast numbers migrated southward in search of food and water. In the mid-1970s the rains returned briefly, but in the 1980s drought conditions resumed. Ethiopia and the Sudan were especially hard hit, and mass starvation was alleviated only by worldwide relief efforts.

Desertification is a natural process, but like other geologic hazards it can be worsened or even initiated by human actions. At a certain point it becomes difficult to distinguish the natural parts of the process from the human impacts. It may be more appropriate to use the term *land degradation* to refer to human activities, such as inappropriate or overly intensive agricultural or forestry practices, that lead to the deterioration and eventual desertification of formerly productive lands.

El Niño, La Niña, and the Southern Oscillation

The intricate interrelationships between oceanic and atmospheric processes are dramatically illustrated by an exceptional weather phenomenon known as **El Niño**. In simple terms, El Niño is an anomalous warming of surface waters in the eastern equatorial Pacific. However, as indicated at the beginning of the chapter, both the causes and effects of El Niño are complicated. They center on the be-

PREDICTING EL NIÑO

A particularly intense El Niño can be a devastating event, upsetting weather patterns, as well as the ecosystems and human industries that depend on them, worldwide. For this reason, scientists who study ocean–atmosphere interactions are devoting a great deal of effort to the prediction of El Niño. A major problem is that the recurrence interval for El Niños is irregular, which makes them particularly difficult to predict using statistical methods. The interval generally ranges from 3 to 7 years, and a single event will typically last 18 to 24 months. But even the length of an event is difficult to quantify and depends on the exact definition chosen for the "beginning" and "end" of the phenomenon.

In general, the first noticeable anomaly associated with the onset of an El Niño is the weakening of the pressure differential between the region centered over Indonesia and the one centered over the southeastern Pacific. This difference is called the *Southern Oscillation Index.* An exceptionally low index is probably the most reliable early indicator of the onset of El Niño, followed by the weakening of trade winds and changes in sea surface temperatures (Fig. B2.1). As of early 1995, one El Niño had recently ended (in 1993) and another had just gotten under way.

The series of atmospheric and oceanic phenomena associated with El Niño is fairly well known, and the sequence of events is usually more or less predictable. However, the processes that produce the anomalies are not understood. Clearly the Southern Oscillation—the seesawing of atmospheric circulation between the two pressure systems—is fundamental to the process. Closely coupled with the changes in atmospheric circulation are changes in sea surface temperature. What is not clear is exactly what initiates the entire process. Research indicates that changes in sea-surface temperature influence the location of atmospheric convection zones because large-scale convection is most vigorous over the warmest waters. This suggests that a thermal disturbance in the ocean may initiate the process—but what, in turn, controls ocean water temperatures?

One intriguing suggestion is that magmatic heat—that is, heat from submarine lava flows or hydrothermal activity at sea floor spreading centers—may affect ocean temperatures enough to begin the El Niño cycle. The input of

magmatic heat to ocean water is comparable to other contributors to the equatorial ocean heat budget. The intermittent time scale and variable size of submarine volcanic eruptions may prove to be consistent with the apparently erratic pattern of variations in the El Niño–Southern Oscillation cycle. It has recently been confirmed that El Niño events correspond to periods of exceptionally intense seismic activity, which may, in turn, be indicative of periods of active submarine volcanism. Although there is much that is not yet understood concerning the contribution of magmatic heat to ocean thermal patterns, the idea is intriguing in that it may represent a coupled process involving the dynamics of the atmosphere, the oceans, *and* the lithosphere of the Earth.

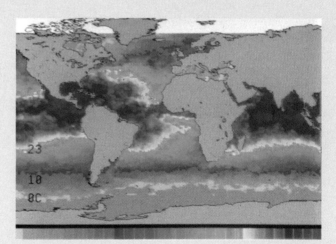

▲ F I G U R E   B2.1

False color, satellite image of the El Niño event of 1983. El Niño events occur when the cold, South Equatorial current fails to reach the equator on the Pacific coast of South America. The image, made July 1, 1983, shows sea surface temperatures, color coded according to the scale on the bottom of the image. A tongue of cold water (blue and green) reaches the equator on the Atlantic coast of Africa, but there is no equivalent on the Pacific coast of South America, as there would be in a normal year.

havior of trade winds and associated currents in the equatorial zone of the Pacific Ocean. The causes of normal annual fluctuations in these processes are not well understood; exceptional El Niños are even more difficult to analyze and predict.

The Normal Scenario

Under ordinary conditions (Fig. 9.31A), trade winds blow from the southeast along the coast of South America, then west along the equator toward Indonesia and Australia. The trade winds create currents that push warm tropical waters

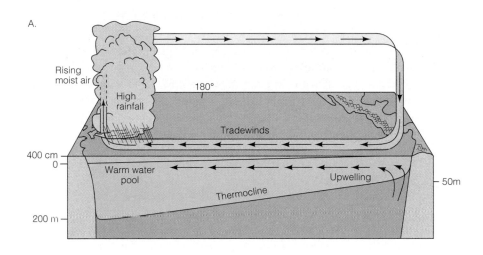

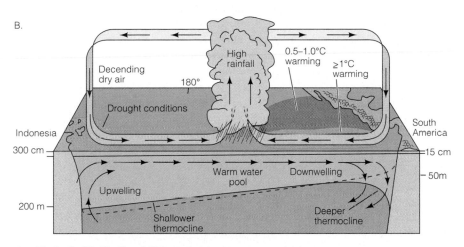

▲ F I G U R E 9.31

El Niño is a manifestation of the Southern Oscillation, an atmospheric pressure seesaw between a high-pressure center in the southeastern Pacific and a low-pressure center over Indonesia. A. Under ordinary conditions the pressure differential between these centers drives easterly trade winds along the equator. The winds pile up warm water, raising the sea level in the western Pacific and depressing the thermocline there. Off South America, where the trade winds drive surface water offshore, the thermocline is shallow and cool water wells up to the surface. Near Indonesia the trade winds converge with westerly winds, producing rising air and heavy rains. The air flows eastward at high altitudes and sinks in the central and eastern Pacific, where the weather is dry. B. During an El Niño the east–west pressure difference becomes so low that the trade winds collapse in the western Pacific. The warm water piled up there flows back toward the east. At the same time, the thermocline is depressed off South America, where the upwelling water becomes warm. Both effects warm the surface of the sea. During the 1982–1983 El Niño, the severest in a century, the wind directions, and hence the weather pattern, were completely reversed.

north along the coast of Peru and then west along the equator. These currents cause warm water to "pile up" in the western Pacific, raising sea levels there by as much as 40 cm. The associated increase in surface water temperature in the western Pacific also causes a depression in the *thermocline,* the boundary between warm surface water and the underlying cooler water. Moisture-laden air in the western Pacific causes seasonal heavy rains to fall; the moisture-depleted air travels back at high altitude to the eastern Pacific, where the resulting weather is normally sunny and dry.

As shown in Fig. 9.31A, the offshore movement of warm surface waters from the coast of South America allows cool water to well up from the depths of the ocean to the surface. These upwelling waters are rich in the nutrients that nourish plankton and ultimately support the Peruvian fishing industry. Every year around Christmastime, warm currents flow southward along the coast of South America and interfere with the upwelling of cold water. This signals an end to the fishing season—but only temporarily. By March or April the situation has usually reversed itself, and the trade winds and water temperatures are back to normal.

The normal pattern is complicated by the *Southern Oscillation,* an atmospheric pressure "seesaw" between a high-pressure system (where air masses are descending) centered in the southeastern Pacific and a low-pressure system (where air masses are rising) centered over Indonesia and northern Australia. Superimposed on the normal annual variations is a back-and-forth, seesaw-like fluctuation of air pressure between these two systems.

The exact causes and timing of the seesaw effect are not well understood, but clearly the two systems are linked. Sometimes the pressure difference between them is very great. In this case the trade winds will be very intense, and the temperature of equatorial coastal waters in the eastern Pacific will be unusually low. An exceptionally intense event of this type is referred to as *La Niña* (the girl). Like El Niño, La Niña can have far-reaching weather effects.

In other years, however, the pressure differential between the two atmospheric systems drops to anomalously low levels. This generally signals the beginning of an El Niño (Fig. 9.31B). The decrease in pressure differential causes the southeasterly trade winds to weaken; sometimes they become so weak that they reverse, blowing from west to east along the equator. The warm water piled up in the

western Pacific flows back toward the east. Along the coast of South America the thermocline is depressed; temperatures of both surface water and upwelling water are unusually warm. The rising, moisture-laden part of the convective pattern in the atmosphere shifts from the western Pacific to the east.

The Impacts of an Exceptionally Intense El Niño

The weakening or reversal of wind and current patterns during El Niño wreaks havoc on normal regional weather systems. In the El Niño of 1982–1983—the worst of this century—the normal weather pattern was completely reversed. The region of rising air (normally situated near Indonesia and Australia) shifted all the way across the Pacific to the coast of South America, causing heavy rains to fall in that normally dry region. Islands in the central Pacific received month after month of record rainfalls, and Peru reported its heaviest rainfall in 450 years. Meanwhile, Indonesia, Melanesia, and eastern Australia suffered exceptionally long droughts. The normal tracking patterns of tropical storms were also disrupted. For example, Hawaii was hit by an unusual northward-moving hurricane in November 1982; the last major storm with similar characteristics had occurred 25 years before, also during an El Niño. Record-breaking weather anomalies developed in other regions as well; the eastern Pacific jet stream current became much more intense than usual, causing extensive wet spells and storminess all the way from California to Cuba.

When temperature zones in the ocean shift, normal zones of marine life also shift; early accounts of El Niños in Peru contained reports of exotic life forms carried south by the warm currents, together with the disappearance of normally abundant bird and marine species. In some El Niños the anomalously warm water temperatures can cause widespread coral mortality and a deadly algal bloom called a *red tide,* which contaminates shellfish and renders them toxic for human consumption. The long-lasting suppression of nutrient-rich cool water upwelling along the coast of South America meant that the 1982–1983 El Niño was also disastrous for the Peruvian anchovy fishing industry which, prior to that time, had accounted for one fifth of the world's marine catch.

SUMMARY

1. The oceans and the atmosphere are complementary parts of a huge, complex, dynamic system. They are fundamental to the functioning of the global hydrologic cycle. The oceans play a critical role in climate and weather systems. Atmospheric circulation, in turn, drives ocean waves and currents.

2. *Weather* refers to the local condition of the atmosphere at any given time. Most weather-related phenomena originate in the troposphere, the lowest layer of the atmosphere. When the characteristic weather patterns for a given region are averaged over a significant time interval, they are referred to as *climate*.

3. Because of uneven heating of the Earth's surface, air is always in motion. The Coriolis effect breaks up the simple convective flow of air between the equator and the poles into a set of beltlike wind systems. The global pattern of air circulation, interacting with oceans and land masses, determines the distribution of climatic zones.

4. Close to 97 percent of the world's surface water is salt water, most of which resides in four great ocean basins: the Atlantic, Pacific, Indian, and Arctic Oceans. The oceans are interconnected to form a single global oceanic system, but there are dramatic local variations in topography and other characteristics.

5. Coastal zones encompass a wide range of ecosystems and landforms. People tend to inhabit coastal zones in order to have access to ocean resources, but this can increase their vulnerability to hazards caused by oceanic and atmospheric processes.

6. Tides are created by the rhythmic rise and fall of water in oceans and other large bodies of water. They result from the gravitational attraction of the Moon and (to a lesser degree) the Sun acting on the Earth. Two tidal bulges, one on either side of the Earth, are raised by gravitational forces. As the Earth rotates, these bulges remain essentially stationary. Each major land mass therefore "runs into" a tidal bulge twice in each rotation. Water-level fluctuations in major bodies of water can also be caused by long-term climatic changes, especially those related to glacial–interglacial cycles.

7. Surface ocean currents are broad, slow drifts of water set in motion by the prevailing surface winds. Both north and south of the equator, currents are driven by the trade winds but are influenced by the Coriolis effect and deflected wherever they encounter a coast. The result is a circular motion of water in each major ocean basin.

8. Like currents, waves receive their energy from the wind. In the open ocean, water moves in a circular motion, the diameter of the circle becoming smaller with depth. When a wave approaches the shore, it begins to "feel" the bottom; its wavelength shortens and wave height steepens until eventually it breaks in the surf zone.

9. Erosional forces tear away at the land, whereas other forces move and deposit sediment, thereby adding to the land. Most erosion is accomplished by wave action in the surf zone. Longshore currents and beach drift move the eroded sediments downdrift along the shoreline. The approximate equilibrium among the erosional and depositional forces that operate along coasts is interrupted by exceptional storms that erode cliffs and beaches at rates far greater than the long-term average.

10. The dynamic interplay between erosion and deposition produces a wide range of coasts. The most common are rocky or cliffed coasts, beaches, and coral reefs. Rocky coasts are characterized by wave-cut cliffs and benches. Depositional landforms associated with beaches include dunes, barrier islands, spits, and tombolos. On the mainland side of barrier islands are found coastal environments such as lagoons and estuaries. Coral reefs, which are particularly susceptible to changes in water temperature and levels of sedimentation, are of two main types: fringing reefs and barrier reefs.

11. Techniques for protecting shorelines from erosion include hard stabilization (engineered structures such as seawalls, breakwaters, groins, and jetties) and soft stabilization (beach nourishment and dune stabilization). Beach stabilization is expensive and may lead to the degradation or destruction of the beach in the long run. Another approach to shoreline protection is nonintervention, that is, protection of coastal areas through zoning laws and other regulations.

12. A cyclone is an atmospheric low-pressure system that gives rise to roughly circular, inward-spiraling wind motion. This broad definition encompasses a variety of intense cyclonic storms, including tornadoes, tropical cyclones, hurricanes, typhoons, and nor'easters. Such storms cause damage through high winds and wave action. A hazard commonly associated with cyclonic storms is storm surges, which often cause coastal flooding.

13. Drought is a condition in which there is little or no rainfall for an extended period. Low rainfall is a characteristic of arid and semiarid regions and the great

deserts associated with them. Dust storms are common in deserts and drought-stricken regions. Desertification occurs when deserts advance onto formerly productive lands. It is a natural process, but it can be worsened or even initiated by human actions.

14. El Niño is an anomalously warm ocean current in the Pacific, with associated weakening of trade winds and widespread meteorologic effects. It is an example of an exceptional weather occurrence driven by complex interactions among the atmosphere, oceans, and even the lithosphere.

IMPORTANT TERMS TO REMEMBER

barrier island (p. 231)
beach (p. 230)
beach drift (p. 228)
climate (p. 218)
Coriolis effect (p. 220)
currents (p. 224)
cyclone (p. 238)
desert (p. 242)
desertification (p. 243)
drought (p. 241)

hurricane (p. 240)
El Niño (p. 243)
longshore current (p. 227)
meteorology (p. 218)
nor'easter (p. 241)
oceanography (p. 222)
reef (p. 234)
storm surge (p. 241)
surf (p. 226)
tides (p. 223)

tidal bore (p. 224)
tornado (p. 238)
trade winds (p. 221)
troposphere (p. 218)
typhoon (p. 240)
wave base (p. 226)
wave-cut cliff (p. 229)
weather (p. 218)

QUESTIONS AND ACTIVITIES

1. Do you live in a coastal zone? If so, what type of coast is it (reef, beach, rocky cliff)? Visit the coast. Can you recognize some of the landforms commonly associated with this type of coast? Which of the landforms appear to be erosional? Which ones are depositional?

2. The next time you are near the shore of an ocean or a large lake, take a close look at the processes of wave action and sediment transport. At what angle are the waves hitting the shore? What is the direction of the longshore currents? Do you observe any signs of coastal erosion (relative to beachfront buildings, for example)? Have attempts been made to stabilize the shore using structures such as groins, jetties, or breakwaters? What effects have these structures had? If the structures are newly installed, start your own program of periodic monitoring to assess their impact on the movement of sediment and the stability of the shoreline.

3. If you can, visit a shoreline before and after a storm. What changes do you notice? What erosional effects did the storm have? Were those effects caused primarily by wave action or did high winds also play a role? What kind of storm was it, what was its intensity, and where did it originate?

4. Investigate the linkages among human activities, land degradation, and desertification (we will also look more closely at this problem in Chapter 14). You may want to carry out your investigation in the context of a case study: the dust bowl years in the American Great Plains or the ongoing drought and desertification in the Sahel are good examples.

5. Do you live near a large lake? Is the lake susceptible to storm surges? What is the orientation of the lake (i.e., what is the map direction of a line drawn through the longest dimension of the lake)? In general, large lakes oriented parallel to the direction of prevailing winds are most susceptible to storm surges. Is this true of your lake?

6. Conduct an experiment to demonstrate the Coriolis effect. Use a large, flat rotating plate (you can put a large, circular piece of cardboard on top of a rotating

cake server, for example). Set the plate rotating, then roll a small rubber ball across it. Observe the ball's motion from above. Does it move directly across the plate in a straight line, or is its path deflected? In what direction is it deflected? Compare your results with what you know about the impacts of the Coriolis effect on air and water circulation.

7. Synthesize what you have learned in this and previous chapters in Part II by creating a table in which you compare and contrast the hazards occurring on the east and west coasts of North America. Which natural hazards are most common on each coast? Are they the same or different? If you wish, you can add the southern Gulf and northern Arctic coasts to your table.

CHAPTER 10

METEORITE IMPACTS

*It is easier to believe that Yankee professors would lie,
than that stones would fall from heaven.*
• attributed to Thomas Jefferson
(upon hearing two Yale professors report a meteorite fall)

Approximately 210 million years ago a large meteorite, probably several kilometers in diameter, tore through the atmosphere and landed in the northern wilderness of Québec with a massive explosion. The impact kicked dust and debris high into the atmosphere and left a scar some 100 km across in the hard rock of the Canadian shield. The ancient, ringlike scar is now filled by the waters of the Lac Manicouagan reservoir. At the center of the lake is a plateau topped by melted rock and a series of uplifted peaks, also remnants of the impact. The entire event probably lasted little more than a minute. At the time, of course, there were no human witnesses, but the event coincides approximately with the mass extinction of marine species at the end of the Triassic Period, one of the greatest extinctions of species in geologic history. Could there be a connection between the meteorite impact and the mass extinction?

On June 30, 1908, a mysterious explosion occurred in the Tunguska region of Siberia. An eyewitness more than 100 km away from the blast site gave the following account of the event:

> Suddenly the whole sky was split in two and above the forest the whole northern part of the sky appeared to be covered with fire. I felt a great heat as if my shirt had caught fire. There was a mighty crash. I was thrown onto the ground about [7 m] from the porch. A hot wind, as from a cannon, blew past the huts from the north. Many panes in the windows [were] blown out, and the iron hasp in the door of the barn [was] broken.

The witnesses closest to the blast were reindeer herders asleep in their tents about 80 km from the site of the explosion. They and their tents were blown into the air, and some of the men were knocked unconscious by the force of the explosion. They reported that "everything around was shrouded in smoke and fog from the burning fallen trees." Even at a distance of 500 km, the sound was described as "deafening," and fiery clouds were observed. Scientists as far away as 3600 km reported anomalous air pressure waves, and seismic vibrations were detected as far as 1000 km from the site. The tremendous blast—approximately equivalent to a 15-megaton bomb—knocked down trees like matchsticks in a circular pattern radiating outward from its center, yet it left no discernible crater. After years of controversy, the consensus among investigators is that a fragment of a comet, probably about 20 to 60 m in diameter (10^7 to 10^8 kg in mass), exploded and shattered in the atmosphere just above the Earth's surface.

History is full of eyewitness accounts of fiery falls of extraterrestrial material from the sky. Fragments of such material have been recovered from all over the globe, and there are documented accounts of people, houses, and cars being struck by falling stones. How common are such events? What is their cause? What are the risks associated with a major event like the meteorite impact that caused Manicouagan Crater? These are some of the questions we will address in this chapter as we examine a phenomenon that has its origin not on Earth but in outer space.

WHAT IS A METEORITE?

A **meteorite** is a fragment of extraterrestrial material that strikes the surface of the Earth. Before the fragment hits the ground, it is referred to as a *meteoroid*—a small (i.e., smaller than a planet) solid body floating in space. The scientific study of meteorites and meteoroids is called *mete-*

◄

Manicouagan Crater in Québec was created 210 million years ago by a large meteorite impact. The original crater, now a ring lake, was 75 to 100 km in diameter.

▲ F I G U R E 10.1
Two meteor tracks (top right and center right) belong
to the Perseid meteor shower, 1993. The shower
reaches a peak about August 12 of each year. Such
regular showers occur when the Earth crosses the
orbit of a comet and its debris enters into the atmosphere. The Perseid shower is associated with the
Comet Swift–Tuttle.

▲ F I G U R E 10.2
Woodcut showing a meteorite fall near the town of
Ensisheim, France, in 1492. The stone was recovered
and preserved in a local church.

oritics. These terms (as well as *meteorology*, the study of
weather phenomena) are all derived from the Greek word
meteōron, meaning "phenomenon in the sky."

When meteoroids enter the atmosphere they are heated
by friction. The surface material of the fragment may become ionized, causing it to glow. Glowing fragments of extraterrestrial material passing through the atmosphere are
meteors, commonly known as *shooting stars;* exceptionally
bright meteors are called *fireballs.* At certain times of the
year, large swarms of meteors, all coming from roughly the
same direction, can be observed; these are called *showers.*
An example is the Perseid shower, which occurs every year
in mid-August (Fig. 10.1).

▼ F I G U R E 10.3
Worshipers circling the Kaaba (rectangular building),
the holy site at the center of the mosque in Mecca,
Saudi Arabia. The stone inside the Kaaba is believed
to be a meteorite.

▲ F I G U R E 10.4
Dr. Ursula Marvin, a scientist at the Smithsonian Astrophysical Observatory, examines a meteorite discovered on the icy surface of Antarctica. The meteorite was discovered in 1981.

Throughout history there have been reports of stones falling from the sky (Fig. 10.2). Such falls were observed and recorded in early Chinese, Greek, and Roman writings. Recovered meteorites were venerated and worshipped in a number of ancient civilizations, including that of the Pueblo Indians in Arizona. The black stone enshrined in the Kaaba, the sacred shrine of Islam in Mecca, is reputed to be a meteorite that was recovered before A.D. 600 (Fig. 10.3). However, the scientific community did not acknowledge the extraterrestrial origin of meteorites until the late 1700s.

Several close encounters with meteorites have been reported in recent times as well. In 1938 a small meteorite smashed through the roof of a garage in Illinois. In 1954 a 5-kg meteorite fell through the roof of a house in Alabama, bounced off the radio, and hit the owner of the house on the head. Another small meteorite demolished a car in a New York suburb in 1992.

Although meteorite fragments have been found all over the world, even in sediments dredged up from the sea floor, Antarctica outstrips all other collection sites in importance. In 1969, Japanese geophysicists discovered that meteorite fragments are scattered over the Antarctic ice sheet—some on the surface, others frozen into the subsurface ice (Fig. 10.4). For the most part, the fragments have been preserved from weathering and have suffered little contamination because of the subzero temperatures of their surroundings. Even the smallest samples are relatively easy to detect. Overall, several times more meteorites have been found in Antarctica than the total amount recovered prior to 1969.

Meteorite impacts have played a major role in the formation and geologic history of the Earth. It now appears that they may also have had significant consequences for the evolution, and possibly even the origin, of life. In this chapter we examine where meteorites come from, what they are made of, and their role in the formation of planetary surfaces. We also consider what happens when a large meteorite strikes the Earth, and the probability of such an event occurring during a human lifetime.

Composition and Classification of Meteorites

Meteorites are commonly named after the location in which they fell. For example, the famous Allende meteorite, the most primitive piece of solar system material known, was recovered near Pueblito de Allende, Mexico. In a very general classification scheme, meteorites can be grouped into three categories, as shown in Table 10.1: (1) *stones;* (2) *irons;* and (3) *stony-irons.*

T A B L E 10.1 • Simplified Classification of Meteorites

Meteorite Type		Description	Probable Origin
Stones	Primitive: chondrites	Unaltered by geologic processes such as metamorphism; contain chondrules and mixtures of high- and low-temperature minerals; some chondrites contain amino acids	Formed from primitive materials in the solar nebula early in solar system history
	Differentiated: achondrites	Silicate minerals similar to those found in terrestrial rocks; alteration and metamorphism evident	Broken off from outer layer of a differentiated parent body such as a large asteroid
Stony-irons		Mixtures of silicate minerals and iron–nickel metal	Contact between outer, rocky layer and metallic core of a differentiated parent body such as a large asteroid
Irons		Primarily iron–nickel metal	Metallic core of a differentiated parent body such as a large asteroid

Stones

The *stony meteorites,* or **stones**, resemble terrestrial igneous rocks. They are made primarily of silicate minerals, including many of the same minerals that are abundant in the Earth's crust. Although they are by far the most common type of meteorite (about 94 percent), because they look so much like terrestrial rocks they often are not recognized as meteorites unless the fall is actually witnessed.

Among the stones, the most abundant are the *chondritic meteorites* or **chondrites**, so named because they contain small, round, glassy-looking spheres called *chondrules,* from a Greek word meaning "seedgrain." Chondrites, especially the *carbonaceous chondrites,* tend to be very old, and most have remained unaltered since the time of their formation (see Box 10.1). Scientists believe that the chondrules in these meteorites are "frozen" droplets of the actual material that condensed from the gaseous solar nebula early in the formation of the solar system.

The other major group of stony meteorites is the *achondrites,* which lack chondrules. These meteorites have been heated, melted, subjected to pressure, and/or fractured to the extent that any chondrules that were present have been destroyed. Many achondrites closely resemble ordinary volcanic rocks of the Earth or Moon.

Irons

The *iron meteorites,* also called **irons**, are alloys of nearly pure metallic nickel and iron, sometimes with enclosed fragments of stony (i.e., silicate) material. They are easily recognizable because they are much denser than the rocks found in the Earth's crust. If you pick up an iron meteorite, it will feel *much* heavier than a terrestrial rock. Iron meteorites also tend to develop a distinctive rusty weathered surface that helps distinguish them from terrestrial rocks. When iron meteorites are cut and polished, they often reveal an intricate interlocking crystal texture called a *Widmanstätten pattern* (Fig. 10.5). These crystal growth pat-

▲ F I G U R E   10.5
Widmanstätten texture in a cut and polished iron meteorite. The crystal-growth texture indicates to scientists that iron meteorites cooled slowly, deep in the core of a large parent body. This photograph is about 10 cm across.

terns indicate that the iron meteorites cooled slowly, probably while buried in the core of a larger body.

Stony-Irons

As the name implies, the **stony-iron** meteorites are composed of intimate mixtures of stony (silicate) and metallic material. Their classification is based primarily on the minerals they contain and the proportions in which those minerals are present. Many of the stony meteorites consist of a type of material called *breccia,* a rock that has been fragmented and then recemented through heat and/or chemical cementation processes. The stony-irons present us with a mystery: How did the stony and metallic materials come to be intimately associated in a single fragment? We will attempt to provide some clues to this question as we consider the origin of meteorites in more detail.

THE ORIGIN OF METEORITES

Most meteorites are fragments of larger bodies, called *parent bodies.* It is believed that all known meteorites come from parent bodies within our own solar system, illustrating how isolated our solar system is from the rest of the vast Milky Way Galaxy.

The Asteroid Belt

Nestled between the orbits of Mars and Jupiter, at the boundary between the inner (terrestrial) and outer (Jovian) planets, is the **asteroid belt** (Fig. 10.6 and Table 10.2). The asteroid belt is a swarm of at least 100,000 **asteroids**—small, irregularly shaped rocky bodies orbiting the Sun (Fig. 10.7). About 4000 of these objects, those with known characteristics and reasonably predictable orbits, have been officially classified. Once the orbit of an asteroid has been determined with some accuracy, the asteroid is given a number and a name. The number indicates the order of discovery, and the name is honorary. Thus, "1 Ceres" was the first asteroid discovered (by an Italian astronomer named Piazzi, on New Year's Day, 1801); it is named after the patron goddess of Sicily. Asteroid "1000 Piazzi" was the 1000th asteroid to be discovered; its name honors the discoverer of Ceres.

The asteroids are probably either remnants of a planet that was broken up or rocky fragments that failed to gather together into a planetary mass. The latter possibility is more likely; in spite of the very large number of objects in the asteroid belt, their total mass is not even equal to that of the Moon. More important for our present discussion are these three characteristics of asteroids: (1) some of them are large enough to have undergone internal *differentiation;* (2) since the formation of the asteroid belt, they have continued to undergo collisions and mutual *fragmentation;* and (3) some of them have *Earth-crossing orbits.* Let's consider each of these in turn.

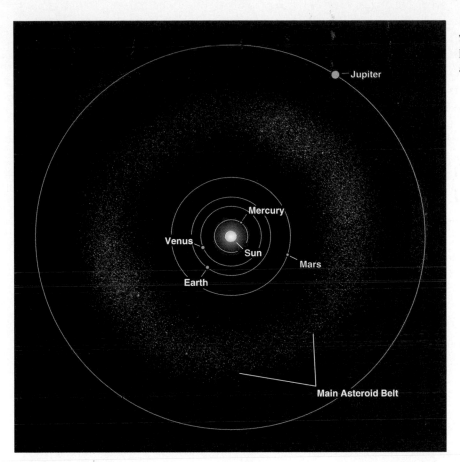

◄ F I G U R E 10.6
The location of the main asteroid belt, between the orbits of Mars and Jupiter.

T A B L E 10.2 • Orbits of the Planets and the Main Asteroid Belt, Stated in Terms of the Semi-major Axis of the Object's Orbit (Its Distance from the Sun)

Planets		Semi-major Axis (AU[a])
Mercury		0.387
Venus	Terrestrial planets	0.723
Earth		1.000
Mars		1.52
Asteroid belt		≈2.8
Jupiter		5.2
Saturn		9.58
Uranus	Jovian planets	19.1
Neptune		30.2
Pluto		39.44

[a]AU = Astronomical unit; 1 AU = 1.496×10^{11} m = mean distance of the Earth from the Sun.

▲ F I G U R E 10.7
Only recently have scientists been able to take high-resolution photographs of asteroids in orbit. This photograph of asteroid 951 Gaspra was taken in 1991 by the Galileo spacecraft. Gaspra is believed to be a fragment of a larger asteroid.

BOX 10.1
•
FOCUS ON . . .

CARBONACEOUS CHONDRITES: THE MOST PRIMITIVE MATERIAL KNOWN

Carbonaceous chondrites are a special group of stony meteorites with some interesting characteristics. As the name implies, they consist primarily of dark, fine-grained carbon-rich material. They are low in density and very easily broken. They contain large amounts of volatile materials and even some complex organic compounds. The ages of carbonaceous chondrites cluster around 4.6 billion years, the age of the solar system. These observations indicate that the carbonaceous chondrites formed in a cold part of the solar system and that they have not undergone significant geologic modification since their origins in the early stages of solar system formation. If they had been modified by geologic processes such as metamorphism, their volatile materials and organic compounds would have been driven off by heat.

The carbonaceous chondrites are scientifically important because they are among the least altered, or most *primitive,* of all known solar system materials. Allende, shown in Figure I.5, is a meteorite of this type. The discovery of protein-related amino acids of confirmed extraterrestrial origin in some carbonaceous chondrites suggests the possibility that incoming meteorites may have contributed some of the first building blocks of life on Earth. Even if

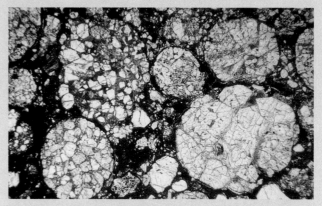

▲ FIGURE B1.1
Round chondrules seen in Tieschitz meteorite that fell on July 15, 1878, in what is now the Czech Republic. The field of view is 2.5 mm across.

the basic organic complexes needed for life were already present in the oceans, it is within the realm of scientific possibility that a meteorite impact supplied the energy needed to catalyze the chemical reactions leading to the formation of amino acids and proteins.

Differentiation

Differentiation refers to processes through which planetary bodies develop concentric layers that differ in their composition. The Earth, for example, differentiated early in its history into a dense, iron-rich metallic core, a silicate mantle, and a thin crust of much less dense silicate material. Some asteroids appear to be homogeneous, but others are large enough to have differentiated into a core, mantle, and crust, like the Earth did. It is probable that the different layers in asteroids have provided the material for the different classes of meteorites.

Fragmentation

With so many solid bodies orbiting in a close swarm, collisions and mutual fragmentation are inevitable. In fact, the size distribution of asteroids in the asteroid belt closely resembles that of fragments in a shattered rock. Asteroids in the main part of the belt encounter one another at velocities sufficient to cause mutual fragmentation upon impact. Figure 10.8 is a painting by astronomer and artist William K. Hartmann that shows the probable appearance of an asteroid collision spraying out meteoritic fragments.

Apollo Objects

Earth-approaching bodies are referred to as **Apollo objects**. Asteroids whose orbits pass within the orbit of the Earth are referred to as **Earth-crossing asteroids;** they are also known as the *Apollo group* of asteroids. About 150 Apollo objects with diameters of 1 km or more are known, but this is only a small fraction of the total number of Earth-crossing objects. The largest known Apollo asteroid is approximately 8 km in diameter. A subset of the Apollo group is the *Aten group* of asteroids, which have orbits that don't just

▲ F I G U R E 10.8
Painting by astronomer and artist William K. Hartmann showing the probable result of a collision between asteroids spraying meteorite fragments in all directions.

cross over but fall completely within the Earth's orbit. Another important orbital grouping of asteroids is the *Amor group* of Mars-crossing asteroids, which approach (but do not quite cross) the Earth's orbit.

Many Apollo asteroids will eventually strike the Earth. In general, these objects have rather unstable orbits because they are under the gravitational influences of Earth, Mars,

and Venus. If the object doesn't hit one of these three planets within 100 million years, it is likely that it will be ejected from its orbit as a result of a near miss with one of them. This means that any asteroid that started out as an Earth-crossing object early in the history of the solar system is probably long gone, so there must be a source of replenishment. The most obvious source of new Earth approachers is fragments that have been ejected from the asteroid belt, where collisions are common.

Asteroids as Parent Bodies of Meteorites

All of the characteristics of asteroids discussed so far coincide rather nicely with the characteristics we would look for in the parent bodies of meteorite samples.

1. We know from the Widmanstätten patterns in iron meteorites that the parent bodies of iron meteorites must have been large enough for minerals to cool very slowly in the core. We also know that the parent bodies of at least some of the meteorites must have been differentiated because of the modifications some of the meteorites have undergone and because of the chemical and mineralogic contrasts among the different types of meteorites.

2. The ongoing collisions and mutual fragmentation of asteroids in the asteroid belt provide a mechanism for breaking apart a differentiated parent body, with the metallic core of the body presumably yielding the iron meteorites, the mantle and crust yielding the stones, and the core–mantle boundary or the recementation of broken fragments yielding the odd mixtures of material found in the stony-irons (Fig. 10.9).

3. Finally, ongoing collisions serve as a mechanism by which some asteroid fragments may be kicked out of their normal orbits within the asteroid belt, yielding the Earth-approaching group of Apollo asteroids. Of

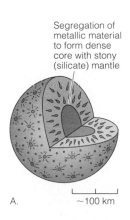
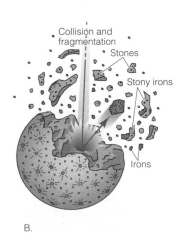

Segregation of
metallic material
to form dense
core with stony
(silicate) mantle

Collision and
fragmentation

Stones

Stony irons

Irons

A.   ~100 km   B.

◀ F I G U R E 10.9
The fragmentation of a differentiated asteroid parent body leads to the formation of three different types of meteorites. A. The parent body differentiates, separating into a metallic (iron–nickel) core. B. The parent body collides with another asteroid in the asteroid belt, breaking into smaller pieces. Iron meteorites are remnants of the parent body's core; stones are remnants of the outer layer; and stony-irons are remnants of the boundary between the layers.

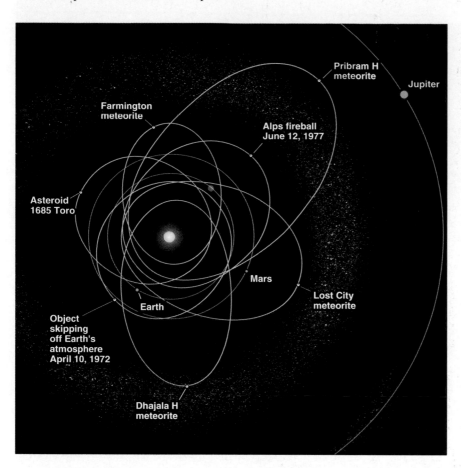

◄ F I G U R E 10.10
Orbits of meteorites whose
incoming trajectories have been
observed. Most cross through or
originate in the asteroid belt.

the few recovered meteorites for which trajectories have been determined, all appear to have originated in or near the asteroid belt (Fig. 10.10).

The natural conclusion is that asteroids in the asteroid belt are the main source of most of the meteorites that strike the Earth. Yet there are some classes of meteorites that do not fit well into this scenario because of their unusual chemical or mineralogic characteristics. Some of these meteorites are virtually identical to samples brought back from the Moon. Other rare meteorites are thought to be derived from Mars. Meteorites of lunar and martian origin must have been broken off from the surfaces of their parent bodies by the impact of another body, which sent them flying toward the Earth. The impact must have been very large for the ejected fragments to exceed the velocities necessary to escape the gravitational attraction of the parent bodies. Another, larger group of unusual meteorites is thought to have been derived from comets.

Comets as Parent Bodies of Meteorites

A **comet** is a bright object with a long, wispy tail that always points away from the Sun (Fig. 10.11). The study of comets has an interesting and colorful history. In European folklore, comets were considered signs of evil and omens of

bad luck and catastrophe. Most of the classical Greek and Roman philosophers believed that comets originated within the atmosphere because the celestial sphere was unchanging and unchangeable. The extraterrestrial origin of comets was not recognized until the 16th century, when the Danish astronomer Tycho Brahe demonstrated that they originate far beyond the Earth's atmosphere.

Comets generally are not visible until they come at least as close as Mars. The closer they come to the Sun, the more they begin to take on their familiar appearance (Fig. 10.12), with a central core called the *nucleus* surrounded by a bright diffuse halo called a *coma*. The nucleus and the coma together make up the comet's *head*. As the comet approaches the Sun, icy material in the nucleus is vaporized; driven by the solar wind, it streams away from the Sun, carrying dust particles along with it. Thus, the comet develops two types of visible tails when it is near the Sun: a *gas* or *plasma tail* made of volatilized icy material from the nucleus, and a *dust tail* made of particles of dust carried along by the solar wind. The two types of tails can be seen in Figures 10.11 and 10.12.

Observations by an international flotilla of research probes during the 1986 appearance of Comet Halley (its thirtieth historically reported appearance and fourth predicted reappearance) revealed it to have an irregularly shaped nucleus about 16 km x 8 km in dimension (Fig. 10.13). Sur-

◄ F I G U R E 10.11
Comet West, showing its dust tail (white) and its gas tail (blue). The photograph was taken in 1976.

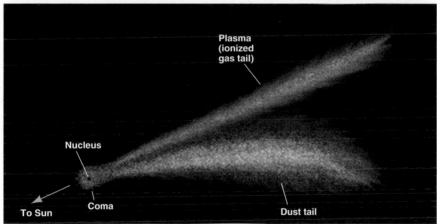

◄ F I G U R E 10.12
The parts of a comet.

Plasma
(ionized
gas tail)

Nucleus

To Sun   Coma

Dust tail

prisingly, it was found to be one of the darkest known objects in the solar system, its dark color resulting from a layer of black carbon-rich material on the outside of the nucleus. The nucleus was also found to have an extremely low density (es-

timated at 0.1 to 0.4 gm/cm³). These observations confirmed scientists' notions of comets as fluffy masses of icy material loosely held together by a matrix of silicate dust—the so-called "dirty snowball" theory.

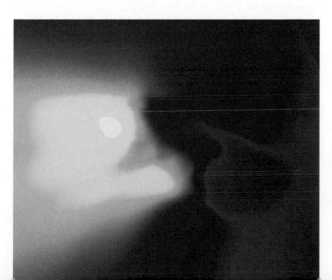

◄ F I G U R E 10.13
Spacecraft observing Halley's Comet during its 1986 trip past the Earth found its nucleus to be an irregular, dark object about 16 km long × 8 km wide. Jets of gas (red and orange), volatilized by the heat of the Sun, can be seen streaming out of the nucleus.

▲ FIGURE 10.14
Destruction caused by the Tunguska event, in Siberia, June 30, 1908, was still very evident when this photograph was taken in 1927. More than 3000 square kilometers of forest were flattened.

The orbits of comets are quite different from those of most other bodies in the solar system. For one thing, they are highly *eccentric,* which means that they are elongated and elliptical, unlike the more nearly circular orbits of the planets. For another thing, comets typically do not orbit within the Sun's equatorial plane, as do most other bodies in the solar system. These observations and the very icy compositions of the comets suggest that they come from the far outer reaches of the solar system.

Comets are almost certainly the parent bodies of some of the objects that strike the Earth's surface. The nucleus of Halley's Comet was found to contain material whose composition is very similar to that of some chondrites. Some comets have been observed to split apart, fragment, and even disintegrate completely upon approaching the Sun. This may be due in part to the effect of jets of gas and volatilized material streaming out of the nucleus. Most periodic meteor showers are thought to consist of swarms of small, glowing fragments of comets that have entered the atmosphere.

The fact that comets are so loosely held together means that they tend to break apart when they enter the atmosphere. Nevertheless, the impact of a large cometary fragment can be devastating, as illustrated by the Tunguska explosion described at the beginning of the chapter (Fig. 10.14). In the Tunguska event, the absence of a discernible crater and the lack of significant amounts of meteoritic material in the vicinity of the blast suggest that the comet exploded above the surface.

IMPACT EVENTS

What happens when an extraterrestrial object strikes the Earth? In the process of *impact cratering,* the planetary surface is deformed as a result of the transfer of kinetic energy from the object to the surface. The result is an impact **crater**, a bowl-shaped depression that is approximately circular. The crater's width and depth and other effects associated with the impact depend primarily on the size and velocity of the incoming body.

Size and Velocity of Incoming Objects

The total *meteorite flux*—that is, the mass of meteorites that strikes the Earth—is about 10^7 to 10^9 kg/year. This amounts to hundreds of tons of meteoritic material a day! Most of this material consists of dust-sized particles, or *micrometeorites.* Because of the Earth's gravitational field, the meteorite flux on Earth is about 50 percent greater than that on the Moon. But friction from the Earth's atmosphere generally causes small objects, anything less than about 1 m in diameter, to disintegrate and burn up before reaching the ground. This, combined with the active modification of the surface by erosional processes, accounts for the fact that the Earth's surface is much less noticeably cratered than that of the Moon.

Objects large enough to make it through the atmosphere and cause significant cratering—that is, objects greater than 1 m in diameter—strike the Earth at a rate of about one a year. Great impact events, which cause craters of several kilometers or more, are much rarer. Meteorites ranging from 100 m to several kilometers in diameter, for example, strike the Earth less than once in every million years (1 in 10^6 years).

The velocities of small meteorites entering the atmosphere have been measured at between 4 and 40 km/s. If a large meteorite struck the Earth at such a velocity, the amount of energy released would be enormous. It has been calculated that the impact of a meteorite 30 m in diameter and traveling at a speed of 15 km/s would release as much energy as the explosion of 4 million tons of TNT. The resulting impact crater would be the size of Meteor Crater (also known as Barringer Crater) in Arizona, 1200 m across and 200 m deep (see Fig. I.6).

Cratered Surfaces

To a large extent, the Earth's surface is shaped by three processes: tectonic processes, magmatic processes, and the surface processes of weathering, mass-wasting, and erosion. To varying degrees, each of these sets of processes has played a role in shaping the surfaces of all the rocky planets and moons in the solar system. In the context of solar system history, however, a fourth process, impact cratering,

BOX 10.2
•

THE HUMAN PERSPECTIVE

JUPITER'S ENCOUNTER WITH COMET(?) SHOEMAKER–LEVY 9

*T*wenty-one pieces of a fragmented object called Shoe-maker–Levy 9 smashed into the turbulent atmosphere of Jupiter over a 1-week period from July 16 to 22, 1994 (Fig. B2.1). The explosion created superheated atmos-pheric pools of magnesium, sulfur, and silicate on the side of Jupiter that was not directly visible from the Earth. At the point of collision the fragments were traveling at 60 km/s and released an amount of energy equal to 6 million megatons of TNT, more than the amount that could be re-leased by all the world's nuclear weapons combined.

Scientists had a long time to prepare for this space spec-tacular. An intact Shoemaker–Levy 9 with an estimated ra-dius of 5 km had approached Jupiter the previous year. At that time the object broke into smaller pieces because of tidal stresses caused by its proximity to the giant planet. More fragmentation resulted from collisions among parti-cles. The comet that returned included 21 fragments rang-ing from 1 km to more than 2 km in radius as well as in-numerable smaller fragments.

With the time and place of the comet's impact pre-dicted a year in advance, the scientists were able to simu-late the event, test theories, and prepare the global scien-tific and public communities to observe an impact similar to the one that could have accounted for mass extinctions on Earth millions of years ago. Amateur and professional astronomers pointed their telescopes at Jupiter. The best view, however, was from NASA spacecraft, which were po-sitioned to have an unobstructed view of the far side of the planet. Signals were sent back from Galileo, an unmanned spacecraft on its way to a rendezvous with Jupiter in 1995, and from the Hubble Space Telescope, which produced the highest resolution images.

Even as Shoemaker–Levy crashed into Jupiter, scientists were scrambling to explain the data they were collecting. The first observations described upwardly expanding dark plumes at each impact site. To the observers' surprise, no water molecules were detected in these plumes. This led some scientists to speculate that Shoemaker–Levy might have been an asteroid instead of a comet. Comets are com-posed of chunks of water ice and other types of ice mixed with silicate dust and dark carbonaceous material; if Shoe-maker–Levy was a comet, why didn't it leave a trail of

water molecules in its impact plume? It may have been a more rocklike object, or possibly a burned-out comet.

Visible-light telescopes picked up dark impact marks in the Jovian clouds. These marks were distorted and merged over time by Jupiter's atmospheric winds, forming a black band that began to appear several weeks after the collision. Ultraviolet spectra taken from equipment on the Hubble Space Telescope revealed sulfur and hydrogen sulfide, two emissions never before detected on Jupiter.

When all the data have been analyzed and reviewed, as-tronomers expect to have a much better understanding of the stratification of Jupiter's lower atmosphere, the planet's magnetosphere, and storms and eddies occurring in its at-mosphere. This information may eventually help scientists comprehend similar features of the Earth's atmosphere.

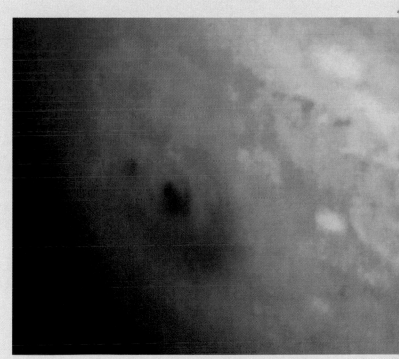

▲ FIGURE B2.1
The impact area of Shoemaker–Levy 9, fragment G, on Jupiter, photographed through the Hubble Space Telescope. The image, which was made about 20 hours after the impact, shows a center of boiling gas the size of the Earth.

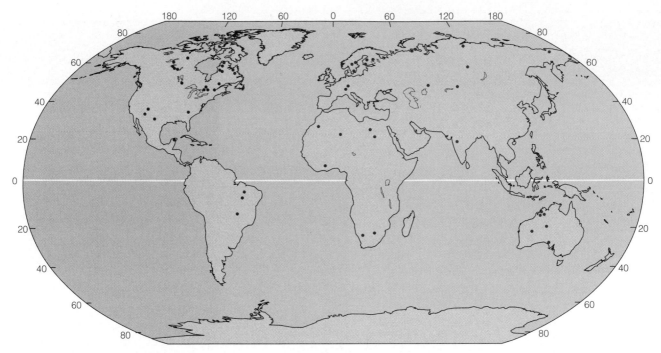

▲ FIGURE 10.15
World map of large impact craters. More than 200 impact craters have been identified on the Earth.

has been equally important in the modification of planetary surfaces, including that of the Earth.

Terrestrial Impact Craters

More than 200 impact craters have been identified on the Earth (Fig. 10.15). However, there must have been many more impacts than this small number implies. We know, for example, that the Moon is heavily cratered as a result of a period of particularly intense meteorite bombardment

early in the history of the solar system. The Earth experienced the same period of intense bombardment, but few craters remain because weathering and the rock cycle continually erase the evidence. Most recognizable terrestrial impact craters have been preserved because they are very large, very recent, or located in very stable geologic environments, or else because they were rapidly buried by protective sediment that has since been removed by erosion.

There are two basic types of craters (Fig. 10.16): simple

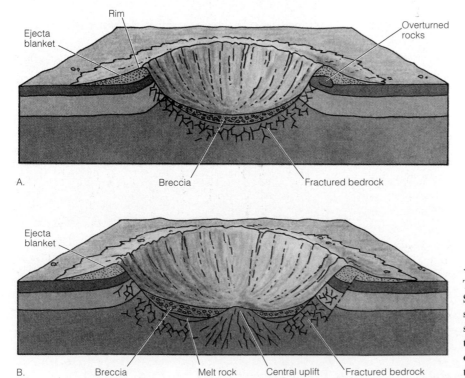

◀ FIGURE 10.16
Two basic types of craters. A. Simple craters are relatively small, with simple, bowl-like shapes. B. Complex impact structures and basins are large (3 km or more in diameter), with central peaks and ringed structures.

craters, which are generally less than 5 km in diameter with a simple, bowl-shaped excavation; and complex impact structures and basins, which are generally 3 km or more in diameter, with a distinct central peak and/or ringed structure. Notable examples of simple craters are Meteor (Barringer) Crater in Arizona; New Québec Crater in Québec; and Roter Kamm in Namibia. Some examples of complex impact structures include Upheaval Dome in Utah; Sudbury in Ontario; and Gosses Bluff in Australia.

The Mechanics of Impact Cratering

Although many ancient impact craters have been discovered, no large, natural impact crater has ever been observed as it was being formed. What is known about the process of impact cratering comes largely from laboratory experiments (see Fig. 10.17). As a high-speed meteorite strikes and penetrates the surface of a planet, it causes a jet of rock and dust, called **ejecta**, to be strewn at high velocity away from the point of impact in all directions. This produces a blanket of ejecta that surrounds the crater and thins out at greater distances from the rim.

At the same time, the impact compresses the underlying rocks and sends intense shock waves outward. The pressures produced by the shock waves from a large meteorite are so great that the strength of the rock is exceeded; the re-

A.

B.

▲ F I G U R E 10.17
The sequence of events in the formation of an impact crater. A. The initial impact ejects a high-velocity jet of material from just below the surface. B. The passage of shock waves through the bedrock produces high pressures and compresses surrounding rock and sediment layers. In places the pressure exceeds the strength of the rock, causing fracturing. C. Rock layers along the rim of the crater are folded back and overturned by decompression. The ejected debris forms a circular blanket around the crater.

▲ F I G U R E 10.18
Meteorite impact craters. A. Meteor Crater, Arizona, is a simple crater. The raised surface surrounding the 1200 m wide crater is a blanket of rock ejected during the impact event. B. Upheaval Dome, Canyonlands National Park, Utah, is a complex crater consisting of more than one ring and a central, uplifted dome.

sult is a large volume of crushed and brecciated material. In very large impact events, vaporization may occur. Once the compressive shock waves have passed, rapid decompression occurs. In some cases the decompression causes melting. Molten rock may then rise along fractures produced by the impact and flood the floor of the crater.

Rock layers adjacent to the rim of the crater are overturned. In the largest impact craters, the central crater is surrounded by one or more raised rings of deformed rock. The outer rings and central peaks of large, complex impact craters (Fig. 10.18) are presumed to have formed as a result of the initial compression.

Cratering is a very rapid process; the event that produced Meteor Crater, for example, is estimated to have lasted only about 1 minute. After the initial impact, however, the crater is modified by several kinds of postimpact events. The walls may slump; rebound may produce changes in the floor and rim; and erosion may fill the crater with debris.

METEORITE IMPACTS AND MASS EXTINCTIONS

Whereas a major meteorite impact would surely ravage local ecosystems, the regional and global effects of such an

event could be equally devastating. Many scientists believe that there is a direct connection between major impact events and the mass extinctions of species that have occurred periodically throughout the history of the Earth. Other scientists disagree vigorously with this theory. The controversy surrounding the impact theory of extinction has generated one of the liveliest multidisciplinary dialogues in recent scientific history.

Regional and Global Effects

First, let's consider whether a major meteorite impact could be devastating to life, in principle. It has been estimated that the impact of an object 10 km in diameter traveling at 20 km/s would instantaneously release about 100 megatons of kinetic energy—a nonnuclear explosion equivalent to about 1000 times the total yield of all existing nuclear weapons. Among the widespread and catastrophic effects of such an impact would be global darkness; extreme cold, followed by enhanced warming of the atmosphere; acidic rain; tsunamis; and global wildfires (Fig. 10.19).

Darkness Fine dust particles created by the impact would be ejected into the upper atmosphere, probably encircling the globe in a matter of weeks. On the basis of modeling of impacts and knowledge about dust injected into the atmosphere by volcanic eruptions, it has been suggested that the

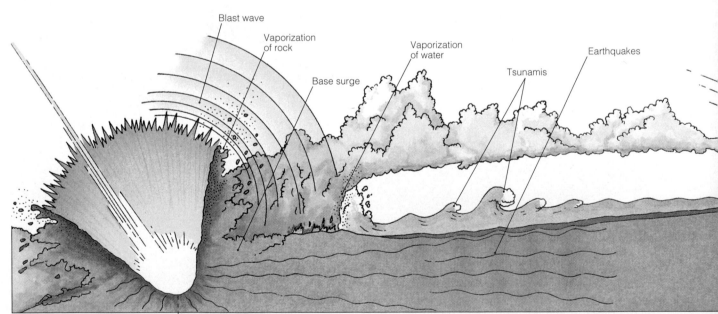

Time ⟶   Immediate regional effects

▲ F I G U R E 10.19
Global and regional effects of a major impact.

entire world could be plunged into darkness for several months following a major impact event. The results would include the suppression of photosynthesis, collapse of food chains, and crop failure. The dust would eventually fall to the ground, depositing a thin blanket of sediment worldwide. This sediment band, preserved in the geologic record, would be the global "signature" of the impact.

Cold The period of global darkness would probably be accompanied by intense cold because the dust would block out incoming solar radiation. (Researchers modeling these effects have noted similarities with the possible aftermath of a nuclear war, the so-called *nuclear winter* scenario.) If the impact occurred in the ocean, water vapor would also be distributed throughout the atmosphere, remaining there long after the dust particles had settled out. Water vapor is an important agent of warming in the Earth's atmosphere, known as the greenhouse effect. The presence of a large quantity of additional water vapor in the atmosphere could enhance this effect, so that the cold period would be followed by a period of extreme global warming.

Other Effects The energy and shock heating resulting from a major explosion could cause nitrogen and oxygen in the atmosphere to combine, creating compounds such as nitric acid that would result in acidification of rain and surface waters. A major impact in the ocean would also cause huge tsunamis. Radiation from the fireball would probably ignite global wildfires, and the impact might even generate earthquakes.

It seems clear that the global effects of a major meteorite impact could be devastating to life. What we must determine next is whether there is any direct evidence for the connection between major impacts and mass extinctions in geologic history.

The Geologic Record of Mass Extinctions

Most people know that dinosaurs became extinct about 65 million years ago, at the boundary between the Cretaceous and Tertiary Periods (commonly referred to as the *K–T boundary*). But many are not aware that other animal and plant species were also affected. Approximately one quarter of all known animal families then alive (including marine and land-dwelling species) became extinct at the end of the Cretaceous Period. This mass disappearance of species is clearly evident in the fossil record; it is the reason that early paleontologists selected this particular stratigraphic horizon to represent a major boundary in the geologic time scale.

The great Cretaceous–Tertiary extinction is not unique in Earth history, nor was it the most devastating of such occurrences; there have been at least five and possibly as many as 12 recognizable episodes of mass extinction during the

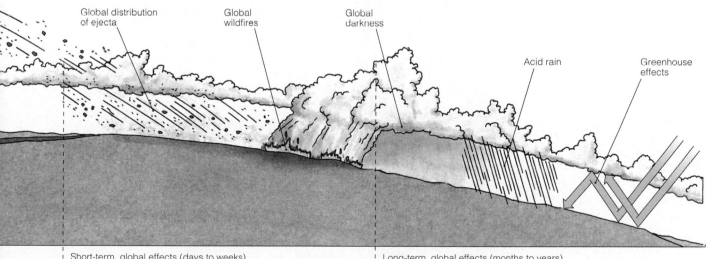

Global distribution of ejecta   Global wildfires   Global darkness   Acid rain   Greenhouse effects

Short-term, global effects (days to weeks)   Long-term, global effects (months to years)

past 250 million years (Fig. 10.20). The most devastating of these occurred 245 million years ago, at the end of the Permian Period. Perhaps as many as 96 percent of all species then alive died out at that time. Another great extinction occurred at the end of the Triassic Period, approximately coinciding with the meteorite impact that created Manicouagan Crater in Québec.

The Iridium Anomaly

The possibility that mass extinctions might be caused by devastating impacts was suggested at least as early as 1970.

In 1978, however, the connection between meteorite impacts and mass extinctions was dramatically substantiated by a piece of evidence uncovered by scientists from the University of California at Berkeley. They found anomalously high concentrations of a rare element, iridium, in clay-rich sediments marking the K–T boundary (Fig. 10.21).

This *iridium (Ir) anomaly* has been found in K–T boundary sediments throughout the world. What is its significance? Only two reasonable explanations have been proposed so far: Either the iridium was brought to the sur-

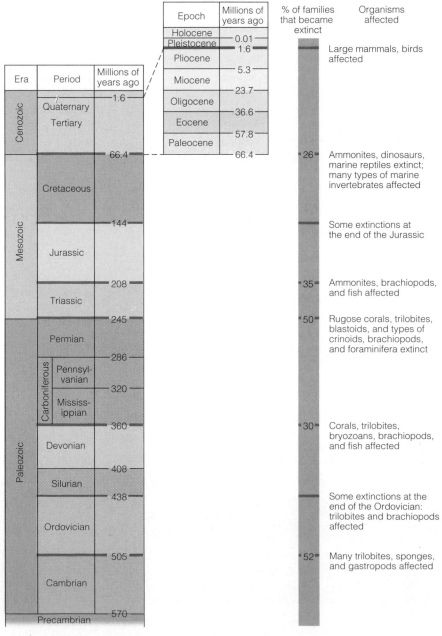

▲ F I G U R E 10.20
Mass extinctions in geologic history.

▲ F I G U R E 10.21
**Geologists climbing a face of eroded limestone
strata to observe a thin, dark layer of rock rich in the
chemical element iridium. The iridium-rich layer,
seen here in New Mexico, has been identified at many
places around the world and is believed to have
formed as a result of a world-encircling dust cloud
formed by a great meteorite impact about 65 million
years ago.**

by the extraordinarily high pressures characteristic of major
explosions. Grains of the mineral *stishovite,* which are
formed only under conditions of intense pressure, have also
been found. Also collected from some of the K–T boundary layers are tiny glass spherules, which are thought to be
the remnants of melt droplets generated by the impact.

Periodicity of Extinctions

The unfolding story of the link between meteorite impacts
and mass extinctions became more complicated when scientists from the University of Chicago proposed that extinctions may actually occur on a regular schedule (Fig.
10.22). Through computer modelling they demonstrated
that extinctions seem to occur at regular intervals of approximately 26 million years. In the absence of an obvious
terrestrial cause for this periodicity, the researchers suggested that it might be caused by the effects of periodic
swarms of comets passing through the inner part of the
solar system. The likelihood of a comet striking the Earth
would be dramatically increased during these encounters. If
this model turns out to be valid, the concept of comet impacts causing mass extinctions on a regular, cyclical basis
would revolutionize our theories of how life on Earth
evolved into its present state.

face of the Earth by a period of exceptionally intense volcanic activity, or it came from a sudden influx of extraterrestrial material. As mentioned earlier, the dust ejected into
the atmosphere by a major impact would eventually settle
out, forming a global layer of fine sediment like the iridium-rich clay layer that marks the K–T boundary. The fact
that iridium is much more abundant in meteorites than in
the crust of the Earth lends support to the hypothesis that
the Ir anomaly resulted from a major meteorite impact.

Soot, Shock, and Spherules

On the basis of evidence presented so far, we cannot rule
out the possibility that the iridium was ejected from the
Earth's interior through volcanism. Nor can we definitively
connect the event that generated the Ir anomaly with the
mass extinction of species at the K–T boundary.

However, other lines of evidence appear to connect the
K–T extinction to an extraterrestrial (i.e., impact) cause.
For example, soot has been discovered in the K–T boundary clay layer at a number of locations. This has been interpreted as evidence of global wildfires ignited by the fireball.
Grains of quartz that have undergone *shock metamorphism*
have also been discovered. Shock metamorphism is a type
of microscopic mineral fracturing that can be caused only

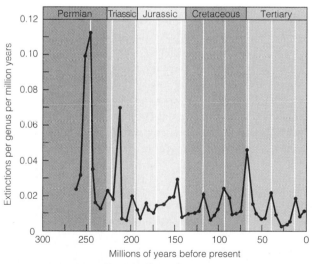

▲ F I G U R E 10.22
**Periodicity in the fossil record of extinctions suggests
that a regular, recurrent event has disrupted the Earth's
biosphere every 26 million years. This diagram is
based on the work of David Raup and John Sepkoski
of the University of Chicago, who analyzed extinctions
in 11,000 genera of marine organisms. The vertical
lines represent intervals of 26 million years, which
seem to be well aligned with the extinction peaks over
the past 150 million years; for earlier times the correlation is not as close.**

The Controversy Continues

To summarize the story so far, it is now generally accepted that:

- Major meteorite impacts can cause devastating world-wide effects.
- Such effects would leave characteristic signatures in the geologic record, such as Ir anomalies, sooty layers, shocked minerals, and glassy spherules.
- Many such pieces of evidence have been located, not only in the clay-rich layer that marks the Cretaceous–Tertiary boundary but in other layers marking major transitions in the geologic time scale.

The suggestion of a 26-million-year periodic cycle of mass extinctions and the significance of this observation are less widely accepted. And the direct cause and effect connection between a major meteorite impact and the extinction of the dinosaurs (and other species) at the end of the Cretaceous Period is still a subject of controversy. Let's take a closer look at this ongoing debate.

Volcanism Versus Impact Theory

A number of experts argue that many of the effects associated with a major impact could have resulted from an extended period of particularly intense volcanic activity. These effects include the Ir anomaly and global climatic changes caused by material ejected into the stratosphere. It has even been suggested that impact-generated volcanic activity could have caused the mass extinctions. Observational evidence seems to favor the impact theory over volcanism as the causal mechanism for the K–T extinction (Table 10.3). However, there is a remarkable correlation between known extinction events and the occurrence of very extensive basaltic lava flows. The largest known mass extinctions (those marking the Permian–Triassic and Cretaceous–Tertiary boundaries) coincide with the formation of the world's largest flood basalt provinces, which are located

in Siberia and India, respectively. These are no ordinary volcanic events; the Deccan Plateau flood basalts in India, for example, represent 1.5 million km³ of lava extruded over a period of 100,000 to 1 million years. The consequences of such an extended period of volcanic activity must have been dramatic and widespread.

The Time Scale of Extinctions

Another major unsolved problem for the impact theory of mass extinctions is related to the timing of the extinctions themselves. The geologic record of sediments, and therefore the fossil record, is incomplete; in any given locality thousands or even millions of years of the record may be missing altogether. Even in the best preserved fossil strata, with extremely tiny and abundant organisms, there will be gaps of thousands of years in the fossil record. The situation becomes even more tenuous when one is dealing with larger, much less common fossils such as dinosaurs; in such cases, there are likely to be gaps of 1 million years or more. This means that what looks like an abrupt extinction on the geologic time scale may turn out to be quite gradual when examined on a finer time scale. Furthermore, it is often difficult to prove that changes observed in a particular species occurred at the same time throughout the world.

Many experts argue that this is the case with the K–T and other mass extinctions—that the decline of many species actually began many thousands of years before the actual transition point and continued for tens of thousands or even millions of years (Fig. 10.23). Mechanisms such as extensive, episodic volcanic eruptions, gradual climatic change, or changes in sea level may offer better explanations for the extended duration of these extinction episodes than an impact scenario.

The "Smoking Gun"

A perennial problem for those who favor the impact theory of mass extinctions has been the lack of a "smoking gun." That is, there was no large impact crater that could be

T A B L E 10.3 • Comparison of Observational Evidence from the K–T Boundary with a Meteorite Impact or Volcanism as the Causal Mechanism

Observational Evidence	Meteorite Impact	Volcanic Eruption Quiet Basaltic	Violent Rhyolitic
Iridium	Yes: all types of meteorites are high in iridium	Possibly: high Ir found in Kilauea gas; no other known case of high Ir	No: high Ir not yet found in rhyolitic volcanics
Spherules	Yes: droplets of impact melt	Yes: drops of liquid basalt	No: violent eruptions produce angular fragments, not droplets
Shocked minerals	Yes: shocked quartz common at known impact sites	No: large explosions do not occur in this setting	Possibly: violent rhyolitic eruptions might produce sufficient pressure for shock metamorphism to occur
Soot (global)	Yes: global wildfires set off by fireball	No: fires only very local	No: fires only very local

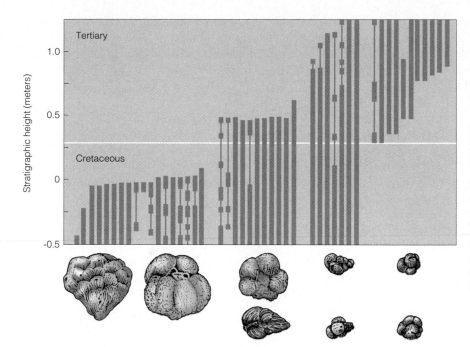

◄ **F I G U R E 10.23**
This graph, based on the work of Gerta Keller of Princeton University, shows a stairstep pattern of extinctions of fossil foraminifera, tiny shelled marine organisms. The important feature of the diagram is that the extinctions began before the Cretaceous–Tertiary boundary (marked by the horizontal line) and continued in stepwise fashion across the boundary. The entire sequence represents roughly a million years of Earth history. This suggests that the K–T extinction may not have been as sudden as the fossil record of larger, rarer fossils (like dinosaurs) seems to indicate.

shown to have formed exactly 65 million years ago, at the time of the K–T extinction. Recently this problem may have been resolved by the discovery of Chicxulub Crater, a

180-km-diameter crater in the northern Yucatán Peninsula of Mexico that appears to have been formed about 65 million years ago (Fig. 10.24). The presence of shocked min-

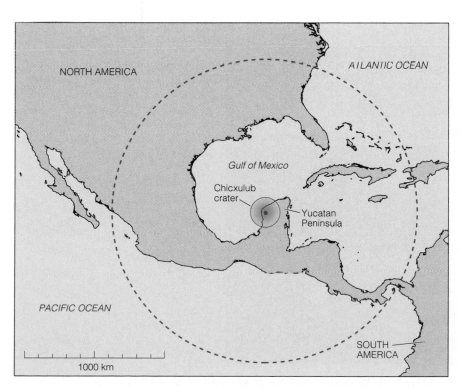

◄ **F I G U R E 10.24**
Recently discovered Chicxulub Crater near the Yucatán Peninsula may be the "smoking gun" sought by impact theorists—the impact made by the meteorite that caused the Cretaceous–Tertiary extinction. Effects of this impact were recorded in the rocks as far away as the red dashed circle.

eral grains in associated ejecta deposits supports the interpretation of the structure as an impact crater. The very large size of the crater is consistent with the production of the global effects discussed earlier—worldwide distribution of iridium dust in the stratosphere, and soot from wildfires. In addition, materials thought to be deposited by impact-generated tsunamis have been found in coastal areas as far away as Texas.

After more than 15 years of debate, the questions linger: What killed the dinosaurs? Was the Earth devastated by the effects of a major impact, or by an extended period of intense volcanism? Could events like these have caused a mass extinction of species, or were they just the final blow in a long, slow decline that had begun thousands of years earlier? Is the apparent cyclicity of mass extinctions related to periodic swarms of extraterrestrial objects? Whether or not these issues are resolved in the near future, the research surrounding these questions has contributed much valuable information about the evolution of life on the Earth. It has also fueled a fascinating and lively scientific debate.

WORLDS COLLIDING

Given the prevalence of impact cratering and the mass of material known to enter the atmosphere each year, what are the chances of a large impact occurring in the foreseeable future? It has been calculated that in a single human lifetime there is about a 1 in 10,000 chance that the Earth will be hit by a meteorite large enough to cause worldwide climatic changes leading to widespread crop failures and, perhaps, mass extinctions of species. These are about the same

as the odds of dying from anesthesia during surgery, dying in a car crash during any 6-month period, or dying of cancer as a result of breathing automobile exhaust on a Los Angeles freeway every day.

Near Misses

Some rather unnerving near-collisions with large extraterrestrial objects have been recorded in the not too distant past. For example, on June 14, 1968, an Apollo object named 1566 Icarus passed within 6 million kilometers of the Earth. If the object had struck the Earth, it would have created a crater 10 km in diameter. The approach of 1566 Icarus had been predicted by astronomers, who assured the public that there was no cause for alarm.

An astonishing near-collision with a slightly smaller body occurred on April 10, 1972, when a rocky object somewhere between a few meters and 80 m in diameter approached the Earth at a velocity of 10 km/s. Impelled by gravitational attraction, it accelerated to 15 km/s and entered the Earth's atmosphere. It traveled over North America from south to north, dipping as low as 50 to 60 km above the surface and becoming visible as a brilliantly glowing object over Utah, Idaho, and Montana. If the object had fallen to the surface it would have struck somewhere in Alberta, causing an explosion about the same size as that of the bomb that destroyed Hiroshima in World War II. Instead, the object glanced off the atmosphere and returned to outer space (Fig. 10.25).

Observers of Apollo objects are identifying Earth approachers at the remarkable rate of 35 new objects each year. In all, there may be as many as 300,000 Earth-cross-

◄ F I G U R E 10.25 On August 10, 1972, a small asteroid passed at an altitude of just 58 km above the western United States. The asteroid is seen as a bright streak in this photo, taken at Grand Teton National Park in Wyoming.

STRATEGIC PLANNING FOR A COSMIC STRIKE

*M*ost of the asteroids in our solar system exist in an orbit between Mars and Jupiter; however, many have orbits dangerously close to the Earth's. Collisions and near-collisions are well documented. If the Earth were hit by a large asteroid or comet, as it has been many times in the past, the results would be catastrophic. The doomsday scenario for a major collision includes a blast large enough to annihilate a significant portion of the Earth's human, plant, and animal population; spew enough Sun-blocking dust into the atmosphere to change the global climate and destroy agriculture; and cause monstrous firestorms and tsunamis. What can we do to prepare for such an event or, better yet, to forestall it?

In 1991, at the request of the U.S. Congress, the National Aeronautics and Space Administration (NASA) presented a proposal to protect the Earth and its inhabitants from such impacts by establishing the Spaceguard Survey. The plan called for a global network of at least six automated wide-field telescopes to map the skies and track the orbits of asteroids and other planetary bodies that could strike the Earth. This plan was designed by experts from Australia, Finland, France, India, Russia, and the United States. NASA's early-warning system would locate 95 per-

cent of potentially deadly large asteroids and predict their paths. In the unlikely case that one of them was on a trajectory for a direct hit on the Earth, the scientists would have a couple of decades to prepare a defense.

Once the trajectory of an Earth-approaching asteroid is known, it may be possible to plan a mission to intercept it. The most realistic proposal is to redirect the asteroid by exploding a nuclear bomb alongside it. The NASA proposal states that very powerful explosive devices would be required to provide enough energy to skew the path of a typical asteroid, an object the size of a mountain traveling toward the Earth at 25 km/s.

The problems associated with the Spaceguard Survey and intercept plans involve cost, technology, and human nature. The costs of the Spaceguard Survey alone are estimated at $50 million initially and $10 million annually for 25 years. Moreover, even after an intense search extending over 25 years, some potentially dangerous Earth-approaching objects might not have been identified. A final argument voiced by critics is that once the technology is in place to modify an asteroid's orbit, the likelihood of its misuse is greater than the probability of an asteroid hitting the Earth.

ing objects with diameters greater than 100 m. As many as 50 objects between 5 and 50 m in diameter regularly pass closer to the Earth than the Moon. Most of the newly identified Apollo objects are asteroids. However, with respect to possible future impacts, the greatest cause for concern comes not from an asteroid, but from a comet.

Comet Swift–Tuttle

Comet Swift–Tuttle is approximately 10 km in diameter, about the same size as the body that could have caused the extinction of the dinosaurs. The comet was named for the American astronomers who observed its passage in 1862. Small fragments of Swift–Tuttle are responsible for the Perseid meteor showers described earlier. The fragments are sprayed out of the comet's nucleus every time it passes the Sun and eventually become dispersed along the comet's orbit. Whenever the Earth crosses the path of the fragments, a meteor shower results.

About 200 comets regularly pass by the Earth as they follow their very elongated orbits from the outer edge of the solar system. Although it is often possible to predict the return of a comet with great accuracy, new (i.e., never before observed) comets have a disconcerting habit of appearing "out of nowhere." As noted earlier, comets can also change their trajectories, split apart, or even disappear without notice. Gas jets streaming out from the core of the comet can change the comet's course or weaken the already loosely bound material in the nucleus, causing it to fragment.

Objects as large as Swift–Tuttle strike the Earth only about once in every 10 to 30 million years. In 2126 Swift–Tuttle will pass within 60 lunar distances (about 23 million kilometers) of the Earth. This is a reasonably safe distance, and the comet probably will not prove to be of concern to our descendants . . . unless the gas jets from its nucleus happen to nudge it slightly off its course. No doubt, astronomers of the twenty-second century will be tracking Swift–Tuttle's course with interest.

SUMMARY

1. A meteorite is an extraterrestrial object that strikes the Earth. Glowing fragments of extraterrestrial material passing through the atmosphere are called meteors. Large meteorite impacts are rare, but such impacts may have had catastrophic effects on the Earth at various times throughout geologic history.

2. Meteorites can be classified into three general categories: stones, irons, and stony-irons. One group of stony meteorites, called chondrites, is thought to represent primitive material that has not been altered since early in the history of the solar system. Another group, the achondrites, resemble terrestrial igneous rocks. The irons are made of an alloy of nickel and iron, while the stony-irons are complex mixtures of stony (silicate) material and metal.

3. The asteroid belt, which contains at least 100,000 asteroids, is located between the orbits of Mars and Jupiter. The asteroids are probably either remnants of a planet that was broken up or rocky fragments that failed to gather together to form a planetary mass.

4. Many of the characteristics of asteroids suggest that they are the most likely parent bodies for most meteorites. Large asteroids have undergone differentiation into metallic cores and rocky outer layers, which would account for the compositional differences among meteorites. Collision and mutual fragmentation are common processes in the asteroid belt; such collisions would produce smaller fragments with erratic trajectories. Asteroids whose orbits cross within the Earth's orbit are called Earth-crossing asteroids or Apollo objects.

5. Comets may be the parent bodies for some types of meteorites. Comets are loosely packed aggregations of ice and silicate dust, resembling "dirty snowballs." When they pass close to the Sun they begin to glow as a result of the vaporization of volatile material, and they take on the characteristic appearance of a comet with a glowing head and long, wispy tails.

6. Impact cratering has played an important part in the modification of planetary surfaces throughout the history of the solar system. The size of the crater that forms, and the nature of the events associated with the impact, depend on the size and velocity of the incoming object. The Earth receives hundreds of tons of meteoritic material per day, but dust-sized particles are far more common than large objects. Meteorites with diameters greater than 100 m, which would cause significant cratering, strike the Earth about once every million years.

7. Impact craters are not as common on Earth as on the Moon because on Earth the processes of weathering and erosion continually wipe out or cover the evidence. The two main types of craters are the simple, small, bowl-shaped crater and the large, complex impact structure. When an impact occurs, material is thrown out of the crater, forming a blanket of ejecta. Fracturing and melting of rocks may occur at the bottom of the crater, and rock layers around the edge of the crater are overturned.

8. A very large impact would have regional or even global effects, including global darkness, extreme cold followed by enhanced warming, acidic rain, tsunamis, and global wildfires.

9. It has been postulated that major impacts might be the cause of mass extinctions of species at different times in geologic history, notably the extinction of the dinosaurs at the boundary between the Cretaceous and Triassic Periods. The presence of a layer of clay marking this boundary, which is anomalously high in the rare element iridium, seems to support this theory. However, an extensive period of widespread volcanism might also have contributed to the extinction of the dinosaurs.

10. During a single human lifetime there is a 1 in 10,000 chance that the Earth will be hit by a meteorite large enough to cause global climatic changes, crop failures, and perhaps mass extinctions. Several near misses have been observed, and a program is under way to identify potentially dangerous Earth-approaching objects. The most significant identifiable danger is associated with Comet Swift–Tuttle, which will pass within 60 lunar distances of the Earth in 2126.

IMPORTANT TERMS TO REMEMBER

Apollo object (p. 256)
asteroid (p. 254)
asteroid belt (p. 254)
chondrite (p. 254)
comet (p. 258)

crater (p. 260)
Earth-crossing asteroid (p. 256)
ejecta (p. 263)
iron (p. 254)
meteor (p. 252)

meteorite (p. 251)
stone (p. 254)
stony-iron (p. 254)

QUESTIONS AND ACTIVITIES

1. It has been suggested that it may one day be possible to land a spaceship on an asteroid and mine it for iron and nickel. Investigate this possibility; do you think it might be feasible? What are the main points in favor of such an expedition? What stumbling blocks might there be?

2. How do you think the inhabitants of the Earth should prepare for the eventuality of a collision with a large extraterrestrial object?

3. Scientists have estimated that the probability of civilization being wiped out by the effects of a large impact is somewhere between 1 in 6000 and 1 in 10,000. Write a fictional account of such an event from the point of view of a survivor. What would the first few days after the event be like? What kind of civilization do you think might emerge afterward?

4. Why are carbonaceous chondrites especially important in terms of our understanding of the origin of the solar system? Investigate the meteorite Allende. When and how was it found? What is so special about it?

5. Do you think a major meteorite impact caused the extinction of the dinosaurs? Do some research to support your opinion. Write a short summary of the theories about the cause of the extinction of the dinosaurs and the evidence in support of each theory.

6. Discuss the pros and cons of the impact theory of mass extinctions versus the volcanic eruption theory. Which of the types of evidence presented by experts do you find most convincing? Which do you find unconvincing? Can you think of a type of evidence that has not yet been uncovered that might settle the controversy one way or another?

7. How do scientists identify a circular structure as having been caused by a meteorite impact? This is not always a straightforward process because impact craters can be partially eroded, covered by sediment, or deformed (folded or faulted). Moreover, circular and ringlike geologic structures (such as volcanic calderas) are common. Do some research to find out how scientists find impact craters and what kinds of evidence they look for to confirm the impact origin of a circular structure.

PART THREE

USING AND CARING FOR EARTH RESOURCES

Resources are like air, of no great importance until you're not getting any.

• Anonymous

★ E • S • S • A • Y ★

THE NATURE OF EARTH RESOURCES

In a geologic context, the term **Earth resources** refers to materials of value to humans that are extracted (or extractable) from the solid Earth. This definition encompasses all minerals, rocks, and metals; energy resources, including both conventional fossil fuels and alternative energy sources such as geothermal or tidal energy; soil; and water, both surface and subsurface. Although it is not extracted from the Earth, air is often considered an Earth resource also, particularly now that this crucial resource is being threatened on a global scale by the chemical byproducts of human activities.

This definition of Earth resources *excludes* living resources, such as forests, fish, livestock, and human resources. Although living resources are affected by many of the same physical and chemical processes that affect solid Earth resources, they usually require different management approaches. The main reason for this is the difference between *renewable* and *nonrenewable* resources.

RENEWABLE AND NONRENEWABLE RESOURCES

A **renewable resource** is a resource that can be renewed, replenished, or regenerated. For example, fish spawn, and new fish hatch each year; trees are cut down, but new ones can be planted; a soil layer that has been stripped off or

eroded away can be regenerated by the physical, chemical, and biologic processes of soil formation; underground water drawn from wells may eventually be replenished by rainwater (Fig. III.1).

▲ F I G U R E III.1
Water is a precious renewable resource. In the desert, as here in Senegal where Fulani women are drawing from a community well, water determines whether land is habitable.

◀

Gold miners working their claims, Serra Pellada, Brazil, sometime in the 1980s. All of the mined ore was carried out in bags on miners' backs. Accidents, deaths, and injury were common.

But what does "eventually" mean? Some of these resources may take a *very* long time to regenerate—longer than humans are able to wait. For example, it can take hundreds, thousands, even tens of thousands of years for a 10-cm layer of arable soil to regenerate from bare rock. Underground water supplies are eventually replenished, but it may take hundreds of years for this to happen. Even some biological communities, such as forests, may take hundreds or even thousands of years to regenerate. So we must modify our definition to say that a renewable resource is one that can be renewed, replenished, or regenerated *on a human time scale.* (Some people prefer to restrict the use of the term *renewable* to the description of living resources, and to refer to nonliving resources such as soil or underground water as *replenishable.*)

In contrast, a **nonrenewable resource** is one that cannot be replenished or regenerated on a human time scale. For example, it can take millions of years for deposits of valuable minerals or oil to form (Fig. III.2). So even though many of the geologic processes that formed such deposits are still occurring, these are considered nonrenewable resources because they are being renewed on a geologic rather than a human time scale.

An **inexhaustible resource** is one that can never be depleted or used up (Fig. III.3). For example, energy derived

from the wind or tides will never be depleted because the ultimate source of that energy is inexhaustible—as long as the Sun shines on the Earth, and as long as we have an atmosphere and a hydrosphere, there will be winds and tides. In principle, the Earth's geothermal energy is also inexhaustible; as we begin to exploit this energy source, however, we are discovering that the supply of geothermal energy in a particular locality can be depleted through overuse.

ISSUES IN RESOURCE MANAGEMENT

With the exception of ancient ecosystems like old-growth forests, most living resources are, in principle, renewable—if they are managed properly. On the other hand, most solid Earth resources are nonrenewable. Some, such as groundwater resources, are borderline. Increasingly, the guiding principle in the management of both renewable and nonrenewable resources is *sustainability*—that is, cautious, planned utilization of Earth resources to meet current needs without degradation of the viability of ecosystems or jeopardization of the future availability of those resources.

Sustainable Management of Renewable Resources

The primary concern in the sustainable management of renewable resources is to ensure that the rate at which the resource is being utilized does not exceed the rate at which it is being replenished or regenerated. This means that fish should not be harvested faster than new fish are hatching; trees should not be cut down faster than new trees are being planted; and water should not be withdrawn from wells faster than rainwater is replenishing underground storage reservoirs.

If a renewable resource becomes too severely depleted, it may never regenerate, no matter how long we wait. This is reasonably obvious in the case of living resources such as fish, where there is often a critical level below which the population simply cannot survive. The fishing industry in many parts of the world, including the eastern coast of Canada and the northwestern coast of the United States, has collapsed (or is on the verge of collapsing) because fish stocks have been harvested beyond the critical level for sustaining the fish population in those areas.

In some cases a population can be renewed artificially. Seedlings can be planted; fish can be spawned in fish farms and then transferred to rivers. However, there is always the possibility that much diversity, either of the species or of the ecosystem as a whole, will be lost. The ability of a fish population to regenerate itself may also be negatively affected by other environmental factors, both anthropogenic and natural. These include accelerated erosion caused by

▼ **FIGURE III.2**
Petroleum is a nonrenewable resource. As accessible oil fields are pumped out new supplies must be sought in increasingly hostile environments. This drilling rig is operating 14 kilometers offshore of the Alaskan coast, in the Beaufort Sea.

▲ F I G U R E   III.3
A wind farm at Gorgonia Pass, near Palm Springs, California. Each windmill drives a generator.

logging, agriculture, or mining; diminished water supply or quality resulting from the withdrawal or contamination of water; and climatic fluctuations.

Interestingly, it is also possible for *nonliving* Earth resources that are potentially renewable to become depleted to the point at which they cannot regenerate naturally. For example, a subsurface water reservoir can be replenished by rainfall, although the process may take many years or even centuries. But if too much water is withdrawn from an underground aquifer the clay particles in the rocks or sediments of the aquifer may become compacted. When this happens, rainwater can't infiltrate and the aquifer can never be replenished, no matter how much it rains and no matter how long we wait.

Sustainable Management of Nonrenewable Resources

The focus of nonrenewable resource management differs from that of renewable resource management. The salient feature of nonrenewable resources is that they can be depleted; the more we use, the less remains. In Chapter 1 we noted that the Earth is essentially a *closed system*. This implies, among other things, that the amount of matter in the Earth system is fixed and finite. The existing solid Earth resources are all we have and probably all we will ever have. Someday it may be possible to visit an asteroid for the purpose of mining nickel and iron; there may even be a mining station on the Moon or on Mars. But for now it is realistic to think of Earth resources as finite.

Nonrenewable resources must be treated with respect and used wisely and sparingly. When we use up nonrenewable resources we do so not just for ourselves but for future generations as well. Wherever possible, we must seek to extend the availability of these resources through conservation, substitution, reuse, or recycling (Fig. III.4). A wise management strategy will also look toward replacing reliance on nonrenewable resources with reliance on renewable or inexhaustible ones, for example, by replacing the use of nonrenewable fossil fuels with the use of solar, geothermal, tidal, or biomass energy (discussed in greater detail in Chapter 12).

RESOURCE LIMITATIONS AND POPULATION GROWTH

We have stated that the Earth is a closed system and that Earth resources are limited. We need to be more specific about the implications of these facts in terms of the number of people the planet can support.

The Growing World Population

One bacterium divides into two, which then divide into four. The four bacteria become eight, then 16, then 32, and then 64. After 10 divisions there are over 1000 bacteria. After 20 divisions there are over 1 million! This kind of growth, in which a population increases by a constant percentage per unit of time, is called **exponential growth.** Populations of living organisms, including humans, tend to grow exponentially when they are able to grow at all. Rates of exponential growth vary tremendously: A population of bacteria can double within half an hour; it is possible for a population of human beings to double within about 20 years.

Exponential growth cannot continue very long within any finite place. If every bacterium really did double every half-hour, it would take only a few weeks for the mass of bacteria to become greater than the mass of the Earth. Most of the time, populations are held in check by some limiting factor—food or water or living space—that keeps birth and death rates equal and populations roughly constant. In the case of human populations, growth is controlled by an incredibly complex variety of factors, some biologic and geologic, others socioeconomic, political, religious, and educational.

▶ F I G U R E III.4
A resource-efficient industrial system. Wastes are captured at every stage and reused, and the end product can be recycled to enter the industrial cycle again as raw material.

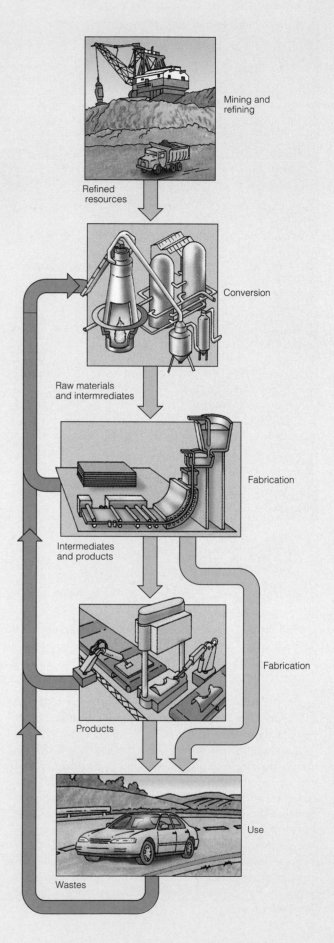

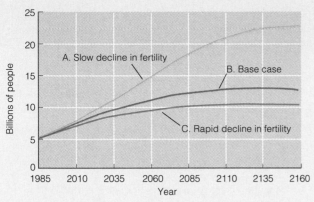

▲ F I G U R E III.5

Three world population projections (A, B, and C). As survival rates have improved with public health, the doubling period for world population has shortened from centuries to decades. All three projections assume that birthrates will fall to zero growth (replacement rate) within the next 50 years but with different rates of change.

Throughout human history population increase has been characterized by exponential growth (Fig. III.5). In some industrialized nations the *rate* of population growth (not the population itself) has begun to level off and even decline. However, in most developing countries birthrates remain high. As a result, the world population is increasing by about 1 billion people every 11 years, or almost 90 million each year (Fig. III.6), and will exceed 6 billion by the year 2000. The "medium" scenario suggested in Fig. III.5 projects a world population reaching 8.9 billion by 2030 and leveling off at 11.5 billion in 2150; some scenarios, such as a worst case scenario proposed by the World Bank, have the global population stabilizing at 23 billion after the

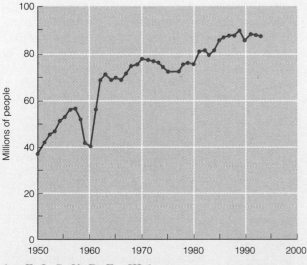

▲ F I G U R E III.6

Annual increase in world population, 1950 to 1993. In 1993 the net population gain was 87 million—roughly equal to the population of Mexico.

year 2135. What will happen after that? Is the Earth capable of sustaining a population of 23 billion, 8.9 billion, or even 6 billion?

Carrying Capacity

The **carrying capacity** of any system or resource base is the number of living things that can be sustained indefinitely by that system. The concept was originally formulated by pasture managers to indicate the highest number of cattle or sheep a given pasture could support without degrading the soil or vegetation. It is extremely difficult to determine carrying capacity precisely, even for a simple system. For a single species (e.g., cattle) eating a single or limited number of things (e.g., grass), the carrying capacity of a pasture depends on the rate at which the grass grows, which in turn depends on uncertain and variable factors such as weather. It also depends on the activities of the grazing species, which may manure the land or tear up the soil. And it depends on human management activities, such as fertilization or irrigation.

In general, the carrying capacity of a system is defined by its most limiting factor. It is like a chain whose strength is determined by its weakest link. For example, corn needs nitrogen in order to grow. You can apply phosphate fertilizers, but that will not make the corn grow if there is insufficient nitrogen in the soil. If you apply nitrogen fertilizers, growth will occur only up to the point at which phosphate, or water, or something else becomes the limiting factor.

The Earth's Carrying Capacity

What, then, is the Earth's carrying capacity? Numerous efforts have been made to identify the factors that limit the planet's carrying capacity in terms of human populations. Different studies have suggested that, judging from the limits on the Earth's mineral and energy resources and biological productivity, the planet theoretically could sustain a human population of 6 billion, or 8 billion, or 12 billion, or 100 billion. Why are the estimates so different?

A significant problem in trying to quantify the Earth's carrying capacity is that different human populations consume different quantities of resources. Different groups place varying amounts of stress on Earth resources. For example, it has been estimated that on a per capita basis Americans use four times as much steel and 23 times as much aluminum as their neighbors in Mexico. The average Dutch person's consumption of food, wood, natural fibers, and other products involves the exploitation of five times as much land outside Holland as inside that country's borders. The average Japanese person consumes nine times as much steel as the average Chinese person. Canadians routinely top the list in per capita energy consumption. And in terms of overall drain on the Earth's resources, the average Swiss person consumes as much as 40 Somalis. The Earth's carrying capacity is therefore 40 times lower for people living a Swiss lifestyle than for people living a Somali lifestyle.

The point here is not that the lifestyle of the Swiss or the Somalis or the Americans is right or wrong, nor is it that everyone's living standard should be reduced to a minimum in order to maximize the Earth's carrying capacity. Rather, the point is that there are many choices to be made, many resource management issues to consider, many ways to use resources wisely, and many ways to do more with less.

*E*ARTH RESOURCES AND SOCIETY

We tend to go through life unaware of our dependence on Earth resources. Yet we are completely dependent on those resources for the maintenance of our lifestyles—and, indeed, for life itself.

Can you imagine a world without machines? Our modern world, with its 5.5 billion inhabitants, couldn't operate without them. Machines are used to produce food, make clothes, transport us, and help us communicate. The metals needed to build machines, as well as the fuels needed to run them, are extracted from the Earth. Without metals we would be unable to build planes, cars, televisions, or computers. Industry would falter, and living standards would decline. The same can be said for many other resources. For example, every part of your car—with the possible exception of leather upholstery—was made from materials derived from mineral and petroleum resources. Even the tires are made from Earth resources (petroleum); the fuel used to run the car is an Earth resource as well.

Look around you now. Even if you are sitting in the middle of a forest, you are probably surrounded by Earth resources that have been made into useful objects. The buckles on your bookbag, the zipper and the rivets on your jeans, even the dyes used to color the fabric are all derived from mineral resources. The shirt you are wearing is probably made from a petroleum-based synthetic fabric. The apple in your bookbag came from a tree that was grown in soil, probably irrigated with groundwater, and nourished by fertilizers made from minerals. Even the paper on which this book is printed probably has some mineral content;

paper is still made from wood fiber, but pulverized minerals are often added to fine papers to produce a specific color or texture.

Our everyday dependence on Earth resources goes deeper than the material products and conveniences on which we have come to depend. The sustenance of all life on this planet depends on the availability of water to drink, air to breathe, and soil in which to grow food.

Support Squares

One way to visualize our dependence on the land is to think of the global population in terms of *density* instead of total numbers. This gives a better idea of how much land area we each have to "call our own." The world's total land area is 153 million km². If we exclude the areas covered by permanent ice and the places above 5000 m in altitude, this leaves about 133 million km² of dry land on which we could conceivably live and from which we could produce or recover resources in one way or another. Some land, of course, is quite rugged and some climates are very inhospitable, but even so, each square kilometer can supply some resources.

In 1895 the global population density was about 13 people per km². This is equivalent to a block of land 280 m square for each person (Fig. III.7). You can think of this block of land as a "support square." Today the support square for each person is a block of land about 160 m square, the size of one city block. This amount of land must supply not only all the Earth resources an individual will use in his or her lifetime but also all the resources that will be used by future generations. Somehow that space must also consume most of the solid wastes created by the use of those resources. By the year 2000, the support square will be 140 m square; by 2095, if the global population continues to grow to 12 billion, the support square will shrink to 100 m on a side (Table III.1).

Population densities in parts of some countries are of interest because they indicate how tightly packed together humans can live. In Japan and Korea, the current support

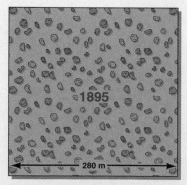
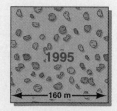

Global population = 1.8 billion	>5.5 billion	12 billion
People / km² = 13	41	90

◄ F I G U R E III.7
Support squares for 1895, 1995, and 2095. The size of the square is calculated by dividing the global population at a given time by the amount of conceivably inhabitable land (133 million km²), which gives population density (No. / km²). In 1995 the square was about 160 m on an edge; by 2095, if global population grows to 12 billion, it will have shrunk to 100 m².

T A B L E III.1 • Approximate Population and Population Density in 1895, 1995, 2000, and 2095[a]

Year	Approximate Global Population (billions)	Population Density (No. / km²)	Support square, meters on an edge
1895	1.8	13	280
1995	5.5	41	160
2000	7.0 (?)	52	140
2095	12.0 (?)	90	100

[a]Densities and individual "support squares" are calculated based on 133 million km² of conceivably inhabitable land.

square is 57 m on an edge; in China, 68 m; in India, 71 m; and in Europe, 100 m. By comparison, the United States seems sparsely populated: each person has a support square 198 m on an edge. Hardly anyone seems to live in Canada, where each person enjoys a support square 1000 m². The support square of Europe today, which is 100 m², is about the size of the anticipated support square for the entire world when the global population reaches 12 billion. Such a densely packed herd of large mammals—the human herd—inevitably must cause massive changes in the environment simply by being alive.

Stressed to the Limit?

Whatever the Earth's carrying capacity may turn out to be, we apparently haven't yet reached its limits. The human population continues to grow, and we continue to find, process, and make use of Earth resources in a wide variety of ways. Perhaps the only definite conclusion we can reach is that the Earth's carrying capacity is large but not infinite. Already there is clear evidence that the carrying capacities of some local ecosystems have been reached and, indeed, greatly exceeded. A particularly stark example is found on the fringes of the Sahara desert and in the semiarid savanna regions of Africa, where population growth and stressful land-use practices have contributed to land degradation and accelerated desertification (Fig. III.8).

Despite our emphasis on the fact that Earth's resources are limited, we will argue in this part of the book that we are *not* in danger of depleting the Earth's energy resources,

nor are we on the verge of running out of mineral resources. These resources can be extended well into the next century by careful use and conservation practices, new extraction and processing technologies, and new discoveries by geologists and geophysicists.

When it comes right down to it, probably the single most important factor limiting the Earth's ability to sustain the human population is the capacity to support food production, which in turn depends on the two Earth resources that we tend to worry about least: water and soil. Of these, soil may turn out to be the most difficult to manage sustainably, partly because its rate of regeneration is so slow. Around the world, soil loss amounts to about 25 billion metric tons per year, or almost 5 metric tons per person per year. Put another way, each person consumes about three quarters of a metric ton of food per year; to produce that food, 5 tons of soil—seven times the mass of the food itself—is lost. This clearly represents unsustainable utilization of a critical Earth resource, and if it continues unchecked it will have serious consequences both for the status of natural ecosystems and for the production of food.

IMPORTANT TERMS TO REMEMBER

carrying capacity (p. 279)
Earth resources (p. 275)
exponential growth (p. 278)
inexhaustible resource (p. 276)
nonrenewable resource (p. 276)
renewable resource (p. 275)

◄ F I G U R E III.8
Attempting to hold desertification at bay in Southern Morocco, on the edge of the Sahara. Palm fronds are woven into fences to slow the drift of sand.

CHAPTER 11

ENERGY FROM FOSSIL FUELS

*It is hard to know which is more remarkable,
that it took 600 million years for the Earth to make its oil,
or that it took 300 years to use it up.*

• M. King Hubbert

*T*he search for oil in Saudi Arabia began May 29, 1933, with the signing of an oil concession agreement between the Saudi government and Standard Oil of California. The geologists immediately headed for the Dammam Dome, an exact copy of the region in Bahrain where oil had been discovered the year before. The discovery well's name could not have been more prosaic: Dammam No. 7. Yet "Lucky No. 7" would become a symbol of success. The well capped a quest that had lasted nearly 5 years, and it made possible the commercial development of oil in Saudi Arabia.

The Dammam Dome tested the patience and expertise of geologists and their drilling crews. An early oil company handbook stated that "not only the drilling rig and equipment, but every item of lumber, hardware, plumbing and steel needed to create and supply living quarters, pipe for water, transportation equipment and spare parts, food and personal requirements had to be brought over in a supply line reaching from the United States." In addition, water wells had to be drilled, roads built, and power provided where none had existed before.

On December 7, 1936, the crew prepared to start drilling a deep-test well at Dammam No. 7. Holes 1 through 6 had been disappointing, but when it came to trouble, No. 7 proved to be in a class by itself. There were delays and stoppages. Drill pipes stuck. Rotary chains broke. Bits were lost down the hole and had to be fished

◄ ───────────────────

The well that first produced oil in Saudi Arabia. After six dry holes, Dammam No. 7 struck oil on March 4, 1938 at a depth of 1440 m. "Lucky No. 7" eventually produced an extraordinary 32 million barrels of oil.

out. Walls caved in. As the steam-driven rotary rig ground downward, the results remained the same: No oil. A year later, after drilling to a depth of almost 1400 m, the crew still hadn't found enough oil to fill the crankcases of their own trucks.

Early in 1938 chief geologist Max Steineke was called to San Francisco. Hard questions were raised about the future of a venture that had cost millions and proved nothing. But even as Steineke argued his case he was dramatically vindicated: the crew found what it was looking for at a depth of 1440 m. Dammam No. 7 produced 1585 barrels of oil on March 4, 3690 barrels on March 7, 2130 barrels 9 days later, 3732 5 days after that, 3810 the next day, and so on, until the head office cabled that there was no reason to continue testing. Jubilation reverberated from Dammam Camp to San Francisco. But even after they struck oil, the first Saudi and United States drillers could not have imagined that Dammam No. 7 would be capable of producing not just for months or years but for several decades, or that this one well would pour out more than 32 million barrels of oil.

THE ENERGY BUDGET

In Chapter 1 we learned that the Earth's energy comes from three sources: solar radiation, geothermal energy, and tidal energy. The energy from these sources circulates through the various pathways and reservoirs of the energy cycle (Fig. 1.10). The total amount of energy in the planet's energy budget is about $174,000 \times 10^{12}$ watts, or 174,000 terawatts.

How does the flow of energy in the Earth's energy cycle compare with the needs of human populations? The annual global demand for energy now exceeds 300 exajoules (300 $\times 10^{18}$ joules), which means that humans are converting

283

T A B L E 11.1 • **World Commercial Energy Consumption, 1990**[a]

Region	Energy Consumption (exajoules)	Per Capita (gigajoules per person)
Developing countries	84	21
Industrial countries	154	185
Centrally planned economies[b]	71	167
World	310	59

[a]British Petroleum, *BP Statistical Review of World Energy*, 1992.
[b]In 1990, the centrally planned economies included the former Soviet Union, China, and other socialist nations.

energy for their own use at an average rate of nearly 10 terawatts (10×10^{12} watts). (To put such a huge number into perspective, consider that a healthy, active human can work at the rate of about 100 watts, but only for 8 hours a day. Therefore, considering our ability to work in terms of a 24-hour day, we humans are at best only 33-watt machines!)

If the Earth's energy budget offers 174,000 terawatts and human populations currently require less than 10 terawatts, it is clear that we are not on the verge of running out of energy in an absolute sense (Table 11.1). What *is* in question is the continued availability of inexpensive, easily accessible energy in forms that are environmentally acceptable and convenient to use.

In this chapter we focus on the energy source that has been the driving force of industrial development throughout modern history and remains the world's primary energy source: *fossil fuels*. We look at the different types of fossil fuels, how they form, where they occur, and how they are found and extracted for human use. The chapter ends with a look at current and future energy needs and the implications of continued dependence on fossil fuels.

Fossil Fuels

Coal, oil, and natural gas, collectively called **fossil fuels**, provide most of the energy that runs modern civilizations. Fossil fuels are sometimes referred to as **hydrocarbons** because they are composed of compounds made up of different amounts of hydrogen and carbon, with varying proportions of other elements present as impurities. They typically occur as local concentrations of organic substances in sediments and sedimentary rocks.

Fossil fuels are one of the storage reservoirs in the energy cycle (Fig. 1.10). Nearly all living organisms derive their energy from the Sun. The chief energy-trapping mechanism is *photosynthesis*, a process by which plants use the Sun's energy to combine water and carbon dioxide to make carbohydrates (organic compounds containing carbon, oxygen, and hydrogen) and oxygen. Animals that consume

plants, therefore, are secondary consumers of trapped solar energy. When plants or animals die and decay, atmospheric oxygen combines with carbon and hydrogen in the organic compounds to form water (H_2O) and carbon dioxide (CO_2) again. Whenever organic matter is buried, a small portion of it escapes decay. In this way some of the solar energy becomes stored in sediments and ultimately in sedimentary rocks.

The total amount of trapped organic matter in sedimentary rocks is far less than 1 percent of the organic matter formed by growing plants and animals. Nevertheless, during the past 600 million years the amount of trapped organic matter has increased steadily and has become very large. The exact nature and occurrence of each of the fossil fuels depends on the kind of sediment, the kind of organic matter trapped in the sediment, and the changes that have occurred during the long geologic ages since the organic matter was first buried.

PETROLEUM

In the ocean, microscopic phytoplankton (tiny floating plants) and bacteria (simple, single-celled organisms) are the principal sources of organic matter trapped in sediment. Most of the organic matter is trapped in clay that is slowly converted to shale. During this conversion, organic compounds are transformed into oil and natural gas. These two products are the main forms of **petroleum**—gaseous, liquid, and semisolid naturally occurring substances that consist chiefly of hydrocarbon compounds.

Petroleum (from the Latin words *petra*, rock, and *oleum*, oil) is referred to as *crude oil* when it emerges from the ground. From this state it must be distilled and refined, or separated into its various hydrocarbon components. The distillation is usually facilitated by the addition of various chemicals that assist in the breakdown of complex hydrocarbons, a process known as *cracking*.

A Brief History of Petroleum Use

Oil—the liquid form of petroleum—is one of the first resources our ancestors learned to use. The earliest people to use oil instead of wood for fuel were the Babylonians, who lived in what is now Iraq about 4500 years ago. They used oil from natural seeps in the valleys of the Tigris and Euphrates rivers. However, the major use of oil really started about 1847, when a merchant in Pittsburgh started bottling and selling rock oil from natural seeps to be used as a lubricant. Five years later, a Canadian chemist discovered that heating and distilling rock oil yielded *kerosene*, a liquid that could be used in lamps. This discovery spelled doom for candles and whale-oil lamps.

Wells were soon being dug by hand near Oil Springs, Ontario, in order to produce oil. In 1856, using the same hand-digging process, workers in Romania were producing 2000 barrels a year. (A barrel is equivalent to 42 U.S. gallons of oil. For a brief summary of energy-related terms and units, see Appendix A.) In 1858, the first oil well was drilled in southern Ontario. On August 27, 1859, at a well in Titusville, Pennsylvania, oil-bearing strata were encountered at a depth of 21.2 m. Soon up to 35 barrels of oil were being pumped each day (Fig. 11.1). Oil was quickly discovered in many other places.

▲ FIGURE 11.1
Reconstruction of the original engine shed of the world's first well drilled strictly for oil, Drakes Well Museum, Titusville, Pennsylvania.

The earliest known use of **natural gas** (naturally occurring hydrocarbon compounds that are gaseous at ordinary temperatures and pressures) was about 3000 years ago in China. Gas seeping out of the ground was collected and transmitted through bamboo pipes to be ignited and used to evaporate salt water in order to recover salt. It wasn't long before the Chinese were drilling wells to increase the flow of gas. Modern use of gas started in the early seventeenth century in Europe, where gas made from wood and coal was used for illumination. Commercial gas companies were founded as early as 1812 in London and 1816 in Baltimore. The stage was set for the exploitation of an accidental discovery at Fredonia, New York, in 1821. A water well drilled in that year produced not only water but also bubbles of a mysterious gas. The gas was accidentally ignited and produced such a spectacular flame that a new well was drilled on the same site and wooden pipes were installed to carry the gas to a nearby hotel, where 66 gas lights were installed. By 1872 natural gas was being piped as far as 40 km from its source.

Today petroleum products are used for a very wide variety of purposes in addition to their main use as fuel for heating and running vehicles and machines. Different components of petroleum are broken down and used in fertilizers, lubricants, asphalt, and an array of synthetic materials (e.g., rubber and plastic) and fabrics (e.g., nylon and rayon).

The Origin of Petroleum

As mentioned above, the term *petroleum* encompasses naturally occurring liquid and gaseous hydrocarbons (oil and natural gas), as well as semisolid hydrocarbons, commonly called **tar.** Petroleum deposits are nearly always found in association with marine sedimentary rocks. Offshore sampling on continental shelves and along the base of continental slopes has shown that fine mud beneath the seafloor can contain up to 8 percent organic matter. Geologists therefore conclude that oil originates primarily as organic matter deposited with marine sediment.

Two kinds of evidence support the hypothesis that petroleum is a product of the decomposition of organic matter: (1) Oil possesses optical properties known only in hydrocarbons derived from organic matter, and (2) oil contains nitrogen and certain compounds that are believed to originate only in living matter. A long and complex chain of chemical reactions apparently is involved in the conversion of organic matter into crude petroleum and natural gas. In addition, chemical changes may occur in oil and gas after they have accumulated, accounting for chemical differences between the oil in one body of petroleum and that in another.

Formation

When marine microorganisms and algae die, their remains settle to the bottom and collect in the mud. Typically this is a fine, dark mud with very little oxygen content, that is, an *anaerobic* environment. As the organisms begin to decay, the decay process uses up the small amount of oxygen that is present. After the oxygen has been used up, the remaining organic material is preserved and covered with more layers of mud and decaying organisms.

Over time, and with continued burial, the muddy sediment with the partially decayed organic material is subjected to heat and pressure. This initiates a series of complex physical and chemical changes that break down the organic material into simpler liquid and gaseous hydrocarbon compounds; this process is referred to as **maturation**.

Meanwhile, the muddy sediment itself turns into a sedimentary rock. The rock containing the organic material that is eventually converted into oil and natural gas is referred to as the **source rock**. Dark marine shales are the most common type of source rock for hydrocarbons.

Migration

The sediment in which organic matter is accumulating today is rich in clay minerals, the type of material that will eventually become a shale. However, most of the rocks that constitute oil or gas *pools*—rocks in which the pore spaces are saturated with hydrocarbon—are sandstones (consisting of quartz grains), limestones and dolostones (consisting of carbonate minerals), and much-fractured rock of other kinds. Long ago, geologists realized that oil and gas must form in one kind of material and travel, or *migrate*, to another at some later time.

Over time, the gas and small droplets of oil in the source rock begin to coalesce, and eventually they are squeezed out by the weight of overlying rocks. For migration to occur, however, the oil must encounter a rock formation through which it can travel easily toward areas near the surface where the pressure is lower. Thus, oil migration requires rock with high *porosity* and *permeability*, that is, a rock with a high proportion of interconnected pore spaces. Such a formation is called a **reservoir rock**. Typically the rocks with the greatest porosity and permeability are sandstones and limestones, and these are the most common types of reservoir rocks.

When oil and gas are squeezed out of the shale in which they have originated and enter a body of sandstone or limestone, they can migrate more easily than before because of the greater permeability of the reservoir rock. The force of molecular attraction between oil and mineral grains is weaker than that between water and mineral grains. Hence, because oil and water do not mix, much of the water remains fastened to the quartz or carbonate mineral grains, while oil occupies the central parts of the larger openings in the porous sandstone or limestone. Because it is lighter than water, the oil tends to glide upward and become partially segregated from the water.

Traps

Most of the petroleum that forms in sediments eventually makes its way to the surface, forming oil *seeps*. It is not sur-

prising, therefore, that the highest ratios of oil and gas pools to volume of sediment are found in rock that is no more than 2.5 million years old, and that nearly 60 percent of all the oil and gas discovered so far has been found in rocks of Cenozoic age, that is, less than 66.4 million years old (Fig. 11.2). This does not mean that older rocks produced less petroleum. It simply means that the oil in older rocks has had a longer time in which to escape. It is estimated that no more than 0.1 percent of all the organic matter originally buried in a sediment eventually becomes trapped underground in an oil or gas pool.

Sometimes a rock with very low porosity and permeability gets in the way of migrating oil and gas and prevents it from going any farther. Such a formation is called a **cap rock**. The most common type of cap rock is shale.

A geologic situation that includes a source rock (to contribute organic material), a reservoir rock (to allow for the migration of oil), and a cap rock (to stop the migration) is referred to as a petroleum or hydrocarbon **trap**. There are several common types of petroleum traps (Fig. 11.3). Both water and natural gas are geologically associated with oil in all of them. All three fluids migrate out of the source rock into the reservoir rock, but because water is the densest of the three fluids, it will wind up on the bottom, with the oil in the middle and the gas, which is lightest, on top.

Structural Traps Traps associated with structural features of rocks, such as folds or faults, are called *structural traps*. Sedimentary rocks form from sediments, which are laid down as more or less flat units one on top of the other, from oldest to youngest. Sometimes these rocks are subjected to the pressures created by plates moving around on the surface of the Earth, and they become folded. A fold in which the rock units are concave downward is referred to as an *anticline*. If the rocks are folded in such a way that a cap rock is on top of a reservoir rock (Fig. 11.3A), the oil will migrate into the top of the fold and become trapped by the

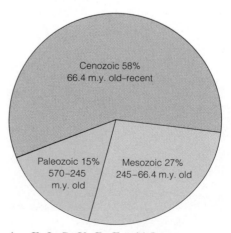

▲ F I G U R E 11.2
Percentages of total world oil production from rocks of different ages. (m.y. stands for million years.)

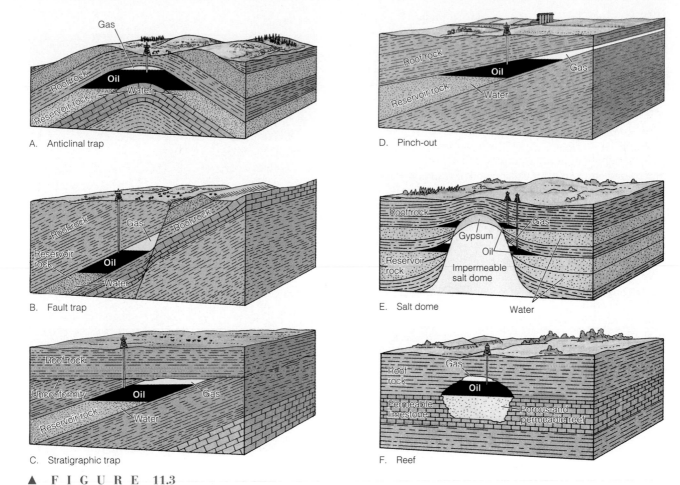

▲ F I G U R E 11.3
Common types of petroleum traps. A. Anticlinal structural trap. B. Faulted structural trap. C. Stratigraphic trap with unconformity. D. Stratigraphic pinch-out with overlying impermeable rock unit. E. Salt dome. F. Coral reef.

anticline. In addition to being folded or tilted, sometimes rocks subjected to tectonic pressures will break or *fault*. If the movement of rocks along the fault places a reservoir rock against a cap rock, it creates a trap (Fig. 11.3B).

Stratigraphic Traps   Traps created by differences in sedimentary layers (strata) are called *stratigraphic traps*. Sometimes flat-lying sedimentary rocks are tilted on their sides as a result of plate tectonic movement. After they have become tilted, they may be eroded; additional flat-lying sediments may be deposited on top of them and eventually turn into rocks. The contact between the tilted, eroded sedimentary rocks and the flat-lying rocks is called an *unconformity*. If the sedimentary rock on top of the unconformity is a cap rock, a stratigraphic trap is formed (Fig. 11.3C). In another type of stratigraphic trap, shown in Fig. 11.3D, a porous reservoir layer thins or pinches out and is overlain by an impermeable cap rock.

Salt Domes   Salt (the mineral halite) belongs to the group of sedimentary rocks called *evaporites*, which are formed as a result of the evaporation of lakewater or seawater. Sometimes shallow marine sediments are laid down on top of salt deposits. These marine sediments eventually turn into sedi-

mentary rocks such as shales and sandstones. However, salt is a very light, buoyant type of rock; that is, the rocks on top of the salt are typically heavier or denser than the salt itself. Over time, and when subjected to pressure, the salt will begin to flow upward after penetrating through the overlying rocks. When it does so it forms an inverted teardrop shape called a *salt dome* (Fig. 11.3E). The rocks around the salt dome become folded and faulted during the process. Salt is highly impermeable, so if the rocks around the dome are reservoirs for oil, the oil will migrate and become trapped along the sides of the salt dome.

Coral Reefs   After the death of the organisms that make up a coral reef colony, the reef becomes a highly porous and permeable form of limestone. Such rocks are ideal reservoirs for liquid and gaseous hydrocarbons. If the limestone reef is surrounded by an impermeable shale unit (Fig. 11.3F), a trap is created.

Overall, structural traps are more common than stratigraphic traps, salt domes, or reefs. The great oil pools of Saudi Arabia, Kuwait, Iraq, and other parts of the Middle East occur primarily in structural traps. The deformation that produced the traps was caused by compressional stresses from the tectonic collision of Africa and Asia.

The Distribution of Petroleum Deposits

Petroleum deposits are common, but they are unevenly distributed. Suitable source sediments for petroleum are widespread; they are as likely to form in subarctic waters as in tropical regions. Yet such rocks yield productive deposits only under very specific circumstances. The critical factors are (1) a supply of heat to convert the solid organic matter into oil and gas, and (2) the formation of a suitable trap before the petroleum has leaked away.

The conversion of solid organic matter into oil and gas happens within specific depth and temperature ranges defined by the conditions shown in Fig. 11.4. If the *geothermal gradient* (change in temperature with increasing depth below the surface of the Earth) in a particular locality is too low—less than 1.8°C/100 m, so that temperature increases very slowly with depth—conversion does not occur. If the gradient is above 5.5°C/100 m, so that temperatures are relatively high at shallow depths, conversion to gas starts at such shallow depths that very little trapping occurs. The depth–temperature "window" within which oil and gas form and are trapped lies between the two thermal gradients. Once oil and gas have been formed, they will accumulate in pools only if suitable traps are present. Most oil and gas pools are found beneath anticlines; the timing of the folding event therefore is a critical part of the trapping process. If folding occurs after petroleum has migrated, pools cannot form. The great oil pools of the Middle East arose through the fortunate coincidence of the right thermal gradient and the development of anticlinal traps.

How Much Oil?

How much oil is there in the world? This is an extremely controversial question. Approximately 600 billion barrels have already been pumped out of the ground. A lot of additional oil has been located by drilling but is still waiting to be pumped. Probably a great deal more remains to be found by drilling. Unlike coal, for which the volume of strata in a basin of sediment can be estimated accurately before mining, the volume of undiscovered oil can only be guessed at. Such guesses are based on the accumulated experience of a century of drilling. Knowing how much oil has been found in an intensively drilled area, such as eastern Texas, experts make estimates of probable oil volumes in other regions where rock types and structures are similar to those in eastern Texas. Using this approach, and considering all the sedimentary basins in the world (Fig. 11.5), experts estimate that somewhere between 1500 and 3000 billion barrels of oil will eventually be discovered.

Exploration and Recovery

A variety of methods are used to locate oil. These include geologic mapping, air photos and remote sensing, geochemical testing of soils, and seismic, magnetic, and gravity

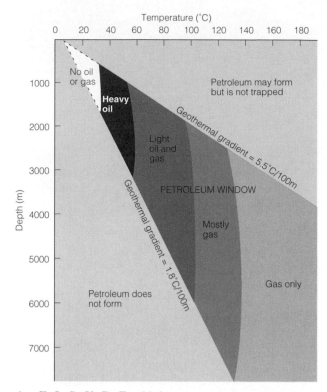

▲ F I G U R E 11.4
The "petroleum window" is the combination of depth and temperature within which oil and gas are generated and trapped.

surveys. The overall goal of petroleum exploration is to recognize and identify rock types and geologic settings that favor the formation, migration, and trapping of hydrocarbons. Computers have become essential in modern petroleum exploration; they are used to design experiments, manage and interpret data, and test new ideas. Computers are also useful in three-dimensional modeling of sedimentary basins.

Often, test wells are drilled in a target area. Rocks taken from the drill core at regular intervals are examined for properties such as porosity, permeability, and water saturation; this is called *core logging*. Test drilling and core logging are done both for exploration purposes and to provide more information about a known oil field.

Early oil drilling was done using a *cable tool rig*, a type of drill that simply lifted a bit and dropped it onto the rock. Today *rotary drills*, which cut directly into the rock, are the most common drilling tools. When an oil well is first drilled, the oil and gas (which have been trapped under pressure) will instantly take the easy way out through the drilled hole, forming a *gusher* (Fig 11.6). If the pressure is too high, the top part of the well may be blown off; modern drill rigs (Fig. 11.7) have special equipment designed to prevent such *blowouts*.

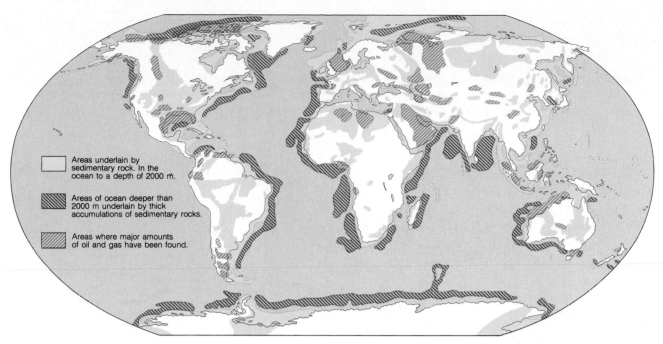

▲ F I G U R E 11.5

Areas underlain by sedimentary rock and regions where large accumulations of oil and gas have been located. Where the ocean is deeper than 2000 m, sedimentary rock has not yet been tested to determine its potential for oil and gas production.

▲ F I G U R E 11.6

A gusher comes in. Texas in the 1920s.

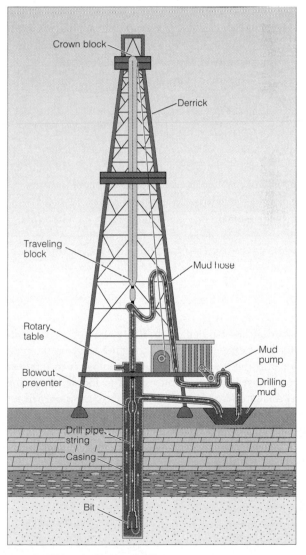

▲ F I G U R E 11.7

A modern petroleum drilling rig.

BOX 11.1

•

THE HUMAN PERSPECTIVE

GEOLOGY AND THE POLITICS OF MIDDLE EAST OIL

*P*etroleum is the premier fossil fuel of twentieth-century civilization. We use it to fuel our cars, trucks, ships, and airplanes; to run generators that produce electricity; and to provide heat for our homes. Industrial societies are so dependent on petroleum that it constitutes a key component of their economies. In the best of all worlds, each nation would have its own reserve of petroleum suitable to meet its present and future needs. Unfortunately, however, geopolitical boundaries and geological resources seldom coincide. A few of the major industrial countries, such as Canada and Russia, have adequate supplies of petroleum for their present needs. The United States can supply most, but not all, of the petroleum it uses. Other countries, such as Japan, France, and Germany, have either no petroleum or miniscule supplies that must import the oil and gas they require. The reason for this disparity is the unequal distribution of petroleum, which is a function of both sedimentary geology and tectonic history.

Sedimentary rocks that contain large quantities of oil are concentrated in a few relatively restricted geologic regions. Their distribution is a function of a long and often complex history of sediment accumulation and tectonic events related to moving lithospheric plates. Such countries as the United States and Russia cover such large geographic areas that the laws of geologic chance make it

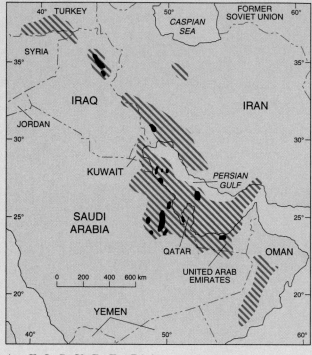

▲ F I G U R E B1.1

Rocks containing oil and gas underlie large areas of the Middle East. The outlined areas enclose the more than 400 oil fields that have been discovered so far. The outlines of the largest individual fields are highlighted.

A recent development in drilling technology is the ability to drill horizontally at depth, rather than just straight down into the ground. This allows geologists to intersect fractures that might transmit oil and to follow oil-bearing units in any direction underground. Drill rigs designed for use on offshore platforms have also been improved; it is now possible to drill for oil in water as deep as 2280 m.

When oil flows from a well under its own pressure or flows easily with pumping, it is referred to as **primary recovery**. The initial surge may continue for many years. For example, the Dammam No. 7 well described at the beginning of the chapter was closed down in 1982 after 45 years of nearly continuous output. The well is still capable of turning out 1800 barrels a day without pumping.

Enhanced Recovery from Existing Oil Wells

Eventually it becomes necessary to pump oil from a well. Sometimes the natural gas retrieved from the well is reinjected into the reservoir rock, facilitating the flow of oil under its own pressure. After a certain amount of pumping, the reservoir rock will have been substantially depleted and the pressure in the well may decrease significantly. At that point it may become necessary to introduce other fluids into the well in order to push the oil out. This is called **enhanced recovery**.

Much of the oil that remains in the reservoir rock after primary recovery has been exhausted is *residual oil*, which is not very mobile because it is held in place by a variety of probable that they will contain at least some large petroleum desposits. For smaller countries, however, the chance is almost like a roll of the dice. Many small countries contain little or no petroleum and so are losers, whereas others are big winners.

The largest known reserve of petroleum happens to lie in a limited geographic region centered around the Persian Gulf (Fig. B1.1). Here the sedimentary and tectonic laws of chance have converged to produce huge underground reservoirs of fossil fuel in a region otherwise blessed with few natural resources. A few relatively small countries surrounding the gulf control more than half of the world's known petroleum resources. As a result, their economic and political significance has skyrocketed in the decades since oil was discovered in this region (Fig. B1.2).

Always a politically sensitive region, the Middle East lies at the center of the world political stage. International conflicts, the legacy of centuries of intertribal and religious disputes, now include petroleum as an added complicating factor, for control of oil resources is associated with both economic and political power. While this oil-rich region may remain politically unstable well into the future, eventually its importance will decline as continued exploitation causes the nonrenewable petroleum resources to diminish and the world turns to alternative sources of power.

▲ F I G U R E B1.2
Geologic exploration teams working in the harsh desert environment of the Persian Gulf region have discovered vast reservoirs of petroleum that comprise more than half the world's known supplies. These discoveries quickly brought the oil-rich countries of the Middle East to economic and political prominence.

physical and chemical forces. Sometimes enhanced recovery can be used to overcome the forces that are immobilizing the oil. The reservoir rock is first flooded with water to increase pressure and force the oil up and out of the well. This is called *secondary recovery*. The next stage of enhanced recovery, called *tertiary recovery*, involves the injection of other chemicals, such as carbon dioxide, surfactants (detergents), or polymerizing gels, which may further increase the reservoir energy and facilitate the flow of oil. Sometimes heated fluids are utilized.

Primary and secondary recovery together generally permit the extraction of about one third of the oil originally present in a given reservoir. It may be possible to extract an additional 40 percent of the oil if tertiary recovery technologies are employed. Enhanced recovery methods can substantially extend the productive life of an existing oil field, but their use adds significantly to the cost of extracting the oil. There are environmental costs as well, such as water pollution, thermal pollution (excess heat), and subsidence.

SOLID HYDROCARBONS

On land, plants such as trees, bushes, and grasses contribute most of the organic matter that becomes trapped in soil and rocks. These plants contain carbohydrates, but they are also rich in resins, waxes, and lignins, which tend to remain in solid form. The most important types of solid hydrocarbons are *peat* and *coal.*

Peat

In water-saturated environments such as bogs or swamps, plant remains accumulate to form **peat**, an unconsolidated deposit of plant remains with a carbon content of about 60 percent (Fig. 11.8). In some parts of the world, such as Ireland and Russia, peat is dried and used as a low-grade fuel for cooking and heating homes. One third of the world's peat is found in Canada, where about 12 percent of the land surface consists of peatlands. Most of the rest of the world's peatlands are located in other circumpolar regions.

Coal

The formation of peat is the initial stage in the development of the most abundant fossil fuel, the combustible sedimentary rock that we call **coal**. If you look at a piece of coal through a microscope, you can sometimes see the shapes of bits of fossil wood, bark, leaves, roots, and other parts of land plants, chemically altered but still identifiable. Decomposed plant matter accounts for more than 50 percent of the content of coal.

The Chinese were probably the first people to mine and use coal, beginning about about 3100 years ago; surface deposits of coal may have been exploited even earlier. Most of the coal mined today is eventually either burned under boilers to make steam for electrical generators or converted into *coke,* an essential ingredient in the smelting of iron ore and the making of steel. In addition to its use as a fuel, coal is a raw material for nylon and other plastics and for a multitude of other organic chemicals.

Coal Formation

The process of **coalification** involves the loss of volatile materials, such as H_2O (water), CO_2 (carbon dioxide), and CH_4 (methane), from peat. As the peat is buried beneath more plant matter and accumulating sand, silt, or clay,

▲ F I G U R E 11.8
A peat cutter harvests peat from a bog in western Ireland. The peat has formed in a cool moist climate that favors the preservation of organic matter in wet environments. When dried, the peat provides fuel for heat and cooking.

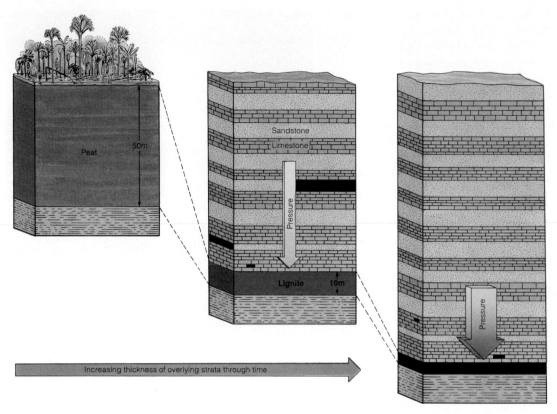

▲ FIGURE 11.9
Plant matter in peat is converted into coal by decomposition, coupled with increased pressure and temperature as overlying sediments build up. By the time a layer of peat 50 m thick has been converted to bituminous coal, its thickness has been reduced by 90 percent. In the process, the proportion of carbon has increased from 60 to 80 percent.

both temperature and pressure rise (Fig. 11.9), causing a series of continuing changes. The peat is compressed, water is squeezed out, and gaseous organic compounds escape, leaving an increased proportion of carbon in the residual material. Through coalification, the peat is converted into *lignite,* and eventually into *subbituminous coal* and *bituminous coal.* Peat is a sediment, whereas these coals are sedimentary rocks. *Anthracite,* a still later phase in the coalification process, is so changed mineralogically that it is essentially a metamorphic rock. Because of its low volatile content, anthracite is hard to ignite, but once alight it burns with almost no smoke. In contrast, lignite is rich in volatile elements, burns smokily, and ignites so easily that it is subject to spontaneous ignition. In regions where metamorphism has been intense, coal has been changed so thoroughly that it has been converted into graphite, which contains no volatile substances. Graphite will not burn in an ordinary fire.

Occurrence of Coal

Coal occurs in layers (miners call them *seams*) along with other sedimentary rocks, mainly shale and sandstone (Fig. 11.10). A coal seam is a flat, lens-shaped body with the same surface area as the swamp in which it originally accu-

▲ FIGURE 11.10
Coal seams in a sequence of sedimentary strata, Healy, Alaska. Coal is a sedimentary rock; the sedimentary layers, once horizontal, have been tilted by tectonic forces.

mulated. Most coal seams are 0.5 to 3 m thick, although some reach more than 30 m (Fig. 11.11). Coal seams tend to occur in groups. For example, 60 seams of bituminous coal have been found in western Pennsylvania.

Coal swamps seem to form in many sedimentary environments, of which two types predominate. One consists of slowly subsiding basins in continental interiors and the swampy margins of shallow inland seas. This is the home environment of the bituminous and subbituminous coal seams in Utah, Montana, Wyoming, and the Dakotas. The second sedimentary environment consists of continental margins with wide shelves that were flooded during times when sea level was high. This is the environment of the bituminous coals of the Appalachian region and the eastern coast of Canada.

Periods of Coal Formation

Although peat can form even under subarctic conditions, it is clear that the luxuriant plant growth needed to form thick and extensive coal seams develops most readily in a tropical or semitropical climate. This implies either that the global climate was warmer when the plant matter of today's coal deposits originally accumulated, or that the wet, swampy environments in which most of the world's coal seams were formed were all located in the tropics, within about 20° of the equator. Probably both conditions were involved. Coal deposits that must have been formed in warm low-latitude environments and now lie in frigid polar lands provide some of the most compelling evidence we have for the slow drift of continents over great distances.

Peat has been formed more or less continuously since land plants first appeared about 450 million years ago. The size of peat swamps has varied greatly, however, and there-

▲ F I G U R E   11.11
On average, about 300,000 tons of coal per day are extracted from a seam 20 to 30 m thick in this strip mine at Wyodak, Wyoming.

▲ F I G U R E   11.12
A modern-day example of a coal-forming swamp, characterized by slow-flowing, near stagnant water and dense vegetation. Great Dismal Swamp, North Carolina.

fore the amount of coal formed has also varied. By far the greatest period of coal swamp formation occurred during the Carboniferous and Permian periods, 360 to 245 million years ago. The great coal beds of Europe and eastern North America were formed at this time, when the swamp plants were giant ferns and scale trees *(gymnosperms)*. The second great period of coal deposition peaked during the Cretaceous Period, 144 to 66.4 million years ago. The plants of coal swamps during this period were flowering plants *(angiosperms)* much like those found in swamps today.

Today, peat formation occurs in wetlands such as the Okeefenokee Swamp in Florida and the Great Dismal Swamp in Virginia and North Carolina (Fig. 11.12). The Great Dismal Swamp, one of the largest modern peat swamps, contains an average thickness of 2 m of peat. However, unless this swamp lasts for millions of years, even that dense growth is insufficient to produce a coal seam as thick as some of the seams in Pennsylvania.

Tar Sands and Oil Shales

Some sedimentary rocks contain a very thick, dense, asphalt-like form of hydrocarbon. This form of solid hydrocarbon is colloquially called *tar,* and the rocks or sediments that contain it are called *tar sands*. Tar sands may be immature petroleum deposits, in which the oil- and gas-forming alterations never reached completion, or they may be petroleum deposits in which the volatile elements have escaped, leaving behind a tarry residue.

Another form of solid hydrocarbon is found in *oil shales.* Shales are sedimentary rocks composed of tiny mineral fragments (mostly clay particles) rather than accumulated plant material. Some shales are unusually high in organic content. The oils and fats contained in dead organisms buried in marine or lake muds may be converted into solid hydrocarbons, which ultimately may be converted into liquid or gaseous petroleum. However, in some shales temperatures never reach the levels required to break down the organic molecules completely. Instead, a waxlike substance called *kerogen* is formed. If a kerogen-bearing rock is heated in a retort, the solid organic matter is converted into liquid and gaseous hydrocarbons similar to those found in petroleum. The use of tar sands and oil shales as alternative energy sources is discussed in more detail in Chapter 12.

THE GLOBAL ENERGY SCENARIO

Since the Industrial Revolution, fossil fuels have been the main energy source for the industrializing world. What is the future of fossil fuels in meeting the world's energy needs? To answer this question we take a broad look at worldwide patterns of energy use, how they have changed in the past, and how they may be expected to change in the future. We also consider some of the environmental and strategic implications of continued dependence on fossil fuels as the dominant energy source in the next century.

Global Patterns of Energy Use

At present, a very small proportion (about 5 percent) of the world's population consumes energy at a rate of more than 10 kilowatts per person; 50 percent of the population consumes energy at an average rate of less than a kilowatt per capita. The significance of this gap, in terms of its impacts on standards of living, is readily apparent:

> In the small Ecuadorean village of Guamalán lives an elementary school teacher named Marco Coloma. Electricity only reached his isolated village in 1990, but most homes are unable to afford the steep hookup fee. Tanks of propane are hauled in occasionally over a rough dirt track, though most of the community's energy is firewood culled from the trees that border local cornfields and pastures. Coloma's energy use—directly or in products he consumes—is equivalent to 4.6 barrels of oil each year, assuming he manages to consume the Ecuadorean average.
>
> Five thousand kilometers away, in the U.S. town of Winsted, Connecticut, his brother uses 48 barrels of oil-equivalent annually, 10 times as much energy. Carlos Coloma owns a car and a pickup truck, as well as a home equipped with an electric cooking range, electric lights, a refrigerator, a washing machine, and two tele-

> visions. These are all things that his brother Marco longs for, a dream likely shared by more than 4 billion people in Africa, Asia, and Latin America.

The world's population is growing at a rate of about 2 percent per year, fast enough to double within 35 years. Even if the rate of growth is reduced, the global population could be close to 7 billion by 2000. Most of the increase will occur in developing regions, which are currently among the lowest in per capita energy use. This means that over the next few decades there will be many more people in the world who, like Marco Coloma, will be increasing their energy consumption while striving to raise their standards of living. Concerns about conserving energy—so prevalent in the industrialized world today—are not shared by the majority of the world's people. While we have been worrying about the rapid rates of population growth since the 1950s, the demand for energy has been escalating at a rate *four times* that of population growth!

Traditionally, there has been a strong correlation between gross domestic product (GDP) or gross national product (GNP)—both indicators of the productivity of an economy—and per capita energy consumption. This correlation, which was established during the Industrial Revolution, can be seen in rapidly industrializing countries today (Fig. 11.13). The correlation between GDP and energy use *suggests* that if nonindustrialized countries want to develop economically, they must increase their energy consumption to avoid becoming poorer relative to industrialized countries. But is this necessarily true?

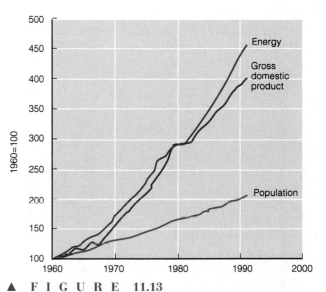

▲ **F I G U R E   11.13**
Increase in energy use, gross domestic product (GDP, a measure of the productivity of an economy), and population in developing countries, 1960–1991.

In fact, energy use is not constant during all stages of economic development. Energy consumption in industrialized nations increased dramatically and steadily for many decades following the Industrial Revolution, but in recent years it has begun to level off and even to decline. This is because the **energy intensity** of economic activity—that is, energy consumption per unit of economic activity—has begun to decrease in the industrialized countries. This has come about as a result of energy conservation, more efficient technologies, and a shift toward more service-oriented industries.

Many developing countries are now at the same energy-intensive stage of development that characterized the industrialized world in the years following the Industrial Revolution. Can the world's energy budget meet the rapidly growing needs of such a large portion of the global population? Can developing countries follow a less energy-intensive development path than the one followed by the industrialized nations? Just maintaining the present level of per capita energy use over the next 20 years will require increasing the global supply by more than 50 percent at an affordable cost both economically and environmentally. How will the world's people meet this shortfall? Part of the answer may lie in exploration and discovery of new fossil fuel deposits. Part of it will have to come from conservation and the development of more energy-efficient technologies, and part will have to come from a shift toward alternative, renewable energy sources.

Implications of Dependence on Fossil Fuels

Fig. 11.14 shows the history of energy consumption and the changing character of the world's energy "mix," including the dramatic increase in demand for energy after World War II (starting in the late 1940s) and the subsequent rise to predominance of the fossil fuels, especially oil, in the energy mix. The switch to coal from wood just before the turn of the century (Fig. 11.14B) came about primarily because of the development of new coal-based technologies such as the steam engine. In later years the shift from coal to oil and gas was also driven by new technologies, primarily the internal combustion engine, which required fuel in liquid form. The ability to discover new fossil fuel deposits to meet these increasing demands was greatly improved by the advent of new technologies for exploration, including seismic studies, organic geochemistry, satellite imagery, computer modeling, and digital recording methods.

Supply and Demand

On the basis of present rates of production from proven reserves, it has been estimated that the United States and Canada have only enough recoverable oil to meet their energy needs for the next 10 years. When such dire predictions have been made in the past, the deadline has always been extended by new discoveries, improved technologies,

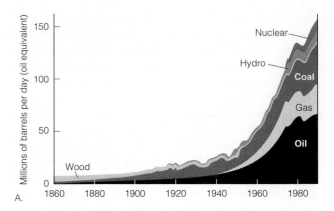

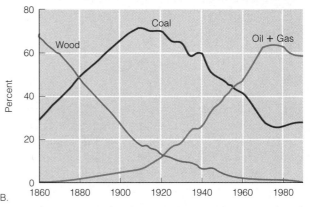

▲ **F I G U R E 11.14**
A. The history of world energy consumption. B. The major types of fuel and their percentages of total worldwide energy consumption.

increased imports, and conservation. But the world's recoverable fossil fuel resources *are* finite, and we must be concerned with what happens when we *do* reach the limits.

Fig. 11.15 shows possible projections of world energy production into the twenty-first century. (The projections assume an increased contribution from so-called *unconventional* oil and gas reserves, including oil shales). A range of possible demand scenarios is shown, all of which assume an eventual leveling off of demand owing to increased energy efficiency and conservation. Coal, the "dirtiest" and therefore least environmentally acceptable of the fossil fuels, is assumed to be limited to current levels of production.

Realistically, it is highly unlikely the demand for fossil fuels will decrease substantially or that an alternative fuel source will supplant the need for oil. Are supplies of fossil fuels adequate to meet future demands? If we use a barrel of oil as our unit of measurement, we can compare quantities of all fossil fuels directly (Table 11.2). Considering the approximate world-use rate for oil (30 billion barrels a year) and the projected changes in global demand shown in Fig. 11.15, and comparing the estimated amounts of fossil fuels remaining, it is apparent that only coal, the most abundant fossil fuel, may have the capacity to meet the world's long-term demands.

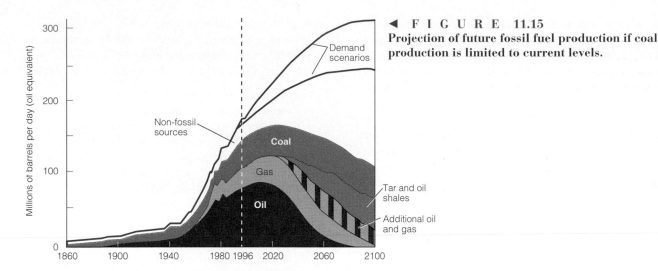

◄ F I G U R E **11.15**
Projection of future fossil fuel production if coal production is limited to current levels.

These are not alarmist scenarios but sober, realistic projections. The world will continue to rely on fossil fuels until well into the next century. To make up for shortfalls in supply, new discoveries will have to be made. Improvements in coal-burning technologies will be needed. New extraction technologies will have to be developed, and unconventional fossil fuel resources will have to be exploited more fully. But the projections also imply that fossil fuels cannot meet global energy needs forever; major advances will be needed in alternative energy sources. And the bottom line is that we must continue to strive for energy efficiency and conservation if we are to close the gap between energy supply and demand.

Environmental Impacts

Continued reliance on fossil fuels has many serious environmental implications. Probably the best publicized of these is its contribution to the buildup of carbon dioxide (CO_2) and other *greenhouse gases* in the Earth's atmosphere, leading to the possibility of global warming. As a result of fossil fuel use, about 1.2 tons of CO_2 is released into the at-

mosphere every year for each person on the Earth. The implications of the CO_2 buildup and its relationship to global warming are discussed in detail in Chapter 18.

The burning of fossil fuels can also result in emissions of sulfur and nitrogen oxides, which combine with water vapor in the atmosphere to create *acid precipitation*. This is particularly true of coal, which has the most impurities (notably sulfur) of all fossil fuels. Acid precipitation is one of the most intractable environmental problems associated with the burning of fossil fuels, primarily because the effects are felt far from the source of the emissions. Tropospheric ozone, volatile organic compounds, and other components of smog are also released by the burning of fossil fuels, mainly in internal combustion engines. These atmospheric problems are also discussed in Chapter 18.

Another environmental concern associated with heavy reliance on oil as an energy resource is the need to transport it over long distances, creating the possibility of spills. One of the reasons that oil has risen to predominance in the world market instead of gas is the relative ease with which oil can be transported from its place of origin to the mar-

T A B L E **11.2** • **Estimates of Amounts of Fossil Fuels Possibly Recoverable Worldwide (Unit of Comparison is a Barrel of Oil)**[a]

Fossil Fuel	*Total Amount in Ground (billions of barrels)*	*Amount Possibly Recoverable (billions of barrels)*
Coal	About 100,000	62,730[b]
Oil and gas (flowing)	1,500–3,000	1,500–3,000
Trapped oil in pumped-out pools	1,500–3,000	0–?
Viscous oil (tar sands)	3,000–6,000	500–?
Oil shale	Total unknown; much greater than coal	1,000–?

[a]0.22 ton of coal = 1 barrel of oil.

ket. Viewed as a whole, the oil industry has a very good safety record in terms of transporting its product; only a small fraction of the total amount of oil transported is accidentally released into the environment. Yet the possibility of spills does exist, and when they occur they can be catastrophic. For example, in March 1989 the *Exxon Valdez* ran aground off the shore of Valdez, Alaska, spilling almost 40 million liters of crude oil into the water. This was the largest oil spill ever in North American waters; it contaminated more than 5000 km of Alaska's shoreline and released many toxic organic compounds. Oil spills and their environmental implications are discussed in more detail in Chapter 17.

Strategic Concerns

In 1991, when the Allied forces went to war in Kuwait, soldiers and civilians gave their lives in what was viewed by many to be a struggle to secure access to a critically important resource. This was not the first time that governments have gone to war over resource-related issues, and surely not the first time lives have been lost in the pursuit or protection of a material commodity. Yet somehow the inability of the world's nations to shake their dependence on cheap oil, together with the continuing failure of most industrialized

countries to meet their own energy needs, have perpetuated the sense of uneasiness and insecurity felt by many during the Gulf War. How did we wind up in this situation?

According to researchers for the WorldWatch Institute,

The forerunners to this crisis were the failed energy policies that allowed oil-consuming nations—both industrial and developing—to greatly increase their dependence on Middle Eastern oil in the late eighties. The world's addiction to cheap oil is as destructive and hard to break as an alcoholic's need for a drink. Since 1986, when oil prices fell back below $20 per barrel . . .,world oil demand shot up by almost 5 million barrels per day, or nearly 10 percent. Virtually all the extra oil now being consumed is supplied by a handful of countries in the Middle East, a region that faces the stresses of rapidly growing populations, autocratic political systems, rampant poverty, and a deadly arms race. Not only is the world addicted to cheap oil, but the largest liquor store is in a very dangerous neighborhood.

The human and environmental tolls of this conflict must surely make us pause to reconsider the strategic implications of continued dependence on cheap oil.

SUMMARY

1. About 174,000 terawatts cycle through the Earth's energy budget. The total required by human populations is less than 10 terawatts; thus, we are not about to run out of energy in an absolute sense. Fossil fuels—mainly oil, coal, and natural gas—have supplied most of the world's energy needs since the Industrial Revolution.

2. Fossil fuels are composed primarily of hydrocarbon compounds in solid, liquid, or gaseous form. They consist of the altered remains of plant and animal life, representing long-term storage of a small fraction of the solar energy originally used in the process of photosynthesis.

3. Petroleum (oil and natural gas) is formed principally in marine environments, where the remains of marine microorganisms accumulate in anaerobic sediments. As the microorganisms decay, they use up the limited oxygen present in the sediments. The remaining organic material is gradually altered by the increasing temperatures and pressures caused by the buildup of overlying sediment. This maturation process eventually results in the formation of oil and natural gas.

4. Oil and natural gas migrate from a source rock to a reservoir rock, which must be porous and permeable

enough to permit the passage and storage of fluids. Because the fluids are buoyant, they tend to migrate toward the surface. If their passage is blocked by an impermeable rock unit or cap rock, a trap is created. The most common types of traps are structural traps (anticlines and faults), stratigraphic traps (unconformities and pinchouts), salt domes, and coral reefs. Saline water is typically associated with the oil and gas found in all kinds of traps.

5. Petroleum deposits are unevenly distributed around the world. This has occurred because the processes leading to the maturation of oil and gas require very specific pressure and temperature conditions, in addition to an appropriate combination of source, reservoir, and cap rocks.

6. A variety of exploration techniques are used to find oil; they include geologic mapping, seismic surveys, core logging, and three-dimensional modeling of sedimentary basins. The main goal of petroleum exploration is to recognize and identify rock types and geologic settings that favor the formation, migration, and trapping of hydrocarbons.

7. Primary recovery occurs when oil gushes out of a well under its own pressure or flows easily with pumping.

Later, when the reservoir is somewhat depleted and pressures have decreased, it may be necessary to inject water or other fluids into the well to facilitate the flow of oil; this is called enhanced recovery.

8. Hydrocarbons also occur naturally in solid forms. Peat and coal are formed from the remains of land plants. In bogs and swamps, plant matter accumulates to form layers of peat. As overlying sediment piles up, the peat is subjected to ever higher temperatures and pressures and is eventually transformed into three kinds of sedimentary rocks: lignite, subbituminous coal, and bituminous coal. In another stage of the coalification process a metamorphic rock, anthracite, is formed.

9. At present, a very small percentage of the world's population accounts for by far the greatest proportion of total energy use. Historically there has been a strong correlation between economic development and energy use, suggesting that developing countries may have to greatly increase their energy consumption if they wish to achieve economic parity with industrialized nations. However, in recent years, energy intensity—the amount of energy used per unit of economic activity—in the industrialized nations has been declining owing to the combined effects of conservation, more efficient technologies, and a shift toward more service-related economic activities.

10. Judging from estimates of the total supplies of extractable oil, natural gas, and coal, it appears that only coal may be present in large enough quantities to meet future demand. Even if sufficient reserves can be found, continued dependence on fossil fuels has serious implications, including environmental impacts and strategic concerns.

IMPORTANT TERMS TO REMEMBER

cap rock (p. 286)
coal (p. 292)
coalification (p. 292)
energy intensity (p. 296)
enhanced recovery (p. 290)

fossil fuels (p. 284)
hydrocarbons (p. 284)
maturation (p. 286)
natural gas (p. 285)
oil (p. 285)

peat (p. 292)
petroleum (p. 284)
primary recovery (p. 290)
reservoir rock (p. 286)
source rock (p. 286)

tar (p. 285)
trap (p. 286)

QUESTIONS AND ACTIVITIES

1. Find out what percentage of the energy consumed in your country is provided by oil, coal, natural gas, and other energy sources (nuclear, hydroelectric, and alternative energy sources). How has this energy mix changed over time? Make a chart or diagram to illustrate your findings.

2. Do some research to answer the following questions about energy consumption: What is the rate of per capita energy consumption in your country? How does it compare to the worldwide average? Has energy intensity risen or fallen in your country in recent years? Can you discover any reasons why this change may have occurred?

3. If you live in a part of the world where oil is found, make a field trip to an oil field. In what kind of geologic environment did the oil form? How was this oil field originally identified and explored? What type of recovery is being carried out? Ask the drillers or geologists to show you the different parts of the drill rig or pump.

4. Make a field trip to an oil refinery. Find out where the oil comes from and how it gets to the refinery. Watch the various steps in the refining process and ask what products will be made from the refined oil. Find out how and where the oil is shipped after it has been refined.

5. If you live in a part of the world where coal occurs, investigate the history and geology of coal and mining in your region. What type of coal mining is usually carried out in your area? (Strip mining? Underground mining?) What are the environmental impacts of the mining? Have there been any major mine disasters in the area? What type of coal is mined there, and when was it formed?

6. What proportion of the fossil fuels used in your country is imported, and what proportion is produced domestically? How have those proportions changed over the past few decades? What factors have produced those changes? Consider a wide range of factors that may influence the price and, therefore, the amount of oil imports. The possibilities include geologic, environmental, economic, historical, societal, technologic, and political factors.

CHAPTER 12

ENERGY ALTERNATIVES

There is always wood—but we've tried that.
• Jonathon Weiner

North Island, one of New Zealand's two main islands, includes a strip of territory more than 250 km long along which there is a volcano, a steam hole, a boiling spring, or a fumarole every 15 km or so. Maori tribes settled here in the fourteenth century, and they have been using the Earth's heat ever since. They prepare vegetable dyes by dropping bark into boiling geothermal pools and letting the bark stew overnight. Children dunk sweet corn into a hot pool, where it cooks in moments. Maori ritual feasts, called *hangi*, are structured around meats wrapped in palm leaves and lowered into steam holes in sturdy baskets.

Since the 1940s this hot strip of Earth has been turned to high-tech applications. Buildings and homes in the area are heated by wells drilled into the hot subterranean reservoirs. The Tasman Pulp and Paper Plant in Kawerau is the largest geothermal plant of its kind in the world. At Wairakei, near Lake Taupo, a large electric power station was completed in 1963. Since that time it has been running almost continuously, a source of energy unaffected by such external factors as coal strikes, oil embargos, or the weather.

The Maori believe that the hot springs were the answer to the prayers of their ancestors. The high priest of the first Maori canoe to arrive had never seen snow before he disembarked on North Island, and he prayed for warmth. According to a Maori elder of Rotorua,

Our people regard this as part of our heritage. My older grandaunts and granduncles and grandparents, when they talked about the legend, would always say, "These are what have been bestowed upon us. These were gifts from the gods. And so, as they were gifts, gifts are to be used, but not to be abused."

Scientists may have a different theory, but ours has lasted 600 years, and as the descendants of the passengers of that canoe, and especially of the high priest, we still feel that ours is probably better than the scientists' theory, anyway.

Geothermal energy is used not only in New Zealand but also in other countries, including Iceland, Mexico, Hungary, Japan, France, Italy, and the United States. As we begin to feel the pinch of dwindling supplies of fossil fuels and increasing concern for environmental impacts associated with their use, alternative energy sources such as geothermal are becoming more attractive—and more economically competitive.

SOME THINGS TO CONSIDER

In the preceding chapter we learned how solar energy can be trapped in the form of partially decayed plant and animal life; subjected to heat, pressure, and chemical changes; and ultimately transformed into fossil fuels. We looked at the history of energy use and at possible future demand scenarios. And we considered whether fossil fuels can continue to meet those needs into the next century. We found that there will be a growing gap between energy supply and demand as both population and per capita energy use continue to increase, especially in parts of the world that are entering an energy-intensive phase of economic development.

Fossil fuels will certainly continue to dominate the world's energy mix until well into the twenty-first century. But continued heavy reliance on fossil fuels has significant environmental, economic, and strategic implications. Moreover, the world's fossil fuel resources are finite and nonrenewable. In the coming decades the shortfall between supply and demand will need to be met by enhanced en-

◀ _____

Geothermal power plant, Wairakei, New Zealand.

▲ F I G U R E 12.1
Oil refinery, Richmond, California. Fossil fuels provide the energy that runs today's societies. Oil, the most important fossil fuel, is refined into different grades of fuel prior to use.

ergy efficiency and conservation measures, by the development of new technologies for exploration and extraction, and by increasing reliance on alternative sources of energy.

In this chapter we build on the discussion of energy resources presented in Chapter 11 by considering some new technologies and alternative energy sources. But before we discuss the details of those alternatives, let's pause to consider what our major concerns should be in planning a sustainable energy future.

Strategies for Planning Our Energy Future

At present, the choice of an energy source for industrial and other applications is controlled primarily by economics: the source that is most competitive economically is most likely to be chosen. In industrialized countries this has made fossil fuels the main source of energy ever since the Industrial Revolution (Fig. 12.1). However, the world is entering a new phase in which technological advances will help make other types of energy more competitive economically. In addition, greater emphasis is being placed on factors other than simple economics, including environmental, social, health, and strategic concerns. The cost of ignoring these concerns is now being acknowledged as a hidden cost of energy use. Let's take a closer look at some of the factors that may shape our energy future.

Conservation Energy conservation is probably the easiest and least expensive means of reducing the gap between energy supply and demand; as such, it should play a major role in any future energy plan. Per capita energy consumption in the industrialized countries still far exceeds that in developing countries, but *energy intensity*—the ratio of energy consumption per unit of economic activity—is leveling off or even declining in most industrialized nations (Fig. 12.2). For example, the member countries of the Organization for Economic Cooperation and Development (OECD), which includes most of the world's industrialized or "developed" countries, have lowered their energy use per unit of production by more than 24 percent since 1973. The future energy demand scenarios discussed in Chapter 11 (see Fig. 11.15) were based on the implicit assumption that global per capita energy demands would decrease substantially by the middle of the next century, mainly

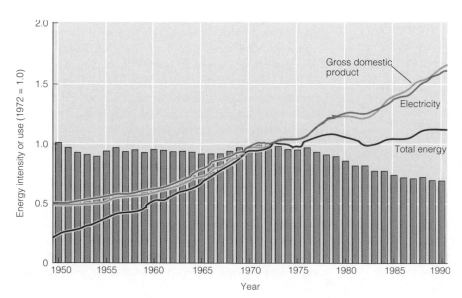

◀ F I G U R E 12.2
Changes in energy intensity—the ratio of energy consumption per unit of economic activity—in the United States from 1950 to 1990. Recently, energy intensities in industrialized countries have begun to level off or even decline, meaning that they are producing more goods while using less energy.

through the development and adoption of new, more energy-efficient technologies.

In developing countries, where economic development and improved standards of living depend on increasing energy use, adopting a more energy-efficient development path may be a particularly strategic move. For example, in 1980 China initiated a program in which 10 percent of the nation's energy investment was directed toward improvements in efficiency. The result was that annual growth in energy use decreased from 7 to 4 percent and energy consumption in China increased at less than half the rate of economic growth during the 1980s. About 90 percent of the savings were achieved through improvements in efficiency, and the rest through shifts to less energy-intensive industries. In addition, the energy gains achieved by investing in more efficient industrial technologies were found to be one third less expensive than comparable investments in coal supplies would have been.

Shift to Renewable and/or Inexhaustible Resources Although supplies of fossil fuels—especially coal—will not be exhausted for some time, it would be a good thing if we could reduce the proportion of fossil fuels in our energy mix, simply because they are nonrenewable and ultimately finite. Oil is the most economical, versatile, and easily transportable fossil fuel, but it will probably be the first to be exhausted. A system based on alternative energy sources will have to be considerably more energy-efficient than our current system because the technologies involved are likely to be more expensive and less readily available than those based on oil.

Environmental Concerns Continued heavy dependence on hydrocarbons is a source of environmental concern, primarily because the products of fossil fuel combustion become pollutants when they are released into the atmosphere. Natural gas is the "cleanest" burning of the hydrocarbons; oil is somewhat more polluting; and coal, which has the most impurities, is the most polluting. The two main environmental problems associated with fossil fuel combustion are the release of carbon dioxide, possibly resulting in global warming, and the release of sulfur and nitrogen oxides, leading to acid precipitation and photochemical smog. Different energy options affect the environment in different ways, but *all* energy sources have some negative environmental impacts. It is increasingly important to design an energy system that minimizes negative environmental effects.

Social, Health, and Regional Considerations Basic human requirements for food and shelter can be met only through the expenditure of energy. A good energy policy must strive to make energy available to all and to ensure an acceptable standard of living. The environmental impacts of energy use—primarily air and water pollution—should also raise concerns about the potential effects of different energy alternatives on human health.

No one energy source or technology is equally applicable to all regions. Alternatives such as solar energy, wind energy, and geothermal energy have different degrees of applicability, depending on the resources and technologies available in each region. The energy sources we will rely on in the future may not be as easily transported as oil. It will be necessary for each region to adapt its own energy policy to meet its specific needs more efficiently.

Strategic Concerns Even though we are not experiencing an energy "crisis" right now, we nevertheless are dependent on oil for most of our energy needs. In North America, we cannot continue to meet the demand for oil with domestic supplies. By far the largest portion of the world's identified oil reserves are in the Middle East. Given the political instability of that region, the strategic disadvantages of heavy reliance on Middle Eastern oil are evident. An energy system based primarily on domestic supplies would make more sense, both politically and economically.

Energy Diversity An energy system based on overdependence on one source of energy is unwise. Although oil will continue to meet the better part of our energy needs for a few decades, we need to begin planning a *sustainable* energy system—one that will meet our energy needs well into the twenty-first century in an environmentally, socially, and strategically acceptable manner. Such a system must maximize energy efficiency, take full advantage of the potential of regional and local energy resources, and be based increasingly on renewable and inexhaustible resources. In addition, it will probably be a system characterized by diversity, both in terms of the technologies utilized and in terms of the energy sources tapped.

The Alternatives

In this chapter we look at two groups of "alternative" energy sources, that is, energy sources other than the conventional fossil fuels. The first group, new fossil fuel technologies, comprises a diverse set of technologies and energy sources with the potential to extend or improve the world's fossil fuel resources. They include techniques for tapping into unconventional hydrocarbon fuels, such as tar sands and oil shales, and technologies for improving the energy efficiency or environmental acceptability of existing resources, such as coal.

The second group is composed of true alternatives to fossil fuels. Some of these, such as hydroelectric and nuclear energy, have been developed more fully than others and might be considered part of the "conventional" energy mix of many countries. Others, such as solar, wind, and geothermal energy, are just beginning to be exploited and may become increasingly important as we begin to shift toward renewable and inexhaustible energy sources.

NEW FOSSIL FUEL TECHNOLOGIES

Clean Coal Technologies

As mentioned earlier, coal is the "dirtiest" of the fossil fuels and therefore is the least acceptable from an environmental point of view. Compared to the other fossil fuels, coal has more impurities, primarily sulfur, and it is therefore more polluting during combustion (Fig. 12.3). The sulfur emissions from coal-fired power plants are the main generators of the sulfur dioxide (SO_2) that causes acid precipitation. Coal burning also releases significant amounts of trace elements into the environment. For some of these elements, including arsenic, beryllium, lithium, selenium, thorium, and vanadium, the emissions resulting from coal burning outweigh emissions from all other sources combined. There are also extensive environmental and health-related problems associated with coal mining and processing. These include acid mine drainage, subsidence, underground fires and explosions, and "black lung" disease.

Yet coal is the most abundant of the conventional fossil fuels, so it may be worth investing in new approaches to render coal more environmentally acceptable and more efficient as an energy source. To that end, a variety of "clean coal" technologies are being investigated.

▲ **FIGURE 12.3**
Smoke plumes from the Four Corners coal-burning power plant, New Mexico. Smoke from this plant is thought by some to be one of the reasons for the haze in the Grand Canyon.

Precombustion Cleaning One approach is to "wash" the coal, that is, to remove as many impurities as possible from the fuel before burning it so that it will emit fewer pollutants into the atmosphere (although the impurities must still be dealt with in other forms). The cleaning is accomplished through a variety of mechanical, chemical, and biochemical techniques. For example, sulfur-digesting microorganisms that thrive in sulfur-rich hot springs can be used to remove both organic and inorganic sulfur.

Combustion-Enhancing Technologies Another approach is to enhance the efficiency with which the coal burns, thereby reducing the sulfur dioxide (SO_2) and nitrous oxide (NO_x) emissions. In one such technique, called *fluidized bed* combustion, air is blown through the bottom of a box of sand or some other type of inert particles (Fig. 12.4). The air lifts the particles and agitates them so that they tumble about like a boiling fluid; this is the fluidized bed. The sand particles are heated until they become incandescent. When the coal is introduced into the bed, it ignites and burn. The amount of nitrous oxides produced in a fluidized bed system is much lower than the amount produced in conventional coal plants because the system runs at much lower temperatures. When limestone is added to the bed *(limestone injection),* most of the SO_2 also is chemically captured before being released into the environment. Fluidized bed technology results in cleaner, more efficient combustion. It has the drawback of leaving behind a solid residue, but in some cases the residue may have commercial value.

Coprocessing and Conversion Technologies One of the biggest problems in considering a switch from petroleum to coal is that many existing industrial technologies rely on a fuel in liquid form. It would be very costly to convert existing machinery or develop new technologies that could accept a solid fuel. Therefore, considerable attention has been paid to the possibility of converting coal into a form that can be used by current industrial equipment. Liquid and gaseous fuels produced synthetically from coal are referred to as *synfuels.* (This term also covers other types of unconventional hydrocarbon resources.)

One method used to produce synfuels is *coprocessing,* whereby finely ground coal is mixed with water or oil to form a liquid mixture called a *slurry.* Existing equipment usually can be converted to accept slurries by adding specially designed burners. Another approach to the production of synfuels is conversion of the coal into gaseous or liquid form, that is, coal *gasification* and *liquefaction.* For example, certain enzymes found in bread mold can liquefy coal. Coal gasification is accomplished by heating the coal in the absence of oxygen and retrieving the volatile fraction, a process called *coking.* In addition to being more readily utilized in existing equipment, gasified coal emits

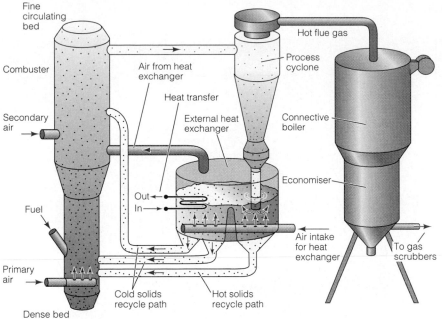

fewer pollutants than normal coal. The residual material from gasification, *coke,* can also be used as a fuel in some industrial processes.

Postcombustion Cleaning Emissions of sulfur dioxide from coal-fired power plants in North America are now limited by law. Many power plants have reduced their emissions by installing *"scrubbers"* that remove sulfur from flue gases just before they are released into the environment. Another approach to postcombustion cleaning is *"in-duct" sorbent injection,* in which sulfur-absorbing chemicals are injected into the exhaust ducts of the power plant; this is typically less expensive than installing scrubbers. Special technologies also exist for reducing nitrous oxide emissions and for removing particulates (tiny solid particles) from gases before their release into the atmosphere.

Unconventional Hydrocarbon Resources

As we saw in Chapter 11, the typical reservoir rock for petroleum hydrocarbons is a rock of high porosity and permeability, such as a sandstone or limestone. However, other types of sedimentary rocks, such as shales, coal seams, and so-called "tight" (low permeability) sandstones, can also contain oil and gas in appreciable quantities. Because of the lack of permeability in these reservoirs, special technologies are required to extract the petroleum from them. There are also large quantities of natural gas at great depths in the pore spaces of basinlike sedimentary rock sequences thousands of meters thick. These high-pressure, high-temperature environments are called **geopressurized zones** (or *geopressured zones*). The broad term **unconventional hydrocarbon** is used in referring to any hydrocarbon fuel

that occurs in an atypical reservoir, such as a tight sandstone or geopressurized zone, or in an unconventional form. The most promising, in terms of the sheer size of the resource and the potential for recovery, are the *tar sands* and *oil shales.*

Tar Sands

Oil that is exceedingly viscous will not flow easily and cannot be pumped. Colloquially called *tar,* viscous oil acts as a cementing agent between mineral grains in an oil pool. **Tar sands** are deposits of dense, viscous, asphalt-like oil that are found in a variety of sedimentary rocks and unconsolidated sediments (not just in sand, as the name implies). Tar sands may be petroleum deposits in which the volatile fraction has migrated away, leaving behind the residual, tarry material. Alternatively, they may be immature deposits in which the chemical alterations that form the lighter liquid and gaseous hydrocarbons have not yet occurred.

The exploitation of tar sands has a number of environmental impacts. Tar can be recovered from the deposits only if the rock itself is mined and heated enough to make the tar flow. The resulting tar must then be processed to recover the usable fuel. Large quantities of waste rock are produced by the recovery process. Because many of the deposits occur near the surface, strip mining is commonly used to extract the material. The process also uses large quantities of water, a cause for concern if the deposit is located in a region where water is scarce. Noise, thermal pollution, and water pollution are additional environmental problems associated with the extraction of petroleum from tar sands.

The largest known occurrence of tar sands is in Alberta, Canada, where the Athabasca Tar Sand covers an area of

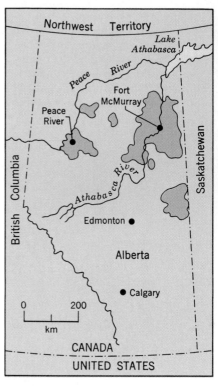

Area underlain
by tar sand

A.

B.

▲ F I G U R E 12.5

A. The part of the province of Alberta, Canada, that is underlain by the Athabasca tar sand. B. Athabasca tar sand being mined in Alberta. After mining, the tar-cemented sands are heated in order to soften and remove the tar before they are processed.

5000 km^2 and reaches a thickness of 60 m (Fig. 12.5). It has been estimated that the Athabasca deposit contains as much as 600 billion barrels of petroleum. Similar deposits almost as large have been identified in Venezuela and in the former Soviet Union. The costs—both economic and environmental—of mining and treating tar sands are high, but the process is technologically feasible.

Oil Shales

Another potential source of petroleum is a waxlike organic substance called *kerogen*, which forms when organic material is buried, compacted, and cemented in very fine-grained sedimentary rocks such as marine and lacustrine (lake) shales. If burial temperatures are not high enough to initiate the chemical breakdowns leading to the formation of oil and natural gas, kerogen may be formed instead. If the kerogen is heated, it breaks down and forms liquid and gaseous hydrocarbons similar to those found in oil and gas. All shales contain some kerogen, but to be considered an energy resource the kerogen in an **oil shale** must yield more energy than is required to mine and heat it. Because the energy needed to mine and process a ton of shale is equivalent to that created by burning 40 liters of oil, only shales that yield 40 or more liters of distillate per ton can be considered.

The world's largest deposit of rich oil shale is located in the United States. Millions of years ago, during the Eocene

Epoch, there were many large, shallow lakes in Colorado, Wyoming, and Utah. In three of them a series of organic-rich sediments was deposited that are known as the Green River Oil Shales (Fig. 12.6). The richest shales were deposited in the Colorado lake, which is now called the Piceance Basin. These shales are capable of producing as much as 240 liters of oil per ton. Scientists of the U.S. Geological Survey estimate that in the Green River Oil Shale alone, oil-shale resources capable of producing 50 liters or more of oil per ton of shale can ultimately yield about 2000 billion barrels of oil.

Although they have not been thoroughly explored, rich deposits of oil shale have been identified in other parts of the world. They include the Irati Shale in Brazil and a large deposit in Queensland, Australia; other deposits have been reported in South Africa and China. Oil shales have been mined and processed on an experimental basis in the United States, but the only countries where extensive commercial production has been tried are China and the former Soviet Union. Thus far, production expenses have made the exploitation of oil shales unattractive compared with that of oil and gas. Most experts believe, however, that large-scale mining and processing of oil shale will eventually happen.

The environmental impacts associated with the extraction of shale oil are similar to those discussed earlier for tar

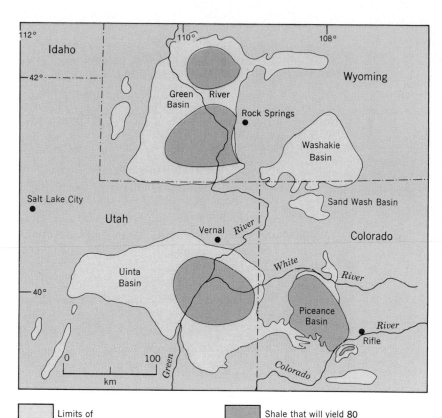

A.

◄ F I G U R E 12.6
A. Vast areas of Colorado, Wyoming, and Utah are underlain by the Green River Oil Shale. The extensive deposits of oil shale were formed as organic-rich sediment that accumulated in ancient freshwater lakes and was buried, compacted, and cemented. When heated, the solid organic matter in the shale is converted into hydrocarbons similar to those in petroleum. B. Bedded shales in this Colorado canyon wall belong to the Green River Oil Shale. C. A polished slab of laminated oil shale clearly shows the thin, dark layers in which the oil is concentrated.

B.

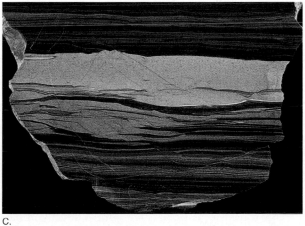

C.

sands. For both types of deposits, some consideration has been given to the possibility of extracting the resource *in situ,* that is, without actually removing the rock that contains the hydrocarbon material. This would involve the use of explosives to fracture the rock while it is still under-

ground, and the injection of heated fluids to extract the kerogen or tar. This type of procedure could minimize land disturbance at the surface, but it would still involve large amounts of water, as well as the potential for noise, thermal, and water pollution.

ALTERNATIVES TO FOSSIL FUELS

Far more energy is available in the Earth's energy budget than humans can use. What is not clear is when—or if—humans will learn how to tap the different energy sources in ways that are economical and don't disrupt the environment. Three sources of energy other than fossil fuels have been developed to some extent in different parts of the world. These are biomass energy, hydroelectric energy, and nuclear energy. Others, including the Sun's heat, winds, waves, tides, and the Earth's internal heat, have been tested and developed only on a limited basis. But the day may not be far off when one or more of these alternative energy sources will become important on more than a local scale.

Biomass Energy

Biomass energy refers to any form of energy that is derived more or less directly from the Earth's plant life. In the form of fuel wood, biomass was the dominant source of energy for millennia; not until the end of the nineteenth century was it displaced by coal-based technologies. Biomass fuels—primarily fuel wood, peat, animal dung, and agricultural wastes in one form or another—are still widely used throughout the world. Scientists working for the United Nations estimate that wood and dung used for cooking and heating fires now account for 4×10^{19} J of energy production annually, or approximately 14 percent of the world's total energy use. The greatest use of wood as a fuel occurs in developing countries, where the cost of fossil fuel is very high in relation to income. Biomass energy supplies about 35 to 40 percent of the energy in developing countries today.

Aside from the burning of fuel wood, one of the main sources of biomass energy is the burning of agricultural wastes. In Brazil, agricultural waste left over from the processing of sugar cane is burned to fuel the sugar extraction process. These wastes, known as *bagasse,* can also be used to run turbines and generate electricity. Bagasse is potentially very important as a biomass energy source because sugar refining operations are found in virtually every developing country. In Hawaii, where there are no conventional fossil fuel deposits, the burning of sugar cane wastes supplies about one third of the state's electricity needs.

Not all biomass energy is obtained directly from the burning of plant matter. Charcoal is a form of biomass energy in which the wood is not burned directly in the stove or fireplace but is first transformed into a more convenient and longer lasting form of fuel. In regions where people rely heavily on charcoal as their primary fuel for cooking, the use of charcoal can contribute significantly to deforestation. Proposed energy and fuel wood conservation schemes focus on more efficient charcoal-producing techniques and the development of more efficient charcoal-burning stoves for home use.

Animal dung is a form of biomass energy in which plant matter is processed in the digestive tracts of animals before it is made available for use as a fuel. Methane (CH_4) gas can also be derived from decaying animal wastes, collected, and used as a fuel, referred to as **biogas**. Systems for recycling animal waste and utilizing biogas to generate electricity have helped many farm families in developing countries improve their standard of living.

Household wastes can be turned into fuel in several ways. The methane that evolves from the biodegradation of organic material in landfills can be tapped and used as a fuel (Fig. 12.7). Some modern landfill operations supplement their operating budgets by selling the biogas as a fuel. Some communities burn household wastes (as well as nonhazardous industrial wastes), partly to reduce their dependence on landfills and partly as a source of energy. This energy source is sometimes called *refuse-derived fuel,* or *RDF.*

Alcohol fuel, generally made from the fermentation of grains such as corn, is another biomass-derived fuel. Alcohol has received considerable attention as an alternative fuel because cars with conventional engines can run on a mixture of 10 percent alcohol fuel with 90 percent gasoline. As oil supplies dwindle, it is likely that the use of "gasohol" will increase.

▲ F I G U R E 12.7
Energy production from waste biomass. This Italian gasification plant produces methane from the pulped waste of sorghum. The methane is used as fuel in an electrical power plant.

◄ F I G U R E   12.8
Hoover Dam on the Colorado River, Arizona. The water impounded by the dam creates Lake Mead and is used to generate electricity. The hydroelectric power stations can be seen at lower left in the canyon below the dam.

How much energy is potentially available from biomass sources? It is estimated that new plant growth on land produces 1.5×10^{11} metric tons of dry plant matter each year. If all of this were burned, or used in some other way as a biomass energy source, it would produce almost nine times more energy than the human population uses each year. Obviously, this is a ridiculous suggestion because it would require that all the forests be destroyed; plants could not be eaten, and agricultural soils would be devastated. Nevertheless, controlled harvesting of fuel plants could probably increase the proportion of the biomass used for fuel without serious disruption of forests or food supplies.

Biomass is, in principle, a completely renewable form of energy. However, as mentioned earlier, *all* energy sources have some drawbacks and environmental impacts. As with all renewable resources, in regions where the rate of use exceeds the rate of replenishment the resource can become effectively depleted. This has happened, for example, in semiarid regions where intensive harvesting of fuel wood has contributed to deforestation. Biomass burning also releases carbon dioxide into the atmosphere and contributes to particulate air pollution. Where charcoal and fuelwood are used for cooking and heating, the smoke can create indoor air pollution and pose a hazard to health.

Hydroelectric Power

Hydroelectric power is generated from the potential energy of stream water as it flows to the sea. In order to convert the power of flowing water into electricity, it is necessary to dam streams (Fig. 12.8). The reservoirs behind the dams eventually fill with silt, so even though water power is continuous, dams and reservoirs have limited lifetimes.

Hydroelectric power is considered a "clean" energy source because it involves no burning and, therefore, no atmospheric emissions. However, hydroelectric installations have a number of environmental impacts. Among the most common impacts are loss of prime agricultural land because of the flooding of large areas behind the dam (recall from Chapter 8 that soils adjacent to rivers are typically very fertile). The filling of large reservoirs has been known to cause earthquakes in seismically inactive areas; this is called *induced seismicity*. Wildlife habitat along a dammed river is lost or degraded both upstream and downstream from the dam. Questions have also been raised about the role of the reservoir-filling process in water pollution. For example, elevated mercury levels have been measured in reservoirs behind some large dams. It has been suggested (though not proved conclusively) that the filling process may release mercury contained in subsurface rocks and soils. Finally, international financial institutions (notably the World Bank) have been severely criticized for their involvement in large hydroelectric projects in developing countries. In the past, such projects have led to the displacement of tens of thousands of local inhabitants, often without compensation. Large reservoirs have also been known to contribute to the spread of water-borne diseases in developing countries.

Water power has been used in small ways for thousands of years, but only in the twentieth century has it been used to any significant extent for generating electricity. The total recoverable energy from the water flowing in all of the world's streams is estimated at 9.2×10^{19} J/year, equivalent to the energy obtained by burning 15 billion barrels of oil per year. Thus, even if all the potential hydropower in the world were developed, it could satisfy only about one third of present energy needs. We have to conclude that although hydroelectric power is very important for countries with large rivers and suitable dam sites, for most countries it holds limited potential for development.

Nuclear Energy

Nuclear energy is the heat energy produced during the controlled transformation of suitable radioactive isotopes. Three of the radioactive atoms that keep the Earth hot by spontaneous decay—^{238}U, ^{235}U, and ^{232}Th—can be mined and used to obtain nuclear energy (Fig. 12.9). Uranium deposits occur primarily in sedimentary rocks and metamorphosed sedimentary rocks. The dense uranium-bearing minerals can be concentrated in sediments by the action of waves and currents. Uranium minerals are also very soluble in oxygen-rich, near-surface environments, which means that they are readily dissolved by groundwater, transported in solution, and eventually precipitated in concentrated layers during weathering processes.

There are two potential sources of nuclear energy, that is, two types of nuclear processes through which power can be generated. They are *fission* and *fusion*.

Fission

Fission is the splitting of an atom into two smaller atoms (Fig. 12.10). The sum of the masses of the fission products

▲ F I G U R E   12.9
Four identical reactors at the Paluel nuclear power plant, Normandy, France. Each reactor can produce 1330 megawatts of power by the fissioning of ^{235}U.

is slightly less than the mass of the original atom; the remaining mass is converted into heat energy. Fission is accomplished by bombarding the radioactive atoms with neutrons, thus accelerating the rate of decay and the release of energy. The device in which this operation is carried out is a *pile*.

When ^{235}U fissions, it not only releases heat and forms new elements but also ejects some neutrons from its nucleus. These neutrons can then be used to induce more ^{235}U atoms to fission, creating a continuous chain reaction. The function of a pile is to control the rate of neutron bombardment so the rate of fission can be controlled. (When a chain reaction proceeds without control, an atomic explosion occurs.) Controlled fission is the method used by nuclear power plants, and it produces a tremendous amount of energy. The fissioning of one gram of ^{235}U produces as much heat as the burning of 13.7 barrels of oil.

However, ^{235}U is the only natural radioactive isotope that will maintain a chain reaction, and it is the least abundant of the three radioactive isotopes that are mined for nuclear energy. Only one out of every 138.8 atoms of uranium is natural ^{235}U. The remaining atoms are ^{238}U, which will not sustain a chain reaction. However, if ^{238}U is placed in a pile with ^{235}U that is undergoing a chain reaction, some of the neutrons will bombard the ^{238}U and convert it into plutonium-239 (^{239}Pu). Under suitable conditions, this new isotope can sustain a chain reaction of its own. The type of pile in which the conversion of ^{238}U takes place is called a *breeder reactor*. The same kind of device can be used to convert ^{232}Th into ^{233}U, which also will sustain a chain reaction. Unfortunately, breeder reactors and nuclear power plants based on them are more complex and less safe than ^{235}U plants, so most operating nuclear power plants still use ^{235}U.

Nuclear power plants utilize the heat energy from fission to produce steam that drives turbines and generates electricity. Approximately 17 percent of the world's electricity is derived from nuclear power plants. In France, more than half of all the electrical power comes from nuclear plants; the proportion is rising sharply in some other European countries and in Japan. The reason for the increase is that Japan and most European countries do not have adequate supplies of fossil fuels to be self-sufficient.

Over 400 nuclear plants are now in operation worldwide. However, the number of new installations per year is decreasing. As operating plants begin to age and, eventually, require decommissioning, the associated costs may mean that nuclear energy will no longer be economically competitive with other forms of energy.

What about the environmental and health concerns associated with radioactivity? In normal operating mode, nuclear power plants contribute only a very small portion of the background environmental radiation that we are exposed to in daily life. In fact, on a fuel equivalent basis, the burning of coal releases more radioactivity *and* more carcinogenic substances into the environment than does the

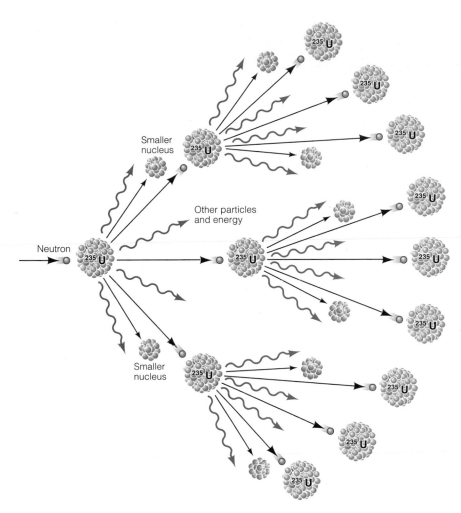

The ^{235}U fission process. When the nucleus of an atom of ^{235}U is bombarded with neutrons, the nucleus splits (fissions) into two smaller nuclei. The sum of the masses of the two smaller nuclei is less than the mass of the original nucleus; the lost mass must be conserved somehow, so it is converted into energy. The fission process also releases other neutrons, which, in turn, cause fission in other ^{235}U nuclei.

generation of nuclear power. Still, many significant problems are associated with nuclear energy. The isotopes used in power plants are the same as those used in atomic weapons, creating a potential threat to security. The possibility of a power plant failing in some unexpected way creates a safety problem. An example is the 1986 Chernobyl disaster, in which the reactor core overheated and exploded, causing a total meltdown accompanied by the release of significant quantities of radioactive material from the reactor (Fig. 12.11). The problem of safe, permanent disposal of dangerous radioactive waste matter must also be faced; this will be discussed in detail in Chapter 16.

▶ F I G U R E 12.11
Nuclear power plant, Chernobyl, Ukraine, site of the world's worst nuclear accident. The photograph shows the fire-damaged remains of reactor unit number three.

Fusion

In principle, nuclear **fusion**—the joining together or fusing of two small atoms to create a single larger atom, with an attendant release of heat energy—is another potential source of nuclear power. Nuclear fusion utilizes hydrogen (actually a heavy isotope of hydrogen, called *deuterium,* is required). The Earth has a virtually endless supply of hydrogen and deuterium in the form of a very common chemical compound—water. The primary byproduct of nuclear fusion would be helium, a nontoxic, chemically inert gas. Although there would be a small amount of mildly radioactive waste, it would be insignificant compared to the waste generated by the fission process.

All of this conjures up images of a cheap, clean power source with a virtually endless fuel supply. So why are we not using energy provided nuclear fusion? Fusion is the nuclear process that occurs in the cores of stars, the process responsible for the tremendous amount of heat energy generated by the Sun. But that, in a nutshell, is the problem. For

two nuclei to overcome their mutual aversion and fuse to form a new nucleus, the ambient conditions must be similar to those at the core of a star—on the order of 100 million degrees.

The possibility that nuclei could be induced to fuse at room temperature, so-called *cold fusion,* represents a utopian dream for society and instant scientific recognition for those who first manage to accomplish it. In 1989 a furor arose when researchers announced—prematurely, as it turned out—that they had achieved cold fusion in a test tube. Unfortunately, the work of those researchers was not substantiated in other laboratories. Ongoing high-temperature fusion experiments at the Plasma Physics Laboratory at Princeton University have made advances in producing energy and, more important, sustaining the energy production for significant periods. However, widespread and economical use of fusion power remains a distant goal.

Geothermal Energy

Geothermal energy, as the Earth's internal energy is called, has been used for more than 50 years in New Zealand, Italy, and Iceland and more recently in other parts of the world, including the United States. The people of Iceland, for example, have found many clever ways to harness volcanic heat to combat the cold climate. Using water warmed by hot volcanic rocks, they heat their houses, grow tomatoes in hothouses, and swim year-round in naturally heated pools. Icelanders also use volcanically produced steam to generate most of the electricity they need (Fig. 12.12).

Broadly, there are three types of geothermal "deposits" or reservoirs: *geopressurized deposits, hydrothermal reservoirs,* and *hot, dry rocks.*

Geopressurized Deposits

Geopressurized deposits are basinlike sedimentary sequences several thousand meters deep that contain huge volumes of warm water (and, in some cases, natural gas). The water associated with geopressurized zones, which is typically salty or briny, is warm by virtue of the geothermal gradient, that is, the normal increase of temperature that occurs with increasing depth in the Earth. Such deposits, which are found in Hungary and the former Soviet Union, among other places, cannot be exploited at the present time, principally because of their great depths.

Hydrothermal Reservoirs

Hydrothermal reservoirs consist of underground systems of circulating hot water and/or steam in fractured or porous rocks near the surface. *Dry-steam* hydrothermal reservoirs, which are relatively rare, produce superheated steam containing very little liquid. These reservoirs are potentially exploitable for geothermal power, because the steam may be directly usable to turn a turbine. The Geysers geothermal installation in northern California—the largest producer of

▲ FIGURE 12.12
Iceland does not have coal or oil, but it has a great deal of geothermal power. At Nesjavellir in southwestern Iceland, steam from geothermal wells drives an electrical power plant. After it passes through the turbine that generates electricity, the steam is used to heat cold water from Lake Thingvallavatna. At a temperature of 83°C the heated water is sent via an insulated 27-km-long pipeline to Reykjavik, Iceland's main city. The hot water is used for space heating and washing.

geothermal energy in the world—is a dry-steam deposit. *Hot-water convecting* hydrothermal systems are much more common, involving a reservoir of circulating water underground that is heated by the hot rocks. The water must be separated from the steam before it can be used to run a turbine. This is the type of deposit found in New Zealand, where geothermal energy is widely used.

To be used efficiently, geothermal steam should be 200°C or hotter. In most parts of the world, holes must be 5 to 7 km deep to reach rock that is 200°C. For development of geothermal energy to be practical, however, rock temperatures of 200°C or higher must be reached within 3 kilometers of the surface. Therefore, most of the world's hydrothermal reservoirs are close to plate margins, where recent volcanic activity has occurred and hot rocks or magma are close to the surface.

Hot, Dry Rocks

Some attention has also been given to the possibility of creating artificial geothermal steam fields using the third type of geothermal deposit—hot, dry rocks. Interesting geothermal experiments have been conducted in New Mexico. In the Jemez Mountains on the edge of an extinct (but still hot) volcano, scientists drilled two holes deep into the hot rock, as shown in Fig. 12.13. Then they shattered the hot rock with explosives to create an artificial reservoir and pumped water through to produce steam. The first tests were only partly successful. A major difficulty was that water did not flow uniformly through the hot rock but instead followed narrow flow paths, and the rocks that lined them soon cooled. Hot, dry rocks are much more abundant than hydrothermal steam fields; if the water flow problems encountered in the Jemez Mountains experiments could be overcome, geothermal energy might someday play a major role in meeting the energy needs of society.

Geothermal Potential

So far, geothermal power is of only local importance. Many volcanoes are too active and too dangerous to be considered as sources of geothermal energy. Some places, like Yellowstone National Park, could produce a large amount of energy, but if we drilled and pumped out the steam and hot water reservoirs beneath them, the famous scenic hot springs and geysers would soon be dry. It is estimated that the world's geothermal reservoirs could yield about 8×10^{19} J of energy—equivalent to burning 13 billion barrels of oil. This estimate incorporates the observation that in New Zealand and Italy only about 1 percent of the energy in a geothermal reservoir is recoverable. If the recovery efficiency were to rise, the estimate of recoverable geothermal resources would also rise. But even if the efficiency rose to 50 percent, geothermal power could satisfy only a small percentage of human energy needs.

The advantages of geothermal energy include its continuous availability and in some situations the ability to use

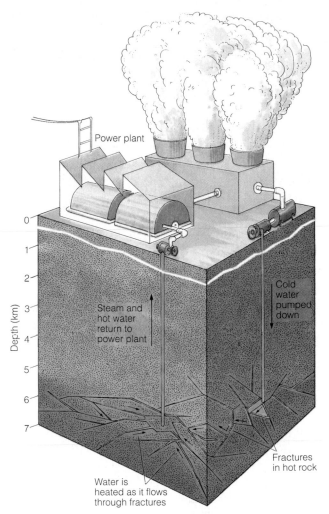

▲ **F I G U R E   12.13**
Experiments with geothermal energy from hot, dry rocks. Cold water is pumped through fractures at the bottom of a 6-km-deep well, becomes heated, and then flows back up to the geothermal power plant, where heat energy is converted into electricity.

the energy directly (that is, without complicated conversions) for certain applications, such as space heating. Geothermal energy occurs in well-defined geologic environments, so potential reservoirs are easy to locate. The technological aspects of geothermal power generation are not highly sophisticated, although difficulties are often encountered in the drilling process because of the high temperatures involved.

Probably the single greatest disadvantage of geothermal power is that it must be exploited at the site of the deposit, as the cost of transporting the steam rapidly exceeds the value of the recoverable energy. We also have little knowledge about the longevity of geothermal reservoirs and the possibility of depletion, although there are indications that some deposits may have useful lifetimes of only 20 or 30 years.

As with other energy sources, the development of geothermal power sources can have negative environmental impacts. These include subsidence associated with the withdrawal of fluids, high noise levels, water pollution, thermal pollution, and some air pollution. Reinjection of geothermal fluids into the producing well might counteract some of these environmental impacts while returning a portion of the heat to the system. However, reinjection has been known to induce earthquakes in some localities.

Solar Energy

The amount of **solar energy** reaching the Earth from the Sun each year is approximately 4×10^{24} J—that is, at least ten thousand times more than the amount humans use. Most of the energy sources discussed in this chapter are actually forms of solar energy that have been altered, stored, or degraded in some way.

We already put some of the Sun's energy directly to work in greenhouses and in solar homes, but the total amount used in these ways is small. Solar energy is best suited to supplying heat at or below the boiling point of water, which makes it especially useful for such applications as home heating and water heating for home use. In Cyprus, Israel, and Jordan, between 25 and 65 percent of the water used in homes is heated by solar panels. Some advanced types of solar collectors can produce water that is hot enough for industrial applications. Large-scale commercial solar concentrators, such as those at the Luz International plants in the Mohave Desert, are used to heat water or another fluid, generating steam to run turbines.

Converting solar energy directly into electricity is a major technological challenge. Devices that effect such a conversion, called **photovoltaic cells,** have been invented (Fig. 12.14). So far their costs are too high and efficiency too low for most uses, although they are widely used in small calculators, radios, and the like. Photovoltaic technology is constantly improving, however, and the cost of energy generated in this manner is decreasing; eventually widespread use of this technology may be more cost effective.

There are some environmental drawbacks to the use of solar energy, although fewer than for most other energy sources. For example, in order to supply solar energy to a large area, a significant amount of land must be devoted to energy "farms." Also, some toxic metals are used in the manufacture of solar panels. Overall, however, the most pressing disadvantage of solar energy is the cost of converting existing equipment to be compatible with a solar power source.

An interesting possible derivative of solar energy is hydrogen fuel, which could be produced from water using solar-generated electricity. Hydrogen fuel burns cleanly and is considered environmentally safe. Existing technologies that rely on the internal combustion engine could be adapted to its use relatively easily. So far, however, solar hydrogen is not competitive with other energy sources in terms of price.

Wind Energy

Winds and waves are both secondary expressions of solar energy. For thousands of years **wind energy** has been used as a source of power for ships and windmills. Today, huge windmill "farms" are being erected in suitably windy places

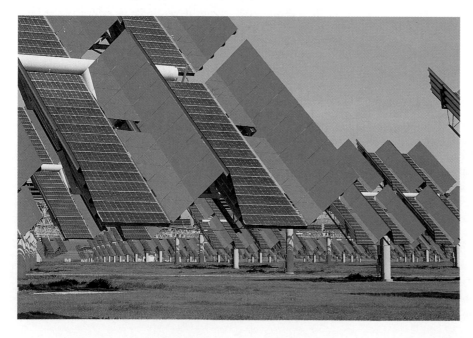

◄ F I G U R E 12.14
Photovoltaic panels convert the energy of sunlight into electricity at a power-generating site in southern California.

◄ F I G U R E 12.15
A field of windmills near Palm Springs, California. The windmills generate electricity using the kinetic energy of wind.

(Fig. 12.15). In Denmark, about 3000 wind turbines supply electricity throughout the country. Although there are some problems with windmill technologies, it seems very likely that windmills will soon be cost competitive with coal-burning electrical power plants.

Unfortunately, much wind energy resides in very high-altitude winds. Steady surface winds can provide only about 10 percent of the amount of energy now used by humans. As with hydroelectric and geothermal power, therefore, wind power may be locally important but probably will not become globally significant. As with solar power, windmill farms require a large amount of land. Critics contend that windmill farms detract from the beauty of the countryside, damage tourism, distract drivers, and create noise pollution.

Ocean Energy

The oceans are an enormous reservoir for energy in the Earth's energy cycle (Fig. 1.10). A large amount of energy is stored in ocean water in the form of heat. Solar energy is cycled through the oceans in the form of waves and currents, and the energy from the Earth's gravitational interaction with the Moon is reflected in the ebb and flow of tides. All of these are essentially inexhaustible energy sources, if the technological means can be found to tap into them effectively. Of the possible means of exploiting ocean energy, we describe three: *wave energy, tidal energy,* and *ocean thermal energy conversion (OTEC).*

Wave Energy

Waves, which are created by winds blowing over the ocean, contain an enormous amount of energy. We can see how powerful waves are by watching any coastline during a storm. For centuries, **wave energy** has been used to ring bells and blow whistles that serve as navigational aids, but so far no one has discovered how to tap waves as a source of power on a large scale.

A number of small-scale wave energy experiments, notably those carried out by Japanese investigators, have been aimed at harnessing wave energy through the use of floating raftlike devices. An alternative technology employs a sluicelike opening in the hold of a ship to permit the passage of water in an incoming wave, trapping the water and then releasing it after the wave recedes. The incoming water forces air to pass through a narrow valve, turning a turbine that drives an electrical generator. Although some ships have been designed that can run on wave energy, the technologies tend to fail because of corrosion, cold water temperatures, or storm damage.

Tidal Energy

Tidal energy is derived from the gravitational forces exerted on the Earth by the Moon and the Sun. If a damlike structure called a *barrage* is constructed across the mouth of a bay so that water is trapped at high tide, the trapped water can be released at low tide to drive a turbine. The use of tidal energy is not entirely new; a site in Britain dating from 1170 still utilizes tidal power to run a mill.

In order for a locality to be suitable for the development of a modern tidal power facility, there must be a substantial *tidal range,* that is, a large difference in height between high and low tides. The bay must also be relatively long and thin, with a narrow opening to the sea, so that the tidal range is amplified as much as possible by the configuration of the shoreline. A bay that is well suited to the development of tidal energy is the Bay of Fundy in Nova Scotia

WIND ENERGY FOR WATER PUMPING IN THE SAHEL

*I*n the mid-1970s the Sahel region of west Africa was hit by a severe drought. Declining annual rainfalls led to widespread erosion and desertification and, in turn, to food shortages. Satellite images from the region revealed that the entire west African landscape was brown and barren (Fig. B1.1). In the mid-1970s a team of Canadian researchers, initially funded by the Institute for the Study and Advancement of Integrated Development (ISAID), traveled to the west African country of Niger to investigate the possibility of establishing a project to fight the encroachment of the desert. The researchers selected a group of 5 villages, known as the Chikal villages, for the pilot project. They worked with researchers from the Nigerian Agriculture Research Center to collect data on soil, water, and wind, as well as information about socioeconomic conditions in the villages. Data from these studies were used to evaluate agricultural practices in the area and their impacts on soil and water.

As the project progressed, it became clear that people in all five communities were very concerned about health and nutrition standards. A survey revealed serious vitamin deficiencies in children, and many deaths of children under the age of 5 were traced to dietary deficiencies. Both the quality and the quantity of available food were inadequate. The researchers determined that vegetable gardens could help meet nutritional needs as well as address some of the ecological concerns about desertification. However, for the gardens to thrive, a source of water had to be found. Technical studies had revealed a significant supply of groundwater in the area, but it was located 10 m below the surface.

In Niger there are two main seasons: a dry season, during which little farming can be carried out, and a wet season, during which the villagers plant and harvest millet,

the staple food in the area. Chikal villagers were very interested in the possibility of dry-season vegetable gardening, but the problems of water supply and appropriate irrigation technologies had to be solved first. Other development programs in the area had experimented with technologies for pumping water during the dry season, such as animal traction pumps and motorized mechanical pumps. Most of these efforts had failed, partly because of the extremely harsh conditions in the area and partly because villagers were not trained in the upkeep of the pumping equipment.

The project team decided that wind-driven pumps would best suit the needs of the Chikal villages. Winds in the Sahel are particularly strong during the dry season (Fig. B1.2). A great deal of emphasis was placed on the selection of a windmill technology that would be simple and easy to maintain, but hardy enough to withstand constant battering by blown dust. Another important aspect of the project at this stage was the selection of a group of villagers to receive training in the installation and maintenance of wind-driven water pumps.

A site was chosen for a pilot community garden, and the first wind pump was installed in 1982. In the first growing season the pattern of declining rainfalls continued, and large amounts of dust blew through the area during the dry season. But the pump provided the water needed to make the garden productive. By the following year, fresh vegetables were available in the local market for the first time. Those who participated in the first garden were pleased with their economic returns, and more villagers became interested in dry-season gardening. There was no shortage of land in the communities; water supply had been the limiting factor in determining the size of

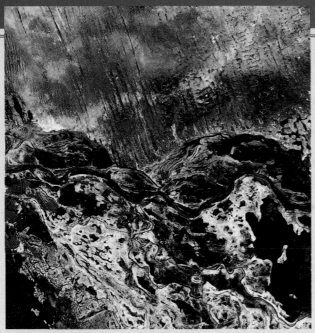

▲ FIGURE B1.1
**Contrasts in Sahel: migrating sand dunes in west
Africa have reached the Senegal River, which marks
the border between Mauretania (top) and Senegal
(bottom).**

▲ FIGURE B1.2
**Groundwater pumped up to surface tanks by wind-
mills is used to grow summer vegetable gardens,
Chikal, Niger.**

community gardens. The project (funded after 1986 by
the Canadian International Development Agency) expanded to nearby villages, including both installation of
new wind pumps and continuation of the training programs.

Today the Chikal gardens continue to thrive. They have
restored large areas of topsoil and green spaces on the edge
of the growing desert. Villagers who participate in the garden projects have improved both their nutritional standards and their incomes. The wind pumps continue to be
managed and maintained by the villagers, and studies suggest that the rate at which water is being withdrawn is

lower than that at which groundwater is being replenished.
The extra water provided by the pumps has also meant
that women and girls in the Chikal villages no longer have
to walk several kilometers each day to haul water for their
families.

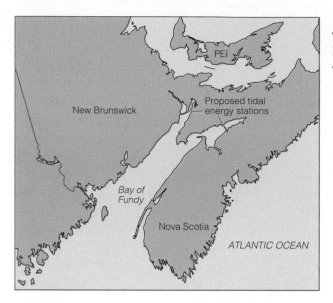

◀ F I G U R E 12.16
The high tidal range and narrow bay inlets at the Bay of Fundy in Nova Scotia are particularly well suited for the exploitation of tidal energy.

(Fig. 12.16), where the maximum difference between high and low tides is close to 9 m. In France, a tidal power project, La Rance, provides 544 million kWh to the French power grid (Fig. 12.17). Other tidal power facilities in the former Soviet Union and China also produce electricity.

Although there are numerous sites on the Earth with potential for development of tidal power sources, a disadvantage is that the power generated is not easily transportable and is thus tied to the locality where it is being harnessed. Tidal energy is attractive because it is inexhaustible, predictable (unlike wind, for example), requires no fuel, has low maintenance costs, and is a nonpolluting

source of energy. However, not enough is known about the possible effects of tidal energy on the ecology of tidal inlets, which tend to be delicately balanced environments. Also, as with solar energy, the development of tidal energy would require a large initial capital investment. Thus, as an alternative energy source tidal power offers some potential but a lot of obstacles.

Ocean Thermal Energy Conversion

Even though the actual temperature of ocean water is not very high, a vast amount of heat energy is stored in the water simply because of its enormous volume. In general,

◀ F I G U R E 12.17
La Rance tidal power plant in southern France. With the tide rising, water is seen flowing through the barrage across the mouth of the River Rance. At high tide the barrage is closed. At low tide the barrage is opened again and the water flowing back to the sea drives electricity-generating turbines.

surface waters are warmer than deeper waters by several degrees. A fundamental principle of thermodynamics is that wherever a *gradient* exists, matter or (in this case) energy will flow along the gradient. Wherever there are flows of this type, energy can be extracted from the system. This is basically the same principle that makes it possible to withdraw energy from the falling water in a waterfall (although in that case, the energy used is *kinetic energy*, or energy of motion). If there is a large enough difference in temperature between surface waters and deeper waters, it should be possible to extract heat energy from the ocean using a simple, widely used technology called a *heat pump*. When heat pumps are used to withdraw energy from ocean water, the technology is called **ocean thermal energy conversion** (abbreviated *OTEC*).

An example of the application of OTEC is found in Hawaii, where the volcanic slopes steepen rapidly just offshore and the temperature gradient in the water is particularly steep. Surface waters are warmed by the Sun, while much colder waters lie a very short distance offshore. The warm surface waters are used to heat ammonia, converting it to its gaseous form. The gaseous ammonia runs a turbine, which generates electricity. After it leaves the turbine, the ammonia is run through a series of pipes cooled by cold offshore water, which is pumped to the power plant through a pipeline. The ammonia cools and condenses very quickly and can then be reused. A large temperature differential is crucial to this process.

Even though the technologies required are relatively simple, in practice the number of potential sites for OTEC is limited. Especially in northern oceans, the temperature difference between surface and deep waters is often too small to be exploited efficiently for power generation. The hostile environment of the ocean setting also leads to frequent equipment failures.

SUMMARY

1. The factors and considerations that will shape our energy future include the continuing need to develop energy conservation methods; the need to shift to renewable and/or inexhaustible resources; environmental concerns; social, health, and regional concerns; strategic concerns; and the need for diversity in the mix of energy sources. It is important to remember that all energy sources, whether conventional or alternative, have some environmental impact.

2. One group of alternative energy sources represents an extension of current energy use patterns; they make fossil fuels longer lasting or more environmentally acceptable. These energy sources include clean coal technologies and extraction of unconventional hydrocarbon resources. A second group of alternative energy sources represents a departure from fossil fuel use. They include solar, wind, hydroelectric, nuclear, biomass, geothermal, and ocean energy.

3. Coal is the "dirtiest" of the fossil fuels, but it is the most abundant. New technologies are aimed at using coal to extend other fossil fuel resources or reducing the negative environmental impacts of coal burning. These technologies include precombustion cleaning, combustion-enhancing technologies, coprocessing and conversion, and postcombustion cleaning.

4. Unconventional hydrocarbon resources include hydrocarbon fuels that occur in atypical reservoirs like "tight" sandstones or geopressurized zones, or in a form other than conventional coal, oil, or natural gas. The most promising of these, in terms of the sheer size of the resources and their potential for recovery, are tar sands and oil shales.

5. Tar sands are sediments or sedimentary rocks that contain a dense, viscous, asphalt-like form of oil. The largest known tar sand deposits are in Canada. Oil shales are organic-rich sediments and sedimentary rocks that contain kerogen. The largest known oil shale deposits are in the United States. The extraction of tar sands and oil shales has many negative environmental impacts, including disruption of land, production of rock waste, intensive use of water, and noise, thermal, and water pollution.

6. A wide range of biomass energy sources have been used throughout history; they include the burning of fuel wood, peat, animal dung, and agricultural wastes. Some promising biomass technologies include production of biogas and alcohol fuel from organic sources, methane from landfills, and more efficient charcoal production technologies.

7. In places where running water is abundant, hydroelectrical power is a clean, inexhaustible energy source. However, dams and large reservoirs have negative environmental impacts, including loss of agricultural land, loss of wildlife habitat, induced seismicity, water pollution, and impacts on siltation rates, as well as displacement of local inhabitants.

8. Nuclear energy can come from fission, the splitting of a nucleus into two smaller parts, or fusion, the joining of two nuclei to make a larger one. Nuclear power generation today is based on the fission process because fusion requires very high temperatures and pressures. The fuel used in nuclear power generation, mostly ^{235}U, occurs naturally in uranium-rich mineral deposits. Nuclear power generation poses little risk under normal circumstances. However, the possibility of accidents, security risks, and the environmental and health impacts of radioactive waste pose formidable problems.

9. Parts of the world with high rates of heat flow near the surface can exploit geothermal energy for power generation. The three types of geothermal deposits are geopressurized zones, hydrothermal reservoirs (including dry-steam and hot-water convecting systems), and hot, dry rocks. The environmental impacts of geothermal energy exploitation include high noise levels, water pollution, thermal pollution, and subsidence associated with the withdrawal of fluids.

10. Most energy sources are actually solar energy that has been stored, altered, and recycled in various forms in the Earth's energy cycle. Solar energy can also be used directly to heat homes and water. With large-scale collectors and photovoltaic cells, solar energy can be used to run turbines and generate electricity. The use of solar energy to produce hydrogen fuel from water is a promising technology.

11. In some areas wind energy is almost cost competitive with more conventional forms of power generation. A drawback is that windmill farms require large amounts of land and may detract from the beauty of the countryside or create noise pollution.

12. Ocean energy encompasses tidal and wave energy as well as ocean thermal energy conversion (OTEC). Tidal energy technologies are similar to those used in conventional hydroelectric installations, requiring the construction of dams across narrow bays with high tidal ranges. Wave energy technologies make use of water or air forced through sluicelike openings in the holds of ships. OTEC relies on the temperature differential between warm surface water and cold, deep water to drive a process in which ammonia is vaporized and used to run a turbine.

IMPORTANT TERMS TO REMEMBER

biogas (p. 308)
biomass energy (p. 308)
fission (p. 310)
fusion (p. 312)
geopressurized zones (p. 305)
geothermal energy (p. 312)

hydroelectric power (p. 309)
nuclear energy (p. 310)
ocean thermal energy conversion (OTEC) (p. 319)
oil shale (p. 306)
photovoltaic cells (p. 314)

solar energy (p. 314)
tar sands (p. 305)
tidal energy (p. 315)
unconventional hydrocarbons (p. 305)
wave energy (p. 315)
wind energy (p. 314)

QUESTIONS AND ACTIVITIES

1. Does your country, region, or municipality have an energy strategy? If so, do you agree with the strategies outlined in the document? How much of the budget of your country, state, or province is spent on research and development of alternative energy resources? Do you think the allocated amount is sufficient? too little? too much?

2. Find out more about an alternative energy source that has been developed in your area. For example, if you live in California you might want to find out about solar energy; if you live in Nova Scotia, find out about tidal energy; if you live in Oregon, find out about geothermal energy.

3. Arrange to visit an alternative energy installation (e.g., a windmill farm, a solar collector, or a geothermal power plant) near where you live. Write a report of your findings.

4. Do any of the landfill sites near your home collect and sell the methane gas generated by the decomposing garbage? Find out how this type of operation works and, if possible, visit a site.

5. Some environmentalists are vehemently opposed to nuclear power generation because of the potential hazards and the problem of disposing of radioactive wastes. Others support nuclear energy as a cleaner alternative to fossil fuels. What are the pros and cons of nuclear energy from the point of view of environmental impacts? What is your position on this issue? What is the position of your community?

6. Biomass, particularly fuel wood, is an important energy source in many developing countries. Research the linkages between fuel wood use and deforestation. Do you think it would be good for people in developing countries to pass quickly out of the fuel wood stage and into a stage in which fossil fuels are used as the primary source of energy? What are the pros and cons of such a transition from the perspective of environmental impacts?

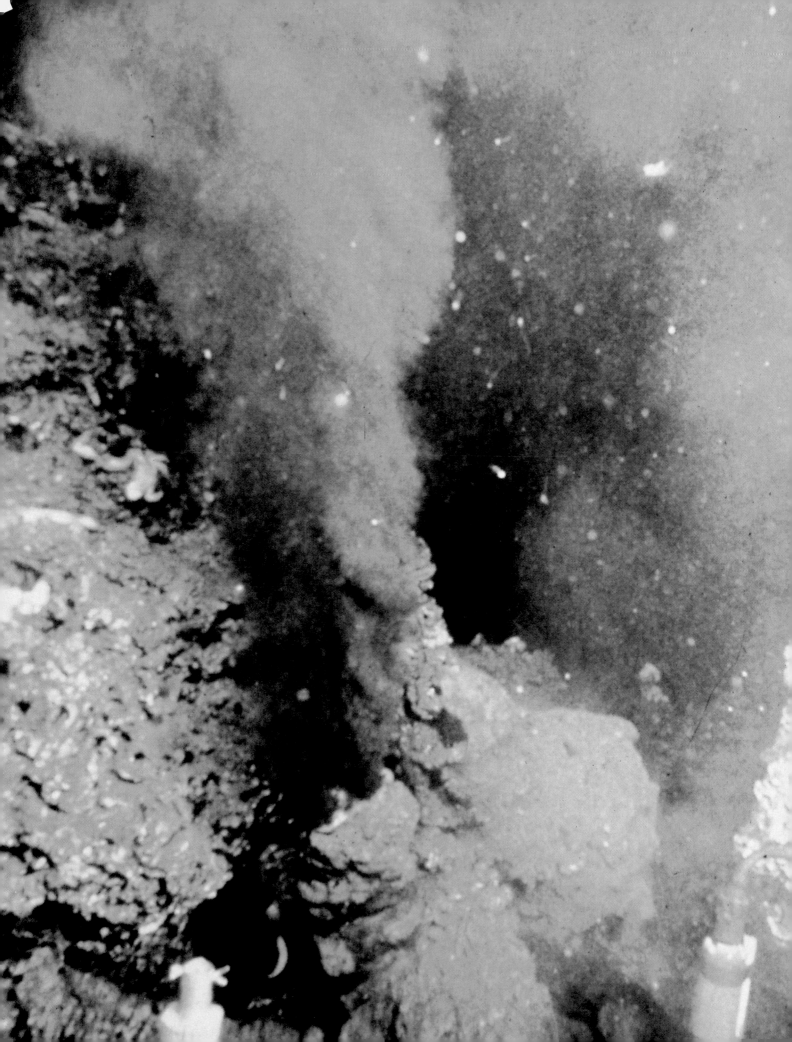

MINERAL RESOURCES

*I*n 1870, science fiction writer Jules Verne wrote a futuristic adventure story, called *20,000 Leagues Under the Sea,* in which an enormous submersible ship, the *Nautilus,* dives to the bottom of the ocean. The voyagers find a fantastic underwater world where food and mineral resources are available in abundance, as reported by the evil genius Captain Nemo: "In the ocean depths there exist mines of zinc, iron, silver and gold which would be quite easy to exploit." Verne was a visionary whose stories anticipated many scientific and technological achievements that he himself did not live to see. Will his vision of boundless mineral resources from the ocean depths become a reality someday, as many of his other fantasies have?

Part of the answer to that question may lie in the story of a modern-day submersible ship, the *Alvin,* and its role in Project FAMOUS (French–American Mid-Ocean Undersea Study). The story of Project FAMOUS began in the early 1970s with a proposal to explore and map an underwater spreading center, part of the midocean ridge near the Azores in the Atlantic Ocean. The project represented the first use of a modern submersible ship for mapping at great

depths. The *Alvin,* which is only 7.6 meters long, is made of a very strong synthetic glass foam mixed with epoxy. It is a highly maneuverable vessel capable of withstanding the extreme pressures of the deep sea.

In the summer of 1974 the *Alvin*'s first dive took place. During this and subsequent dives the "aquanauts" discovered a bizarre world of extreme temperature variations, previously unknown life forms, great fissures in the ocean floor marked by newly formed lava flows, and streams of gases coming from sources deep within the Earth. But their most amazing discovery was the existence of underwater thermal hot springs, which they called "black smokers." These take the form of tall chimneys of rock where heated seawater—black and smoky looking because it is loaded with fine particles of sulfide minerals—is vented from cracks in the ocean floor. The temperature of the water can be as high as 350°C, and the chimneys themselves are composed of solid ore minerals, mainly iron, zinc, and copper sulfides deposited by the heated water.

Scientists now believe that the black smokers represent ore formation in progress. Essentially the same process created many of the rich ore deposits that were formed millions of years ago at the bottoms of ancient oceans and are now found on land. Whether the newly forming mineral deposits deep beneath the sea or any other oceanic mineral resources will ever be exploitable remains uncertain; the technological, environmental, and political challenges are daunting. Jules Verne's bountiful harvest of underwater mineral resources may never be fully realized, but already the scientific harvest from the *Alvin* expeditions has been invaluable in furthering our understanding of the processes involved in ore formation.

◄

An ore-forming hot spring on the floor of the Pacific Ocean, at a depth of 2500 m. Heated as a result of submarine volcanism, water in the hot spring is at 350°C and appears black because fine particles of iron sulfide and other minerals are precipitated from solution as the plume is cooled through contact with cold seawater.

MINERAL RESOURCES AND MODERN SOCIETY

The term **mineral resources** encompasses any element, compound, mineral, or rock that can be extracted from the ground and is of potential value as a commodity. (This definition is meant to exclude energy resources, soil, and water, as well as living resources.) From mineral resources we get the metals to make machines, the ingredients for chemicals and fertilizers, and the materials to construct roads and buildings. Many industrialized nations are rich in **mineral deposits**—volumes of rock that contain enrichments of one or more minerals. Yet no nation is entirely self-sufficient in mineral supplies, so each must trade with other nations to meet its needs.

Most minerals that are very abundant in the Earth's crust have no commercial value; the minerals that are used in industrial processes tend to be harder to find. Geologic processes such as weathering, sedimentation, and volcanism can, under suitable conditions, form valuable mineral deposits. The "suitable conditions" are not common, however, and that is why mineral resources are limited in quantity and difficult to find. No geologic challenge is more difficult or more rewarding than the search for, and discovery of, new resources of minerals. In this chapter we focus on how and where the different kinds of valuable mineral deposits form and the role they play in modern societies. The field of study devoted to these topics is **economic geology.**

Our Dependence on Minerals

The number and diversity of mineral and rock substances that provide materials used by humans are so great that to make a simple classification scheme is almost impossible.

Nearly every kind of rock and mineral can be used for something, although those that are most valuable tend to be rare. An industrial society not only requires a diverse group of metals for machines but also demands a host of nonmetallic mineral products, such as shale and limestone for making cement, gypsum for making plaster, salt for making chemical compounds, and calcium phosphate (the mineral apatite) for making fertilizer.

Each of us uses, directly or indirectly, a very large amount of material derived from minerals (Fig. 13.1). The amount of material consumed per person each year is much greater in industrialized countries such as the United States or Canada than in most of the rest of the world. Table 13.1 lists just a few of the products made from mineral resources that are found in a typical North American home.

In the past, the consumption of mineral resources has been equated with wealth and a high standard of living. Today economic growth and development worldwide continue to feed the demand for mineral resources. Therefore, it is worthwhile to consider where the world's mineral deposits are located and whether remaining supplies of those minerals will be sufficient to meet the needs of a growing global economy in the coming decades.

SUPPLIES OF MINERALS

Mineral resources have three distinctive aspects. First, economically exploitable occurrences of minerals are distinctly localized within the Earth's crust. Second, unlike plants and animals, which can be harvested annually or seasonally and then replenished, mineral deposits are depleted by mining and eventually exhausted. Third, the quantity of a given material available in any one country is rarely known with

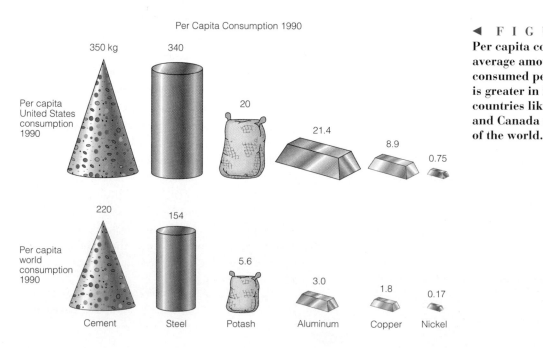

Per Capita Consumption 1990

Per capita United States consumption 1990: 350 kg, 340, 20, 21.4, 8.9, 0.75

Per capita world consumption 1990: 220, 154, 5.6, 3.0, 1.8, 0.17

Cement Steel Potash Aluminum Copper Nickel

◀ F I G U R E 13.1
Per capita consumption, or the average amount of material consumed per person each year, is greater in industrialized countries like the United States and Canada than in the rest of the world.

TABLE 13.1 • Mineral Products Found in a Typical Home in an Industrialized Country

Building materials	Sand, gravel, stone, brick (clay), cement, steel, aluminum, asphalt
Plumbing and wiring materials	Iron and steel, copper, brass, lead, cement, asbestos, glass, tile, plastic
Insulating materials	Rock, wool, fiberglass, gypsum (plaster and wallboard)
Paint and wallpaper	Mineral pigments (e.g., iron, zinc, titanium) and fillers (e.g., talc, asbestos)
Plastic floor tiles, other plastics	Mineral fillers and pigments, petroleum products
Appliances	Iron, copper, and many rare metals
Furniture	Synthetic fibers made from minerals (principally coal and petroleum products); steel springs; wood finished with rottenstone polish and mineral varnish
Clothing	Natural fibers grown with mineral fertilizers; synthetic fibers made from minerals (principally coal and petroleum products)
Food	Grown with mineral fertilizers; processed and packaged by machines made from metals
Drugs and cosmetics	Mineral chemicals
Other items	Windows, screens, light bulbs, porcelain fixtures, china, utensils, jewelry: all made from mineral products

From: U.S. Geological Survey Prof. Paper 940, 1975.

accuracy, and the likelihood of new deposits being discovered is difficult to assess. Let's look more closely at each of these aspects.

The Distribution of Mineral Resources

The uneven distribution of material in the Earth's crust is a natural consequence of processes in the rock cycle and the tectonic cycle. This is the main reason that no nation is self-sufficient in mineral supplies. Because usable minerals are localized, they must be searched out, and the search ranges over the entire globe. The branch of geology concerned with discovering new supplies of usable minerals is **exploration geology.** Exploration geologists refer to regions in which mineral deposits occur in unusually large quantities as *metallogenic provinces.*

To a certain extent, the distribution of different mineral resources and metallogenic provinces can be linked to specific geologic environments, both past and present, within the tectonic cycle. This connection makes sense because the processes of the rock cycle and the tectonic cycle are ultimately responsible for forming mineral deposits. The identification of specific tectonic environments permits geologists to identify locations that will be more or less favorable for specific types of mineral deposits. A striking example is the metallogenic province shown in Fig. 13.2, which runs along the western side of the Americas. Within that province is the world's greatest concentration of large copper deposits. These deposits, called *porphyry copper deposits,* were formed as a consequence of subduction and are associated with the subduction boundaries of the South and North American plates.

Depletion of Mineral Resources

Unlike living resources, deposits of minerals are depleted by mining; they are nonrenewable Earth resources. The rich

mineral deposits that are now being exploited were formed millions, even billions of years ago. Many of the same geologic processes that created these deposits are still operating today, and new mineral deposits are constantly forming. But the renewal time for mineral deposits is exceedingly long compared to the time it can take to deplete a deposit;

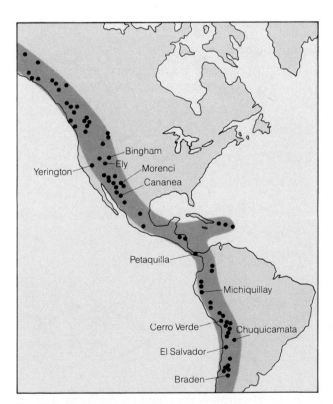

▲ FIGURE 13.2
Copper deposits associated with subduction form a well-defined belt parallel to the subduction edges of the South and North American plates.

on a human time scale, mineral resources are truly nonrenewable. This disadvantage can be offset only by the discovery of new occurrences, the use of alternative materials, or repeated use of the same material—that is, recycling.

Assessing Reserves

A third distinctive aspect of mineral deposits is that it is difficult to assess the quantity of a given material available in any one country with accuracy, and it is even more difficult to anticipate whether new deposits will be discovered. As a result, it is exceedingly difficult to predict production and supplies over a period of years. A country that can meet its needs for a given mineral substance today may face a future in which it will become an importing nation (Fig. 13.3).

England is an example of a country that once was able to meet most of its own mineral needs but can no longer do so today. A little more than a century ago, England was a great mining nation, producing and exporting such materials as tin, copper, tungsten, lead, and iron. Today, the known deposits have been exhausted. The pattern followed by England—a pattern of intensive mining and metals production, followed by depletion of the resource, declining exports and production, and eventually increasing dependence on imports—can be applied in a local, regional, or global context to estimate the remaining effective lifetime of a given mine or mineral resource. (This pattern, which was first demonstrated in 1929 by a geologist named D. Foster Hewett, also applies to energy resources.)

Resources and Reserves

It is important to understand that a *resource* is not the same as a *reserve*. The term **resource** refers to any potentially valu-

able mineral compound, including hypothetical or undiscovered deposits of the material. On the other hand, the term **reserve** refers specifically to *identified resources;* that is, that portion of the resource for which there is detailed information concerning the quality and quantity of material available. In addition, for the material to be called a reserve, it must be technologically, economically, and legally feasible to extract it from the ground at the present time.

Figure 13.4 illustrates the difference between a resource and a reserve. The vertical boundaries on the diagram, such as the boundary between *identified reserves* and *hypothetical resources,* are movable. If exploration geologists find new, economically extractable deposits of a given material, the boundary will shift to the right (i.e., reserves will grow). The horizontal boundaries are also movable. For example, a new technology might be developed that would make it cheaper to extract a particular material from the ground. In that case the horizontal boundary would move down, so that a *marginally economic* reserve might become economically extractable.

Where Are Mineral Deposits Now?

Over the past few decades there has been a slow but steady shift in the emphasis of mineral exploration and production away from the industrialized nations and toward the developing nations. In the industrialized world the geologic locations that are most favorable for conventional mineral exploration have already been prospected, assessed, and in some cases depleted. This doesn't mean that there are no more mineral deposits to be found in these areas; it just means that geologists will have to look harder, utilize innovative exploration techniques, and learn to look for mineral deposits in unconventional locations.

Still, many of the obvious sites remaining for mineral exploration are located in developing countries. A great deal of vigorous exploration is being carried out in South America and Southeast Asia, for example. Other factors have also contributed to this shift in emphasis, including the cheaper cost of labor in less developed regions and the

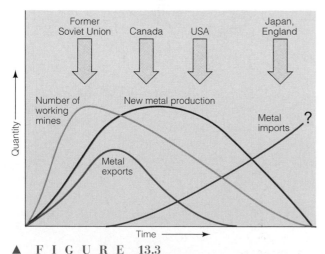

▲ F I G U R E 13.3
Curves demonstrating the evolution of mining and metal production in five industrial countries. The arrows indicate the estimated position of each country on the evolutionary curve. Note that as time passes, a country moves from left to right on the diagram.

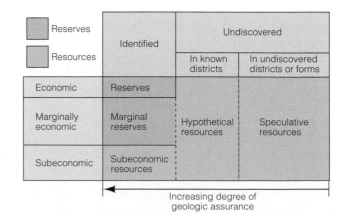

▲ F I G U R E 13.4
Classification of mineral resources and reserves.

increasing cost and rigor of environmental regulations on mining activities in many industrialized countries.

The Challenge of Exploration Geology

The challenge of exploration geology has traditionally been twofold: (1) to find deposits of valuable minerals and (2) to mine the minerals and get rid of the waste rock as cheaply as possible. To these challenges we have recently added a third: to extract the minerals with as little permanent environmental disruption as possible.

The mineral industry places a premium on the skills of exploration geologists and engineers, who play the essential roles in finding and mining mineral deposits. Much ingenuity has been expended in bringing the production of minerals to its present state. Because known deposits are being rapidly exploited while demands for minerals continue to grow, we can be sure that even more ingenuity will be needed to extend mineral reserves in the future.

We turn now to a brief discussion of what it is that distinguishes an economic mineral deposit from the rocks that surround it. After that we will look in more detail at the geology of different types of mineral deposits and the environmental impacts of mining.

ORE

Minerals for industry are sought in deposits from which the desired substances can be recovered least expensively. To distinguish between profitable and unprofitable mineral deposits, we use the word **ore,** meaning an aggregate of minerals from which one or more minerals can be extracted profitably. All ores are mineral deposits because each ore deposit is a local enrichment of one or more minerals. The reverse is not true, however; not all mineral deposits are ores. *Ore* is really an economic term, while *mineral deposit* is a geologic term. A mineral deposit forms as a result of one or more geologic processes, but whether or not a given mineral deposit is an ore is determined by how much humans are prepared to pay for its content.

Grade and Enrichment Factor

For every desired mineral substance there is a **grade,** or level of concentration, below which the deposit cannot be worked economically. The more concentrated the desired minerals, the more valuable the deposit. It is not always possible to say exactly what the grade must be, or how much of a given mineral must be present, for the substance in a given deposit to be considered an ore. Two deposits may have the same grade and be the same size, but one may be identified as an ore and the other may not. There are many possible reasons for the difference. For example, the uneconomic deposit could be buried too deeply or located in so remote an area that the costs of mining and transport are too high relative to other deposits.

The extent to which a material must be concentrated in order to be economically recoverable, over and above its average abundance in the Earth's crust, is called its **enrichment factor.** Each mineral or metal has its own enrichment factor, which represents a balance between the price of the material and its average abundance in the crust. In general, the enrichment factor is greatest for metals that are least abundant in the crust, such as gold and mercury (Fig. 13.5). If the price of gold were to plummet, it would re-

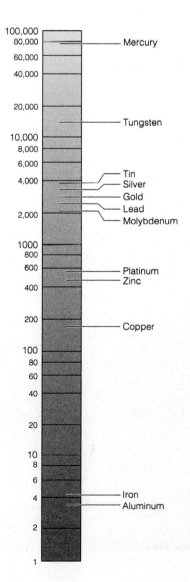

▶ F I G U R E 13.5
Before a mineral deposit can be worked profitably, the percentage of valuable metal in the deposit must be greatly enriched above its average percentage in the Earth's crust. The amount of enrichment required is greatest for metals that are least abundant in the crust, such as gold and mercury. As mining and mineral processing become more efficient and less expensive, it is possible to work leaner ore, and enrichment factors decline. Note that the scale is a logarithmic scale, in which the major divisions increase by multiples of ten.

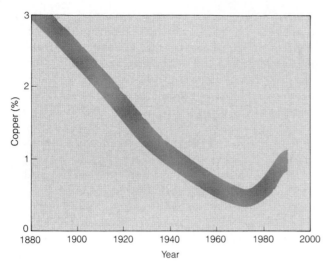

▲ F I G U R E 13.6
The declining percentage of copper needed for a mineral deposit to be an ore. The large reduction before the 1970s occurred because large-volume mining with efficient machines led to a steady reduction in mining costs.

quire an even higher enrichment factor—that is, a greater concentration of gold in one locality—for the gold to be economically extractable. In contrast, aluminum, which is very abundant in the Earth's crust, doesn't require a very high enrichment factor in order to be economically recoverable.

Because of technical advances in finding and extracting ores, some deposits that are now considered ore are only one-sixth as rich as were the lowest grade ores 100 years ago, as shown in Fig. 13.6 for the case of copper. Notice, however, that the curve in Fig. 13.6 has risen from its lowest value, reached during the 1970s. The reason is that overproduction of copper around the world, combined with an economic recession, has produced a glut of newly mined copper. This, in turn, drove down the price of copper and led to the closing of many mines, especially those with the lowest grades. In order for a mineral deposit to be profitably mined as copper ore, the concentration now must be higher than it was in the 1970s.

Gangue

Ore minerals from which desired metals can be extracted are usually mixed with other, nonvaluable minerals, collectively termed **gangue** (pronounced "gang"). Familiar minerals that commonly occur as gangue are the silicate minerals quartz, feldspar, and mica, and the carbonate minerals calcite and dolomite.

If you visit an old mine site you will probably see piles of gangue, the waste rock that was discarded in the search for high-grade ore. Some prospectors have succeeded in utilizing new extraction techniques to turn these old waste rock piles into profitable low-grade ores (Fig. 13.7). Piles of waste rock raise some environmental concerns: they are eyesores, and the minerals in them may combine with rainwater to form acidic runoff. Mining operations deal with waste rock much differently today than in the old days of mining. Waste rock piles are carefully managed and monitored, and the landscapes at mine sites are restored to their premining condition if possible.

◄ F I G U R E 13.7
Taking a second look. Miners in Llallagua, Bolivia, reworking the waste left from tin mining. The mineral they seek, cassiterite (SnO_2), is the main ore mineral of tin.

USEFUL MINERAL SUBSTANCES

For the purpose of discussion it is convenient to group mineral resources according to how they are used (Table 13.2). There are two broad groups of mineral resources: (1) *metallic* and (2) *nonmetallic*.

Metallic Mineral Resources

Metallic mineral resources are mined specifically for the metals that can be extracted from them. Metallic resources are often divided into **base metals,** which includes any of the common metals such as copper and lead, and **precious metals**—gold, silver, and metals of the platinum group.

Metals also can be usefully subdivided on the basis of their average percentage in the Earth's crust. Without exception, the useful metals are present in the crust in such small amounts that rich ore deposits must be located before mining can be carried out economically. Metals that are present in such abundance that they make up 0.1 percent by weight or more of the crust are considered *geochemically abundant*. They include iron, aluminum, manganese, magnesium, and titanium. Geochemically abundant metals require comparatively small enrichment factors to form large deposits. The minerals that are concentrated in deposits of these metals tend to be oxides and hydroxides.

T A B L E 13.2 • Principal Mineral Substances, Gouped According to Use

Metals

Geochemically abundant metals
   Iron, aluminum, magnesium, manganese, titanium
Geochemically scarce metals
   Copper, lead, zinc, nickel, chromium, gold, silver, tin, tungsten, mercury, molybdenum, uranium, platinum, palladium, and many others

Nonmetallic substances

Used for chemicals
   Sodium chloride (halite), sodium carbonate, sulfer, borax, fluorite
Used for fertilizers
   Calcium phosphate (apatite), potassium chloride (sylvite), sulfur, calcium carbonate (limestone), sodium nitrate
Used for building
   Gypsum (for plaster), limestone, clay (for brick and tile), asbestos, sand gravel, crushed rock of various kinds, shale (for cement), building stone
Used for jewelry
   Diamond, corundum (sapphire, ruby), garnet, amethyst, beryl (emerald), and many others
Used for ceramics and abrasives
   Ceramics: Clay, feldspar, quartz
   Abrasives: Diamond, garnet, corundum, pumice, quartz

Metals that make up less than 0.1 percent by weight of the crust, such as nickel, cobalt, or copper, are said to be *geochemically scarce*. Most ore minerals of geochemically scarce metals are not found in common rocks. However, chemical analysis of any rock will reveal that even though the minerals are absent, the metals themselves are present. This is because atoms of the scarce metals can substitute for more common atoms (such as magnesium or calcium) in the crystal structures of common rock-forming minerals. In order for a mineral deposit to form, some gathering and concentrating agent must react with the rock-forming minerals and remove the scarce metals from them. This agent must then transport the metals and deposit them as separate minerals, concentrated in a specific location. With such a complicated chain of events, it is hardly surprising that deposits of the geochemically scarce metals are rarer and very much smaller than deposits of the abundant metals.

Nonmetallic Mineral Resources

Nonmetallic mineral resources, such as salt, gypsum, and clay, are used not for the metals they contain but for their properties as chemical compounds. In other words, they are not mined for metallic properties such as conductivity or ductility but for a wide range of other chemical and physical characteristics. The nonmetallic substances can be subdivided according to their specialized uses (Table 13.2), of which the most important are in construction and as raw materials for chemicals and fertilizers. Although such metals as gold, silver, and platinum may sound more alluring, worldwide production of nonmetallic minerals far outweighs metal production. The reason for this is obvious if you consider the importance of products such as plaster, cement, building stone, ceramics, table salt, and chemical fertilizers, all of which are derived from nonmetallic minerals.

FORMATION OF MINERAL DEPOSITS

For a mineral deposit to be formed, some process or combination of processes must produce a localized enrichment of one or more minerals. Minerals can become concentrated in a number of ways:

1. Concentration may result from the flow of hot, aqueous solutions through fractures and pore spaces in crustal rock, producing *hydrothermal* mineral deposits.

2. Concentration may result from metamorphic banding or recrystallization, producing *metamorphic* mineral deposits.

3. Concentration may be caused by igneous processes within a body of magma, producing *magmatic* mineral deposits.

4. Concentration may result from precipitation from lake water or seawater, producing *sedimentary* mineral deposits.

5. Concentration may be caused by the action of waves or currents in flowing surface water, producing *placers.*

6. Concentration may result from weathering processes, producing *residual* mineral deposits.

Hydrothermal Mineral Deposits

Many famous mines contain ores that were formed when ore minerals were deposited from **hydrothermal solutions**—hot, water-rich fluids (*hydro* is the Greek word for water, and *thermal* is derived from *therme,* the Greek word for heat). Hydrothermal solutions alter the mineralogy of the rocks they pass through, forming *disseminated* ore deposits. They can also deposit dissolved constituents in cracks, forming mineral *veins* (Fig. 13.8). Hydrothermal deposits often consist of common gangue minerals such as quartz and calcite in combination with the valuable materials copper, lead, zinc, and molybdenum sulfides, as well as gold, silver, mercury, and other substances.

Origins of Hydrothermal Solutions

It is likely that more mineral deposits have been formed by deposition from hydrothermal solutions than by any other mechanism. However, it is often difficult to determine the origins of hydrothermal solutions. Some solutions originate when water dissolved in a magma is released as the magma rises and cools. Other solutions are formed from rainwater or seawater that circulates deep within the crust.

An example of the formation of a hydrothermal solution from deeply circulating seawater is shown in Figure 13.9.

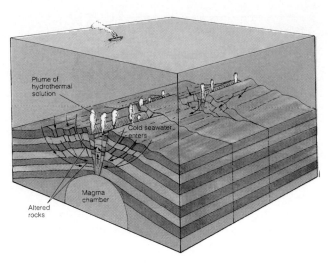

▲ F I G U R E 13.9
Seawater penetrates volcanic rocks on the ocean floor. Heated by an underlying magma body, seawater becomes a hydrothermal solution, alters the chemistry and mineralogy of the rocks, and rises at a midocean ridge as a hydrothermal plume.

This is how the "black smokers" described at the beginning of the chapter were formed. Because volcanism is the heat source for seawater hydrothermal solutions, and because the ore minerals deposited are always sulfides, mineral deposits of this type are called *volcanogenic massive sulfides.*

The ore minerals in volcanogenic massive sulfide deposits originate in the igneous rocks of the oceanic crust. Heated seawater reacts with the rocks, causing changes in the composition of both the minerals and the solution. As the minerals are transformed, metals such as copper and zinc are released and become concentrated in the slowly evolving hydrothermal solution. The kinds of rocks involved in the formation of the solution determine which ore constituents will be concentrated in the solution. For example, copper and zinc are commonly present in small amounts in the mineral pyroxene, so pyroxene-rich rocks of the oceanic crust yield hydrothermal solutions charged with copper and zinc. As a result, most volcanogenic massive sulfide deposits are rich in copper and zinc.

Causes of Precipitation

What makes the solutions precipitate their soluble mineral load and form a mineral deposit? When a hydrothermal solution moves slowly upward through the crust, the solution cools very slowly. If dissolved minerals were precipitated from such a solution, they would be spread over great distances and would not be concentrated enough to create an ore. But when a solution flows rapidly, as in an open fracture in a mass of shattered rock or through a layer of porous rock where flow is less restricted, cooling can occur suddenly over short distances. The result is rapid precipitation and a concentrated mineral deposit. Other effects—such as

▲ F I G U R E 13.8
Metasomatically altered marble. The white mineral is calcite, brown is garnet, green is a pyroxene, and purple is a fluorite. Materials needed to form the garnet, pyroxene, and fluorite were added to the marble by metasomatic fluids. The sample is 8 cm wide and comes from King Island, Australia.

boiling, a rapid decrease in pressure, changes in the composition of the solution caused by reactions with adjacent rock, or cooling as a result of mixing with seawater—can also cause rapid precipitation and form concentrated deposits.

When hydrothermal solutions deposit minerals in open fractures, hydrothermal veins are formed. Many such veins are found in regions of volcanic activity (Fig. 13.10). The famous gold deposits at Cripple Creek, Colorado, were formed in fractures associated with a small caldera, and the huge tin and silver deposits in Bolivia occur in fractures localized in and around stratovolcanoes. In each case the fractures were formed as a result of volcanic activity, and the magma chambers that fed the volcanoes served as the source of the hydrothermal solutions that rose up and formed the mineralized veins.

Like the magma chamber beneath a volcano, a cooling plutonic rock body is a source of heat and can also be a source of hydrothermal solutions. Such solutions move outward from the cooling rock and flow through any fracture or channel they encounter, altering the chemistry and mineralogy of the surrounding rock and depositing valuable minerals. Many famous ore bodies are associated with plutonic igneous rocks; example include the tin deposits of Cornwall, England, and the copper deposits at Butte, Montana, and Bingham, Utah.

Metamorphic Mineral Deposits

The alteration of minerals by interaction with hydrothermal solutions is an example of *metasomatism,* a form of

▲ F I G U R E 13.11
Pronounced compositional banding in a metamorphic rock. The black mineral is graphite, the light-colored mineral is siderite ($FeCO_3$). The sample is from Eisenerz, Austria.

metamorphism in which solutions are responsible for moving around the chemical constituents of the rocks. Hydrothermal alteration and metasomatism are usually associated with *contact metamorphism,* the mineralogical and chemical alteration of cool rocks adjacent to bodies of hot rock or magma. *Regional metamorphism*—the alteration of large areas of rock as a result of the high temperatures and pressures associated with the tectonic movement of crustal plates—can also concentrate minerals, creating mineral deposits.

In response to the differential stresses (i.e., directed pressure) that occur during regional metamorphism, the minerals in a rock commonly segregate, or become separated and concentrated into distinct bands or layers (Fig. 13.11). A wide variety of economically valuable, nonmetallic mineral resources are concentrated in banded metamorphic rocks. They include micas, asbestos, graphite, vermiculite, and some gemstones.

Another common process in regional metamorphism is the recrystallization of minerals, usually into coarser grained, interlocking aggregates. For example, the pure

▲ F I G U R E 13.10
A rich vein in Potosi, Bolivia, containing chalcopyrite, sphalerite, and galena cutting andesite. The andesite has been altered by the hydrothermal solution that deposited the ore minerals.

◄ F I G U R E 13.12
Large garnet crystals (red), surrounded by a black halo of amphi-
bole, Adirondack Mountains, New
York. At one location in the Adiron-
dacks, Gore Mountain, garnets of
this kind are mined and used for
their superior abrasive properties.

white marbles from Carrara, Italy, which are highly valued
by sculptors, began as porous aggregates of shells and other
organic material. Through the process of metamorphic recrystallization, all traces of the original fossil shells were
obliterated and the rocks were transformed into seamless
crystalline marbles composed entirely of interlocking grains
of the mineral calcite.

Mineral assemblages change during metamorphism.
The minerals that were originally present in the rock are replaced by a different assemblage that may contain ore minerals. An example of a deposit in which the mineral assemblage has been changed and coarsened through
metamorphic recrystallization is the famous garnet deposit
of Gore Mountain, New York, where enormous blood-red
garnets are embedded in a matrix of black hornblende; the
result is a spectacular looking rock (Fig. 13.12). Garnets are
used in the manufacture of abrasive products such as sand-
paper; some garnets are used as gemstones.

Magmatic Mineral Deposits

A number of processes that occur during the cooling and
crystallization of magmatic bodies can lead to the separation
and concentration of minerals. We will look briefly at three
of the most important types of magmatic mineral deposits:
(1) *pegmatites;* (2) *layered intrusions;* and (3) *kimberlites.*

Pegmatites

When a magma crystallizes, the earliest formed crystals
preferentially incorporate the elements that fit into their
crystal structures most easily. This means that as crystallization proceeds, the remaining melt will become increasingly
enriched in the *other* elements—those that do *not* fit easily

into crystal structures. Typically, these elements are large
(e.g.,uranium) or have high electrical charges (e.g., niobium, Nb^{5+}). Solidification eventually produces an igneous
rock called a *pegmatite,* which is highly enriched in the elements that had been concentrated in the last fraction of
melt. Pegmatites are often very coarse grained, partly because the magma is cooling off so slowly at this stage that it
has time to grow very large crystals, and partly because
water, which promotes the growth of large crystals, is also
concentrated in the last melt fraction (Fig. 13.13).

▲ F I G U R E 13.13
Huge single crystals of feldspar (white) in the King's
Mountain pegmatite, North Carolina. All pegmatites
are coarse-grained, but the individual mineral grains
shown in this pegmatite are unusually large.

Rocks of this type are most commonly formed in the last stages of crystallization of magmas of granitic composition. Pegmatites contain rich concentrations of such elements as lithium, beryllium, cesium, and niobium. Much of the world's lithium is mined from pegmatites such as those at King's Mountain, North Carolina, and Bikita, Zimbabwe. The huge Tanco pegmatite in Manitoba, Canada, produces much of the world's cesium.

Layered Intrusions

Under some circumstances magma solidifies underground to form large rock bodies called *layered intrusions*. These are plutonic rock bodies of basaltic composition that are layered or banded; even though they are entirely of magmatic origin, they look very much like banded or bedded sedimentary rocks (Fig. 13.14). When a large chamber of basaltic magma crystallizes, one of the first minerals to form is chromite. Chromite is the main ore mineral of chromium, a constituent of steel. As shown in Fig. 13.15, the chromite crystals, which are denser than the liquid around them, settle and then accumulate at the bottom of the magma chamber like particles of sediment settling in water. This process can produce almost pure layers of chromite. Many of the world's principal chromite ores,

▲ F I G U R E 13.14
Layers of chromite (black) and plagioclase (white) formed by crystal settling during the crystallization of the Bushveld Igneous Complex. This unusually fine outcrop is located at the Dwars River in South Africa.

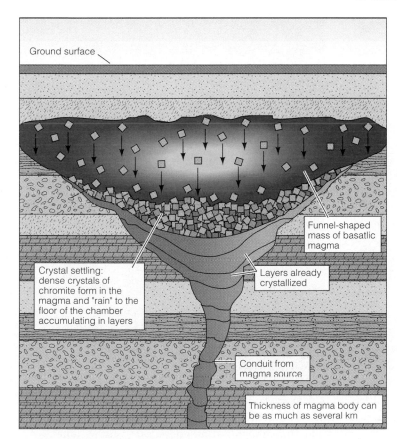

Ground surface

Funnel-shaped mass of basaltic magma

Crystal settling: dense crystals of chromite form in the magma and "rain" to the floor of the chamber accumulating in layers

Layers already crystallized

Conduit from magma source

Thickness of magma body can be as much as several km

▲ F I G U R E 13.15
Chromite that crystallizes from a melt is denser than the surrounding melt and therefore will sink to the bottom and accumulate there in a process called crystal settling.

including the Bushveld Igneous Complex in South Africa and the Great Dike of Zimbabwe, occur in layered intrusions.

An unusual example of a layered intrusion is the Sudbury Complex in Ontario, which contains the world's greatest known concentration of nickel. In this case, a somewhat different settling process led to the formation of the ore. Certain magmas separate—as oil and water do—into two *immiscible* (nonmixing) liquids, one of sulfide composition and the other of silicate composition. Sulfide liquids, which typically are rich in copper, nickel, and other metals, sink to the floor of the magma chamber because they are denser than silicate liquids. After cooling and crystallization, the resulting igneous rock has a copper or nickel ore at the base like that found at Sudbury. The Sudbury Complex has another very unusual feature: the melt that eventually solidified into the ore deposit is thought to have been generated by the impact of a very large meteorite (see Chapter 10). This theory is supported by the fact that the rock body was originally circular and is surrounded by broken-up rocks called *breccias* and conical deformation features called *shatter cones,* both of which are characteristic features of impact craters.

Kimberlites

Another important type of ore deposit that results from an unusual combination of igneous processes is *kimberlite.* Kimberlite consists of a long, thin, pipelike body of igneous rock that originates deep in the mantle—150 to 200 km or more (Fig. 13.16). How and why kimberlite magma forms remains a puzzle, but when it does form it rises explosively upward, punching a circular hole in the crust and carrying broken fragments of mantle rock (called *xenoliths*) upward as it does so.

One of the mineral constituents of the mantle fragments brought up by kimberlites is diamond (Fig. 13.17). Diamond is a high-pressure mineral that is formed only at depths greater than 150 km. The only way diamonds can reach the surface of the Earth is through kimberlite pipes. Even the largest kimberlite pipes are no more than a few hundred meters in diameter, and the quantity of diamond present is never large. For example, the rich diamond pipes

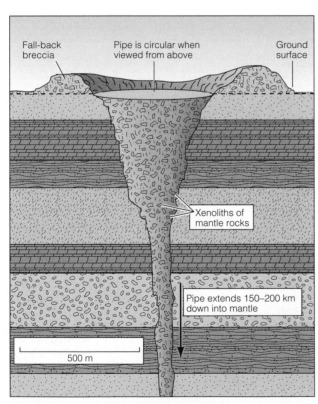

▲ **F I G U R E 13.16**
Kimberlites are unusual igneous rocks that take the form of long, funnel-shaped pipes. The pipes originate at great depths in the Earth and move upward explosively, carrying with them minerals that have been formed under high pressure, such as diamonds.

▲ **F I G U R E 13.17**
The Oppenheimer diamond, a rare yellow stone weighing 253 carats, in the collection of the Smithsonian Institute. The octahedral shape is the characteristic crystal form of diamond. The specimen is 4.5 cm long.

at Kimberley, South Africa, yield on average only one carat of diamond for every 5 m³ of kimberlite mined.

Sedimentary Mineral Deposits

The term *sedimentary mineral deposit* can be applied to any local concentration of minerals that has been formed through processes of sedimentation. However, it is most commonly applied to mineral deposits that have been formed through the precipitation of substances carried in solution.

Evaporites

The most direct way in which sedimentary mineral deposits are formed is through evaporation of lake water or seawater. The layers of salts that precipitate as a consequence of evaporation are called **evaporite** deposits.

Lake-Water Evaporites   Evaporation of an inland water body can occur when the water becomes trapped in closed drainage basins. Examples of salts that precipitate from lake waters of suitable composition are sodium carbonate (Na_2CO_3), sodium sulfate (Na_2SO_4), and borax ($Na_2B_4O_7$)·[$10\ H_2O$]. These and other salts have many applications; they are used in the production of paper, soap, detergents, antiseptics, and chemicals for tanning and dyeing. Huge evaporite deposits of sodium carbonate were laid down in the Green River Basin of Wyoming during the Eocene Epoch. This is the same lake in which the rich Green River Oil Shales were deposited (Chapter 12). A modern-day analogue would be a setting like Great Salt Lake, Utah. Borax and other boron-containing minerals are mined from evaporite lake deposits in Death Valley (Fig. 13.18) and Searles and Borax lakes, all in California, and in Argentina, Bolivia, Turkey, and China.

Marine Evaporites   Much more common and important than lake-water evaporites are the marine evaporites formed by the evaporation of seawater. Evaporation can occur when a portion of shallow sea becomes cut off from the main part of the ocean by a reef or other barrier. This will eventually happen in the part of the ocean cut off by the Great Barrier Reef in Australia. The most important salts that precipitate from seawater are gypsum ($CaSO_4{\cdot}2\ H_2O$), halite ($NaCl$), and carnallite ($KCl{\cdot}MgCl_2{\cdot}6\ H_2O$). Common table salt, the gypsum used in plaster and construction materials, and the potassium used in fertilizers are recovered from marine evaporites.

Marine evaporite deposits are widespread; in North America, for example, strata of marine evaporites underlie as much as 30 percent of the land area (Fig. 13.19). When ancient evaporite deposits became covered by later sediments, an interesting phenomenon occurred. Rock salt is less dense than most other types of rock, and it deforms plastically. The evaporites typically flow upward over many, many years through the overlying layers of denser sedimentary rocks, eventually forming large, upside-down teardrop-shaped bodies called *salt domes*. Because oil and natural gas

◀ F I G U R E   13.18
Evaporite salts encrust the surface of the desert on the floor of Death Valley in California. A shallow lake is created during rainy periods. As the water evaporates and the lake dries up, salts crystallize out of the brine. Polygonal fractures form as the drying sediment contracts.

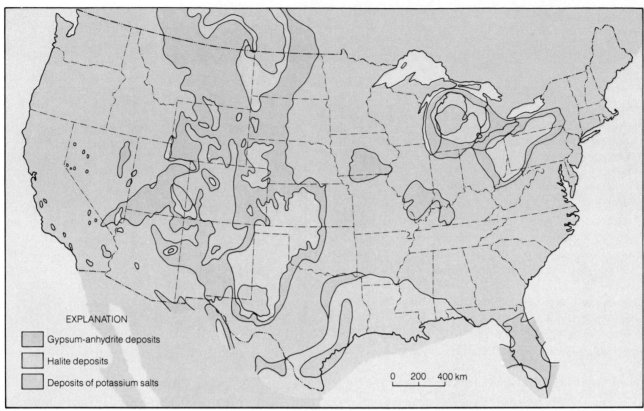

▲   F I G U R E   13.19
Marine evaporite deposits in the United States and southern Canada. The areas underlain by gypsum and anhydrite do not contain halite. The areas underlain by halite are also underlain by gypsum and anhydrite. The areas underlain by potassium salts are also underlain by halite and by gypsum and anhydrite.

also form in shallow marine settings, they tend to be associated with salt; for this reason, salt domes are a good target for petroleum exploration (Chapter 11).

Chemical Precipitates

Phosphates   Sedimentary deposits of phosphorus minerals, which are a primary source of fertilizer, are formed as a result of the precipitation of the mineral apatite $[Ca_5(PO_4)_3(OH,F)]$ from seawater. The surface waters of the ocean are low in phosphorus because fish and other marine animals extract phosphorus to make bone, scales, and other body parts. When the animals die and sink to the ocean floor, their bodies slowly decay and release phosphorus to the deep ocean water. If the resulting phosphorus-rich waters are carried to the surface by upwelling currents, precipitation of apatite can occur. Phosphorus-rich sediments are forming today off the western coasts of Africa and South America, but the process was much more common in the past, especially when shallow seas inundated large portions of the continents.

Iron Deposits   Some of the world's most important concentrations of iron are found in sedimentary rocks that are billions of years old. These deposits, called *iron formations,* are among the most unusual kinds of chemical sedimentary rocks (Fig. 13.20). They consist of iron-rich sediment interbedded with siliceous and/or clastic sedimentary rocks. They originated through chemical precipitation in a marine environment. All sedimentary iron deposits are tiny by comparison with the class of deposits characterized by the *Lake Superior-type iron deposits.* These remarkable deposits, mined principally in Michigan and Minnesota, were long the mainstay of the steel industry in the United States but are now being replaced by imported ores. The deposits are about 2 billion years or older and are found in sedimentary basins on every stable continental shield, particularly in Labrador, Venezuela, Brazil, the former Soviet Union, India, South Africa, and Australia.

Lake Superior-type iron deposits are not ores. The grades of the deposits range from 15 to 30 percent Fe by weight, and the deposits are so fine grained that the iron minerals cannot be easily separated from the gangue. Two additional processes can produce ore. Removal of silica during weathering can lead to iron enrichment and produce ores containing as much as 66 percent Fe. Compare Figure 13.20, a Lake Superior-type iron deposit in the Hamersley

▲ F I G U R E 13.20
Iron-rich sediments of the Brockmans Iron Formation in the Hamersley Range of Western Australia.
The white layers are composed largely of silica in the
form of chert, whereas the darker bluish and reddish
layers consist mainly of iron-rich silicate, oxide, and
carbonate minerals.

▲ F I G U R E 13.21
Leaching of silica during weathering of a Lake Superior–type iron deposit leads to the formation of a sec-
ondarily enriched mass of iron minerals that is rich
enough to be an ore. This sample, from the Hamersley
Range in Western Australia, was developed by secondary enrichment from the kind of iron-rich sediment shown in Fig. 13.20.

Range, Western Australia, with Figure 13.21, a sample of
ore developed by secondary enrichment in the Hamersley
Range. The rocks in Figure 13.20 contain about 25 percent
Fe, while those in Figure 13.21 have had most of the silica
leached out and contain about 60 percent Fe.

The second way in which a Lake Superior-type iron deposit can become an ore is through metamorphism. In this
process, grain sizes increase so that it becomes easier (and,
therefore, cheaper) to separate ore minerals from gangue.
New mineral assemblages may also be formed, and the iron
silicate and iron carbonate minerals that were originally
present may be replaced by magnetite or hematite, both of
which are desirable ore minerals. The grade of the ore is not
increased by metamorphism; it is the changes in grain size
and mineralogy that change the sedimentary rock into an
ore. Iron ores formed as a result of metamorphism are
called *taconites* and are now the main kind of ore mined in
the Lake Superior region.

Stratabound Deposits

Some of the world's most important ores of lead, zinc, and
copper occur in sedimentary rocks. The ore minerals—
galena, sphalerite, chalcopyrite, and pyrite—occur in such
regular, fine layers that they look like sediments (Fig.
13.22). The sulfide mineral layers are enclosed by and are
parallel to the sedimentary layers or *strata* in which they
occur, and for this reason such deposits are called
stratabound mineral deposits. They look like sediments but
are not sediments in the true sense of the term.

▲ F I G U R E 13.22
Stratabound ore of lead and zinc from Kimberley,
British Colombia. The layers of the ore minerals
pyrite (yellow), sphalerite (brown), and galena (grey)
are parallel to the layering of the sedimentary rock in
which they occur. The specimen is 4 cm across.

Stratabound deposits are formed when a hydrothermal solution invades and reacts with a muddy sediment. Reactions between the grains of sediment and the solution result in deposition of the ore minerals. Deposition commonly occurs *before* the sediment has become a sedimentary rock.

The famous copper deposits of Zambia, in central Africa, are stratabound ores, as are the great Kupferschiefer deposits of Germany and Poland. The world's largest and richest lead and zinc deposits, found at Broken Hill and Mount Isa in Australia and at Kimberley in British Columbia, are also stratabound ores.

Placers

A deposit of heavy minerals that have been concentrated by mechanical processes is called a **placer.** The concentration usually happens as a result of the winnowing action of waves or currents. The most important minerals concentrated in placers are gold, platinum, cassiterite (SnO_2), and diamond. More than half of the gold recovered throughout human history has come from placers.

Stream Placers

The famous California Gold Rush of 1849 followed the discovery that sand and gravel in the bed of a small stream contained bits of gold. Similar gold-bearing gravels are found in many other parts of the world. The gravels themselves are sometimes rich enough to be mined, but even when they are too lean, the gold is a clue that a source must lie upstream. In fact, many mining districts have been discovered by following trails of gold and other minerals upstream to their sources.

Because pure gold is a heavy mineral, it is quickly deposited from a stream's sediment load. As most silicate min-

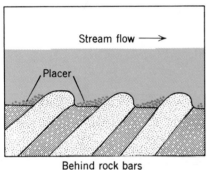

Below waterfalls

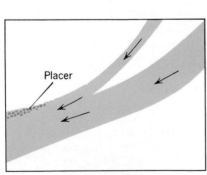

Downstream from a tributary

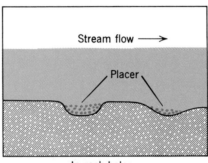

Behind rock bars

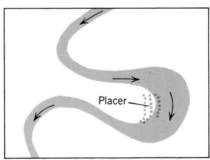

Inside meander loops

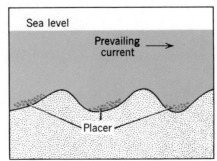

In rock holes

Behind undulations on ocean floor

◀ F I G U R E 13.23
Placers occur where barriers allow flowing water to carry away the suspended load of lightweight particles while trapping heavier particles. Placers can form whenever water moves but are most commonly associated with streams or longshore currents.

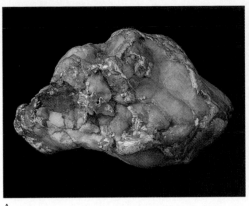
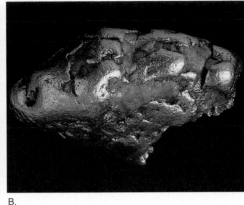
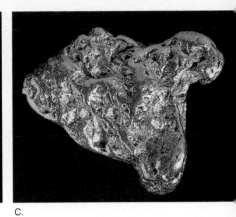

A. B. C.

▲ FIGURE 13.24

Formation of a nugget. A. A vein of metallic gold cuts through a pebble of quartz. Stream abrasion causes the brittle quartz to chip and become smaller, while the malleable gold deforms but is not reduced in size. B. The ratio of gold to quartz increases as the quartz is abraded away. Eventually a nugget of almost solid gold is formed. C. A nugget of gold from California. No quartz remains. Each of the specimens is about 4 cm in diameter.

erals are lighter than gold, grains of gold become mechanically concentrated in places where the velocity of stream flow is high enough to remove the light particles but not high enough to remove the heavy ones. Such concentration occurs, for example, behind rock bars or in bedrock holes along the channel, below waterfalls, on the inside of a meander bend, and downstream from the point where a tributary enters a main stream (Fig. 13.23).

Every phase of the conversion of gold from a hydrothermal vein to placer gold has been traced. Chemical weathering of the exposed vein releases the gold, which is moved slowly downslope by mass-wasting. In some places mass-wasting alone concentrates the gold sufficiently to justify mining. More commonly, however, the mineral particles enter a stream, which concentrates them more effectively than mass-wasting. Most placer gold occurs as grains the size of silt particles, the "gold dust" of miners; some of it is coarser. A lump the size of a pebble or cobble is a "nugget" (Fig. 13.24); the largest nugget ever recorded weighed 80.9 kg. In following placers along stream channels, prospectors have learned that the rounding and flattening (by pounding) of nuggets increases downstream. Therefore, when they find rough, angular nuggets, prospectors know that the primary source is nearby.

Many heavy, durable metallic minerals other than gold also form placers. These include minerals that occur as pure metals, such as platinum and copper, as well as cassiterite (SnO_2) and nonmetallic minerals such as diamond, ruby, emerald, and sapphire. Even if a vein contains a low percentage of a mineral, the placer it yields may be quite rich. In order to become concentrated in placers, the minerals must be not only dense but also resistant to chemical weathering and not readily susceptible to cleaving as the grains are tumbled in the stream.

Beach Placers

Gold, diamonds, and several other heavy minerals can also be concentrated in beach sands by surf and longshore currents. If you dig down into the sand at a beach, you will probably find layers of dark minerals. These are layers of magnetite and other heavy minerals that have been concentrated by the winnowing action of waves. Magnetite-rich placer sands occur on the coasts of Oregon, California, Brazil, and New Zealand, and chrome-rich sands are mined in Japan. Ilmenite, a primary source of titanium, is highly concentrated (50 to 70 percent) along several beaches in India. Nome, Alaska, has become famous as the finish line of the Iditarod Dogsled Race, but the town initially attracted world attention because of its six beaches, four above sea level and two below, which are rich in placer gold. Diamonds are being obtained in large quantities from gravelly beach placers, both above and below sea level, along a 350-km strip of coastal Namibia in southwestern Africa. Weathered from deposits in the interior, the diamonds were transported by the Orange River to the coast and spread southward by longshore drift. Later, some beaches were raised tectonically and others were drowned as a result of a rise in sea level.

"Fossil" Placers

The reason that so much of the world's gold comes from placers is the huge gold deposits found in South Africa. The South African gold deposits are "fossil" or ancient placers, and they have many unusual features. Most placers are

BOX 13.1
•
THE HUMAN PERSPECTIVE

GOLD AND CYANIDE

*C*yanide is one of the few chemical compounds that reacts with gold. This well-known fact is applied in the mining and mineral processing industries to extract gold from ore. When gold ore is soaked in a cyanide solution the gold dissolves, leaving behind all the other materials in the ore. (*Leaching* is the process in which a solution dissolves soluble material. The process of dissolving gold in cyanide is referred to as *cyanide leaching*.) After the gold is dissolved in cyanide it can be withdrawn from solution by adding zinc shavings, which bond with the gold. If heat is then applied, the zinc will vaporize, leaving behind a residue of gold.

Cyanide leaching was discovered in 1887 by a Scottish chemist, John MacArthur, and two physicians, John and William Forrest. The process quickly came into worldwide use. The MacArthur-Forrest technique was inexpensive and could be expected to remove much of the gold (and silver) in a volume of milled ore. Although small changes have been made in the process, the basic chemistry is the same today as it was a century ago.

Today cyanide is sometimes used in *heap leaching,* a technique in which gold ore is piled on large pads and drenched with a cyanide solution. After the solution has drained through the pile, it is retrieved and the gold is recovered from it; the solution can then be recycled through the pile. Heap leaching, if properly carried out, is an effective, low-cost technique that, through the recycling process, effectively limits the amount of cyanide used and reduces its impact on the environment.

Although cyanide is widely used in agriculture, photography, and many other industries, its use entails some risks. Cyanide is toxic when inhaled, absorbed through the skin, or swallowed. In addition, ore bodies containing gold often contain other potentially harmful elements, such as arsenic, which may remain in the mine spoils after the leaching process has been completed. The main challenge in cyanide leaching is to prevent negative environmental and health impacts. Workers must wear eye protection and impenetrable clothing when dealing with cyanide, and work areas must be well ventilated. Properly managed cyanide leaching facilities are carefully engineered and continually monitored. Any cyanide and cyanide-metal complexes must be removed from effluents before they are released into the environment. Several treatment technologies are available that are over 95 percent effective in removing cyanide from solutions.

The environmental problems of a gold mine in Summitville, Colorado, provide an example of cyanide leaching at its worst. In 1985 a young entrepreneur with no formal business or mining training persuaded investors to support an experimental heap leaching operation. The project went bankrupt and was abandoned, a fate shared by many thousands of "orphaned" mine sites throughout North America. At the Summitville site, millions of liters of cyanide-contaminated water and mine tailings were left to be cleaned up by the U.S. Environmental Protection Agency. The deadly poison still poses a threat to wildlife and water supplies in the area. The cost of the cleanup—which is expected to last longer than the mine was in operation—is estimated at $100 million.

▲ FIGURE 13.25
Gold is recovered from the ancient fossil deposits of the Witwatersrand in South Africa. The gold is found at the base of conglomerate layers interbedded with finer grained sandstone, here seen in a weathered outcrop at the site where gold was first discovered in 1886.

found in stream gravels that are geologically young. The South African fossil placers are a series of gold-bearing conglomerates (Fig. 13.25) that were laid down 2.7 billion years ago as gravels in the shallow waters along the edge of a marine basin, now called the Witwatersrand Basin. Associated with the gold are grains of pyrite and uranium minerals. In terms of size and richness, nothing like the Witwatersrand deposits has been discovered anywhere else. Nor has the original source of all the placer gold been discovered, so it is not possible to say why so much of the world's minable gold is concentrated in this one sedimentary basin.

Mining in the Witwatersrand Basin has reached a depth of 3600 m, the deepest mining carried out anywhere in the world, and there are plans to continue mining to depths as great as 4500 m. Despite these ambitious plans, the heyday of gold mining in South Africa has probably passed because the deposits are running out of ore.

Residual Mineral Deposits

Weathering occurs because newly exposed rock is not chemically stable when it is in contact with rainwater and the atmosphere. *Chemical weathering,* in particular, leads to concentration of minerals through the removal of soluble materials. The soluble materials are carried downward with the infiltrating solution, leaving behind a residual concentration of less soluble material. This is the process that caused the leaching of silica and the resulting iron enrichment in the Lake Superior-type iron deposits discussed earlier. (Weathering is discussed in detail in Chapter 14.)

Laterite

A common example of a deposit formed through residual concentration is *laterite,* a hard, highly weathered subsoil that is rich in insoluble minerals. Under conditions of high rainfall in a warm, tropical climate, water-soluble minerals are slowly leached out of a soil. Because the iron-rich mineral limonite is among the least soluble of the many minerals formed during chemical weathering, the leaching process leaves an iron-rich limonitic crust at the surface (Fig. 13.26A). In a few places laterites can be mined for iron. Some laterites are also enriched in nickel, and are mined in such places as New Caledonia, Cuba, and Oregon.

A.

B.

▲ FIGURE 13.26
Residual mineral deposits that are rich in iron and aluminum are typically formed under tropical or semitropical conditions. A. Red laterite enriched in iron, near Djenné, Mali. B. Bauxite from Weipa in Queensland, Australia. Nodules of the aluminum-rich mineral gibbsite have formed by repeated solution and redeposition. The Weipa bauxite deposits are among the largest and richest in the world.

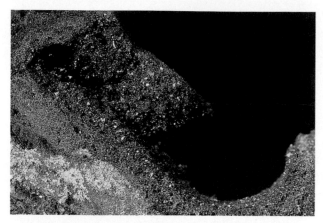

▲ F I G U R E 13.27
A sample of the blanket of sulfide minerals that forms around the seafloor vents on the East Pacific Rise at 21°N latitude. This specimen is part of a system of black smokers. Lining the vent or "chimney" of the black smoker is the sulfide mineral pyrite (FeS_2), and scattered through and around the pyrite are grains of other sulfide minerals. The specimen is 28 cm across.

Bauxite

Although iron-rich laterite is by far the most common kind of residual mineral deposit, the most important deposits for human exploitation are the aluminous laterites called *bauxites,* the source of aluminum ore. Bauxites are widespread but are concentrated in the tropics (Fig. 13.26B). Where bauxites are found in present-day temperate conditions, such as France, China, Hungary, and Arkansas, it is clear that the climate was tropical when the bauxites were formed.

All bauxites are vulnerable to erosion. They are not found in glaciated regions, for example, because overriding glaciers scrape off the soft surface materials. The vulnerability of bauxites means that most deposits are geologically young. More than 90 percent of all known deposits were formed during the last 60 million years, and all of the very large deposits were formed less than 25 million years ago.

Mineral Resources from the Oceans

Some of the ore-forming processes discussed earlier occurred in ancient oceans. Ore formation also occurs in present-day oceans. Seawater contains a variety of chemical constituents besides common salts. Some of them, such as magnesium, are economically extracted from seawater as byproducts of the recovery of halite and other minerals. Another ore-forming process associated with the oceans is the concentration of sulfide minerals through hydrothermal vents ("black smokers") at seafloor spreading centers (Fig. 13.27). This is an example of active hydrothermal ore formation; the deposits are still being formed, and most likely the metals are not economically extractable. At any rate, the technological challenges and political complexities of exploiting hydrothermal mineral deposits in the deep ocean create major barriers.

One of the most promising mineral deposits of the ocean floor may be the least well understood. Metal oxides, notably those of manganese, iron, cobalt, nickel, and copper, cover large areas of the deep-ocean floor in the form of rounded concretions called *nodules* (Fig. 13.28). As a group, they are usually referred to as *manganese nodules* because they typically contain 10 to 35 percent Mn. The exact source of the metals, and therefore the process by which the nodules are formed, is something of a mystery. The most likely source is material that has been weathered from continents and carried by rivers and currents to the deep ocean, where it is precipitated as nodules. Alternative theories suggest that the material for the nodules could be derived from seafloor volcanism, or even from biochemical processes near the water-sediment interface on the ocean bottom. Mining of seafloor manganese nodules appears to be technologically feasible, but it remains to be seen whether it will ever be profitable.

◄ F I G U R E 13.28
Manganese nodule recovered from the deep ocean floor in the center of the Pacific Ocean. The bottom of the nodule is seen on the left, the top on the right. The specimen has a maximum dimension of 5.5 cm.

ENVIRONMENTAL IMPACTS OF MINING

The production of metals and minerals involves many stages, from exploration to the extraction, milling and smelting, refining and processing, fabrication, use, and eventual disposal of products (Fig. 13.29). Wastes and environmental impacts can occur at each stage in the process; so can the recycling and reuse of materials. In general, the exploration phase involves little or no permanent environmental damage. The real disruption begins when commercial production is undertaken.

Mineral Extraction and Processing

The form in which a mineral or metal occurs in nature determines the types of mining operation needed, the cost of extraction, and the amount of waste produced in the process of extracting, separating, and concentrating it. In sand and gravel mining, very little material is discarded. In other cases, however, huge quantities of rock and ore must be removed and discarded to obtain a relatively small amount of mineral or metal. For example, a typical copper grade of 2 percent will produce 20 kg of pure metal from each ton of ore processed.

Once the ore has been extracted, it is crushed and concentrated. The material that is discarded at this point is called *tailings*. For many of the nonmetallic minerals and construction materials (such as sand, limestone, and gravel), basic processing is all that is needed before the substance can be transported and used. Other nonmetallic minerals (such as asbestos, talc, and potash) require further processing of the concentrate. Metallic minerals must be smelted or refined using high-temperature or chemical processes. These operations create wastes that are the source of some of the most significant environmental impacts of mining.

Impacts on Land

Mine site development can involve the construction of roads and buildings, the stripping of surface vegetation and soil, and the sinking of mine shafts or excavation of large open pits (Fig. 13.30). Storage of rock waste, tailings, and other discarded materials is another significant impact of mining, in terms of the amount of land area affected. These changes can be long-term or even permanent, depending

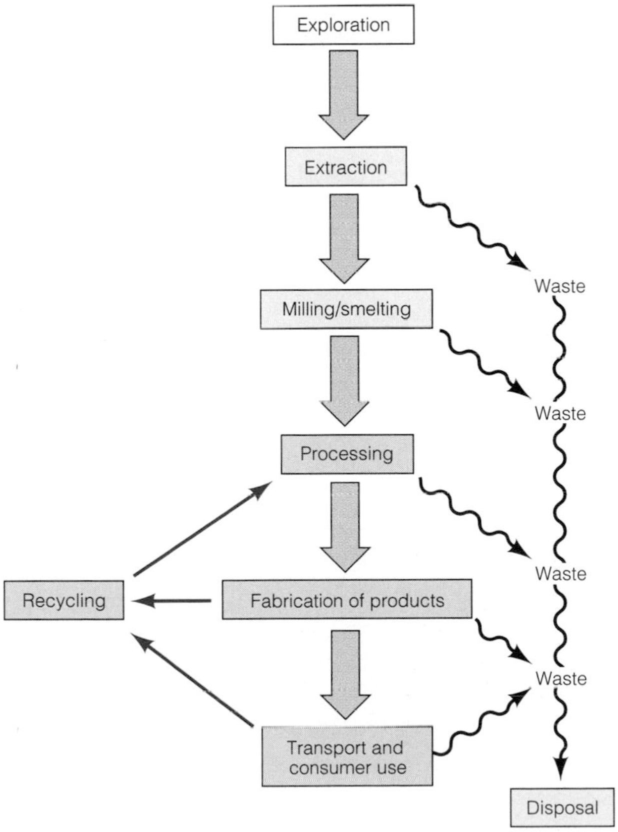

▲ FIGURE 13.29
The resource cycle, showing the stages involved in exploration, processing, use, and disposal of resources and the potential for generation of wastes at each stage.

▲ FIGURE 13.30
The copper mine at Bingham Canyon in the Oquirrh Mountains, not far from Salt Lake City, Utah, is the largest open cut copper mine in North America. The oval-shaped pit is now 4.5 kilometers wide, 7.5 kilometers long, and nearly a kilometer deep. This view looks northeasterly to the Wasatch Mountains on the horizon.

BOX 13.2
•
THE HUMAN PERSPECTIVE

ROCKBURSTS: AN UNDERGROUND HAZARD

A rockburst is a sudden and violent failure of rock that occurs in an underground mine. Rockbursts are caused by the buildup and sudden release of stresses in the rocks. They can cause serious injuries and fatalities among miners, as well as damage underground mine workings (Fig. B2.1) The term *rockburst* was originally used to describe the shattering of support pillars underground. It is a twentieth century phenomenon, simply because earlier mine workings were not deep enough to encounter the large stresses that cause rockbursts. Rockbursts were first observed in the early 1900s in gold mines on the Witwatersrand in South Africa and the Kolar Gold Field in India. In 1908 a South African Commission established that tremors felt on the surface were induced by mining rather than by natural forces (i.e., earthquakes).

Over the years considerable effort has been devoted to defining rockbursts. Most definitions refer to rockbursts as sudden releases of energy or stress underground or at the opening of a mine. During the rock failure, excess energy is liberated as seismic energy, which causes the surrounding rock mass to vibrate as it would during an earthquake. These vibrations are felt both underground and on the surface. Defining a rockburst as a seismic event causes some ambiguity, however, because the term *rockburst* is used to describe the failure process and associated damage as well as the sounds and seismic vibrations emanating from the source. During the past two decades seismic detection systems installed underground have shown that only a very small proportion of the seismic events that occur in underground mines actually cause damage. This has led to the classification of rockbursts as a subset of seismic events; in other words, all rockbursts are seismic events but not all seismic events are rockbursts.

There are several types of rockbursts, classified according to their causes. *Pillar bursts* are caused by the sudden failure of underground rock pillars that have become overloaded. Multiple pillar failures are called *crush bursts;* although they are infrequent, they cause serious damage. Pillar and crush bursts are also called *induced bursts,* emphasizing the fact that the mining operation transferred and concentrated the stresses onto the pillars, thereby inducing the failure. Bursts that occur at the openings of

mines are called *inherent bursts,* reflecting the fact that rock stresses were already high before the development of the mine. Rockbursts can also be caused by slippage along preexisting planes of weakness; this mechanism is essentially the same as the one that causes earthquakes.

Rockbursts occur only in "hardrock" mines, that is, mines situated in igneous or metamorphic rocks. They are common in the metal mines of the Canadian Shield of Ontario and Québec, the Cordilleran silver mines of northern Idaho, and the Homestake Gold Mine in South Dakota, among others. "Softrock" mines, such as potash, gypsum, and halite mines, do not suffer from rockbursts because those rocks deform in a more ductile manner and do not build up stresses. Open-pit mines also do not have rockbursts in the technical sense, although a similar phenomenon occurs when the floor of the mine pit buckles or pops up as a result of the removal of overlying rock.

Rockbursts and the fatalities and damage associated with them became a particular problem in the 1930s and 1940s as mining technologies drove underground mines deeper into the ground, where stresses are high (Figure B2.2). In subsequent decades several methods were developed to alleviate the problem. All hardrock mines now stabilize underground workings with rockbolts and other types of structural supports. Experts on rock stress can assist in planning the layout of a mine in order to minimize the occurrence of rockbursts. Another technique that has proved useful is *preconditioning* or *destressing* the rocks in a mine. The normal technique for excavating tunnels in underground mines involves drilling closely spaced holes into a rock wall (called a *face*) to a depth of about a meter. Explosives are placed in the holes and detonated, advancing the tunnel and moving the rock face ahead. In destressing, the miners drill a few extra holes into the rock face, these holes are deeper than the other holes and are angled outward into the material that will be left behind to form the sides of the tunnel. Blasting in these extra holes fractures the rocks just enough to release some of the built-up stress but not enough to cause them to fail. These approaches have been successful in decreasing the frequency of rockbursts.

A.

B.

◀ F I G U R E B2.1
Rockbursts, one of the dangers of deep mining. A. Mine tunnel in a deep South African gold mine. The wire mesh and cable lacing are designed to contain rock fragments ejected during a rockburst. B. After a violent rockburst. Even though the floor has heaved and the walls collapsed, the broken rock is contained by the wire mesh.

◀ F I G U R E B2.2
Reported rockbursts and associated fatalities in Ontario mines, 1928 to 1990.

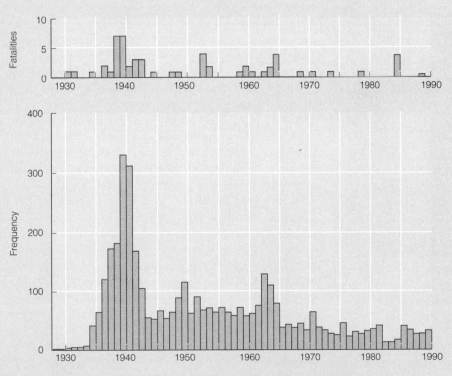

on the scale of the project, the characteristics of the minerals being extracted, and the mining methods used.

Impacts on Air

Smelting and refining processes can emit pollutants into the air, unless the substances are captured by pollution control equipment. The pollutants from these operations include *particulates* (smoke and fine particles); nitrogen oxides and sulfur oxides (which play a role in the production of acid precipitation); vaporized metals; and volatile organic compounds. Any of these may be either deposited locally or transported over long distances. Another airborne problem is dust blown from piles of waste rock and tailings; sometimes the dust contains potentially hazardous materials, such as asbestos fibers. Mining companies today generally seal their tailings by either spraying on a latex cover, maintaining a water cover, or revegetating the area.

Impacts on Water

Liquid effluents containing a wide range of metals (such as arsenic, cadmium, copper, iron, lead, and nickel) and chemical reagents (such as cyanide, kerosene, organic flota-

tion agents, activated carbon, and sulfuric acid) can also be generated during milling and processing. Liquid wastes from mines containing sulfide minerals are often acidic. *Acid mine drainage* forms when water in the hydrologic cycle interacts chemically with sulfide minerals in waste rock and tailings, creating acids such as sulfuric acid (H_2SO_4). The resulting acidic runoff can be devastating to the surrounding ecosystem. Mine drainage associated with evaporite deposits is sometimes saline. In some cases, such as the mining of uranium for nuclear reactor fuel, liquid mine wastes may contain radioactive materials.

Mining companies deal with the problem of liquid wastes by designing the operation to function as much as possible as a closed system. Wastes are generally treated and stored on site until they are clean enough to be released into the environment. Any effluents leaving the mine site are closely monitored.

Mine Site Decommissioning

When the ore has been depleted, the mine is closed. Steps must then be taken to ensure that any discarded materials are properly contained and monitored on an ongoing basis. Once a mine is closed permanently, laws usually require that the mine site be cleaned up and, if appropriate, reclaimed for other uses. Cleaning up and rehabilitating a contaminated site after industrial use has ended is called **decommissioning.** In most industrialized countries, mine site decommissioning is a strictly regulated procedure.

The problem of contamination from old mine sites remains, however. There are many abandoned or "orphaned" mine sites in North America. Strict environmental regulations on mining are relatively recent, and mines that were abandoned before the 1970s may not have received the kind of attention that is now given to waste control during either production or decommissioning. One of the main problems associated with abandoned mines is acid mine drainage (Fig. 13.31). Additional hazards are posed by other types of water pollution, underground fires, and collapse, subsidence, or failure of mine structures. It can be especially difficult to deal with the environmental impacts of abandoned mines from a legal and financial standpoint because it is often impossible to find anyone with legal responsibility for the site.

Mining is of critical importance to industrial societies. It is also a controversial industry that suffers from a negative public image resulting in part from many years of neglectful practices. Since the 1970s the mining industry has undergone enormous changes in both attitudes and practices. It is now tightly regulated in terms of its environmental impacts. Many large mining companies are committed to employing the same environmentally sound practices in their operations in other countries that they use at home, even if environmental regulations in other countries are less stringent.

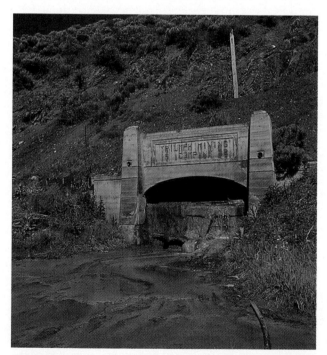

▲ **F I G U R E 13.31**
Pollution problems can continue long after mining has ceased. Water from the entrance tunnel of the old Triumph Mine, Idaho, is acidic as a result of the oxidation of pyrite (FeS), and carries iron in solution. When the water is exposed to the atmosphere the iron is oxidized and precipitated as a reddish-brown ooze of ferric hydroxide.

BOX 13.3

•

THE HUMAN PERSPECTIVE

RECYCLING PLATINUM

*P*latinum provides an interesting example of a material that was recycled efficiently until environmental regulations undermined the recycling mechanism in an unexpected way. Platinum is rare, expensive, and difficult to mine because it occurs at very low concentrations. The high price of platinum had always dictated a high rate of recycling for industries using it. Until the mid-1970s platinum was recycled at rates of 85 percent or better. However, with the advent of catalytic converters for automobiles the platinum recycling situation changed dramatically.

About 60 percent of the platinum mined is made into jewelry, ingots, and chemical-reaction vessels. The remainder is used in chemicals and catalysts for chemical plants and petroleum refineries. Most of these products are recycled with almost complete efficiency. However, platinum is an essential component of catalytic converters, in which it acts as a catalyst to promote chemical reactions that reduce noxious emissions in automobile exhaust. Catalytic converters are now required by law in many parts of the world and represent the fastest growing use of platinum and the major source of permanent (i.e., nonrecycled) platinum use.

The rate of recycling of platinum from catalytic converters is only about 12 percent (Fig. B3.1). The main reason for this is that there is no effective mechanism for recovering platinum from discarded catalytic converters. The problem does not stem from the technology for recovering platinum from the converter; this technology is well understood, and as much as 90 percent of the platinum in used converters is recoverable. However, along with used cars, most used catalytic converters wind up in junkyards and scrapheaps. The costs of locating and collecting the converters and recovering and reprocessing the platinum are so high that recycling is not cost effective unless the price of platinum is higher than $500 per ounce.

The outlook for platinum recycling from catalytic converters is improving somewhat. Some Japanese companies have set up organizations in the United States to acquire used converters and ship them to Japan for reprocessing. Also, as emissions standards become more stringent throughout the world, the demand for platinum will increase, making recycling more profitable.

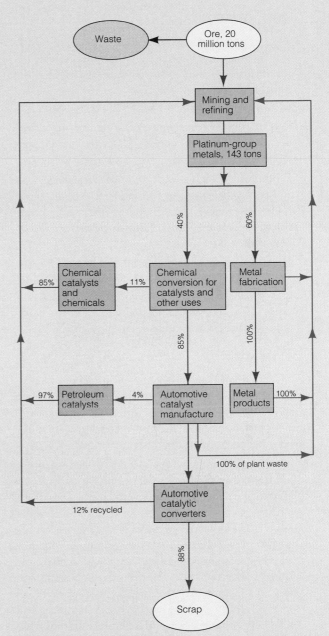

▲ F I G U R E   B3.1
Platinum is recovered efficiently from jewelry and other products, which account for about 60 percent of total consumption. Industrial catalysts and chemicals, which are also recycled efficiently, account for another 6 percent. The fastest-growing use of platinum is in automobile catalytic converters, an application marked by low recycling rates.

SUMMARY

1. Mineral resources are elements, compounds, minerals, or rocks that can be extracted from the ground and are of potential value as commodities. Modern industrial societies depend on mineral resources, yet no single country is completely self-sufficient in mineral supplies.

2. Mineral resources have three important characteristics that influence their supply: (1) they are unevenly distributed; (2) they are nonrenewable; and (3) it is very difficult to assess how much of any particular mineral resource remains to be discovered or exploited.

3. Reserves differ from resources in that reserves are well documented and economically, technologically, and legally extractable.

4. Economic geologists and exploration geologists are constantly challenged to find mineral deposits and to mine the minerals and dispose of the waste rock as cheaply as possible, with little or no permanent disruption of the environment.

5. Ore deposits are mineral deposits in which useful minerals are sufficiently concentrated or enriched to be economically extractable. In general, the scarcest minerals—those that are the least abundant in the Earth's crust—require the greatest amount of enrichment to be economic. Gangue is the unwanted rock associated with ores.

6. There are two main groups of mineral resources. Metallic resources contain ore minerals that are mined specifically to obtain metals, including base and precious metals. Nonmetallic mineral resources are mined for other chemical or physical properties.

7. Hydrothermal mineral deposits are formed when hot, mineral-bearing aqueous solutions precipitate minerals in a rock or sediment, often in systems of fractures called veins. Some hydrothermal solutions are created when water dissolved in a magma is released as the magma rises and cools; others are produced from rainwater or seawater circulating deep within the crust. When a migrating hydrothermal solution encounters a zone in which the temperature, pressure, or chemistry changes rapidly, the dissolved minerals can precipitate suddenly, forming a concentrated mineral deposit.

8. The hot fluids associated with hydrothermal mineral deposits are characteristic of the processes of metasomatism and contact metamorphism, in which rocks are heated and their chemistries changed through contact or close proximity to a body of hot rock or magma. Regional metamorphism, which is caused by the large-scale differential pressures and high temperatures associated with tectonism, can also concentrate

minerals into rich deposits, mainly through the processes of segregation and recrystallization.

9. Magmatic mineral deposits are formed during the cooling and crystallization of magma bodies. Three important types of magmatic mineral deposits are pegmatites, or coarse-grained rocks with concentrations of rare elements; layered intrusions, or plutonic rocks with mineral layering that forms through the process of crystal settling; and kimberlites, rocks confined to long, thin, funnel-shaped pipes that originate deep in the mantle and carry minerals formed under high pressure, such as diamond, up to the surface.

10. Sedimentary mineral deposits are local concentrations of minerals that are formed through the processes of sedimentation, especially the precipitation of material carried in solution. Common types of sedimentary deposits are evaporites, the salts that precipitate when lake water or seawater evaporates; chemical precipitates, including phosphates and iron formations; and stratabound ore deposits, in which hydrothermal solutions deposit ore minerals between sedimentary layers before the sediment has completed the process of becoming a sedimentary rock.

11. Placers are deposits of heavy minerals, such as gold, diamonds, and uranium-bearing minerals, that are concentrated as a result of the action of waves or currents in streams or along shorelines. Most placers are geologically young, but a unique "fossil" placer gold deposit in Witwatersrand, South Africa, supplies much of the world's gold.

12. Residual mineral deposits are formed when soluble materials are leached out of rocks and carried away in solution, leaving behind a residue that is enriched in insoluble materials. Iron-rich laterites and aluminum-rich bauxites are the most common types of residual mineral deposits.

13. Many rich mineral deposits were formed in ancient oceans, and mineral-forming processes are still at work in the oceanic environment. Hydrothermal "black smokers" deposit ore minerals at seafloor vents, and nodules of manganese, iron, cobalt, nickel, and copper oxides are formed as precipitates on the ocean floor.

14. Every stage in the mining process has impacts on the environment. In general, extraction and processing have the greatest impacts. Mining typically involves construction of roads and buildings, stripping of surface vegetation and soil, sinking of mine shafts or excavation of large open pits, and storage of rock waste, tailings, and other discarded materials. Mining processes may also release particulates, nitrogen oxides

and sulfur oxides, vaporized metals, and volatile organic compounds, as well as dust blown from piles of waste rock and tailings. Bodies of water may be polluted by effluents that are contaminated with metals, acids, chemical reagents, salts, or radioactive materials.

15. The mining industry is strictly regulated. In most industrialized countries proper mine site decommissioning—the cleanup and rehabilitation of former mine sites—is required by law. Nevertheless, many abandoned mine sites cause serious environmental problems.

IMPORTANT TERMS TO REMEMBER

base metals (p. 329)
decommissioning (p. 346)
economic geology (p. 324)
enrichment factor (p. 327)
evaporite deposits (p. 335)
exploration geology (p. 325)

gangue (p. 328)
grade (p. 327)
hydrothermal solution (p. 330)
metallic mineral resources (p. 329)
mineral deposit (p. 324)
mineral resources (p. 324)

nonmetallic mineral resources (p. 329)
ore (p. 327)
placer (p. 338)
precious metals (p. 329)
reserve (p. 326)
resource (p. 326)

QUESTIONS AND ACTIVITIES

1. Have you ever noticed how many of the objects around you are made from mineral resources? Are there metals in your immediate surroundings? Synthetic fibers? Plastics? Paints? Building or insulating materials? Glass? These and other common products all use mineral resources. Make an inventory of your use of mineral resources in everyday life.

2. Is any mining carried out near where you live? If so, what kinds of deposits are mined? Do some research to find out more about the economic geology of your area.

3. Arrange to visit an operating mine. You will probably need to call ahead, and you may have to arrange for a special tour with the help of your instructor. When you visit the mine site, ask what measures are taken to ensure that the environmental impacts of the mining operation are kept to a minimum.

4. The locations of mines and quarries often determine the locations of towns and villages. In what ways were early settlement patterns in your area influenced by the locations of exploitable mineral deposits? Are there any towns that still rely heavily on mining as a source of employment?

5. Choose a specific mineral resource, such as diamond, gold, emerald, platinum, or copper, to investigate in detail. Find out what types of processes form ore deposits of this material, where the largest ore deposits are located, and how the material is mined.

6. An interesting long-term project is to monitor the price of a mineral commodity. If there is a major shift in the price, try to find out why it has occurred. Has a new technology been developed to aid in extracting or processing the material, making it less expensive? Has there been a major new discovery? Has a new technology been developed that replaces the material in its present uses? Or is there political instability in a major producing area, making it more difficult to obtain the commodity? Think about other types of economic, technological, geologic, or regulatory developments that might have an impact on the price of a mineral commodity.

CHAPTER 14

SOIL RESOURCES

When you get right down to the bare facts and weigh all the odds, I myself would say that soil is more valuable than oil or gold.

• J. Bilyeu

*S*tarting in the late 1950s, world food production experienced a dramatic and steady increase, outstripping even population growth. For example, world grain production nearly tripled between 1950 and 1990, increasing at an average rate of 29 million tons per year. A number of interrelated factors contributed to this achievement, which has been labeled the "Green Revolution." Those factors include increases in the overall area of cultivated land and the amount of irrigated land; the use of new, high-yield and disease-resistant types of seeds developed through biotechnology; and a significant increase in the use of chemical fertilizers, herbicides, and especially pesticides.

The Green Revolution was most successful in Asia, enabling such countries as India to become self-sufficient in food production. But progress toward global food security is now threatened by a widespread environmental problem: loss of topsoil through contamination, degradation, and human-induced erosion. In India, for example, an estimated 4.3 billion tons of fertile soil are lost each year— more than in any other country. In the United States, where the problem of erosion has been addressed very aggressively, soil loss is still estimated to amount to 1 billion tons per year *in excess* of the rate of soil regeneration.

The real problem with soil loss is that soil, like most other Earth resources, is nonrenewable on a human time scale. Although new soils are constantly being created as a natural part of the rock cycle, it can take 100 to 10,000 years or more to produce a 10-cm-thick layer of fertile soil

from hard bedrock. Unfortunately, the converse is not true; it takes only a short time to turn fertile farmland, forest, or vegetated hillslopes into bare rock.

Since the late 1980s the rate of increase in global agricultural production has slowed; from 1990 to 1993 world grain production per hectare actually declined. It appears increasingly likely that the best years of the Green Revolution may be over. Moreover, the heavy reliance on chemicals associated with the modernization of agriculture may have significant environmental consequences that are just now being recognized, including the salinization and chemical pollution of soils and irrigation waters. To maintain agricultural productivity and ensure global food security in the future, it will be necessary to learn how to manage soils properly.

SOIL: A CRITICAL RESOURCE

At present rates of population growth, the world's farmers must feed an additional 90 million people *every single year.* To meet these demands, food production must increase as rapidly as population. Global food security is influenced by a complex web of technological, socioeconomic, and environmental variables. But the limiting constraint on agricultural productivity is the availability of fertile land, which in turn depends on the status of a single, finite, critically important resource: *soil.*

There are almost as many definitions of the term **soil** as there are people who work with soils. An engineer, for example, might define soil as unconsolidated overburden, or Earth material that can be removed without blasting. A soil scientist more likely would define soil as Earth material that has been broken down and altered in such a manner that it can support rooted plant life. The latter definition, which we will use in this chapter, tells us why soil is such an important resource: Plant growth is directly dependent on

Agriculture in marginal lands. Tuareg families cultivating a market garden maintained by irrigation on the edge of the Sahel in Niger.

soil quality. Soils help maintain equilibrium in the Earth's surface environment by supporting vegetation, an integral part of the hydrologic cycle. In addition, soils store organic matter, thereby influencing the quantity of carbon that is cycled to the atmosphere as carbon dioxide. Soils also trap pollutants and participate actively in the oxygen and nitrogen biogeochemical cycles.

In this chapter we look at the processes involved in the formation of soils as well as at the degradation, contamination, and loss of soils through both natural and anthropogenic causes. We also consider the human impacts of soil loss.

WEATHERING

Soil formation begins when solid bedrock is altered and broken down through a variety of processes. Those processes are encompassed by the term **weathering,** which is defined as the chemical alteration and physical breakdown of rock and sediment when exposed to air, moisture, and organic matter. Rock that has been weathered and broken down so that it consists of unconsolidated particles of rocks and minerals is referred to as **regolith.**

The physical and chemical alteration of rock and the formation of regolith take place throughout the zone in which the lithosphere, hydrosphere, biosphere, and atmosphere mix. This zone extends downward into the ground as far as air and water can penetrate. Rock in this zone contains numerous fractures, cracks, and other openings that are continually being attacked, primarily by water solutions. Given sufficient time, the result is conspicuous alteration of the rock.

In Figure 14.1, for example, loose, unorganized earthy material in which the texture of the bedrock is no longer apparent grades downward into rock that has been altered yet retains its organized appearance, and then into unaltered bedrock. It is evident that alteration of the rock progresses downward from the surface. The soft, earthy regolith near the top of the outcrop in Figure 14.1 no longer

resembles the original minerals in the bedrock, which have largely decomposed. The mineralogical changes that have occurred result mainly from **chemical weathering**—the decomposition of rocks and minerals as chemical reactions transform them into new chemical combinations that are stable at or near the Earth's surface.

Sometimes regolith consists of fragments that are mineralogically identical to the adjacent bedrock. When compared with bedrock, the fragments show little or no evidence of chemical weathering, implying that bedrock can be broken down not only chemically but also mechanically. **Mechanical weathering** is the disintegration or physical breakup of rocks with no change in their chemistry or mineralogy. Although mechanical weathering is distinct from chemical weathering, the two processes generally work together, with effects that are difficult or impossible to separate.

Mechanical Weathering

Mechanical weathering of rock is common in nature. Rock masses buried deep beneath the surface are subjected to enormous pressures by the weight of overlying rock. As erosion wears down the surface, the weight of the overlying rock, and hence the pressure, is reduced, and the rock responds by expanding upward. As it does so, fractures typically develop (Fig. 14.2). These are *joints,* which are defined as fractures in a rock along which no observable movement has occurred. Joints act as passageways through which rainwater can enter a rock and promote both mechanical and chemical weathering. Mechanical weathering takes place whenever crystals of ice or salts grow in rock fractures, whenever rock is heated by fire, and whenever the growth of plant roots disrupts rock.

Frost Wedging Wherever temperatures fluctuate around the freezing point for part of the year, water in the ground periodically freezes and thaws. When water freezes, its volume increases by about 9 percent. The high stresses that result from this volume increase lead to disruption of rocks.

◄ F I G U R E 14.1
A roadcut on the island of Hawaii exposes a weathering profile through a thick lava flow. Loose, earthy regolith grades downward into discolored, weathered rock that retains its organized appearance. Fresh, virtually unweathered rock is seen at the base of the exposure near where the man is standing.

◄ F I G U R E 14.2
Three sets of joints, one horizontal and two vertical, intersect at nearly right angles to form a spectacular rocky vantage point, called the Pulpit, overlooking Lysefjord in southwestern Norway. A widening crack along a vertical joint suggests that the Pulpit will eventually collapse and plunge into the icy waters of the fjord far below.

As freezing occurs in the pore spaces of a rock, water is strongly attracted to the ice and freezes, thereby increasing the stress on the rock. The process leads to a very effective type of mechanical weathering known as *frost wedging,* the formation of ice in a confined opening within rock, thereby causing the rock to be forced apart. Frost wedging is strong enough to force apart not only tiny particles but huge blocks weighing many tons (Fig. 14.3).

▲ F I G U R E 14.3
An extensive field of blocks covers the summit of Mount Whitney, the highest peak in California's Sierra Nevada. Frost wedging, caused by melting snow percolating downward into cracks where it refreezes, has disrupted the bedrock, producing a vast litter of angular boulders.

Crystal Growth Groundwater moving slowly through fractured rocks contains ions that may precipitate out of solution to form salts. The force exerted by salt crystals growing either within rock cavities or along grain boundaries can be enormous and can cause rupturing or disaggregation of rocks. The effects of such mechanical weathering can often be seen in desert regions, where salt crystals grow as rising groundwater evaporates and dissolved salts are precipitated.

Effects of Heat Some geologists have speculated that daily heating of rock by bright sunlight followed by cooling each night should cause mechanical breakdown of the rock because minerals in the rock expand by different amounts when heated. Surface temperatures as high as 80°C have been measured on desert rocks, and daily temperature variations of more than 40°C have been recorded on rock surfaces. Despite a number of careful laboratory experiments, no one has yet demonstrated that daily heating and cooling have noticeable physical effects on rocks. However, these experiments have been carried out only over relatively brief

intervals. It is possible that thermal fracturing occurs only after repeated, extreme temperature fluctuations over long periods.

Fire, on the other hand, can be very effective in disrupting rocks, as anyone knows who has seen a rock beside a campfire become overheated and shatter explosively. Because rock is a relatively poor conductor of heat, an intense fire heats only its thin outer shell, which expands and breaks away as a *spall*. The heat of forest fires and brush fires can lead to the spalling of rock flakes from exposed bedrock or boulders (Fig. 14.4). Over long periods, fires may contribute significantly to the mechanical breakdown of surface rocks.

Plant Roots When a seed germinates in a crack in a rock, the plant extends its roots into the crack. As trees grow, their roots wedge apart adjoining blocks of bedrock. In much the same way, roots also disrupt stone pavements, sidewalks, garden walls, and even buildings (Fig. 14.5). Large trees swaying in the wind can cause cracks to widen, and when they are blown over they can pry rock apart. Al-

▲ F I G U R E 14.4
A large forest fire moving rapidly through a pine forest in Yellowstone National Park, Wyoming, has caused weathering of a surface boulder. Fresh, light-colored spalls, which flaked off as heat caused differential expansion of the rock surface, litter the fire-blackened ground.

▲ F I G U R E 14.5
Tree roots force apart the walls of ruined buildings of Ta Prohm at Angkor in the Cambodian jungle. In the same manner, roots growing in cracks in bedrock force the rock apart.

though it would be difficult to measure, the total amount of rock breakage caused by plants must be very large. Much of it is obscured by chemical weathering, which takes advantage of the new openings as soon as they are created.

Chemical Weathering

Minerals in igneous and metamorphic rocks that form at high temperatures and pressures are chemically unstable when exposed to the lower temperatures and pressures at the Earth's surface. Such minerals break down chemically, and their components form new, more stable minerals. The effects of chemical weathering are most pronounced in regions where both precipitation and temperature are high, two conditions that promote chemical reactions.

The principal agents of chemical weathering are water solutions that behave as weak acids. As rainwater falls through the atmosphere, it dissolves small quantities of carbon dioxide, producing carbonic acid. The slightly acidic water moves downward and laterally through the soil, dissolving additional carbon dioxide from decaying vegetation. The carbonic acid ionizes to form hydrogen and bicarbonate ions (Table 14.1, reaction 1). The hydrogen ions (H^+) are so small that they can enter a crystal and replace other ions, thereby changing the composition of the mineral.

The effectiveness of the H^+ ion in decomposing minerals is illustrated by the way potassium feldspar, a common rock-forming mineral, is decomposed by carbonic acid (Table 14.1, reaction 2). The H^+ ions enter the potassium feldspar and replace potassium ions, which then leave the crystal and pass into solution. Water combines with the remaining aluminum and silica to create the mineral kaolinite. The kaolinite was not present in the original rock; it was created by chemical weathering. A common member of the group of very insoluble minerals that constitute *clay*, kaolinite forms a substantial part of the regolith. The chemical reaction in which the H^+ or OH^- ions of water replace the ions of a mineral is called *hydrolysis*. It is one of the main processes involved in the chemical breakdown of common rocks.

Iron is a constituent of many common rock-forming minerals. When iron-bearing minerals are chemically weathered, iron is released and rapidly oxidized from Fe^{2+} to Fe^{3+} if oxygen is present. Typically this results in the growth of a yellowish mineral, goethite (Table 14.1), through a combination of *oxidation* and *hydration*, the incorporation of water into a crystal structure. Goethite may later be dehydrated (meaning that it loses water) to form hematite, a brick-red mineral (Table 14.1). The intensity of red and yellow colors in weathered rocks and soils can provide clues to how much time has elapsed since weathering

T A B L E 14.1 • Common Chemical Weathering Reactions

1. Production of carbonic acid by solution of carbon dioxide:

$$H_2O + CO_2 \rightleftharpoons H_2CO_3 \rightleftharpoons H^+ + HCO_3^-$$

Water   Carbon   Carbonic   Hydrogen   Bicarbonate
         dioxide   acid   ion   ion

2. Hydrolysis of potassium feldspar:

$$4KAlSi_3O_8 + 4H^+ + 2H_2O \rightarrow 4K^+ + Al_4Si_4O_{10}(OH)_8 + 8SiO_2$$

Potassium   Hydrogen   Water   Potassium   Kaolinite   Silica
feldspar   ion   ion

3. Oxidation of iron (Fe^{2+}) oxide to form goethite:

$$4FeO + 2H_2O + O_2 \rightarrow 4FeO{\cdot}OH$$

Iron   Water   Oxygen   Goethite
oxide

4. Dehydration of goethite to form hematite:

$$2FeO{\cdot}OH \rightarrow Fe_2O_3 + H_2O$$

Goethite   Hematite   Water

5. Dissolution of carbonate minerals by carbonic acid:

$$CaCO_3 + H_2CO_3 \rightarrow Ca^{2+} + 2(HCO_3)^-$$

Calcium   Carbonic   Calcium   Bicarbonate
carbonate   acid   ion   ions

▲ F I G U R E 14.6
An oxidized soil in Hawaii, showing the typical deep red color of these soils. Sugar cane is being cultivated here.

began and the degree or intensity of weathering that has occurred (Fig. 14.6).

Another common chemical weathering process is *leaching*—the removal of soluble matter from bedrock or regolith by water solutions. For example, when silica is released from rocks by chemical weathering, some of it remains in the clay-rich regolith and some is slowly taken into solution by water moving through the ground. Many of the potassium ions weathered from the rock also escape in solution. Soluble substances leached from rocks during weathering are present in all surface water and groundwater. Sometimes their concentrations are high enough to give the water an unpleasant taste.

Factors That Influence Weathering

Several factors influence the susceptibility of a rock to chemical or mechanical weathering. They include the rock type and structure; the steepness of slopes in an area; climate, especially levels of heat and moisture; the activity of burrowing animals; and time.

Rock Type and Structure Because different minerals react differently to weathering processes, the type of rock

subjected to weathering clearly must influence the nature of the resulting decomposition. Quartz is so resistant to chemical breakdown that rocks rich in quartz are also resistant. In the Appalachian Mountains, for example, layers of quartz-rich rocks form ridges that stand prominently above valleys underlain by more erodible rocks containing less quartz. The rate of weathering of a rock is also influenced by its structure. Even a rock that consists entirely of quartz may break down rapidly if it contains closely spaced joints or other partings that make it susceptible to the action of frost.

Slope A mineral grain loosened by weathering may be washed downslope by the next rain. On a steep slope the solid products of weathering move away quickly, continually exposing fresh bedrock. As a result, weathered rock seldom extends far beneath the surface. On gentle slopes, however, weathering products are not easily washed away and therefore may accumulate, forming deep deposits of regolith.

Climate Moisture and heat promote chemical reactions. Not surprisingly, therefore, weathering is more intense and generally extends to greater depths in a warm, moist climate than in a cold, dry one (Fig. 14.7). In moist tropical lands, such as Central America or southeast Asia, obvious effects of chemical weathering can be seen at depths of 100 m or more. By contrast, in such cold, dry regions as northern Greenland and Antarctica, chemical weathering proceeds very slowly. The effects of mechanical weathering are generally obvious in these areas, for bedrock surfaces typically are littered with rubble dislodged by frost action.

Burrowing Animals Burrowing animals such as rodents and ants bring partly decayed rock particles to the surface, where they are exposed more fully to chemical action. More than 100 years ago Charles Darwin calculated that earthworms bring particles to the surface at the rate of more than 2.5 kg/m² (25 tons per hectare) per year. After a study in the Amazon River basin, geologist J. C. Branner wrote that the soil there "looks as if it had been literally turned inside out by the burrowing of ants and termites." Although burrowing animals do not break down rock directly, the amount of disaggregated rock moved by them over many millions of years must be enormous.

Time Hundreds or even thousands of years are required for hard rock to decompose to depths of only a few millimeters. For example, granite and other hard bedrock surfaces in New England, the Canadian Shield, and Scandinavia still display polish and fine grooves made by glaciers of the last ice age more than 10,000 years ago. In such regions, where cool climate and successive glaciations have significantly reduced the rate and effectiveness of chemical weathering, it takes many tens of thousands of years to create weathered regolith like that shown in Figure 14.1. However, in regions that were not repeatedly glaciated and have been continuously exposed to weathering, the zone of weathering often extends to great depths. In some tropical areas, where weathering has continued for millions of years, mining operations have exposed bedrock that has been thoroughly decomposed to depths of 100 m or more.

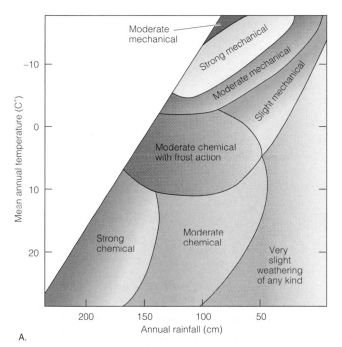

A.

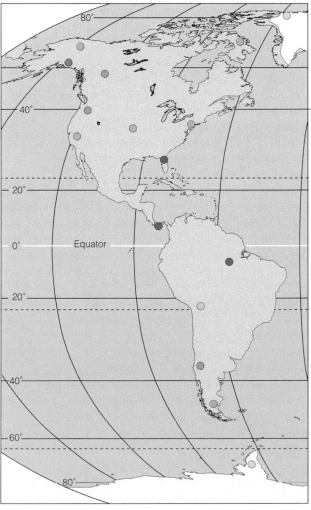

B.

▲ F I G U R E 14.7
A. Climate plays a major role in controlling the type and effectiveness of weathering processes. Mechanical weathering is dominant in areas where rainfall and temperature are low. High temperature and precipitation favor chemical weathering. B. Places where different weathering regimes predominate are shown on the map.

SOIL FORMATION AND CLASSIFICATION

The physical breakdown and chemical alteration of solid rock by weathering is the first step in the formation of soils. However, soils also contain organic matter mixed with the mineral component. The presence of organic matter is crucial to soil fertility, and its formation is a fundamental stage in the processes leading to the development of a *soil horizon*.

Biological Processes in Soil Formation

Partially decomposed organic matter in soils is called **humus.** The organic matter is derived from the decay of dead plants and animals. Humus performs the function of retaining the chemical nutrients released from decaying organisms, as well as those released by the weathering of mineral matter. Thus, humus is critical to **soil fertility**—the ability of a soil to provide the nutrients needed for plant growth, especially nitrogen, phosphorus, and potassium.

Throughout their life cycle, plants are directly involved in producing the fertilizers that will nourish future generations of plant life. Nutrients in water solutions are drawn upward through their roots. When plants die, the organic matter in them decays, contributing to the formation of new humus. Many animal species also live in soil and play an important role in soil formation, from bacteria and fungi to earthworms and small mammals. Earthworms continually rework the soil, not only by burrowing, as discussed earlier, but also by passing soil through their intestinal tracts. They ingest large amounts of decaying leaf matter and carry it down from the surface, where it is incorporated into the mineral soil horizon.

All of the functions performed by organisms in the soil are an integral part of the continuous cycling of nutrients between regolith and the biosphere. With its partly mineral, partly organic composition, soil forms an important bridge between the lithosphere and the biosphere.

The Development of Soil Profiles

As bedrock and regolith weather, soil gradually evolves, extending downward from the surface. Eventually, an identifiable succession of subhorizontal weathered zones is formed. Each of those zones, or **soil horizons**, has distinctive physical, chemical, and biologic characteristics. Taken together, the soil horizons constitute a **soil profile**—the succession of soil horizons lying between the surface and the underlying **parent material,** the rock and mineral regolith from which soil develops (Fig. 14.8). It is difficult to generalize about soil profiles because of the diversifying influence of such factors as climate. However, certain types of horizons are common to many soils.

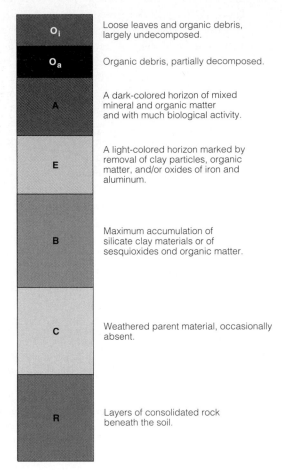

▲ FIGURE 14.8
A sequence of soil horizons—that is, a soil profile— which might appear in a forest soil that developed in a cool, moist climate.

O_i — Loose leaves and organic debris, largely undecomposed.

O_a — Organic debris, partially decomposed.

A — A dark-colored horizon of mixed mineral and organic matter and with much biological activity.

E — A light-colored horizon marked by removal of clay particles, organic matter, and/or oxides of iron and aluminum.

B — Maximum accumulation of silicate clay materials or of sesquioxides ond organic matter.

C — Weathered parent material, occasionally absent.

R — Layers of consolidated rock beneath the soil.

Types of Soil Horizons

The uppermost horizon in a soil profile may be a surface accumulation of organic matter (*O horizon*) that overlies mineral soil (Fig. 14.8). The organic matter (primarily litter, such as leaves) is in various stages of decomposition. Underlying the O horizon is the *A horizon,* which is typically dark in color because of the presence of humus mixed with the mineral matter. The term **topsoil,** used to describe a fertile soil layer near or at the surface, is essentially a synonym for A horizon. The A horizon has lost some of its original substance through the downward transport of clay particles and the chemical leaching of soluble minerals.

The *E horizon,* which is sometimes present beneath the A horizon, typically is grayish or whitish in color. The light color results mainly from the lack of a darker oxide coating on light-colored mineral grains. E horizons are commonly found in acidic soils that develop beneath forests of ever-

▲ F I G U R E 14.9
A soil profile in semiarid central New Mexico includes whitish caliche forming a prominent K horizon between a yellowish-brown C horizon (beneath) and a reddish-brown B horizon (above).

green trees. A *B horizon,* which is usually brownish or reddish in color, underlies the E horizon. The B horizon is enriched in clay and/or iron and aluminum hydroxides that are produced by the weathering of minerals within the horizon and are transported downward from overlying horizons. Although the B horizon is generally penetrated by plant roots, it contains less organic matter than the humus-rich A horizon. The *C horizon,* consisting of parent material in various stages of weathering, is the deepest. Oxidation of parent material in the C horizon may give the soil a yellowish-brown color.

In other climatic zones different types of soil profiles and horizons are formed. For example, the *K horizon,* which is present beneath the B horizon in some arid zones, is densely impregnated with calcium carbonate that coats the mineral grains and constitutes up to 50 percent of the volume of the horizon. When carbonates build up in this way, the soil profile may develop a hard, impervious, insoluble layer of whitish calcium carbonate known as *caliche* or *hardpan* at or just below the surface (Fig. 14.9). In dry, cold polar regions, such as Greenland, northern Canada, and Antarctica, soils may lack well-developed horizons altogether. In tropical regions, where rainfall is high and temperatures warm, soils form deeply weathered horizons that are rich in iron. These soils are typically infertile because essential nutrients have been leached away, leaving the reddish iron-rich residue that geologists call *laterite.* If lateritic soil dries out, it becomes very hard and bricklike (Fig. 14.10). The name *laterite* is derived from the Latin word for "brick" *(latere).*

▼ F I G U R E 14.10
Laterite, formed by the hardening of deeply weathered, iron-rich soils in tropical climatic zones, is being quarried for building stone in this scene from India.

Rates of Soil Formation

A soil profile can form rapidly in some environments. With moderate temperatures and high rainfalls, an A horizon may begin to develop within a few years on newly exposed parent material. As plant cover becomes denser, carbonic and organic acids acidify the soil and leaching becomes more effective. After about 50 years a B horizon may appear and the combined thickness of the A and B horizons may reach 10 cm.

In less humid climates soil is formed more slowly. It may take thousands of years for a detectable B horizon to appear. In the midcontinental United States, for example, B horizons that have developed during the last 10,000 years contain little clay and lack structure, whereas those dating back about 100,000 years generally are rich in clay and have a distinctive structure. An extreme example of the effect of wetness on the rate of soil formation can be seen in the glacier-free deserts of Antarctica, which are so dry and cold that sediments more than a million years old have only weakly developed soils.

Classification of Soils

The soil that is formed under prairie grassland differs from soil formed in an Arctic tundra, boreal forest, or tropical rainforest. The character of a soil may change abruptly as we move from one type of rock to another or from a gentle slope to a steep one, and it also will change with the passage of time. Over the years, soil scientists have developed systems for classifying different types of soil. Soils are classified according to their physical and chemical properties, making it easier to map, study, and understand how soils are distributed on the landscape and how their properties reflect soil-forming factors (Table 14.2 and Fig. 14.11).

The soil classification scheme used in the United States and many other parts of the world is a hierarchical scheme,

▼ F I G U R E   14.11
Soils of the world.

SOILS OF THE WORLD

U.S. Comprehensive Soil Classification System.
Based on data of Soil Conservation Service,
U.S. Dept. of Agriculture.

S Spodosols
A Alfisols
A1 Boralfs
A2 Udalfs
A3 Ustalfs
A4 Xeralfs
U Ultisols
O Oxisols
V Vertisols
M Mollisols
D Aridisols
T Tundra soils
H Highland (I Icesheet)

Kilometers
0   1000   2000   3000

0   1000   2000
Miles

(True distances on mid-meridians and parallels 0 to 40)
Interrupted homolosine projection
Based on Goode Base Map

T A B L E 14.2 • Orders of the U.S. Soil Classification System

Soil Order (Meaning of Name)	Main Characteristics
Alfisol (Pedalfer[a] Soil)	Thin A horizon over clay-rich B horizon, in places separated by light-gray E horizon. Typical of humid middle latitudes.
Aridosol (Arid Soil)	Thin A horizon above relatively thin B horizon and often with carbonate accumulation in K horizon. Typical of dry climates.
Entisol (Recent Soil)	Soil lacking well-developed horizons. Only a thin incipient A horizon may be present.
Histosol (Organic soil)	Peaty soil, rich in organic matter. Typical of cool, moist climates.
Inceptisol (Young soil)	Weakly developed soil, but with recognizable A horizon and and incipient B horizon lacking clay or iron enrichment. Generally occurs under moist conditions.
Mollisol (Soft soil)	Grassland soil with thick dark A horizon, rich in organic matter. B horizon may be enriched in clay. E and K horizons may be present.
Oxisol (Oxide soil)	Relatively infertile soil with A horizon over oxidized and often thick B horizon.
Spodosol (Ashy soil)	Acidic soil marked by highly organic O and A horizons, an E horizon, and iron/aluminum-rich B horizon. Occurs in cool forest zones.
Ultisol (Ultimate soil)	Strongly weathered soil characterized by A and E horizons over highly weathered and clay-rich B horizon. Characteristic of tropical and subtropical climates.
Vertisol (Inverted soil)	Organic-rich soil having very high content of clays that shrink and expand as moisture varies seasonally.
Andisol (proposed) (Dark soil)	Soil developed on pyroclastic deposits and characterized by low bulk density and high content of amorphous minerals.

[a] Soils rich in iron and aluminum

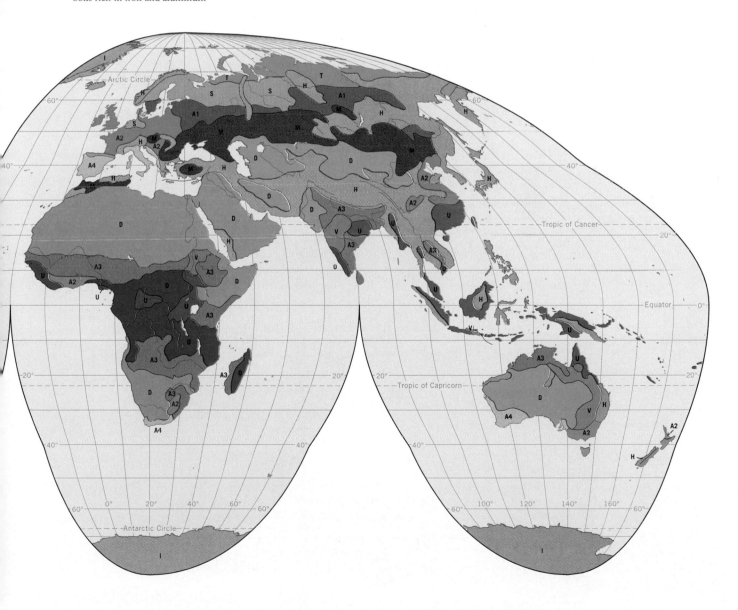

THE CANADIAN SYSTEM OF SOIL CLASSIFICATION

*T*he formation of the National Soil Survey Committee of Canada in 1940 was a milestone in the evolution of soil classification and of soil science in general in Canada. Before that time, the United States system was used in Canada. Although that system was useful in the western plains, it proved to be less applicable in eastern Canada, where parent materials and relief factors dominated soil properties and development in many areas.

Canadian soil scientists ultimately adopted several important features of the U.S. Comprehensive Soil Classification System (CSCS). Nevertheless, special needs required

the establishment of a completely independent classification system. Because Canada lies entirely north of the 40th parallel, there was no need to incorporate soil orders found only in lower latitudes (Figure B1.1). Moreover, the vast expanse of Canadian territory consisting of boreal forest and tundra necessitated the recognition at the highest taxonomic level of soils of cold regions, which appear only at lower taxonomic levels in the CSCS.

The present Canadian system of soil taxonomy is more closely related to the United States system than to any other. For example, major mineral horizons (A, B, C) are

T A B L E B1.1 • **Orders of the Soil Classification System Used in Canada**

Order	Main Characteristics
Brunisolic	Soils under forest with brownish-colored B horizons; occur in a wide range of climatic and vegetative environments, including boreal forest; mixed forest, shrubs, and grass; and heath and tundra.
Chernozemic	Well- to imperfectly drained soils with surface horizons darkened by the accumulation of organic matter from the decomposition of grasses; frozen during some period each winter and dry at some period in summer; semiarid climate.
Cryosolic	Occupy much of northern Canada, where permafrost remains close to the surface of both mineral and organic deposits; common in subarctic forest, extend into boreal forest and alpine areas of mountainous regions; disturbance by freeze–thaw activity common.
Gleysolic	Features indicative of periodic or prolonged saturation with water and reducing conditions; associated with high water tables or temporary saturation above a relatively impermeable layer.
Luvisolic	Light-colored A horizons and clay-rich B horizons; develop in well- to imperfectly drained sites under forest vegetation in subhumid to humid, mild to very mild climates.
Organic	Composed largely of organic materials; include peat, muck, and bog soils; occur in poorly and very poorly drained depressions and level areas in regions of subhumid to humid climate and are derived from vegetation.
Podzolic	B horizon in which the dominant accumulation is composed mainly of humus with Al and Fe; occur under forest and heath vegetation in cool to very cold humid to very humid climates.
Regosolic	Weakly developed horizons, which may be caused by youthfulness of the parent material, instability of the slope, nature of the material, or dry, cold climate; occur in a wide range of vegetation and climates.
Solonetzic	B horizons that are very hard when dry and swell to a sticky mass with very low permeability when wet; occur in semiarid to subhumid plains; associated with vegetative cover of grasses; developed from parent materials that have been salinized through leaching.

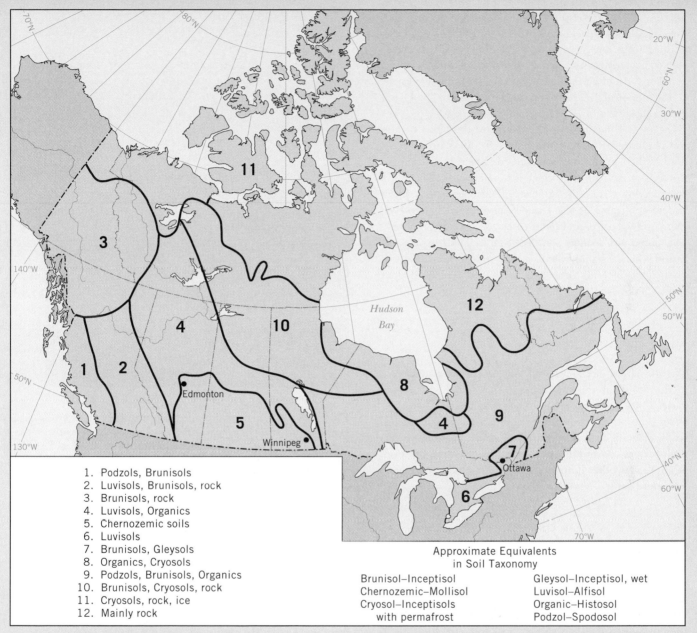

1. Podzols, Brunisols
2. Luvisols, Brunisols, rock
3. Brunisols, rock
4. Luvisols, Organics
5. Chernozemic soils
6. Luvisols
7. Brunisols, Gleysols
8. Organics, Cryosols
9. Podzols, Brunisols, Organics
10. Brunisols, Cryosols, rock
11. Cryosols, rock, ice
12. Mainly rock

Approximate Equivalents in Soil Taxonomy

Brunisol–Inceptisol	Gleysol–Inceptisol, wet
Chernozemic–Mollisol	Luvisol–Alfisol
Cryosol–Inceptisols with permafrost	Organic–Histosol
	Podzol–Spodosol

▲ F I G U R E B1.1
Generalized map of soil regions of Canada.

defined in much the same way as in the United States system. Both systems are hierarchical, and definitions of soil types are based on measurable soil properties (Table B1.1). However, the Canadian system is designed to classify only soils that occur in Canada and hence is not a comprehensive system. The Canadian system is pragmatic; it is designed to organize the knowledge of soils in a reasonable and usable way. The classes are based on properties of the soils themselves rather than on how the soils are used.

or *taxonomy*. Soils are classified into *orders,* which are divided into *suborders,* and then subdivided into a hierarchy of descriptive classifications. The resulting hierarchical levels, in increasing order of detail, are *great groups, subgroups, families, series,* and *types.* The orders—the highest and, therefore, most general taxonomic grouping are distinguished from one another on the basis of easily recognizable characteristics such as lack of well-developed horizons; accumulation of aluminum, iron, clay, or carbonate in the B horizon; or high acidity and organic content. The 11 soil orders are shown in Table 14.2.

Many soil scientists would like to see the United States soil classification system adopted as an international standard. Others prefer to use a system designed specifically for the soils of the region in which it is being applied. The soil classification system used in Canada, for example, is not as widely applicable as the United States system. However, many soil scientists consider it easier to use and particularly well suited to Canadian soils (see Box 14.1). Other classification schemes group soils according to characteristics such as their engineering properties or the grain sizes of their component particles.

EROSION BY WIND AND WATER

The formation of regolith and soils *in situ* is a natural part of the set of interrelated processes that make up the rock cycle. Once formed, soils and regolith in temperate latitudes become susceptible to removal, transport, and deposition through erosion by wind and water. Erosion is a natural, ceaseless part of the rock cycle; as we will see, however, land use and other human activities can sometimes greatly accelerate natural rates of erosion.

Erosion by Running Water

Erosion by water can begin even before distinct stream channels are formed. It occurs in two ways: (1) through impact as raindrops hit the ground and (2) through overland flow during heavy rains, a process known as **sheet wash** (also called *sheet erosion*). As raindrops strike bare ground they dislodge small particles of loose soil, spattering them in all directions; this is called **splash erosion** (Fig. 14.12). On a slope, the result is the displacement of particles downhill and eventually the erosion of a thin layer of topsoil over the entire surface of the slope.

The effectiveness of raindrops and overland flow is greatly diminished if the soil is covered by vegetation. The leaves and branches of trees break the force of falling raindrops and cushion their impact. More important, the intricate network of roots forms a tight mesh that holds the soil in place, greatly reducing erosion. The root network also holds water and allows it to percolate slowly down through the soil. In vegetated areas, therefore, there is less runoff than in areas of bare ground. Organic litter (i.e., O horizon

▲ F I G U R E 14.12
Erosion by raindrops. When a raindrop (upper photograph) lands on bare, wet soil, small droplets of water carrying soil particles are thrown into the air and a tiny crater is formed.

material) in the soil can cushion the impact of raindrops, thereby diminishing the effects of splash erosion. Organic material also absorbs water, thereby limiting runoff.

As discussed in Chapter 8, rainfall on a fairly uniform regolith slope will quickly begin to organize itself into a drainage system with defined stream channels. If furrows are already present in the soil (e.g., plowed rows, tire tracks, or natural lines of drainage), these will concentrate the surface runoff into **rills,** the small, narrow channels that first begin to cut into the regolith. As erosion progresses, the running water will begin to develop **gullies,** distinct, narrow stream channels that are larger and deeper than rills (Fig. 14.13). Gullies and rills are unlike streams in that they typically carry water only during and immediately after rainstorms or during periods of snowmelt.

Erosion by Wind

Wind is an important agent of erosion and sediment transport. Its effects are most visible in semiarid and arid regions (Fig. 14.14). Wind erosion is referred to as **eolian erosion** after Aeolus, the Greek god of the wind.

▲ F I G U R E 14.13
Gully erosion due to overly heavy grazing, Shawnee, Oklahoma. Vegetation, contour ter-
races, and dams (center right) have halted the headward extension of the gullies. (Mark A.
Melton, Ph.D.)

▲ F I G U R E 14.14
Sand dunes advance from right to left across irrigated fields in the Danakil Depression,
Egypt.

▲ **F I G U R E 14.15**
A cave excavated by hand in compact loess provides a roomy and comfortable home for a Chinese family in the loess region near Xian.

sists largely of silt but is commonly accompanied by some fine sand and clay. Loess is an important resource in countries where it is thick and widespread, because of the productive soils developed on it. The rich agricultural lands of the upper Mississippi Valley, the Columbia Plateau of Washington State, the Loess Plateau region of central China, and much of Eastern Europe have loess soils that provide food for millions. Loess deposits can also provide shelter; in central China, caves carved in loess house thousands of families (Fig. 14.15).

The Mechanisms of Wind Erosion

Flowing air erodes the land surface in two ways. The first process, **abrasion,** results from the impact of wind-driven grains of sediment. Airborne particles act as tools, chipping small fragments off rocks, which acquire distinctive, curved shapes and a surface polish as a result. A bedrock surface or stone that has been abraded and shaped by wind-blown sediment is called a *ventifact.*

The second process, **deflation** (from the Latin word meaning "to blow away"), occurs when the wind picks up and removes loose rock fragments, sand, and dust. Deflation on a large scale takes place only where there is little or no vegetation and loose particles are fine enough to be picked up by the wind. Areas of significant deflation are found mainly in deserts; nondesert sites where deflation occurs include ocean beaches, the shores of large lakes, and the floodplains of large glacial streams. Of greatest economic importance, however, is the deflation of bare plowed fields in farmland (Fig. 14.16). Although deflation may occur seasonally when land is plowed, it is especially severe

Sediment carried by the wind tends to be finer than that moved by water or ice. Because the density of air at sea level ($1.22 kg/m^3$) is far less than that of water ($1000 kg/m^3$), air cannot move as large a particle as water flowing at the same velocity. In extraordinary windstorms, when local wind speeds can reach or exceed 300 km/h, coarse rock particles up to several centimeters in diameter may be lifted to heights of a meter or more. In most regions, however, wind speeds rarely exceed 50 km/h and the largest particles of sediment that can be suspended in the airstream are grains of sand. Larger particles settle out too quickly to remain aloft. At lower wind speeds, sand moves along close to the ground surface and only fine grains of dust move in suspension.

Although most regolith contains a small proportion of wind-laid dust, the dust is thoroughly mixed with other fine sediments and is indistinguishable from them. However, in some regions wind-laid dust is so thick and uniform that it constitutes a distinctive deposit and may determine the primary characteristics of the landscape. Known as *loess* (German for *loose*), this wind-deposited dust con-

▲ **F I G U R E 14.16**
Deflation in a plowed field, eastern Colorado. Plowing removes vegetation, loosens and dries out the soil; a dust storm and deflation is the result.

◄ F I G U R E 14.17
The Qattara Depression in the desert of western Egypt. This huge basin, with a floor more than 100 m below sea level, was created by intense deflation in an arid climate.

during times of drought, when no moisture is present to hold soil particles together.

Generally, deflation lowers the land surface slowly and irregularly, making it difficult to measure wind erosion. However, in the Dust Bowl of the 1930s (see Chapter 9), deflation in parts of the western United States amounted to 1 m or more within only a few years. This is a tremendous rate compared with the long-term average rate of erosion for the region as a whole, which is only a few centimeters per thousand years.

Small saucer- or trough-shaped hollows and larger basins created by wind erosion are among the most conspicuous signs of deflation. Tens of thousands of these basins are found in the semiarid Great Plains region of North America from Canada to Texas. Most are less than 2 km long and only a meter or two deep. In wet years the larger ones are carpeted with grass and may contain shallow lakes. However, in dry years moisture evaporates from the soil, the grass dies away, and wind deflates the bare soil. Where sediments are particularly susceptible to erosion by wind, deflation basins can be excavated to depths of 50 m or more. The immense Qattara Depression in the Libyan

Desert of western Egypt (Fig. 14.17), whose floor lies more than 100 m below sea level, has been attributed to intense deflation. In any basin, the depth to which deflation can extend is limited only by the water table or bedrock. As deflation lowers the land to the level at which the ground is saturated, the surface soil becomes moist, encouraging the growth of vegetation that inhibits further deflation.

SOIL LOSS: A GLOBAL CONCERN

Soil loss is a worldwide problem of massive proportions. As discussed in the essay for Part III, the estimated global rate of soil loss through erosion now exceeds 25 billion tons per year. Over the past 20 years an estimated 480 billion tons of topsoil has been lost, more than the amount on all the cropland in the United States. Because it takes so many years for topsoil to form, this loss of soil is tantamount to mining a nonrenewable resource, instead of remaining a continuously productive, renewable resource; soil becomes an ever-diminishing and degraded resource.

Rates of Soil Loss

Soil loss affects everyone. A person eats about 750 kg of food a year, and there are now more than 5.5 billion people on the Earth. Soil erosion leads to an annual loss of 25 billion tons of topsoil, or 4.5 tons (4500 kg) per person. This means that for every kilogram of food eaten in a given year, 6 kg of soil are lost. At the current rate of loss, the world's most productive soils are being depleted at the rate of 7 percent each decade. One recent estimate projected that as a result of excessive soil erosion and rapidly increasing populations, only two thirds as much topsoil will be available to support each person at the end of the century as was available in 1984.

T-Value

Soil scientists often talk about soil loss in terms of a *tolerable annual rate* of soil loss, or **T-value.** The T-value refers to losses that are balanced or offset by natural regeneration of the soil. Of greatest concern is soil loss that occurs *in excess* of the rate at which topsoil is being regenerated in a given region. In the United States, for example, the amount of farmland soil lost to erosion at the beginning of the 1990s exceeded the amount of newly formed soil by almost 2 billion tons, or about 5 tons of soil for every ton of grain produced. This rate of loss has been reduced substantially by the implementation of aggressive soil erosion control programs. However, rates of soil loss in the other major food-producing countries—China and the former Soviet Union—are at least as high, whereas in India the rate of soil erosion is estimated to be more than twice as high.

Sediment Yield

One way of quantifying rates of soil loss is by measuring the amount of sediment being transported and deposited by streams. Different land areas are eroded at different rates. We can see this by comparing streams: some streams run clear nearly all the time, whereas others are constantly muddy and obviously are transporting a considerable load. These differences are related to geologic, climatic, and topographic factors such as rock type and structure, local climate, vegetation, and topography. Together, these factors control the **sediment yield,** the amount of sediment eroded from the land by runoff and transported by streams (Fig. 14.18).

We might guess that the greater the amount of precipitation, the greater the rate of erosion and, therefore, the higher the sediment yield. Globally, the highest sediment yields do occur in tropical areas, where high precipitation and temperature combine to promote rapid chemical weathering and effective transport of sediment (Fig. 14.18). However, the influence of climate on sediment yield is more complicated than that. In moist regions plant

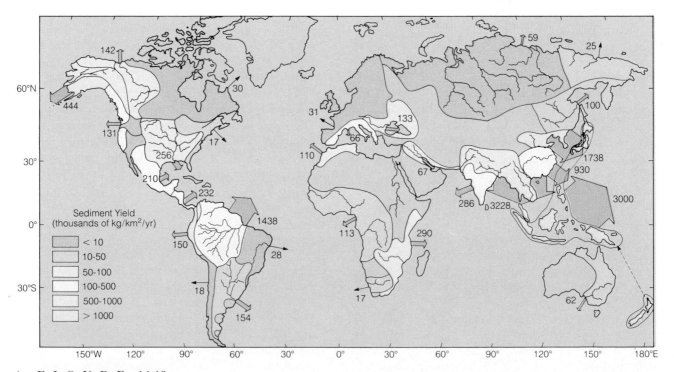

▲ **F I G U R E   14.18**
Average yields of sediment from various drainage basins. The width of the arrows is related to the amount of sediment entering the ocean; the numbers represent annual input in billions of kilograms. Annual sediment yields (thousands of kg/km²) for major basins are indicated by different colors. Green areas contribute minimal sediment to the oceans.

roots tend to anchor the soil, thereby curtailing erosion. In temperate eastern North America and Western Europe, for example, vegetation cover is more or less continuous and rates of erosion are low. In drier regions, minimal precipitation limits the growth of vegetation, making the land more vulnerable to erosion.

Clearing of forests, cultivation of land, damming of streams, construction of cities, and other human activities also affect erosion rates and sediment yields. Sometimes the results are dramatic: In parts of the eastern United States, for example, areas that have been cleared for construction produce between 10 and 100 times more sediment than comparable rural or natural areas that remain vegetated. Let's examine the impacts of land use on soil erosion in greater detail.

Soil Loss and Land Use

Although soil loss results naturally from erosion, the impacts of human activities have overwhelmed natural systems in many parts of the world. Activities that can produce significant soil loss include removal of vegetation, stressful or inappropriate agricultural practices, and a range of activities that contaminate the soil. Topsoil is also lost at an alarming rate through the conversion of prime agricultural land for urban and suburban development.

Removal of Vegetation

Removal of vegetation through practices such as clearcut logging, slash-and-burn agriculture, and overly intensive grazing can lead to accelerated rates of surface runoff and destabilization of soils owing to the loss of anchoring roots. Soils in the humid tropics quickly lose their fertility when they are stripped of their natural vegetation (Fig. 14.19). It may seem strange that soils in tropical rainforests should be so susceptible to degradation; however, most of the nutrients in a rainforest are found in the lush, dense vegetation, not in the soil layers. If the vegetative cover is stripped away, the chemically barren soils are exposed and become vulnerable to erosion.

Removal of natural vegetation in sensitive regions with relatively low precipitation, such as the African Sahel, the North American Great Plains, and the Brazilian Caatinga, can expose vulnerable soils to damaging atmospheric effects such as those caused by strong winds and intense sunlight. Vegetation may be stripped for agricultural purposes, in the gathering of fuelwood, or by intensive grazing of cattle, goats, or sheep, leading to land degradation and desertification (see Chapter 9).

◀ F I G U R E 14.19
Deforestation in Rondonia, Brazil, has led to accelerated surface runoff and turned luxuriant rainforest into an eroded, devastated landscape. Stripped of their natural vegetation cover and planted with crops, these soils quickly lose their fertility.

Inappropriate Agricultural Practices

With the world's population now exceeding 5.5 billion, competition for agricultural land is increasing steadily. In the past, pressures to increase agricultural production have led to the use of stressful or inappropriate farming techniques, which have caused disastrous soil erosion (Fig. 14.20). In many developing nations, farmers have been forced to move beyond traditional farm and grazing lands onto steep, easily eroded slopes or into semiarid regions where periodic crop failures are a fact of life and plowed land is subjected to severe wind erosion. Fallow periods, in which fields are allowed to lie dormant with a mulching cover of crop residues or farmyard manure, have become shorter. Traditional practices such as interplanting of different crops are being abandoned in favor of *monoculture,* that is, fields devoted to a single crop. This can strip the soil of certain nutrients and lead to soil degradation as well as increased susceptibility to pests. At the same time, economic pressures in the more developed countries have led farmers

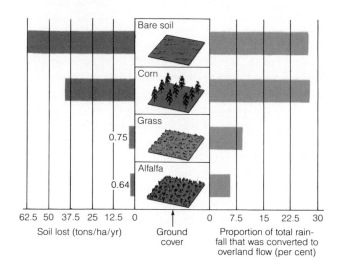

▲ F I G U R E 14.21
Effect of plant cover on rates of soil erosion measured over 4 years at Bethany, Missouri. The soil is silty, the slope is 4.5°, and the annual rainfall is 1000 mm. The measurements show that grass and alfalfa, with their continuous network of roots and stems, are nearly 300 times as effective as "row crops" such as corn in holding the soil in place. Erosional loss from bare soil in this area occurs at a rate of about 45 cm/100 years.

to plant profitable row crops that often leave the land more vulnerable to erosion (Fig. 14.21).

Contamination of Soil

Although the principle cause of soil loss is accelerated erosion, soil degradation resulting from chemical changes is a growing problem. Soil contamination encompasses a variety of destructive processes, including salinization, contamination by pesticides and other chemicals, waterlogging, and acidification. In Europe, industrial and urban pollution, pesticides, and other chemicals have contaminated more than 34 million acres of land.

Human-induced chemical changes can have far-reaching consequences, affecting soil quality on a regional scale. For example, acid rain can cause regional acidification of soils. Alterations in the flow patterns of freshwater bodies can lead to excess sedimentation, salinization, and alkalinization of soils and anaerobic (oxygen-starved) conditions. Chemical changes in the atmosphere caused by the burning of fossil fuels may cause changes in the way the soil retains certain chemical components, which in turn may cause changes in crop yields. And heavy reliance on agrochemicals such as pesticides can lead to chemical pollution of soils and of irrigation water, groundwater, and surface runoff from agricultural land.

Soil Resources and the World Economy

Soil can be viewed as a renewable resource only on a geologic time scale. Over the lifetime of individuals and even of nations, soils should be considered nonrenewable resources that must be carefully utilized and preserved.

▲ F I G U R E 14.20
Soil erosion in a thin layer of soil as a result of overly intensive irrigation farming of peanuts in Oklahoma.

T A B L E 14.3A • **Reduction in Yields of Key Crops as Soil Erodes (Southern Iowa)**

Change in Erosion Phase	Reduction in Yield Per Acre (bushels)		
	Corn	Soybeans	Oats
Slight to moderate	16	5	9
Moderate to severe	7	3	4

T A B L E 14.3B • **Increase in Amount of Fertilizer Required to Grow Corn as Soil Erodes (Southern Iowa)**

Change in Erosion Phase	Nitrogen (pounds per acre)	Potash (pounds per acre)
Slight to moderate	10	6
Moderate to severe	30	7

Source: Paul Rosenberry, Russell Knutson, and Lacy Harmon, "Predicting Effects of Soil Depletion from Erosion," *Journal of Soil and Water Conservation,* May/June, 1980.

Effects of Soil Loss on Agricultural Productivity

Because agriculture is the foundation of the world economy, progressive loss of soil signals a potential crisis that could undermine the economic stability of many societies. The upper layers of a soil contain most of the organic matter and nutrients that support crops. When the A and B horizons are eroded away, not only the fertility but the water-holding capacity of a soil diminishes. The soil that is left behind after the topsoil has eroded away consists of less fertile subsurface layers. This infertile soil lacks the ability to recycle nutrients, and it may become dry, compacted, and even more vulnerable to erosion by wind and water. The relationships between increasing severity of topsoil erosion, decreasing soil fertility, and diminishing crop productivity are shown in Table 14.3, A and B.

Indirect Effects of Soil Erosion

Aside from the direct effects on soil fertility and agricultural productivity, the indirect and often unanticipated impacts of soil erosion on society can be significant in terms of property damage and even loss of life. Much of the topsoil that is eroded from agricultural lands is transported down rivers and deposited along valley floors, in marine deltas, or in reservoirs behind large dams. For example, the designers of a major dam and reservoir in Pakistan expected the reservoir to be usable for at least a century. However, population increases in the region above the dam have led to greatly increased soil erosion, which in turn has caused such a high rate of sediment production that the reservoir is now expected to be filled with eroded soil within 75 years, making it unusable (Table 14.4). Sedimentation can also

T A B L E 14.4 • **Rates of Siltation in Selected Reservoirs**

Country	Reservoir	Annual Siltation Rate (metric tons)	Time to Fill with Silt (years)
Egypt	Aswân High Dam	139,000,000	100
Pakistan	Mangla	3,700,000	75
Philippines	Ambuklao	5,800	32
Tanzania	Matumbulu	19,800	30
Tanzania	Kisongo	3,400	15

Source: El-Swaify, S.E. and E.W. Dangler, "Rainfall Erosion in the Tropics: A State of the Art," in *Soil Erosion and Conservation in the Tropics,* American Society of Agronomy, Madison, Wisconsin: 1982; cited in Brown and Wolf (1984) *Soil Erosion: Quiet Crisis in World Economy;* Worldwatch Institute Paper.

clog waterways, sometimes necessitating expensive dredging operations. Excess sediment is considered to be the second most problematic type of pollution in coastal waters (the first is excess nutrients from sewage and fertilizers).

Blowing sand and dust can severely damage crops and other vegetation, as well as damage equipment and cause health hazards for people and livestock. In one severe storm in California in 1977, cattle were asphyxiated by dense dust, and hair and skin were sandblasted from their hindquarters. A strong, dense dust storm can quickly turn a windshield into a sheet of frosted glass, and blowing sand can pit and strip the paint from a car. The engines of vehicles operating in dusty areas are particularly susceptible to damage; dust can contaminate fuel, clog air filters, abrade cylinders, and cause electrical short circuits. During the Persian Gulf War dust created a problem for military vehicles operating in the deserts of Saudi Arabia and Kuwait. Airborne dust also can pose a hazard to aircraft. In 1973 a Royal Jordanian Airlines plane crashed at an airport in northern Nigeria while flying through dense dust, and in 1979 an attempt to rescue United States hostages in Iran was aborted after engine failure attributed to dense airborne dust caused a fatal helicopter crash in the desert staging area. Inhalation of dust can also lead to various medical problems, such as emphysema and silicosis. In addition, disease-causing organisms such as anthrax and tetanus are carried in dust, where they may survive for long periods.

PROTECTING SOIL RESOURCES

Halting soil loss will require effective programs at both the national and international levels. The good news is that soil management techniques are known and readily available;

they just need to be applied more widely and more aggressively. These techniques include a wide range of agricultural approaches, as well as identification and mapping of vulnerable soils, avoidance of heavy dependence on agrochemicals, and reforestation.

Agricultural Approaches to Erosion Control

In places where crops are grown, the ground surface is bare during part of each year. On unvegetated sloping fields, on pastures that are too closely grazed, and in areas planted with widely spaced crops such as corn, rates of erosion can be high (Fig. 14.21). To reduce the rate of erosion, farmers limit areas of bare soil and prevent overgrazing of pastures. If "row crops" such as corn, tobacco, or cotton must be planted on a slope, they are planted in strips alternating with grass or similar plants that help resist soil erosion.

Crop rotation is another method of reducing soil loss. A study in Missouri showed that land that lost 49.25 tons of soil per hectare when planted continually in corn lost only 6.75 tons per hectare when corn, wheat, and clover crops were rotated. The bare land exposed between rows of corn is far more susceptible to erosion than land planted with a more continuous cover of wheat or clover.

Erosion is greatest on steep hillslopes. In Nigeria, for example, land with a gentle 1 percent slope that was planted with cassava (a staple food source) lost an average of 3 tons of soil per hectare each year. On land with a 5 percent slope, however, the annual rate of soil loss increased to 87 tons per hectare. At this rate, 15 cm of topsoil would disappear in a single generation (about 20 years). On a 15 percent slope, the annual rate of erosion increased to 221 tons per hectare, a rate that would remove all topsoil within a decade. In many parts of the world farmers build low walls

▼ F I G U R E 14.22

A terraced hillside near Lanzhou, China, creates productive agricultural fields from steep hillslopes carved in deposits of windblown dust. Hillslopes that are not terraced are rapidly eroding and deeply gullied.

BOX 14.2
·
THE HUMAN PERSPECTIVE

ASSESSING GLOBAL SOIL DEGRADATION:
A CHRONOLOGY OF MILESTONES

1974   At the World Food Conference (Rome) it was decided that the Food and Agricultural Organization (FAO), the United Nations Educational, Scientific, and Cultural Organization (UNESCO), and the United Nations Environment Programme (UNEP) would prepare an assessment of lands that could still be brought under cultivation, taking into account the hazards of irreversible soil degradation.

1977   The United Nations Conference on Desertification adopted a Plan of Action to Combat Desertification (PACD), which was endorsed by the U.N. General Assembly the same year. The worldwide program was directed at stopping the process of desertification and rehabilitating affected lands.

1979   An interagency group published a provisional methodology for assessing soil degradation and prepared a first approximate identification of areas with the potential for degradation through salinization and erosion. Maps at a scale of 1:5,000,000 were prepared for Africa north of the equator and for the Middle East.

1981   The FAO adopted the World Soils Charter, a global action plan for soil conservation.

1982   The International Society of Soil Science established a Subcommission on Soil Conservation and Environmental Quality.

1982   The UNEP Governing Council adopted a World Soils Policy, an element of which was the development of methodologies to monitor global soil and land resources.

1987   UNEP convened an expert meeting to consider the possibility of preparing, even on the basis of incomplete knowledge, a scientifically credible global assessment of soil degradation.

1987   Global Assessment of Soil Degradation (GLASOD) began when UNEP signed an agreement with the International Soil Reference and Information Centre (ISRIC) at Wageningen, the Netherlands. GLASOD information has been digitized in a computer database.

1990   ISRIC/UNEP published the GLASOD *World Map of Human Induced Soil Degradation* at a scale of 1:10,000,000.

1992   UNEP published a *World Atlas of Desertification,* which examines interactions among the variables that determine desertification.

along the contour lines of slopes to impede the flow of water and, thus, the downslope movement of soil particles.

Steep slopes often can be utilized safely and effectively through contour plowing or terracing. In central China, where steep hillslopes are underlain by erodible deposits of wind-blown silt, terracing is a major factor in reducing soil loss (Fig. 14.22). In the arid regions of western China, where strong winds, limited vegetation, and highly erodible soils can lead to rapid soil loss, rows of trees planted as wind screens substantially reduce erosion on the downwind side.

Reducing Dependence on Agrochemicals

Reducing the use of agrochemicals such as chemical fertilizers, pesticides, and herbicides can contribute significantly to the stabilization and protection of soil resources. More-

over, it is clear that heavy reliance on chemicals has outlived its usefulness in many areas; for example, increases in the number and variety of pesticide-resistant insect species threaten productivity in some areas despite continued heavy use of chemicals. One alternative to excessive dependence on chemical fertilizers is the use of unprocessed crushed rocks containing potassium and phosphate to replenish degraded soils. Other promising efforts are focusing on how geochemical and biogeochemical processes in soils affect the nutritional quality of foods and on the ways in which various agricultural practices influence these processes. If scientists can more fully understand the processes by which plants take up nutritional elements from the soil and convert them into organic nutrients, then it may be possible to enhance these processes and improve the nutritional quality of the product.

Another approach, *integrated pest management (IPM)*, utilizes information about pests and their natural enemies to design effective and ecologically sound insect management programs. In some areas IPM has been shown to reduce the need for pesticides by as much as 90 percent. Practices as simple as plowing crop residues into the soil, rotating crops, and alternating rows of different crops can also reduce the need for chemical fertilizers.

Reforestation

Still another approach to the stabilization of soil resources is restoration of deforested land. Reforestation can help preserve soil by reducing rainfall runoff, increasing the soil's capacity for water storage and reducing its vulnerability to wind erosion. For these and many other reasons, many countries have launched ambitious tree-planting programs. For example, in 1990 the United States announced a plan to plant 1 billion trees each year during the 1990s. If the plan is successful, forest cover would be restored on 16 million acres.

Identifying and Mapping Vulnerable Soils

The search for new approaches to preserving global soil resources depends on knowledge about the nature and pre-

sent condition of those resources. As William Fyfe, a prominent Canadian geologist, puts it:

We need new ways to describe soils. This description must focus on sustainable biological productivity and the [agricultural] potential for given climatic and geographic situations. The soil map of our future must say: good for careful farming, good for trees only, don't touch! We know now how to do this, but we rarely do it! Is it too simple? We are moving into the decade of natural disaster reduction. For our future, one of the greatest natural disasters that can be reduced and can be controlled is soil erosion.

Addressing this need, the United Nations Environment Programme (UNEP) has initiated a global assessment of the status and trends of soil degradation. Its goal is to produce a world atlas of soil "health" indicators, including regional, national, and local land assessments. In 1990 a *World Map of Human Induced Soil Degradation* was completed and published by UNEP in association with the International Soil Reference and Information Centre (ISRIC) in the Netherlands. These efforts and others like them represent important steps toward the identification and preservation of vulnerable soils.

SUMMARY

1. Soil is a critical resource because it supports rooted plant life and is a fundamental part of the geochemical cycling of nutrients. With the world population increasing rapidly, there is pressure to expand the amount of land under cultivation and to increase agricultural productivity.

2. Soil formation begins when rock is broken down by weathering. Factors that influence susceptibility to weathering include the rock type and structure; the steepness of the slope; climate, especially levels of heat and moisture; the activity of burrowing animals; and time.

3. In mechanical weathering, a rock is broken down by physical processes with no change in chemistry or mineralogy. Mechanical weathering takes place when crystals of ice or salts grow in rock fractures, rock is heated by fire, or rock is disrupted by the growth of plant roots.

4. Chemical weathering, in which the chemistry and mineralogy of the rock are altered, is most pervasive in areas with high temperatures and levels of precipitation because these conditions promote chemical reac-

tions. The main agents of chemical weathering are weakly acidic water solutions. Among the most common types of chemical weathering are hydrolysis, oxidation, hydration, and leaching.

5. The formation of a soil profile begins with chemical and mechanical weathering, but soils also contain organic material—humus—that is critical to soil fertility, the ability of soil to provide nutrients for plant life. The organic matter in soil comes from the decay of dead plants and animals. Other types of biological activity, such as the activity of burrowing animals, are also important in soil formation.

6. Different soil horizons and profiles are formed in different climatic zones. Among the most common soil horizons are the O horizon, which consists of organic litter at the surface; the A horizon, which is rich in organic matter and has lost soluble minerals through leaching; the B horizon, where clay accumulates along with the substances leached from the A horizon; and the C horizon, which consists of slightly weathered parent material.

7. Rates of soil formation are highly dependent on such

factors as climate, biologic activity, type of parent material, and topography. In many parts of the world rates of soil regeneration are so slow that soil must be considered a nonrenewable resource.

8. The classification system used in the United States and in many other parts of the world is a taxonomic (i.e., hierarchical) system that places soils in 11 orders based on their physical characteristics.

9. Once they have been formed, soils and regolith become susceptible to removal, transport, and deposition through erosion by wind and water. Erosion by running water begins with splash erosion and sheet wash, which move topsoil downslope even before stream channels have been formed. Eventually, narrow rills and deeper, wider, steep-sided gullies form, carrying water and sediment downslope.

10. Eolian (wind) erosion is most effective in arid and semiarid regions. However, a major cause of concern is eolian erosion of topsoil from agricultural land. Wind acts as an agent of erosion through the processes of abrasion and deflation.

11. Soil erosion and degradation are global problems that have been increasing as the world population has grown. At current rates of loss, the world's most productive soils are being depleted by 7 percent each decade.

12. The formation of soil is a natural part of the rock cycle, and so is erosion. However, activities related to land use, such as deforestation, inappropriate agricultural practices, and overly intensive grazing, often accelerate natural rates of erosion. This can lead to soil losses at rates greater than the T-value, that is, rates of loss that exceed the rate of regeneration of soil in a given area.

13. In addition to reducing agricultural productivity, soil erosion has many indirect negative effects, such as shortening the lifespans of large dams, clogging waterways, damaging machinery, and creating health hazards.

14. Halting the widespread loss of soils is a formidable challenge. Effective soil control measures include agricultural approaches such as terracing and crop rotation as well as identification and mapping of vulnerable soils, avoidance of heavy dependence on agrochemicals, and reforestation.

IMPORTANT TERMS TO REMEMBER

abrasion (p. 366)
chemical weathering (p. 352)
deflation (p. 366)
eolian erosion (p. 364)
gullies (p. 364)
humus (p. 358)
mechanical weathering (p. 352)

parent material (p. 358)
regolith (p. 352)
rills (p. 364)
sediment yield (p. 368)
sheet wash (p. 364)
soil (p. 351)
soil fertility (p. 358)

soil horizon (p. 358)
soil profile (p. 358)
splash erosion (p. 364)
T-value (p. 368)
topsoil (p. 358)
weathering (p. 352)

QUESTIONS AND ACTIVITIES

1. Can you identify any land-use practices that are contributing to accelerated soil erosion in your area? What might be done to modify or eliminate those practices? Are any soil conservation techniques being employed? What are they?

2. Find out whether the soils in your area have ever suffered from problems of chemical degradation, such as acidification, salinization, or waterlogging. How were the problems addressed (or how should they be addressed, if they still exist)?

3. Do you know where to obtain data and information concerning the status of soil resources in your area? Are any organizations carrying out monitoring programs or scientific studies of soils near you? Have the results of such studies been integrated into soil conservation programs or land use planning?

4. The next time you go grocery shopping, pay attention to the sources of the food you buy. Do you buy only produce grown in your immediate area? Investigate agricultural practices and regulations concerning the use of agrochemicals in the regions or countries where your groceries were grown.

CHAPTER 15

WATER RESOURCES

When the well's dry, we know the worth of water.

• Benjamin Franklin

*I*n 1990 a crack almost 1 km long and 3 m deep opened in the ground along a runway at Edwards Air Force Base in California. The fissure appears to have resulted from compaction and subsequent land subsidence associated with excessive withdrawal of groundwater. At the time, supplies of surface water had been seriously depleted by an extended period of drought (1987–1993), during which Californians turned to groundwater for irrigation, municipal, and industrial uses. In California nearly 18 million people—more than in any other state—rely on groundwater to meet their water needs, withdrawing a total of almost 60 billion liters per day. The increased pumping of groundwater during the drought depleted pore waters that maintained the separation of clay particles in the subsurface. When the water between them was withdrawn, the particles collapsed upon one another and became compacted. The result was land subsidence accompanied by fissuring around the edges of the compacted area (see Chapter 7).

Groundwater depletion is a serious issue in California, but the use of surface water has also caused problems. Since 1941 Los Angeles has diverted water from four of the streams feeding Mono Lake in the foothills of the Sierra Nevada. Mono Lake is an inland saline lake with an unusual, picturesque landscape and unique wildlife habitats. For many years the city of Los Angeles derived be-tween 10 and 20 percent of its water supply from the Mono Lake feeder streams via an extensive system of aqueducts. This diversion of streamflow caused a drop of more than 10 m in the level of the lake, endangering the lake ecosystem. In 1989, a court order temporarily prevented the city from diverting any more water from the Mono Lake feeder streams. Then, in 1993, the California legislature appropriated $60 million to purchase water rights to supply water to Los Angeles while ensuring continued protection of the lake's unique ecosystem and wildlife habitats.

California is an example of an area that is heavily settled yet lacks an abundant natural supply of water. The settlement and industrialization of Southern California have been made possible by extensive transfers of water from the Sierra Nevada and Colorado River drainage basins. In this region there is a sharply defined conflict between water as a commodity, needed to support expanding agricultural, municipal, and industrial needs, and water as a basic ecological necessity, needed to support wildlife habitats, replenish groundwater supplies, dilute and disperse pollutants, and maintain the ecological integrity of wetlands and other bodies of water. As we begin to recognize the ecological value of water left "instream," water users are increasingly being required to utilize water resources in the most efficient way possible. Conservation, rationing, and efficient water resource management are becoming a way of life for Californians, as they are for people in other parts of the world where water is scarce.

◄ ——————————————

Mono Lake, California. Drainage of glacial meltwater into a volcanic crater and removal of water by evaporation has left a very saline lake. The tufa spires are composed of soft, porous calcium carbonate deposited as a consequence of evaporation when the lake level was higher.

WATER FOR LIFE

Access to water is a vital human need. Most early cities and towns were founded close to streams that would serve as a reliable source of water. As the populations of these settlements grew, the amount of water supplied by the streams

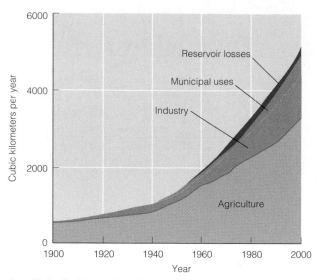

▲ **F I G U R E   15.1**
Estimated annual water use by sector, 1900-2000.

often became insufficient. People then resorted to bringing water from a more distant source through canals, or dug wells to obtain water from underground.

A reliable water supply is critical not only for survival and health, but also because of its role in industry, agriculture, and other economic activities. Urban development creates a spatially concentrated need for resources, especially water. Unfortunately, heavy population concentrations and abundant natural freshwater supplies are sometimes poorly matched. In water-rich, population-poor countries like Canada, New Zealand, and Iceland, many thousands of cubic meters of water are available per person per year; by contrast, in Egypt the amount of available fresh water may run as low as 20 cubic meters per person per year.

Fresh water is used mainly for agriculture, industry, and household consumption. Globally, crop irrigation accounts for 73 percent of the demand for water, industry for 21 percent, and domestic use for 6 percent, although these proportions vary from one region to another. Demand in each of these sectors has more than tripled since 1950 (Fig. 15.1). Population growth is partly responsible for the increasing demand, but improvements in standards of living have also contributed to the large increase in per capita water use over the past few decades. The result is that global water use is increasing at a rate that is much greater than the rate of growth of the world population. The total amount of water being withdrawn or diverted from rivers, lakes, and groundwater for human use is now about 4,340 cubic kilometers per year—eight times the annual streamflow of the Mississippi River!

As we begin to understand the complex interconnections between various Earth systems and their environmen-

tal functions, we also begin to appreciate the broader economic and ecologic functions performed by water, both on the surface and underground. We rely on natural waters to support wildlife habitats and fish populations, dilute and disperse pollutants, maintain soil moisture, generate electricity, transport goods, and provide recreational sites. Increasingly, water managers are being called upon—by laws or court rulings, in some cases—to allocate water specifically for the fulfilment of environmental needs.

Water resource management focuses on two major concerns: (1) how to preserve the *quality* of water resources and (2) how to ensure that an adequate *quantity* of water will be available to meet present and future needs. In Chapter 17 we will deal with the issues surrounding the quality of water resources and the problem of contamination of both surface and subsurface waters. In this chapter we concentrate primarily on the question of quantity.

Although water is, in principle, a renewable resource, the time it takes to replenish a particular water source can be much longer than a human lifetime. Balancing the rate of withdrawals and diversions of water with the rate of replenishment is therefore a central goal of sustainable water resource management. Because of the importance of subsurface water, both in terms of human use and in the hydrologic cycle as a whole, much of the discussion in this chapter focuses on groundwater.

WATER UNDER THE GROUND

Less than 1 percent of all water contained in the Earth is **groundwater,** which may be loosely defined as all subsurface water, that is, the water contained in spaces within bedrock and regolith. Although the percentage of groundwater sounds small, its volume is 40 times larger than that of all the water in freshwater lakes or flowing in streams, and nearly a third as large as that of the water contained in all the world's glaciers and polar ice. More than 98 percent of the unfrozen fresh water in the hydrologic cycle occurs as groundwater.

Most groundwater originates as rainfall, which soaks into the ground and becomes part of the groundwater reservoir in the hydrologic cycle (Fig. 1.11). The water then moves slowly toward the ocean, either directly through the ground or by flowing out onto the surface and joining streams. The fact that groundwater comes from rain was established in the seventeenth century, when Pierre Perrault, a French physicist, measured mean annual rainfall and mean annual stream runoff for part of the drainage basin of the Seine River in eastern France. After allowing for loss by evaporation, Perrault concluded that over a period of years the difference between the amounts of rainfall and runoff was sufficient to account for the amount of water in the ground.

Depth to Groundwater

Water is present everywhere beneath the Earth's surface, but more than half of all groundwater—including most of what is readily obtainable—occurs within about 750 m of the surface. The volume of water in this zone is estimated to be equivalent to a layer of water approximately 55 m deep spread over all of the world's land areas. At greater depths the amount of groundwater gradually, though irregularly, diminishes. Holes drilled in search of oil have found water as deep as 9.4 km, and one experimental hole drilled on the Kola Peninsula in Russia encountered water at more than 11 km. However, even though water may be present in crustal rocks at such depths, the pressure exerted by overlying rocks is so great and the openings in rocks are so small that it is unlikely that much water is present or that it can move freely through the enclosing rocks. Thus, for our purposes here we will think of groundwater as the water found between the land surface and a depth of about 750 m.

The Water Table

Much of what we know about the occurrence of groundwater has been learned from the accumulated experience of generations of people who have dug or drilled millions of wells. This experience tells us that a hole penetrating the ground ordinarily passes first through a layer of moist soil and then into a zone in which open spaces in regolith or bedrock are filled mainly with air. This is the **zone of aeration,** or *vadose zone* (Fig. 15.2). The zone of aeration is also called the *unsaturated zone,* for although water may be present, it does not completely saturate the ground. After passing throughout the vadose zone, the hole enters the **saturated zone,** in which all openings are filled with water. The saturated zone is sometimes called the *phreatic zone.* (Note

that some scientists prefer to define *groundwater* as water in the saturated zone, thus excluding subsurface water occurring in the zone of aeration.)

The upper surface of the saturated zone is the **water table.** Just as water is present at some depth everywhere beneath the land surface, so too is the water table always present. Normally the water table slopes toward the nearest stream or lake, where it intersects the ground surface. In deserts, however, the water table may lie far underground. In fine-grained sediment, a narrow fringe immediately above the water table is kept wet by capillary attraction, which pulls water from the top of the water table a small distance (usually less than 60 cm) into the zone of aeration. This is the *capillary fringe* shown in Fig. 15.2. (*Capillary attraction* is the adhesive force between a liquid and a solid that causes water to be drawn into small tubelike openings. It is the same force that draws ink through blotting paper and kerosene through the wick of a lamp.)

In humid regions the water table is a subdued imitation of the land surface above it (Fig. 15.2). It is high beneath hills and low beneath valleys because water tends to move toward low points in the topography, where the pressure is lowest. If all rainfall were to cease, the water table would slowly flatten and gradually approach the levels of the valleys. Seepage of water into the ground would diminish and then cease, and streams would dry up as the water table beneath the valleys fell. During droughts, when rain may not fall for several weeks or even months, the depression of the water table is evident from the drying up of wells. When a well becomes dry, we know that the water table has dropped to a level below the bottom of the well. Repeated rainfall, which soaks the ground with fresh supplies of water, maintains the water table at a normal level.

Whatever its depth, the water table represents the upper limit of readily usable groundwater. For this reason, a major

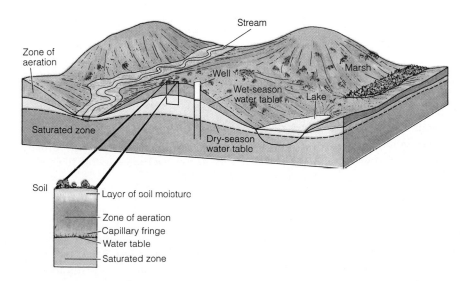

◄ **F I G U R E 15.2**
In a typical groundwater system the water table separates the zone of aeration from the saturated zone and its level fluctuates with seasonal changes in precipitation. Corresponding fluctuations in the water level are seen in wells that penetrate the water table. Lakes, marshes, and streams occur at points where the water table intersects the land surface. A layer of moisture coincides with the surface soil, and a thin capillary fringe lies immediately above the water table. In shape, the water table is a subdued imitation of the overlying land surface.

aim of groundwater specialists (*hydrologists* and *hydrogeologists*) is to determine the depth and the shape of the water table. To do this they must first understand how groundwater moves and what forces control its distribution underground.

HOW GROUNDWATER MOVES

Most of the groundwater within a few hundred meters of the land surface is in motion. Unlike the swift flow of rivers, however, which is measured in kilometers per hour, groundwater flows so slowly that its movement is measured in centimeters per day or meters per year. The reason for this contrast is simple. Whereas the water of a stream flows unimpeded through an open channel, groundwater must move through small, constricted passages, often along a tortuous route. Therefore, the rate of movement of groundwater is largely dependent on the nature of the rock or sediment through which the water moves, especially its porosity and permeability.

Porosity and Permeability

Porosity is the percentage of the total volume of a body of regolith or bedrock that consists of open spaces, or *pores*. Porosity determines the amount of fluid a given volume of sediment or rock can contain. The porosity of sediments is affected by the sizes and shapes of the rock particles as well as by the compactness of their arrangement (Fig. 15.3A and B). In sands and gravels with fairly uniform grain sizes, porosity may reach 20 percent, and some very porous clays have porosities as high as 70 percent. The porosity of a sedimentary rock is affected not only by the shapes, sizes, and arrangement of the particles but also by the extent to which the pores have been filled with cement (Fig. 15.3C). The porosity of igneous and metamorphic rocks, which consist of many closely interlocking crystals, is generally much lower than that of sediments and sedimentary rocks. However, if crystalline rocks have many joints and fractures, their porosity will be higher.

Permeability is a measure of how easily a solid allows fluids to pass through it. A rock with very low porosity is likely also to have low permeability. However, a high porosity does not necessarily mean a correspondingly high permeability, because the size and continuity of the pores (i.e., the extent to which they are interconnected) also influence the ability of fluids to flow through the material.

The relationship between pore size and the molecular attraction of rock surfaces plays a large part in determining the permeability of a rock. *Molecular attraction* is a binding force that causes a thin film of water to adhere to a solid surface despite the pull of gravity. If the open space between adjacent particles in a rock is small enough, the films of water that adhere to the particles will come into contact

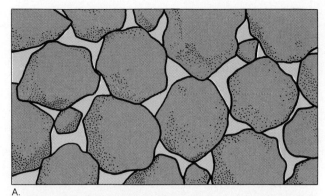
A.

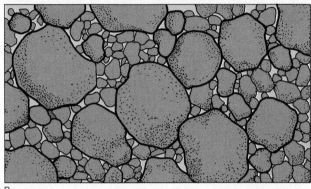
B.

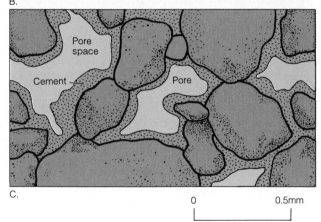
C.

0   0.5mm

▲ FIGURE 15.3
Porosity in different sediments. In these examples all of the pore spaces are filled with water, as they would be in a sample from the saturated zone. In contrast, in the zone of aeration at least some of the pore spaces would be filled by air. A. A porosity of 30 percent in a sediment with particles of uniform size. B. A porosity of 15 percent in a sediment in which fine grains fill spaces between larger grains. C. Reduction in porosity in an otherwise very porous sediment owing to cement that binds grains together.

with each other. The force of molecular attraction therefore extends right across the pore (Fig. 15.4). At ordinary pressures this water is held firmly in place, so permeability is low. In clay, for example, the particles have diameters of less

In very small spaces water is held
by molecular attraction.

Water can move through larger spaces,
although some is held.

▲ F I G U R E 15.4
**Effect of molecular attraction on permeability in the
intergranular spaces of a fine, silty sediment (top) and
a coarse, sandy sediment (bottom) of equal porosity.**

than 0.005 mm. Although clay may have a high porosity,
its permeability is typically low because the pores are very
small. In contrast, in a sediment with grains at least as large
as sand grains, which have diameters of 0.06 to 2 mm, the
pores commonly are wider than the combined thickness of
the films of water adhering to adjacent grains. The water in
the centers of the pores is therefore free to move (Fig. 15.4),
and the sediment is permeable. As the diameters of the
pores increase, permeability increases. Gravel, with very
large pores, is more permeable than sand and can yield
large volumes of water to wells.

Movement in the Zone of Aeration

Water from a rain shower soaks into the soil by **infiltration.** Soil usually contains clay created by the chemical
weathering of bedrock. The fine clay particles may make
the soil less permeable than underlying coarser regolith or
rock. The low permeability and molecular attraction between water and fine clay particles cause part of the water
to be retained in the soil. This is the layer of **soil moisture**
shown in Fig. 15.2. Some of the moisture evaporates directly into the air, but much of it is absorbed by the roots of
plants that later return it to the atmosphere through *transpiration* (see Fig. 1.11).

Because of the pull of gravity, water that cannot be held
in the soil by molecular attraction seeps downward until it
reaches the water table. More water enters the ground with
every rainfall. However, apart from soil moisture and the
capillary fringe, the zone of aeration is likely to be nearly
dry between rains.

Movement in the Saturated Zone

The movement of groundwater in the saturated zone—
termed **percolation**—is similar to the flow of water that

occurs when a saturated sponge is squeezed gently. Water
percolates slowly through very small pores along parallel,
threadlike paths. Movement through the central parts of
the spaces occurs easily. However, immediately adjacent to
the sides of each space the movement diminishes because
the water is held in place by molecular attraction.

Responding to gravity, water percolates from areas
where the water table is high toward areas where it is lower.
In other words, it generally percolates toward surface
streams or lakes (Fig. 15.5). Only part of the water moves
directly down the slope of the water table by the shortest
route. Much of it flows along curving paths that go deeper
within the ground. Some of these paths turn upward and
enter the stream or lake from beneath, seemingly defying
the force of gravity. The upward flow is possible because
water in the saturated zone is under greater pressure beneath a hill than beneath a stream or lake. Since water
tends to flow toward points where pressure is lower, it flows
from the higher pressure areas toward bodies of water at the
surface, where the pressure is lower.

Recharge and Discharge

Recharge, or replenishment of groundwater, occurs when
rainfall and snowmelt enter the ground in *recharge areas* or
recharge zones, places where precipitation infiltrates and
percolates downward through soil layers to reach the saturated zone (Fig. 15.5). The water continues to move slowly
along its flow path toward zones where **discharge** occurs,
that is, where subsurface water leaves the saturated zone. In
discharge zones or *discharge areas,* subsurface water either
flows out onto the ground surface as a *spring* or joins bodies
of water such as streams, lakes, ponds, swamps, or the
ocean. The pumping of groundwater from a well also creates a point of discharge.

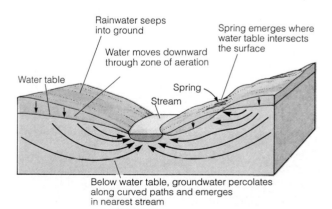

Rainwater seeps
into ground

Spring emerges where
water table intersects
the surface

Water moves downward
through zone of aeration

Water table

Spring

Stream

Below water table, groundwater percolates
along curved paths and emerges
in nearest stream

▲ F I G U R E 15.5
**Paths of groundwater flow in a humid region in uni-
formly permeable rock or sediment. The long, curved
arrows represent only a few of many possible paths.
Surface bodies of water are located wherever the
water table intersects the land surface.**

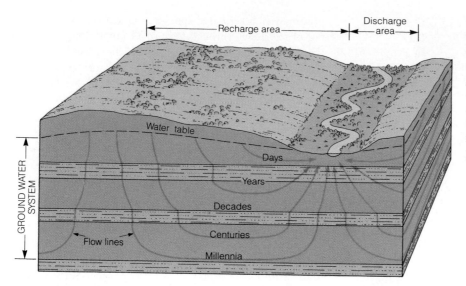

Distribution of recharge and discharge areas in a humid landscape. The time required for groundwater to reach the discharge area from the recharge area depends on the path and distance of travel. Percolation is faster and more direct in the most porous strata.

The surface extent of natural recharge areas is invariably larger than that of discharge areas. The time water takes to move through the ground from a recharge area to the nearest discharge area depends on distance and rates of flow. It may take as little as a few days, or as much as thousands of years (Fig. 15.6). In humid regions, recharge areas encompass nearly all of the landscape beyond streams and their adjacent floodplains (Fig. 15.6). In more arid regions, recharge occurs mainly in mountains and the sedimentary deposits that border them. In such regions recharge also occurs along the channels of major streams that are underlain by permeable sediment; the water leaks downward through the sediment and recharges the groundwater (Fig. 15.7).

SOURCES OF GROUNDWATER

People generally obtain groundwater either from natural springs or from wells reaching down to a body of water un-

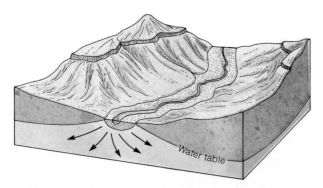

▲ F I G U R E 15.7
In arid regions direct recharge is minimal. During times of low flow, large through-flowing streams and intermittent streams lose water, which seeps downward to resupply groundwater in the saturated zone.

derground. Let's briefly consider the nature of each of these water sources.

Springs

A **spring** is a flow of groundwater that emerges naturally at the ground surface. The simplest kind of spring is one that issues from a place where the land surface intersects the water table (Fig. 15.5). Small springs are found in all kinds of rocks, but almost all large springs issue from limestone, gravel, or volcanic rocks. A vertical or horizontal change in permeability often gives rise to a spring or line of springs (Fig. 15.8). Such a change may be due to the presence of an **aquiclude,** a body of impermeable or distinctly less permeable rock adjacent to a permeable one. In Fig. 15.8A, for example, a porous limestone overlies an impermeable shale, and a spring discharges at the point of contact between the two types of rock with different permeabilities. If a porous sand overlies a relatively impermeable clay aquiclude, water percolating downward through the sand will flow laterally when it reaches the underlying clay and will emerge as a spring at the point where the stratigraphic boundary between the two units intersects the land surface, for example, along the side of a valley or a coastal cliff (Fig. 15.8B). Springs may also issue from lava flows, especially where a jointed lava bed overlies an aquiclude (Fig. 15.8C), or along the trace of a fault (Fig. 15.8D).

Wells

A well will supply water if it intersects the water table (Fig. 15.2). As shown in Fig. 15.9, a shallow well may become dry during periods when the water table is low, whereas a deeper well may yield water throughout the year. When water is pumped from a new well, the rate of withdrawal may initially exceed the rate of local groundwater flow. This imbalance creates a conical depression, called a **cone of depression,** in the water table immediately surrounding the

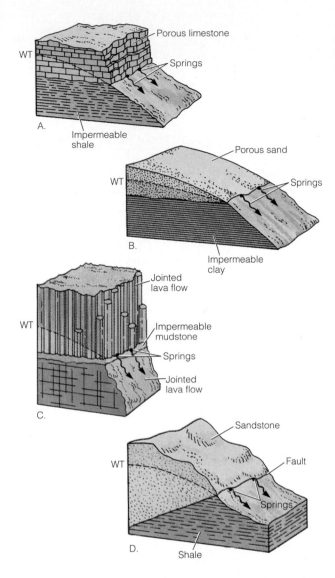

A.

B.

C.

D.

Examples of springs formed under different geologic conditions (WT is water table). A. A spring discharges water at the contact between a porous limestone and an underlying impermeable shale. B. Springs lie along the contact between a porous sandy unit and an underlying impermeable clay. C. Springs issue from the contact between a highly jointed lava flow and an underlying impermeable mudstone. D. Springs discharge along the trace of a fault where it intersects the land surface.

well (Fig. 15.9). The locally steepened slope of the water table increases the flow of water to the well, consistent with Darcy's law (see Box 15.1).

When the rate of groundwater inflow balances the rate at which water is withdrawn from the well, the hydraulic gradient stabilizes, but it will change if either the rate of pumping or the rate of recharge changes. In most small domestic wells the cone of depression is hardly discernible. Wells pumped for irrigation and industrial uses, however, withdraw so much water that the cone can become very wide and steep and can lower the water levels in surrounding wells. When large cones of depression from closely spaced pumped wells begin to overlap, regional depression of the water table may result.

If the source of a groundwater supply is inhomogeneous rock or sediment, the amount of water yielded by wells may vary considerably within short distances. For example, igneous and metamorphic rocks generally are not very permeable because the spaces between the mineral grains in these types of rock are extremely small and constricted. However, what is true for small samples does not necessar-

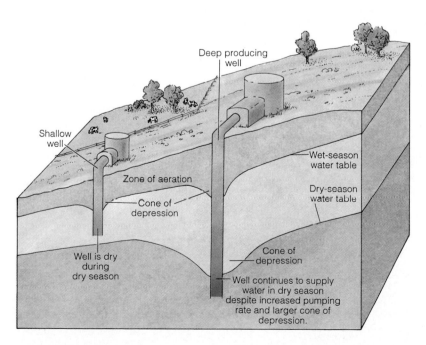

Effects of seasonal changes in precipitation on the position of the water table. During the wet season, the rate of recharge is high and the hydraulic gradient is relatively steep, so water is present both in a shallow well and in a deeper well upslope. During the dry season, the water table falls, the hydraulic gradient decreases, and the shallow well is dry. The deeper well continues to supply water, but increased pumping during the dry season enlarges the cone of depression.

BOX 15.1

•

FOCUS ON . . .

QUANTIFYING GROUNDWATER FLOW

*G*roundwater does not flow at the same rate everywhere. This can be demonstrated by injecting colored dye into a recharge area and measuring the time it takes for the dye to appear in nearby springs or wells. Experiments conducted with materials of uniform permeability have shown that the velocity of groundwater flow increases as the slope of the water table increases; in other words, the steeper the slope, the faster the flow. As shown in Fig. B1.1, the slope can be determined by measuring the altitude of two points (h_1 and h_2) on the water table, subtracting the lower figure from the higher one, and dividing this figure by the horizontal distance (*l*) between the points. The resulting slope value is essentially equivalent to the **hydraulic gradient.** Thus, the velocity of groundwater (*V*) is proportional to the hydraulic gradient:

$$V \propto \frac{h_1 - h_2}{l}$$

In 1856 Henri Darcy, a French engineer, concluded that the velocity of groundwater must be related not only to the slope of the water table (the hydraulic gradient) but also to the permeability of the rock or sediment through which the water is flowing. He proposed an equation in which permeability, together with the acceleration resulting from gravity and the viscosity of water, is expressed as a coefficient (*K*). This coefficient, referred to as the *coefficient of permeability,* is simply a measure of the ease with which water moves through a rock or sediment. The equation Darcy proposed can thus be expressed as follows:

$$V = \frac{K(h_1 - h_2)}{l}$$

In Chapter 8 we learned that discharge (*Q*) in streams varies as a function of both stream velocity (*V*) and cross-sectional area (*A*). The discharge of groundwater through a rock or sediment also depends on the velocity of flow and cross-sectional area of flow; in this case, however, the cross-sectional area is not that of an open channel but that of an interconnected system of pores. We can express discharge with the equation given in Chapter 8:

$$Q = AV$$

Combining this equation with the previous one, we arrive at a new equation:

$$Q = \frac{AK(h_1 - h_2)}{l}$$

This is known as *Darcy's law.* If we take the cross-sectional area (*A*) as constant for any given situation, by measuring any two of the remaining three variables [discharge (*Q*), coefficient of permeability (*K*), and hydraulic gradient ($(h_1 - h_2)/l$)], we can calculate the value of the third.

Because percolating groundwater encounters a large amount of frictional resistance, flow rates tend to be very slow, ranging from 0.5 m per day to several meters per year. The highest rate yet measured in the United States, in exceptionally permeable material, is only about 250 m/year.

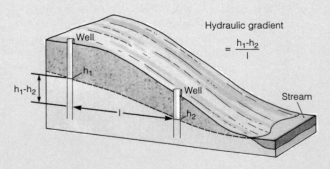

◀ F I G U R E B1.1
The hydraulic gradient is the slope of the water table. In the example illustrated, the hydraulic gradient is found by subtracting the altitude of h_2, the level of water in the downslope well, from h_1, the level of water in the upslope well, and dividing the difference by the distance (*l***) between the two wells.**

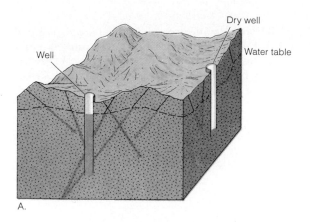

A.

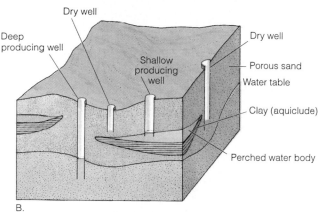

B.

◄ F I G U R E 15.10
The yield of water to wells drilled into inhomogeneous rocks can be highly variable. A. Wells that penetrate fractures in metamorphic and igneous rocks produce water. Dry wells result if no water-bearing fractures are encountered. B. Perched water bodies above the main water table are held up by aquicludes and provide shallow sources of groundwater. Wells that miss the perched water body and do not reach the deeper water table are dry.

from wells. An impermeable layer of rock or sediment in the zone of aeration can produce a *perched water body,* a small saturated region atop an aquiclude that lies above the main water table (Fig. 15.10B). The impermeable layer catches the water reaching it from above. A well drilled into this unit may produce water, whereas an adjacent well drilled to the same depth may be dry.

Aquifers

When we wish to find a reliable supply of groundwater, we search for an **aquifer** (Latin for *water carrier*), a body of highly permeable rock or regolith lying in the zone of saturation. Bodies of gravel and sand generally are good aquifers, for they tend to be highly permeable and often have large dimensions. Many sandstones are also good aquifers. However, the presence of a cementing agent between grains of sandstone may reduce the diameter of the openings and thus reduce the permeability of the rock.

Unconfined and Confined Aquifers

An aquifer whose upper surface coincides with the water table and therefore is in contact with the atmosphere (through the porosity of overlying soil layers) is an **unconfined aquifer** (Fig. 15.11). About 30 percent of the

ily apply to large masses of the same rock. Many massive bodies of igneous and metamorphic rock contain numerous fissures, joints, and other openings that permit free circulation of groundwater. A well that reaches such a conduit may produce water, whereas one that does not intersect any openings may be dry (Fig. 15.10A).

Discontinuous bodies of permeable and impermeable rock or sediment can also result in widely differing yields

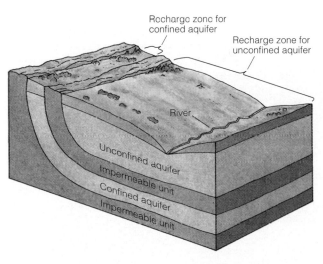

◄ F I G U R E 15.11
A confined aquifer is bounded by impermeable units (aquicludes). An unconfined aquifer is in contact with the atmosphere through the pore spaces of overlying permeable rocks. Note that the recharge area is much more restricted for the confined aquifer than for the unconfined aquifer.

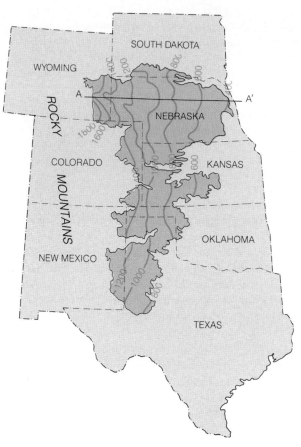

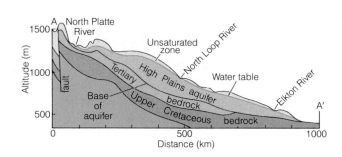

◀ F I G U R E 15.12
The High Plains Aquifer, an example of an unconfined aquifer. A. Regional extent of the aquifer and contours (in meters) on the water table. The flow of water is generally eastward, perpendicular to the contour lines. B. Cross section along profile A–A′ showing the slope of the water table and the relation of the High Plains Aquifer to underlying bedrock.

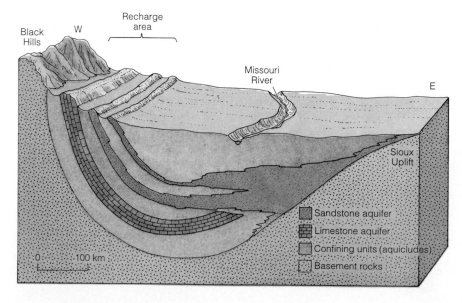

◀ F I G U R E 15.13
The Dakota Aquifer system in South Dakota, an example of a confined aquifer. The aquifer units consist of porous sandstones and a permeable limestone that reach the surface in the recharge area as linear ridges lying beyond the central core of the Black Hills. The intervening layers are the confining aquicludes.

groundwater used for irrigation in the United States is obtained from the High Plains Aquifer, an unconfined aquifer lying at shallow depths beneath the High Plains (Fig. 15.12). The aquifer, which averages about 65 m in thickness and is tapped by about 170,000 wells, consists of a number of sandy and gravelly rock formations. The water table at the top of this aquifer slopes gently from west to east, and water flows through the aquifer at an average rate of about 30 cm/day. Recharge comes directly from precipitation.

An aquifer that is bounded by aquicludes is a **confined aquifer** (Fig. 15.11), of which the Dakota Aquifer system in South Dakota is a good example. The Dakota Aquifer lies to the east of the Black Hills, an elongate dome of uplifted rocks around which permeable strata and bounding aquicludes form the land surface (Fig. 15.13). Rain falling on the permeable rock at the surface recharges the groundwater, which flows down the inclined rock layers toward the east.

Artesian Systems

Water that percolates into a confined aquifer flows downward under the pull of gravity. As it reaches greater depths, the water comes under increasing hydrostatic pressure. If a well is drilled down to the aquifer, the difference in pressure between the water table in the recharge area and the level of the well intake will cause water to rise in the well. Potentially, the water could rise to the same height as the water table in the recharge area. If the top of the well at ground level is lower than the potential height of the water table, the water may flow out of the well without having to be pumped. Such an aquifer is called an **artesian aquifer,** and the well is an *artesian well* (Fig. 15.14). Similarly, a freely flowing spring supplied by an artesian aquifer is an *artesian spring*. The term *artesian* comes from the name of a town in France, Artois (called Artesium by the Romans), where artesian flow was first studied. Under some conditions artesian water pressure can be great enough to create fountains rising as much as 60 m above ground level.

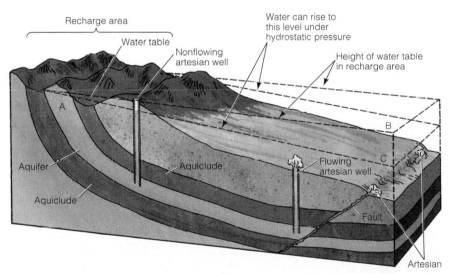

▲ **F I G U R E 15.14**
Two conditions are necessary for an artesian system to exist: a confined aquifer and sufficient water pressure to make the water in a well rise above the aquifer. Water in an artesian well may or may not flow from the well without having to be pumped; this depends on the specific conditions present. The water rises to the same height as the water table in the recharge area (line AB) minus an amount determined by the loss of energy caused by friction. Thus, the water can rise only to line AC, which slopes downward and away from the recharge area. In the nonflowing artesian well, the water rises to line AC but does not reach the ground surface. In the flowing artesian well downslope, water flows out at the surface without being pumped because the top of the well lies below line AC.

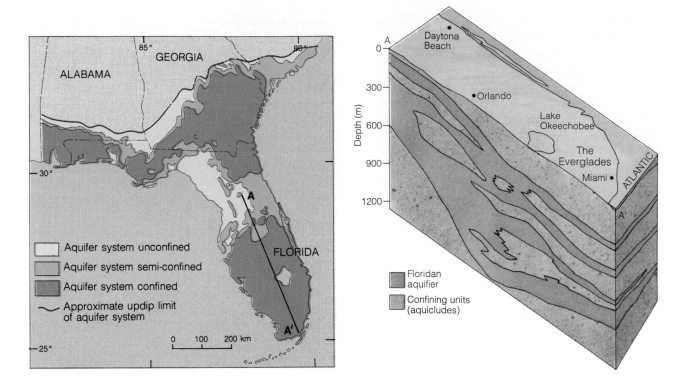

▲ F I G U R E 15.15
**The Floridan Aquifer system. A. Map showing the distribution of the system and areas
where it is unconfined, semiconfined (overlying confining unit is less than 30 m thick), and
confined (overlying confining unit is more than 30 m thick). B. Cross section along line A–A′
showing the relationship of the aquifer to overlying, underlying, and intervening
aquicludes.**

A vertical sequence of highly permeable carbonate rocks along the Florida peninsula creates a complex regional aquifer system in which both confined and unconfined units are present and water reaches the surface by artesian flow (Fig. 15.15). The aquifer system is restricted mainly to limestones, in the upper part of which are numerous caves and smaller openings. The permeable beds are interconnected, and their permeability is at least 10 times greater than that of impermeable layers above and below. The aquifer gives rise to many springs, whose concentration and discharge probably exceed those of any country in the world. Most of the major springs are artesian and occur where the overlying impermeable beds have been breached and the water pressure is high enough to allow water to rise above the land surface.

MINING GROUNDWATER

In dry regions where streams are few and average discharge is low, groundwater is a major source of water for human consumption. Sometimes the rate of withdrawal exceeds the rate of natural recharge; when this happens, the volume of stored water is steadily diminished. Just as petroleum is being withdrawn from the most accessible oil pools and minerals are being mined from the most accessible rocks of the upper crust, groundwater is being pumped from the best aquifers. We regard fossil fuels and minerals as nonrenewable resources, for they form only over geologically long time intervals, but we don't often realize that groundwater can also be a nonrenewable resource. Removal of groundwater from aquifers at a rate greater than the rate of natural recharge is sometimes referred to as *groundwater mining.*

Impacts of Excessive Groundwater Withdrawal

In some regions it would take so long for natural recharge to replenish a depleted aquifer that formerly vast underground water supplies have essentially been lost to future generations. Even where the problem has been recognized and measures have been taken to stem the loss, centuries or millennia of natural recharge may be required to return aquifers to their original state. Sometimes it is possible to recharge an aquifer by artificial means. In other cases, however, the results of excessive withdrawal—which include lowering of the water table, compaction, and subsidence—may be permanent.

Lowering of the Water Table

When the rate of groundwater withdrawal exceeds the rate of recharge, the water table falls. Lowering of the level of the water table can lead to the drying up of springs and streams if the water table no longer intersects the land surface. It can also cause shallow wells to run dry, necessitating the drilling of deeper wells. Under these conditions the cost of pumping groundwater increases.

For example, the natural rate of recharge of the High Plains Aquifer from precipitation is much lower than the rate at which water is being withdrawn. The inevitable result has been a significant drop in the level of the water table. In parts of Kansas, New Mexico, and Texas, the water table has dropped so much over the past half century that the thickness of the saturated zone has declined by more than 50 percent. The resulting decreases in water yield and increases in pumping costs have given rise to concerns about the future of irrigated farming on the High Plains.

Sometimes urban development can aggravate the depletion of groundwater, not only by increasing the demand for water but also by increasing the amount of impermeable ground cover in the area. When roads, parking lots, buildings, and sidewalks cover an aquifer's recharge area, the rate of recharge can be substantially reduced. This is especially true of confined aquifers, which often have limited recharge areas (as shown in Fig. 15.11).

On islands and in coastal zones, lowering of the water table and depletion of the saturated zone can result in a type of groundwater contamination called *saltwater intrusion,* in which saline water is drawn inland by the flow of groundwater in the direction of a pumped well. Eventually the presence of saline water may render the groundwater supply unusable.

Compaction and Subsidence

The water pressure in the pores of an aquifer helps support the weight of the overlying rocks or sediments, keeping the particles from collapsing upon one another. When groundwater is withdrawn, the pressure is reduced and the particles of the aquifer shift and settle slightly. This process, discussed in Chapter 7, is called *compaction.* When an aquifer compacts, it is permanently damaged and may never be able to hold as much water as it originally held.

Another result of compaction is that the land surface may subside. The amount of subsidence resulting from the withdrawal of groundwater depends on the extent of the reduction in water pressure and on the thickness and compressibility of the aquifer. Subsidence is widespread in the southwestern United States, where withdrawal of groundwater has caused disruptions of the land surface; structural damage to buildings, roads, and bridges; damage to buried cables, pipes, and drains; and an increase in the amount of land susceptible to flooding. Subsidence due to groundwater withdrawal is a major problem in many urban areas, such as Mexico City, Bangkok, and Venice, where the weight of buildings contributes to the compaction of compressible sediments.

Artificial Recharge

To halt compaction and the fall of the water table, groundwater can sometimes be artificially recharged. For example, artificial recharge can be accomplished by spraying nonhazardous biodegradable liquid wastes from a food-processing or sewage-treatment plant over the land surface. Pollutants in the water are removed by biologic processes as the liquid percolates downward through the soil, and the purified water recharges the groundwater system. Runoff from rainstorms in urban areas can be channeled and collected in basins, from which it will seep downward into permeable strata below the surface and raise the water table. In addition, groundwater that has been withdrawn for nonpolluting industrial use may be pumped back into the ground through injection wells, thereby recharging the saturated zone.

Allocation of Groundwater Rights

Law and policy relating to groundwater rights are very complicated. This is partly because water law, in general, has a complicated history, but it is also because groundwater is largely hidden from view. It is difficult to monitor the flow of groundwater and regulate its use. Some very complex issues are associated with the allocation of groundwater rights among potentially competing interests. For example, if you drill a well into an aquifer that underlies your property, are you entitled to withdraw as much water as you need from that well? Should you withdraw water only for your own purposes, or should you be permitted to withdraw the water for the purpose of selling it elsewhere? What happens if withdrawing the groundwater depletes the aquifer and your neighbor's well runs dry? Similar problems arise when a landowner's activities cause an aquifer to become contaminated; in some areas, groundwater contamination is not considered to be a problem *unless* the contaminant migrates underground and crosses a property boundary. In most parts of the world, groundwater use is unlimited and unregulated, which can lead to serious conflicts between adjacent landowners.

COMPETING FOR SURFACE WATER

People in many parts of the world derive their water supplies from surface freshwater bodies, which usually are more accessible than groundwater. Often water can be drawn from a nearby lake or stream to serve the local population. However, the demand for water varies substantially from one region to another. Sometimes, because of patterns of population growth and development, regions with the greatest demand for water do not have an abundant and readily available supply of surface water. For this reason, surface water is often transferred from one drainage basin to another, sometimes over long distances.

Interbasin Transfer

Interbasin transfer is the diversion of surface water from one drainage basin to serve users in a different drainage basin. Usually the transfer is accomplished by diverting water from a large stream via pipelines, reservoirs, canals, or aqueducts. The movement of water over long distances to service users is not a new phenomenon; extensive aqueduct systems were used to transport water thousands of years ago. Ancient aqueduct systems called *qanats* (Fig. 15.16) are believed to have been built more than 2000 years ago in Iraq (ancient Persia) and are among the earliest examples of large-scale transfer of water for human use.

New York City provides a modern example of an urban center that relies heavily on interbasin transfer of surface water for its water supply. New York boasts what is probably the world's most elaborate system of dams, aqueducts, and tunnels, which carries almost 6 billion liters of water per day to the city from the Catskill Mountains. The main components of the system are two huge tunnels buried deep beneath the city; they were constructed in 1917 and 1936. A third tunnel, still under construction (Fig. 15.17), will eventually be connected to the nearly 10,000 km of pipes that distribute water to the city's households and businesses.

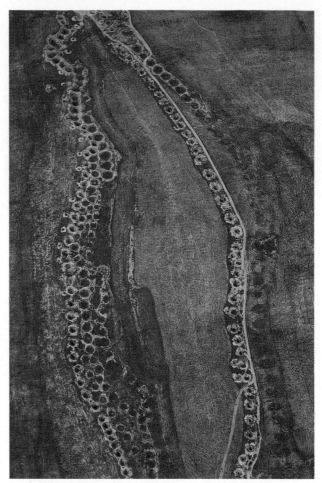

▲ **F I G U R E 15.16**
Aerial view of an ancient qanat system in southern Iraq that was built to carry water toward agricultural fields in a desert basin.

Environmental Impacts of Surface Water Diversions

Major surface water diversions often have unexpected and unwelcome environmental consequences. For example, over the past few decades the Amu Dar'ya and Syr Dar'ya rivers, which feed the Aral Sea on the border between Kazakhstan and Uzbekistan, have been diverted for irrigation purposes. Thirty years ago the Aral Sea was the world's fourth largest lake. It covered 68,000 km², had an average depth of 16 m, and yielded 45,000 tons of fish a year. As a result of the water diversions, the two feeder rivers were reduced to a trickle. The Aral Sea has shrunk so drastically that once-prosperous fishing villages are now 50 km from the shore (Fig. 15.18). The lake now covers only 40,000 km², with an average depth of only 9 m; its fishing industry is dead. Because the lake is shrinking, local rainfall is declining, the average temperature is rising, and wind velocities are increasing. Most of the newly exposed lake bottom is covered with salt, which is picked up by winds and depos-

ited on surrounding areas. Supplies of potable water have decreased, and various diseases, especially intestinal diseases, are afflicting the local population at alarming levels.

A proposed plan calls for the diversion of water from the Boro River, which feeds Botswana's Okavango Delta, a world-renowned wetland and wildlife habitat. The water would be used to increase irrigation, provide drinking water for nearby cities, and supply water for industrial use in the Orapa diamond mine. The plan would cause the wetland habitat of the Okavango Delta to shrink and therefore has been vociferously opposed by conservationists as well as by local herders and fishers, whose livelihoods depend on the ecological integrity of the wetland. As of 1992, the government of Botswana had put the diversion plan on hold.

Two major diversion schemes have been proposed for North America. One would involve damming James Bay in northern Québec and diverting water into the Great Lakes and farther south. The plan could create enormous environmental impacts, ranging from regional climatic change to the invasion of the Great Lakes by nonindigenous species. The marine fauna that now inhabit the saline bay would be affected, since the bay would be turned into a huge freshwater lake. Analogous problems have occurred in the Netherlands, where smaller diversions for the purpose of flood control have turned the Zuider Zee from a freshwater habitat into a saltwater one. The other North American proposal would divert water from rivers in northwestern Canada, including the Mackenzie and Yukon rivers, transforming the Rocky Mountain Trench into a huge reservoir and canal and eventually funneling the water to the semiarid southwestern United States. Because of the role of the Mackenzie River in controlling climate, regional climatic changes would be among the expected environmental impacts of such a massive diversion scheme.

Allocation of Surface Water Rights

Like groundwater law, the laws and policies governing surface water rights are complex. However, it is somewhat easier to monitor and regulate withdrawals and diversions of surface water, since the water source is in plain view. The two doctrines on which surface water law and policy are commonly based are the doctrines of *prior appropriation* and *riparian rights*.

Under the principle of **riparian rights,** which is based on English legal tradition, bodies of water are viewed as a *common property resource,* that is, a resource owned in common by all who have access to the resource. In the case of surface water, this would apply to anyone who owns property along the shores of the lake or river in question. In principle, each property owner would be granted equal access and equal rights to withdraw and use the water. Riparian rights generally include a requirement that the water

▲ FIGURE 15.17
Eighty-five meters below the surface in an access shaft being sunk to connect with a water tunnel 185 meters below Brooklyn.

▲ FIGURE 15.18
Fishing boats stranded in the desert by the shrinkage of the Aral Sea.

must be used for "natural" or "beneficial" purposes, but the precise meaning of these terms is not always clear.

The doctrine of **prior appropriation,** which is based on Spanish law and practice, grants water use rights on the basis of historical precedent. In other words, whoever was the first to draw water from a particular body of water takes precedence over latecomers. Once one has used a certain amount of water for so-called "beneficial purposes," one has established a claim to future rights to the same amount of water.

The Colorado River is an example of a river with a long and complex history of conflicts over water rights; indeed, the Colorado has been characterized as "the most fought-over river in the world." It flows through arid and semiarid lands and is an important source of water for seven states in the southwestern United States, as well as northwestern Mexico. Since 1922 the river water has been heavily allocated among the seven bordering states and Mexico by international treaties and United States laws based mainly on the doctrine of prior appropriation. In the 1960s the flow of Colorado River water into Mexico was so reduced by upstream diversions, and the salinity of the water was so high (three times the normal level), that the water could not be used for agriculture. An international incident ensued. The problem was eased temporarily when salty irrigation waters in the southwestern United States were diverted instead of being dumped into the river upstream. This allowed the river to return to its normal salinity. But because water rights were allocated at a time when flows in the river were much higher than now, future conflicts will arise if the upper states claim their full, legally designated allocations.

FACING UP TO WATER SHORTAGES

Hydrologists use the term *water-stressed* to designate a country or region with annual supplies of 1000 to 2000 m^3 of water per person. If the supply drops below 1000 m^3 (about 2700 liters per person per day), the country is said to be *water-scarce*—that is, lack of water places serious constraints on agricultural production, economic development, and environmental protection. Some countries in arid and semiarid climatic zones, such as Niger and Sudan, are not water-scarce, simply because their populations are still very small. Twenty-six countries, with a combined total population of 232 million people, are currently designated as water-scarce; many of them have rapidly increasing populations, foreshadowing ever-increasing water problems (Table 15.1).

Water-Conservation Measures

Mismatches between supply and demand can produce local and regional water shortages even in a country such as the United States, which does not fall into the water-scarce cat-

egory. Rates of water use in some regions far outstrip the rates at which resources are being recharged. In this final section we briefly examine some of the conservation measures that have been taken to bring rates of water use and withdrawal into line with the rates of replenishment of surface and groundwater resources.

Households

Household water conservation is taken very seriously in water-stressed parts of the United States. In some states and municipalities, for example, people can be fined for such infractions as overwatering their lawns. During the 6-year drought described at the beginning of the chapter, Californians became adept at devising household water-conservation measures. They resorted to "painting" their brown lawns green with biodegradable dyes. Landscaped yards with drought-resistant plant species (called *xeriscaping*) began to replace traditional grass lawns, which can require twice as much water. Household "gray" water—left over from bathing, toothbrushing, and dishwashing—was used to water houseplants and lawns. In southern California water reclaimed from sewage treatment facilities was, and remains, an important source of water for dust control, watering, and recharge of groundwater. Californians were also offered a substantial rebate, $100, on traditional toilets as an incentive to replace them with more water-efficient models.

Agriculture

Throughout the world the largest user of water is the agricultural sector, primarily for the purpose of irrigation (Fig. 15.19). Some of this water use is related to the cultivation of highly water-intensive crops in dry areas. For example, farmers in California have been criticized for growing rice in what is essentially a desert climate. Californian rice growers retaliate by claiming that the seasonal wetlands cre-

▲ FIGURE 15.19
Irrigation of alfalfa, Saskatchewan, Canada. The sprinklers rotate around a central supply, like the minute hand on a giant clock.

T A B L E 15.1 • **Water-Scarce Countries—Those with Annual Renewable per Capita Freshwater Supplies of Less Than 1000 cm³— 1992, with Projections for 2010**

Region/Country	Annual Renewable Water Supplies (cm³ per capita)		Change (percent)
	1992	*2010*	
Africa			
Algeria	730	500	−32
Botswana	710	420	−41
Burundi	620	360	−42
Cape Verde	500	290	−42
Djibouti	750	430	−43
Egypt	30	20	−33
Kenya	560	330	−41
Libya	160	100	−38
Mauritania	190	110	−42
Rwanda	820	440	−46
Tunisia	450	330	−27
Middle East			
Bahrain	0	0	0
Israel	330	250	−24
Jordan	190	110	−42
Kuwait	0	0	0
Qatar	40	30	−25
Saudi Arabia	140	70	−50
Syria	550	300	−45
United Arab Emirates	120	60	−50
Yemen	240	130	−46
Other			
Barbados	170	170	0
Belgium	840	870	+4
Hungary	580	570	−2
Malta	80	80	0
Netherlands	660	600	−9
Singapore	210	190	−10
Additional Countries by 2010			
Malawi	1030	600	−42
Sudan	1130	710	−37
Morocco	1150	830	−28
South Africa	1200	760	−37
Oman	1250	670	−46
Somalia	1390	830	−40
Lebanon	1410	980	−30
Niger	1690	930	−45

ated in their rice fields serve as important habitats for migrating waterfowl and facilitate the migration of salmon.

Globally, irrigation is not only the largest *user* but also the largest *consumer* of water. Water that is *consumed* is used and not returned to the cycle as wastewater. Most consumed water is simply lost through evaporation; some leaks from underground pipes. (In the United States, industry is more important than agriculture as a user of water, but not as a consumer, because a large proportion of the water used in industrial processes is eventually returned to the source as treated wastewater.) Irrigation is a particular problem because a large proportion of the water is lost through evaporation and transpiration.

Modern water-saving irrigation technologies have been pioneered in semiarid parts of the world, such as Israel and the southwestern United States. Drip systems, precision

BOX 15.2

•

THE HUMAN PERSPECTIVE

WATER CONTROVERSIES IN THE MIDDLE EAST

*U*rlama, King of Lagash, diverted water through canals to intercept Mesopotamian water supplies from Umma in 2450 B.C. Hammurabi wrote irrigation guidelines and water ownership provisions in Hammurabi's Code in 1790 B.C. Alexander the Great destroyed the Persian dams on the Tigris River in 350 B.C. These and many other examples document more than 4000 years of conflict over water rights in the Middle East.

Today, as in the past, issues involving water rights in the Middle East center on four rivers: the Euphrates (with a drainage basin covering 44,000 km²), Tigris (378,850 km²), Jordan (19,850 km²), and Orontes (13,300 km²) (Fig. B2.1). Through centuries of negotiations, treaties, discussions, multinational ventures, and wars, political boundaries and water rights have been superimposed on the natural courses and flows of these rivers and their tributaries. Each of the major rivers in the region now crosses at least one international boundary. The rights to these waters continue to generate conflict in the area. As Israeli professor of hydrology Uri Shamir concludes, "If you want reasons to fight, water will give you ample opportunities."

The water of the Jordan River, for example, is shared by four nations—Jordan, Syria, Israel, and Lebanon—that are in continuing conflict over the appropriation of river water and groundwater in the drainage basin of the Jordan. During the 1950s, despite strong protests by Syria, Israel built its north-to-south National Water Carrier. In retaliation,

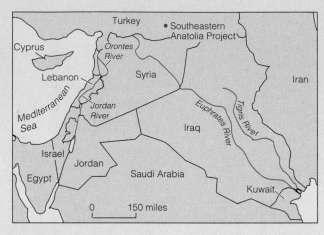

▲ F I G U R E B2.1
Political borders superimposed on natural drainage basins contribute to conflict in the Middle East.

Syria attempted to divert the headwaters of the Jordan River away from Israel. These conflicts ended in the Arab–Israeli War, the occupation of the West Bank, and Israeli control of the Jordan River headwaters. Today Israel derives two thirds of its fresh water from the occupied territories of the Jordan River Basin.

Similar conflicts exist over the Tigris River, which has some tributaries located between Iraq and Iran, and the Euphrates River, with much of its basin in Saudi Arabia

sprinklers, and other types of *microirrigation* systems are often costly to install, but they can substantially reduce the amount of water consumed for irrigation. In Texas, for example, the use of water-conserving irrigation methods by farmers has led to a decrease in the rate of depletion of the High Plains Aquifer from an average of nearly 2 billion m³ per year in the late 1960s to less than 250 million m³ per year. Such systems also can be more beneficial to crops than traditional methods. In Israel many farms are equipped with automated drip systems, which use computerized sensors to monitor the exact amount of moisture needed by the plants. In some systems the water is delivered directly to the plant's roots via a network of perforated piping installed at or just below the surface of the soil. The world leaders in

the use of microirrigation techniques (in terms of percentage of irrigated land under microirrigation) are Cyprus, Israel, Jordan, South Africa, and Australia.

Water supplies for irrigation can also be extended by recycling domestic wastewater. In Israel, 70 percent of domestic sewage is treated and reused to irrigate 19,000 hectares of agricultural land. There are plans to expand this recycling program by the end of the decade.

Industry

Industrial processes can be astonishingly water intensive. It can take as much as 700 kg of water to make 1 kg of paper, and the production of 1 ton of steel can require 280 tons of

▲ F I G U R E B2.2
Precious water from the Yarmuk River in Jordan irrigates a banana field that is 800 feet below sea level, in the Jordan River Valley. Irrigation by flooding, as in the photograph, is wasteful of water and is being replaced by more conservative drip–irrigation methods.

and headwaters in Turkey and Syria. Turkey has attempted to maintain hydrologic control of the Euphrates with the Atatürk Dam (Fig. B2.2), part of a huge water management plan called the Southeast Anatolia Development Project. The dam detains the water of the Euphrates upstream, severely reducing the flow into Syria and Iraq. Negotiations and conflicts among these three countries over the quality and quantity of their shared water resource have continued without a formal tripartite agreement for 3 decades.

As water use in the Middle East evolves from small-scale agrarian to large-scale irrigated farming and industrial uses, the controversies become more difficult to resolve. Limited water supplies are already putting constraints on development; levels of per capita water availability are very low, and the situation will worsen as populations increase. Erratic rainfall in the region contributes to problems of water management. For example, the Euphrates has a fairly reliable flow, but half of its annual recharge comes during a 2-month period in the spring, making it difficult to predict streamflow and water supplies.

No single solution will resolve the water conflicts and shortages in the Middle East. Considerable effort has been focused on the development of new water conservation technologies. Some economists have suggested creating a water reserve market, similar to a successful venture undertaken in California during the recent long drought. The idea behind such a market is that water can be "banked" during periods of excess or sold by countries with a surplus to countries in need. The extra cost of water is an incentive to recycle or to invest in water-conservation technologies such as desalinization plants and aqueducts. Politicians and international agencies have focused their efforts on the establishment of principles governing shared water rights and on negotiated agreements regulating access to available supplies. In 1991 the International Law Commission drafted 32 general principles for international water distribution; they include sharing of hydrologic data, equitable distribution, and riparian rights, that is, equal access for all those whose property borders on a given body of water.

water. But water use in industrial applications tends to differ from agricultural water use, because only a small amount of the water used in industry is actually consumed. Many industrial activities heat or pollute the water but do not actually cause it to evaporate or to leave the system; this means that industries can be quite efficient in recycling their wastewater. For example, through recycling, steelmakers have managed to reduce the average water intake from 280 to 14 tons per ton of steel produced.

In many industrialized countries, such as Japan, Germany, and the United States, industries are managing to cut their water use dramatically by employing a wide range of conservation measures. California's industries are world leaders in water efficiency—another legacy of the 6-year

drought. A comparison of several industries showed that water-conservation measures introduced during the drought not only resulted in dramatic reductions in water use but also were cost effective. Many of the conservation techniques employed were simple ones, such as fixing leaks, decreasing rates of flow, and treating and recycling wastewater.

Interestingly, much of the new efficiency in industrial water use has been an offshoot of compliance with pollution-control laws. To meet legislated water quality standards, many industries find it more economical to treat and recycle their water than to release it as effluents. Water quality legislation thus has had the dual impact of reducing pollution and leading to more efficient water use.

RESTORING THE FLORIDA EVERGLADES

*T*he Everglades is a unique cypress and mangrove wetland that is home to a number of endangered species, including the southern bald eagle, American crocodile, loggerhead turtle, and roseate spoonbill. It is part of an enormous freshwater drainage system that originally covered a large part of southern Florida. Everglades National Park was dedicated in 1947 in an effort to preserve this fragile ecosystem; at 1.5 million acres, it is the second largest national park in the lower 48 United States (Fig. B3.1). Nevertheless, encroaching development has left its mark. The disruption of southern Florida's natural drainage system for purposes of flood control, water supply, irrigation, and urban development has resulted in the loss of more than 50 percent of the original wetlands and left the rest vulnerable to salinization and other forms of degradation.

Efforts to manage the Everglades began in 1905, when the governor of Florida, Napoleon Bonaparte Broward, proposed to drain the entire area. Engineers from the Everglades Drainage District built four canals to carry fresh water from Lake Okeechobee to urban centers on the Atlantic Ocean (Fig. B3.2). This system was later enlarged by the Army Corps of Engineers and is still in use. After a major hurricane in 1947 and unusually severe floods in 1948, the Central and Southern Florida Project for Flood Control and Other Purposes was initiated. This massive federal project used levees, water storage areas, channel improvements, and large-scale pumps for drainage. All of the Everglades region north of the park was encompassed by this project, which grew in scope for more than 40 years. The ultimate effect was to open the area to urban and agricultural development.

For many years, water resources in southern Florida have been managed in ways designed to maximize crop production. Water for irrigation is imported from the north during dry periods and discharged to the south during rainy periods. However, phosphorus in drainage water discharged from farms has contributed to algal blooms and other problems in Lake Okeechobee and the Water Conservation Areas to the north of the park and in Florida Bay to the south. The decreasing flow of fresh water through the park has also caused the salinity of Florida Bay to increase dramatically. Tolerance to salinity is species specific; some species flourish in low salinity ranges, whereas others tolerate higher ranges. When the salinity of a body of water fluctuates too quickly, the result may be a loss of species diversity.

◄   F I G U R E   B3.1
Everglades National Park, Florida is home to a large number of plant, fish, and bird species, all of which depend on water conditions. The Park and its wildlife are endangered by human disruptions of the water flow.

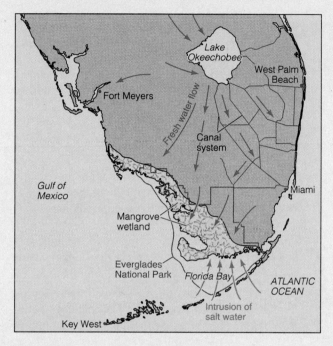

Today the management of the Florida Everglades is divided among several agencies, often with competing constituencies. Agencies representing the water management needs of agricultural and urban development are often pitted against agencies with a mandate to preserve and protect the fragile Everglades ecosystem. But sometimes the goals of conservation agencies themselves appear to be contradictory. For example, the mission of the National Park Service is to protect the integrity of the ecosystem as a whole, whereas the U.S. Fish and Wildlife Service is dedicated to the protection of individual species. Thus, the latter agency recently objected to a Park Service plan that would restore the flow of fresh water to parts of the park on the grounds that it would disrupt one of the main nesting grounds of an endangered species, the snail kite.

Although many conflicts remain to be resolved, the primary objective of water management officials in Florida seems to have changed over the years, from flood control to water supply to environmental restoration. In 1983 the governor of Florida initiated a Save Our Everglades program dedicated to the remediation of the state's water quality problems. Efforts are under way to improve water quality and provide more natural habitats for Everglades biota. Preventing fluctuations in salinity by managing discharges will be a critical step toward improving habitat in Florida Bay. In 1994 the Florida legislature adopted a plan to build large marshes at the border of the agricultural areas. These marshes are intended to filter pollution out of water seeping toward the Everglades. Urban runoff will be delivered to the marshes to recharge and maintain groundwater levels. In addition, the marshes will reduce the consumption of water by taking some agricultural lands out of production.

SUMMARY

1. A reliable water supply is critical not only to meet demands for water for household consumption but also because of the role of water in industry, agriculture, and other economic activities. Unfortunately, areas with the greatest demand do not always correspond with areas where water supplies are abundant; this mismatch creates problems and conflicts in regulating water use. Over the past few decades improvements in standards of living have led to a large increase in per capita water use, and as a result global water use is increasing at a rate much faster than the rate of population growth.

2. The total percentage of groundwater in the hydrologic cycle (less than 1 percent) sounds small, but its volume is 40 times larger than that of all the water in freshwater lakes or flowing in streams, and nearly a third as large as that of the water contained in all the world's glaciers and polar ice. More than 98 percent of the unfrozen fresh water in the hydrologic cycle occurs as groundwater.

3. Just below the Earth's surface there is an aerated zone (or vadose or unsaturated zone) in which some water exists in the form of soil moisture but at least some of the pore spaces are filled with air. Below that is the saturated (or phreatic) zone, in which all pore spaces between the grains in sediment or rock are filled with water. The top of the saturated zone is the water table, whose surface tends to be a subdued imitation of the land surface above it.

4. Groundwater moves much more slowly than surface water. Its movement is controlled to a great extent by the porosity (percentage of pore space) and permeability (size and interconnectedness of pore spaces) of the material through which it is flowing. In the zone of aeration, groundwater infiltrates and moves downward under the influence of gravity, until it joins the saturated zone. In the saturated zone, groundwater percolates along a curved path, flowing toward regions where pressure is lower. Thus, it may move downward under the influence of gravity or upward toward a stream or lake where the pressure created by overlying material is low.

5. In zones of recharge, groundwater is replenished by precipitation. In zones of discharge, water flows out from beneath the surface into springs or wells or joins bodies of water at the surface such as streams, rivers, lakes, marshes, or oceans.

6. Groundwater comes from aquifers, bodies of permeable rock or regolith lying within the zone of saturation. Unconfined aquifers are in contact with the atmosphere, whereas confined aquifers are bounded at the top and bottom by impermeable layers (aquicludes). An artesian aquifer is one in which the poten-

tial level of the water table is above the level of the confined aquifer; if a well is drilled into the aquifer, water will rise to this potential level, sometimes flowing or gushing out onto the surface under its own pressure.

7. When groundwater is withdrawn from a pumped well, a cone of depression is created in the surrounding water table. If water continues to be withdrawn at a faster rate than the rate of recharge, the results may include thinning of the saturated zone, regional lowering of the water table, compaction of aquifer sediments, and subsidence of the ground surface. Sometimes these changes can be reversed by artificially recharging the aquifer, but sometimes they are permanent.

8. The use of surface water often involves interbasin transfer, the transport of water from one drainage basin to another. Diversion of surface water has a variety of environmental impacts, including changes in rates of flow and siltation, drying up of wetlands and loss of habitat, and even changes in local climatic conditions.

9. Water law has a long and complicated history. The allocation of surface water rights is often based on either the doctrine of prior appropriation, which gives priority to users according to the order in which they began using the resource, or riparian rights, which grant equal access to anyone whose property borders on a given water body. Groundwater rights, where specified, are typically based on similar principles, but regulation and monitoring are complicated by the fact that groundwater sources are largely hidden from view.

10. Twenty-six countries are now classified as water-scarce; that is, lack of fresh water places serious constraints on agricultural production, economic development, and environmental protection. Because of continuing population growth, several other countries are expected to join this list within the next decade.

11. Even in countries where water is not scarce, high levels of demand and irregularities in supply create shortages in some regions. Water conservation measures undertaken in households include reuse of "gray" water; limitations on lawn watering; adoption of water-saving technologies such as more efficient toilets; and landscaping with drought-resistant plants. Industries can be very efficient at reducing water use, mainly through recycling and treatment of wastewater and adoption of water-saving technologies. Throughout the world agriculture is both the largest user and the largest consumer of water, primarily for the purpose of irrigation. Microirrigation techniques—water-saving devices that deliver precisely measured volumes of water directly to the roots of plants—can significantly reduce water use in the agricultural sector.

IMPORTANT TERMS TO REMEMBER

aquiclude (p. 382)

aquifer (p. 385)

artesian aquifer (p. 387)

cone of depression (p. 382)

confined aquifer (p. 387)

discharge (p. 381)

groundwater (p. 378)

hydraulic gradient (p. 384)

infiltration (p. 381)

percolation (p. 381)

permeability (p. 380)

porosity (p. 380)

prior appropriation (doctrine of) (p. 392)

recharge (p. 381)

riparian rights (p. 391)

saturated zone (p. 379)

soil moisture (p. 381)

spring (p. 382)

unconfined aquifer (p. 385)

water table (p. 379)

zone of aeration (p. 379)

QUESTIONS AND ACTIVITIES

1. Find out whether the water laws in your region are based primarily on the doctrine of prior appropriation or on that of riparian rights. What is the history of water use policy in your area?

2. Where does the water that you use in your household originate? Does it come from a groundwater source or from a surface water source? How far must the water be transported before it reaches your home?

3. Do you live in a water-stressed area? Are there strict regulations on water use (e.g., specific times when lawn watering or car washing are limited)? Are there limitations on agricultural or industrial use? Are any water conservation programs in effect? Is there any provision in your area for reusing treated sewage (e.g., for irrigation of municipal landscaping or for dust control)?

4. Select and investigate an area with an interesting history of water use. Possibilities include the High Plains Aquifer, Mexico City, Bangkok, the Colorado River, the Colombia River, the Aral Sea, the Middle East, New York City, and Los Angeles. Is water use in the area dominated by groundwater withdrawal or by diversion and interbasin transfer of surface water? What problems or conflicts are associated with water use policy in the area?

5. Investigate one of the proposed schemes for diversion of surface water in North America, such as the plan to dam James Bay in northern Québec. Who would benefit from the plan? Who might suffer negative impacts? Are there other goals besides the transfer of fresh water? What are some of the potential environmental impacts of such a plan?

★ E • S • S • A • Y ★

MANAGING WASTES

It has been said that pollution is "resources out of place." In some respects this is a pretty good definition; it emphasizes the fact that *all* Earth materials, even wastes and unwanted byproducts, have potential value as resources. However, although it is true that pollution is sometimes caused by potentially useful materials accumulating in inappropriate places, this definition isn't entirely correct. The term **pollution** usually refers to materials with harmful impacts on the natural environment (or the act of releasing those materials).

It may be more accurate to say that **waste** is resources out of place. Waste consists of the residual materials and byproducts that are generated by human use of Earth resources and wind up unwanted and unused. According to this definition, only some kinds of wastes are **pollutants** or **contaminants**—materials that have harmful impacts and degrade the environment. Other kinds of wastes are not necessarily destined to become pollutants; they are merely "leftovers" that haven't yet found their way to an appropriate user.

*R*ETHINKING THE CONCEPT OF WASTE

Used tires are a classic example of waste as a potential resource. There are *millions* of discarded tires in North America. Tires are made from petroleum, and it is possible to re-cycle this petroleum for use as a fuel. Used tires can also be retreaded or utilized (either whole or in ground-up form) as components of highway barriers, landscape borders, asphalt paving, and carpet backings. A few places recycle tires today, but more often used tires are piled up in junkyards, becoming an unwanted waste and a resource that is definitely out of place. Piles of used tires are highly susceptible to fires, which can rage out of control for weeks or even months, causing billowing clouds of black smoke and serious contamination of groundwater in surrounding areas (Fig. IV.1).

A vast pile of used tires near Westley, California. The 6 million tires are burned as fuel in a power plant that provides electricity for 3500 homes.

▲ FIGURE IV.1
A pile of used tires catches fire and burns out of control, Hagersville, Ontario, 1990.

401

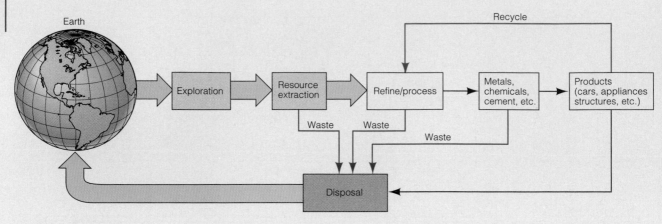

▲ FIGURE IV.2
The cycling of resources through the Earth system, and the generation of wastes.

Today humans are generating wastes of all kinds at an unprecedented rate. We must learn to deal with harmful, contaminating, polluting wastes as well as with the unwanted but potentially useful materials that for one reason or another have accumulated in the wrong place in the resource cycle (Fig. IV.2). The *resource cycle* is a complicated, interconnected set of processes through which useful materials are transferred from the environment to the human user and then back into the environment. The cycle as it now operates is "leaky" and inefficient. Materials often turn up where they weren't intended to be; or they fail to reenter the cycle and languish somewhere as wastes; or they have negative environmental side effects that were unanticipated or underestimated.

In order to reduce the production of wastes and decrease their harmful, polluting impacts, we have to close the resource cycle and stop its leaks. The ideal solution would be to mimic nature by becoming highly efficient users, reusers, and recyclers of Earth materials. We can do this only by learning to think of *all* wastes as potential resources, and, more importantly, by learning to guide these materials to an appropriate setting so that they can be utilized as resources.

GENERATING WASTE

Industrial economies are based on the extraction, transformation, and consumption of raw materials. This causes problems at the extraction end of the resource cycle because of the disruption of land; the generation of wastes, effluents, and emissions; and other environmental impacts. At the other end of the cycle, the materials that feed a consumption-based society are eventually discarded, and this ever-growing output of waste also poses a serious problem.

Production of Raw Materials

The extraction and processing of mineral and energy resources creates more waste than does the actual use of these resources. For example, to extract the materials and energy needed to construct a typical building requires a hole the size of the building; about half of what comes out of the hole is waste (Fig. IV.3). Much of the waste associated with mining is not necessarily pollution, just piles of waste rock. When a pile of rock is mixed with rainwater, however, many harmful contaminants can be dissolved and carried away from it. The processing, manufacture, and transportation of products and raw materials also generate wastes and harmful emissions.

Over the last century, per capita consumption of raw materials increased dramatically throughout the world, and it is still increasing. It has been estimated that from 1940 to 1976 the United States alone consumed more minerals than did all of humanity up to 1940. Since the 1970s, however, the need for raw materials has begun to level off somewhat in most industrialized nations. This has occurred partly because of a shift toward industries, such as pharmaceuticals and computer technologies, that use fewer raw materials and less energy than traditional industries. This trend will continue to gain momentum if industrial recycling programs are improved and greater attention is paid to efficient use of materials at all stages of the cycle (Fig. III.4).

An innovative approach to reducing both wastes and raw material inputs in industrial usage is *waste exchange* programs, in which the waste materials of one industrial process become the raw materials for another. Such a program has been established in southern Ontario, where the Ontario Waste Management Corporation essentially acts as a waste "broker," matching the resource needs of one company with the waste disposal needs of others.

◄ F I G U R E IV.3
Waste rock removed in order to get at mineralized ore, Bingham Canyon, Utah. In many mining operations 3 or 4 tons of waste rock must be removed for every ton of ore mined.

The Throwaway Society

The output of waste by consumers is actually less of a problem, in terms of both volume and associated hazards, than the waste generated by the extraction and processing of resources and the manufacture of products. However, it is a more visible problem and one that is particularly challenging for many communities. The amount of solid waste generated per person has been rising sharply in most industrialized countries and doesn't show any signs of leveling (Table IV.1). Unfortunately, it is hard to imagine any sizable municipality without a solid-waste management problem or controversy over landfill location.

T A B L E IV.1 • Change in Municipal Solid-Waste Generation in Selected Countries over a 5-Year Period in the 1980s

Country	Total (% change)	Per Person (% change)
Ireland	+72	+65
Spain	+32	+28
Canada	+27	+21
Norway	+16	+14
United Kingdom	+12	+11
Switzerland	+12	+9
Denmark	+6	+6
Sweden	+6	+5
France	+7	+5
Italy	+7	+4
Portugal	+13	+4
United States	+8	+3
Austria	+3	+3
Luxembourg	+2	+2
Japan	0	−3
West Germany	−10	−9

Source: From Young, John E., "Reducing Waste, Saving Materials" in *State of the World 1991,* Worldwatch Institute & W. W. Norton, 1991, NY.

THE NATURE OF WASTE

Because wastes come from so many different sources and take so many forms, it is virtually impossible to come up with a classification scheme that fits all kinds of waste and serves all waste management purposes. Sometimes it is useful to classify wastes according to their *source:* industrial waste, municipal or household waste, agricultural waste, and so on. This approach is useful for regulatory purposes or for setting limits on the amounts of waste that may be generated by different sectors of society. In other cases it is more useful to classify wastes according to their *physical form:* solid, liquid, or gaseous. Such an approach is helpful because it tells us something about how wastes can be transported for disposal and how they move about in the environment: gaseous wastes are transported in the atmosphere; liquid wastes come out of pipelines and are transported by streams; and so on.

In still other cases it is more useful to classify wastes according to their *special properties.* For example, ordinary municipal solid waste requires an entirely different treatment approach from that required by radioactive waste, toxic waste, or medical waste. Usually the special properties of wastes—the properties that determine what kinds of treatment are necessary—have a lot to do with their chemical characteristics. These characteristics, in turn, determine how the wastes behave in the natural environment and how they respond to different treatment and management procedures. For example, petroleum behaves very differently in the environment from dry-cleaning fluid (trichloroethylene, or TCE, one of the most troublesome of the toxic chemicals commonly involved in groundwater contamination). Petroleum floats on top of groundwater at the level of the water table, with little intermixing, whereas TCE, which is denser, sinks and mixes with the groundwater. Both of these chemicals are hazardous liquid wastes; they come from similar sources and occur in similar settings as contaminants (usually in decommissioned industrial sites or poorly designed hazardous-waste disposal sites). Yet because of their different chemical characteristics they behave very differently in the environment, cause entirely different problems, and require different treatment approaches (which are discussed in some detail in Chapter 17).

APPROACHES TO WASTE MANAGEMENT

In countries with high population densities, such as Japan, the problems of waste, pollution, and lack of space for waste disposal were of necessity faced much earlier than in North America. As a result, Japan and many European countries have much lower rates of per capita waste genera-

tion, as well as much higher rates of incineration and recycling, than do Canada and the United States.

The Three "R"s

But which management approaches are really best? Is burning garbage the best answer to waste disposal problems? Is recycling the solution? How can we stop the flow of hazardous industrial wastes into waterways? And—the bottom line—can we ever stop generating so much waste?

The United Nations Environment Programme, along with many governments throughout the world, endorses an integrated, hierarchical approach to waste management that begins with **source reduction,** that is, cutting back on waste generation in the first place. Methods of source reduction include reducing the volume of products, increasing the useful lifetime of products, reducing the amount of packaging, and decreasing consumption. The waste management hierarchy continues, in order of decreasing desirability, with direct **reuse** of products (using a product repeatedly instead of throwing it away), **recycling** (taking apart an old product and using the material it contains to make a new product), incineration, and—the last resort—landfilling. Some people like to add a fourth "R"—*refuse*—suggesting consumers have the power to refuse to purchase items with excessive or environmentally harmful packaging.

A new approach to responsible waste management and informed consumerism is *cradle-to-grave product analysis,* also called *lifecycle analysis.* This approach takes into consideration the Earth resources utilized over the entire lifetime of a product. Cradle-to-grave analysis has both industrial and consumer applications. For example, a few years ago disposable diapers became unpopular among environmentalists because they take up space in landfills. However, cradle-to-grave analyses have shown that, by comparison, cotton cloth diapers generate much more air pollution, water pollution, and sewage sludge over their useful lifetimes than do disposable diapers for a comparable number of uses. Moreover, as an agricultural product, cotton requires heavy use of chemical fertilizers and pesticides. It has also been shown that, contrary to popular belief, disposable diapers take up less than 1 percent of the space in the average landfill. On the other hand, disposable diapers are made from plastic, which is not biodegradable, and wood pulp, which uses up trees. They may also contribute pathogenic materials such as harmful bacteria to landfills, where they are not always properly treated. Thus, in the cloth versus disposable diaper controversy some environmentalists have decided to call it a draw.

Sociological Aspects of Waste Disposal

As challenging as the technical aspects of waste disposal are, the sociological barriers may be even more difficult to overcome. We have a longstanding cultural tradition of shying away from garbage and anything having to do with it. In part, this tradition is based on common sense built from many years of learning that wastes can be hazardous to health. Historical patterns of health and disease, particularly in urban areas, have revealed the links between disease transmission and have shaped our attitudes toward waste disposal. Another source of our unwillingness to deal with wastes is knowledge about the effects of exposure to deadly chemicals. Incidents such as those described in Table IV.2 form part of our collective consciousness and contribute to our fear of wastes and hazardous substances.

Challenging the "NIMBY" Attitude

As a result of these historical developments, we have evolved a societal attitude toward waste disposal that is best summarized by the acronym **NIMBY,** which stands for

"not in my back yard." No one wants to live next door to a waste disposal facility or even to acknowledge the need for such a facility.

Or do they? Some regions and communities have solved their waste problems by challenging prevailing attitudes toward waste. Large, meticulously engineered waste disposal facilities on carefully selected sites can safely and effectively solve regional garbage problems for decades while providing employment for nearby residents. In some cases governments have turned the tables: instead of forcing such a facility on a community, they offer tax incentives and other benefits to communities that are willing to provide waste disposal sites.

The Traditions of Reuse and Recycling

Efficient reuse and recycling of products is a way of life in many parts of the world (Fig. IV.4). In Cairo, for example,

T A B L E IV.2 • Major Toxic Pollution Incidents in the Past Five Decades

1956	Widespread mercury poisoning was discovered in the Japanese fishing village of Minamata.
1960s	The drug thalidomide was withdrawn from the market in several countries after its use by pregnant women was shown to be instrumental in causing serious birth deformities.
1968	A mass poisoning incident took place in Yusho, Japan, after cooking oil was contaminated with polychlorinated biphenyls (PCBs) and other toxic chemicals.
1970s	A mysterious disease began killing horses at three riding stables in the area of Times Beach, Missouri. Before the epidemic was over, more than 60 horses had died and several people had become ill. The cause of the epidemic was traced to a tank of contaminated industrial wastes.
1976	An explosion at a chemical factory showered the town of Seveso, Italy, with toxic chemicals, killing thousands of birds and small animals.
1978	A health emergency was declared near Niagara Falls, New York, when it was discovered that highly toxic chemical wastes from the nearby Love Canal disposal site were seeping into residential backyards.
1979	Cooking oil contaminated with PCBs and other toxic chemicals was once again the cause of mass food poisoning, this time in Yu-Cheng, Taiwan.
1979	Mercury discharges from a pulp and paper plant forced the closing of the commercial and recreational fisheries on the Wabigoon–English River system in northern Ontario, causing considerable damage to the economic and social stability of local Native Canadian bands.
1979	Chlorine leaking from damaged tank cars forced the evacuation of 200,000 people after a train carrying toxic chemicals derailed in Mississauga, Ontario.
1979	An accidental release at the *Ixtoc I* offshore oil platform in the Gulf of Mexico was the largest unintentional oil spill ever.
1986	Meltdown of the core of the nuclear reactor at Chernobyl, in the former Soviet Union, released large quantities of radioactive material, causing deaths and as yet unknown health impacts in the immediate area. Some of the radioactive material was spread globally by atmospheric transport.
1984	An accidental discharge of deadly methyl isocyanate from a chemical plant in Bhopal, India, left 4000 people dead and thousands of others with lingering and incapacitating illnesses.
1989	The *Exxon Valdez* oil tanker ran aground and released approximately 10 million gallons of crude oil into Prince William Sound, the largest oil spill ever in North American waters.
1991	More than 600 burning oil wells and tanker spills in Kuwait and the Persian Gulf made the Persian Gulf War the largest oil spill ever and probably the greatest intentional environmental disaster of all time.

Based on Box 21.1 from *The State of Canada's Environment, 1991,* Minister of Supply & Services, Canada. Major Toxic Chemical Incidents.

◀ F I G U R E IV.4
Informal waste management. Picking over garbage at the main municipal trash heap, Cairo, Egypt.

the *Zabbaleen* or "garbage sifters" earn a living by collecting, sorting, and recycling trash. The Zabbaleen live and work in settlements on the outskirts of Cairo and sell recycled materials to small businesses throughout the city. The environmental conditions in which these people live and work are terribly unhygienic, but the Zabbaleen play a crucial role in solid waste management in Cairo. This type of informal waste management and recycling of materials is characteristic of many cities in less developed parts of the world. Informal waste management can be very important in such areas because they lack the money to construct a "modern" waste management facility; materials and products also tend to be in very short supply, and the income provided by recycling provides a livelihood for some of the poorest residents.

Efficient reuse and recycling were also characteristic of industrialized nations before the advent of convenience products and a consumption-oriented, throwaway lifestyle. Recently, however, new attitudes toward waste disposal have produced changes in the way industrialized societies view the materials and resources contained in both garbage and industrial waste. Our new understanding of waste as "resources out of place" has led to the establishment of resource recovery programs. The term **resource recovery** refers to any means of extracting economically valuable components, whether materials or energy, from wastes. Landfill operations can be subsidized by selling as fuel the methane gas generated by decomposing garbage (Fig. IV.5). Small businesses can profit from the "harvesting" of resources contained in trash. Large corporations can offer

▲ FIGURE IV.5
Power generation from a municipal landfill, Enderby, Leicestershire, England. A network of pipes, one of which is seen here, extracts methane gas from decomposing refuse. The gas is used to run a turbine generator.

substantial financial rewards to employees who suggest less wasteful ways of carrying out everyday procedures.

As long as we use Earth resources, we will never be able to avoid generating some waste. Solutions to waste management problems are likely to be arrived at in stepwise fashion through a web of incentives—in the form of information, education, taxes, fines, pricing policies, laws, and regulations—designed to encourage producers and consumers to close the resource cycle by using both materials and energy more efficiently.

IMPORTANT TERMS TO REMEMBER

contaminant (p. 401)
NIMBY (p. 405)
pollutant (p. 401)
pollution (p. 401)
recycling (p. 404)

resource recovery (p. 406)
reuse (p. 404)
source reduction (p. 404)
waste (p. 401)

WASTE DISPOSAL

Garbage isn't generic gunk; it's specific elements
of our behavior all thrown together.

• William Rathje

*J*ust over 20 km from Manhattan in New York City lies an enormous monument to the throwaway lifestyle. Fresh Kills (from the Dutch word *kil,* meaning "stream") is the world's largest landfill and one of the largest structures made by humans. It covers more than 3000 acres of Staten Island and stands more than 150 m high. About half of New York City's garbage—more than 100,000 tons each week—winds up at Fresh Kills. The pile of trash housed in this landfill amounts to more than 25 times the volume of the Great Pyramid at Giza.

The site for the landfill was chosen in 1948. It would not be considered geologically appropriate for a modern landfill. The facility is located on what is essentially a salt marsh, with several small streams running through it (between the hills of garbage), and a water table that is very close to the surface. The facility itself was designed without the benefit of pollution control technologies (such as liners, drainage systems, and monitoring stations) that are routinely installed in modern landfills.

The high water table under the landfill is problematic, especially in the vicinity of the small streams. The garbage soaks up water, creating a very wet environment within the landfill. The moisture enhances the growth of *anaerobic* microorganisms (microorganisms that thrive in nonoxygenated environments), which in turn promotes the biodegradation of organic materials in these sections of the landfill. However, it is extremely difficult to determine the chemical composition of the fluids resulting from the decomposition process, and to monitor and control their flow. Before a recent cleanup project was undertaken, it is estimated that more than a million gallons (almost 4 million liters) of contaminated fluids were seeping from the landfill into nearby water bodies each day.

Fresh Kills is destined to be filled to capacity by about the year 2000, at which point it will be almost 200 m high. New York City residents will then have to find somewhere else to put their garbage, and try to figure out what to do with the small mountain of urban waste on Staten Island.

DISPOSING OF WASTE

Virtually every human activity generates waste. Even the simple act of eating results in one of the most challenging waste problems of modern society: sewage. And this doesn't even take into account the wastes generated in the production and transportation of the food that was eaten. Waste disposal is a problem that increasingly demands the attention of scientists, engineers, policy makers, and the general public. This is partly because the volumes of waste are increasing at an alarming rate, and partly because our understanding and appreciation of the hazards associated with improperly handled wastes are growing.

The Geology of Waste Management

An understanding of geology is fundamental to the management of waste for three reasons: (1) Wastes are generated as a consequence of the use of Earth materials; it is important to understand the properties of these materials, both individually and in combinations. (2) Wastes and contaminants occur in and travel through the geologic environment; we need to understand the nature of this environment and the transport processes. And (3) most waste management schemes involve the use of geologic reservoirs for containment or the use of geologic processes for dilution and dispersal of the waste.

◄ ——————————————

Fresh Kills, New York City, is the world's largest landfill. The bulldozer is leveling and flattening the garbage prior to covering it with a layer of soil.

PROBLEMS WITH SUPERFUND

*T*he Superfund program was created by the U.S. Congress in 1980 to provide funds for cleaning up the worst of the nation's contaminated sites. However, there is growing concern that Superfund may not be achieving its goal. Originally, $8.6 billion was allotted for the program. As of June 1993, only 161 sites had been cleaned up, leaving another 1256 sites on the National Priority List (NPL) of contaminated sites. The cost of cleanup per site has averaged $30 million, and the program has already cost a total of $15 billion.

One of the problems encountered by Superfund has been the extremely high cost of proper treatment and disposal of toxic materials. Materials removed from Superfund sites are commonly incinerated, but in May 1993 President Clinton imposed a moratorium on the development of new hazardous waste incinerators to allow some time to review the procedures involved in the incineration of 4 million tons of toxic waste each year. With many environmentalists strongly opposed to incinerators, the costs of dealing with toxic wastes—already as much as $1000 per ton for some materials—are likely to remain high.

Many Superfund sites are located in or near heavily populated metropolitan areas (Table B1.1), which makes their cleanup more urgent. For example, the New York metropolitan area has 10 Superfund priority sites (most of which are actually located in northern New Jersey). Superfund was created to deal with past emissions or improper disposal of hazardous materials. In some cases, states with a large number of severely contaminated sites are not emitting large quantities of hazardous materials at the present time. For example, New Jersey currently has the highest number of Superfund priority sites but releases a relatively small amount of toxic materials—11,600 tons per year in 1991, compared with 229,300 t/year in Louisiana and 205,300 t/year in Texas, for example.

How much will the cleanup of contaminated sites cost? It is difficult to say; estimates range from $750 billion to $1.5 trillion. Federal legislation has created other programs besides Superfund to deal with some hazardous waste sites; these cover approximately 40,000 former dump sites that must be cleaned up by states or private individuals. In addition, there are as many as 2.7 million underground storage tanks containing petroleum products that need to be removed. The cleanup will certainly continue until well into the next century. At this point it is not clear whether Superfund will continue to be the main mechanism for handling the remediation of these sites or whether another, more effective process can be found.

T A B L E B1.1 • **Superfund Sites:**
The 16 Worst Metropolitan Areas

Metro Area	Number of Sites
New York	56
Philadelphia	44
San Francisco	29
Chicago	22
Minneapolis	22
Los Angeles	21
Seattle	18
Boston	16
Miami	16
Detroit	11
Houston	11
Grand Rapids	11
Milwaukee	9
Denver	9
Portland	8
Saint Louis	8

As discussed in the essay at the beginning of Part IV, the range of wastes is so vast in terms of sources, chemical characteristics, and physical properties that it is virtually impossible to come up with a simple, all-encompassing categorization scheme. In this chapter we focus on a few types of waste that present significant disposal problems, either because of their sheer volume (e.g., municipal solid waste and sewage) or because of their hazardous characteristics (e.g., toxic and radioactive wastes). Fortunately for this discussion, the range of waste treatment and disposal methods is fairly limited.

SOLID WASTE

Solid waste, which can be defined very broadly as waste material that cannot be easily passed through a pipe, comprises a very wide range of materials that come from a variety of sources. The primary sources of solid waste are agriculture and mining, followed by industrial and municipal sources. The sources of solid waste, the amount generated at each individual source, and its physical and chemical characteristics are important factors in determining how the waste is handled.

Although the difference between a solid and a liquid seems clear enough, the distinction between solid and liquid waste is not always quite so obvious. When solid wastes accumulate, water may pass through and pick up soluble components; as a result, the distinction between solid and liquid waste may become blurred. On the other hand, a waste that is principally in liquid form, such as sewage or fine suspended sediment, may become concentrated into a

▲ F I G U R E 16.1
Agricultural waste being destroyed in the Snake River Valley, Idaho. Wheat stubble is raked into rows and burned.

more or less solid form as a result of treatment procedures or natural settling processes.

Agricultural Solid Waste

More than half of all solid waste is generated by the agricultural sector, which includes sources like farms, orchards, and animal feedlots (Fig. 16.1). Sediment eroded from fields is one important type of agricultural solid waste. Sediment actually starts out as a liquid waste when it is washed off a field in the form of surface runoff. The eroded sediment can settle and clog adjacent waterways. Also, the soils lost from agricultural lands often carry with them a wide range of chemicals such as pesticides, fungicides, herbicides, and chemical fertilizers, all of which can have damaging effects.

Aside from sediment and the associated chemicals, agricultural solid wastes are almost entirely organic. This waste consists primarily of such materials as animal excrement, dead animals, and offal; stubble from fields; prunings from orchards; and other types of plant and animal debris. They are usually disposed of by being allowed to accumulate onsite or washed off into surface waters; only a small fraction of agricultural waste is collected and burned or treated in waste disposal facilities. The disposal of agricultural wastes in surface water bodies, which may become contaminated as a result, is a source of concern, especially in areas where agricultural land is located near population centers.

Because of its organic nature, agricultural waste can biodegrade if it accumulates in the right type of environment. *Biodegradation,* which will be discussed in further detail in Chapter 17, is the process whereby the decomposition of a substance is promoted by the action of microorganisms. Biodegradation uses up oxygen, so the main problem associated with organic wastes is the demand placed on the oxygen content of the system in which the material is decaying. If oxygen becomes depleted, it can lead to an excess of organic material in the system, a type of water pollution known as *eutrophication.*

Mine Waste

Mining generates a very large volume of waste, virtually all of which is disposed of at the mining or processing site (Fig. 16.2). Some of the solid waste is in the form of piles of discarded rock, or *gangue.* The volume of a pile of gangue is typically much greater than the original volume of the material underground because the rock is much more compact when it is underground. Another major component of mining wastes is *tailings,* the slags and sludges that are left behind after processing or smelting. Piles of waste rock and tailings are an eyesore, and the minerals in them may combine with rainwater to form acidic runoff. Dust can blow from the piles and create problems for neighboring communities.

◀ F I G U R E 16.2
Waste from a large mining operation. The tailings deposit after the ore mineral, molybdenite (MoS$_2$), has been removed. Climax molybdenum mine, Colorado.

The proportion of waste rock to ore depends on the type of mining and the material being sought. For example, relatively little waste is generated in the mining of sands and gravels; most of the material removed is utilized. In the mining of most metallic ores, however, huge quantities of rock must be removed and discarded to obtain a relatively small amount of useful material. For example, a typical copper grade of 2 percent will produce 20 kg of pure metal from each ton of ore processed.

In most modern mining operations, piles of waste rock are carefully managed and monitored. If possible, the waste rock is used to fill and grade the mined area and restore the landscape to its premining status. When this is not possible, the piles of waste rock are stabilized by the planting of vegetation, and the land may be used for other purposes, such as forestry or recreation. However, in spite of recent changes in minesite management and site decommissioning, piles of waste rock and tailings at thousands of "orphaned" minesites pose a major environmental concern.

Industrial Solid Waste

Industries other than agriculture and mining also generate solid waste, mostly in the form of paper and cardboard products, scrap metal, wood, plastics, glass, yard waste, tires, and rags (Table 16.1). Much of this waste comes from small industries, which tend to burn their waste in small incinerators, dispose of it on the site of the plant or factory, or transport it to municipal landfills. Often a substantial tipping fee is charged for disposing of industrial solid wastes in municipal landfills. This is especially true of materials such as wood, which is—in principle—reusable or recyclable. Over the past few years the construction industry has improved recycling rates for wood products, but wood is still a significant component of the rubbish in municipal landfills.

Municipal Solid Waste

Like most types of waste, municipal solid waste overlaps to a certain extent with other categories of waste. *Municipal waste* is generally considered to include any solid waste generated by households, small commercial institutions, and light industries. Most of this waste is collected and disposed of by the municipality in which it is generated (or by a contracted collector on behalf of the municipality). In other words, if you put garbage and trash at curbside on garbage collection day and it disappears, or if you haul it to a local disposal facility yourself, it falls in the category of municipal solid waste.

T A B L E 16.1 • **Industrial Solid Waste: A Breakdown of Solid Waste from Commercial Establishments, Institutions, and a Mixture of Light and Heavy Industries[a]**

Component	Percentage
Old corrugated cardboard	8
Office paper	7
Other paper	7
Wood	21
Glass	5
Plastic	3
Plant and yard wastes (organics)	11
Metal	11
Tires	2
Other	24

[a]From *The State of Canada's Environment,* Minister of Supply & Services Canada, 1991, Table. 25.1. Based on data for the province of Ontario.

Solid waste produced in homes, institutions, and small commercial enterprises actually accounts for a relatively small proportion of the total volume of solid waste, but it presents a major disposal problem for several reasons. Municipal waste is the fastest growing category of solid waste; studies suggest that in some areas the amount of garbage generated per capita has increased by as much as 58 percent since 1960! Another reason for concern is that the bulk of household and commercial waste is generated in areas of high population density, where little land is available on which to dispose of the waste. Finally, environmental and health problems can be caused by municipal wastes that are disposed of improperly, whether in open dumps, inactive landfills, or active landfills that are inadequately engineered.

What's in It?

Not only the volume—on the order of 400,000 tons *per day* in the U.S.—but also the character of the solid waste generated in towns and cities has changed over the past few decades (Fig. 16.3). Paper, cardboard products, and other packaging materials account for the largest proportion of most landfills (on the order of 35 percent) and are by far the fastest growing component (Fig. 16.3A). Household organic material (yard waste and kitchen garbage) is another major component of solid waste in terms of volume. The volumes of nonbiodegradable products, such as glass, steel, aluminum, and plastics, have also increased. Used tires, wood, and other waste materials from construction are also found in municipal landfills. Another solid material that must be disposed of is the ashy residue that is left after waste has been burned in an incinerator.

Note that some materials take up a lot of room but are quite light (Fig. 16.3B). Plastic, for example, constitutes about 18 percent by volume but only 7 percent by weight of municipal solid waste. Other materials, such as glass and metals, are denser, so they weigh more but take up less space. Therefore, if the space remaining in a landfill is the main concern, it may make sense to look at the percentages of each component by volume. If the actual amount of wasted resources is the main concern, the percentages by weight are a better indicator.

A potentially serious problem is the presence in municipal landfills of a wide range of toxic chemicals derived from household products. These products range from nail polish to batteries, household and garden pesticides and fertilizers, paints, cleaners, and solvents—many of the products that we rely on and casually throw away in our everyday lives (Fig. 16.4). When many small, seemingly harmless batches of such chemicals accumulate in a municipal landfill, they can add up to a very big toxic waste problem. This is illustrated by the fact that about one fifth of the hazardous waste sites on the Superfund cleanup list are municipal landfills. (Superfund is the United States Environmental Protection Agency's emergency hazardous waste cleanup fund for abandoned or inactive disposal sites. See Box 16.1)

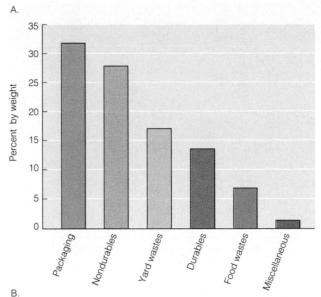

A.

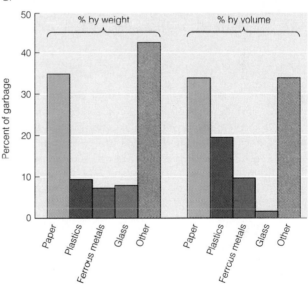

B.

▲ F I G U R E 16.3
A. Percentages by weight of nondurables, durables, packaging, yard wastes, and miscellaneous in United States municipal waste. Nondurables include newspapers, books, magazines, tissue paper, office and commercial paper, clothing, footwear, and the like. Durables includes items such as major appliances, furniture, and rubber tires. Data are for 1988 from the U.S. Environmental Protection Agency. B. A comparison of the relative importance, by weight and by volume, of the various types of garbage in U.S. municipal solid waste. Some materials, such as plastic, are quite light, so they take up a lot of room in landfills, while materials such as glass and metals weigh more but take up less room. Based on 1990 data from the Environmental Protection Agency.

▲ F I G U R E 16.4
Hazardous household wastes—paints and chemicals of various kinds—separated for disposal so that they do not find their way into a municipal dump.

Disposing of Solid Waste

We have mentioned some of the more commonly used methods of solid waste disposal. We now take a closer look at the available options. Note that in addition to the disposal options discussed here, a wide variety of wastes (solid and liquid, some hazardous, some not) are disposed of by being dumped into the ocean. Ocean dumping and pollution of the marine environment create special problems that are discussed in Chapter 17.

Open Dumps and Nondisposal

In the past most towns disposed of their garbage in an open dump—generally little more than a hole in the ground where garbage was allowed to accumulate. Open dumps still exist in some rural areas. They tend to be eyesores, attracting birds and vermin, causing unpleasant smells, and creating health hazards.

Sometimes people dispose of solid waste by *not* disposing of it properly; this informal, uncontrolled dumping of wastes is sometimes called *nondisposal*. Informal dumping is usually done to avoid the costs of proper disposal. If you take a walk in a wooded area you may see evidence of informal dumping of a load of garbage off the edge of a cliff or into a roadside ditch. Problems can also arise when landowners permit the dumping of materials in a corner of their land; materials disposed of in this manner are usually unmonitored and inadequately contained. The landowner may have little idea of the content of the waste and the hazards involved in its mismanagement.

Landfills

Landfills are the most common approach to waste management and are used by most municipalities in North Amer-

ica. Landfill types range from the old-fashioned open dump to highly engineered, carefully sited disposal facilities. Modern landfills are designed to confine the waste and prevent it from causing environmental or health problems in nearby areas. The typical dumping procedure in a modern landfill is to compact the waste as much as possible and periodically (usually daily) cover it with a compacted layer of soil. Because the soil layer isolates the waste from birds, insects, and rodents and minimizes the amount of precipitation that is able to infiltrate into the refuse pile, this type of disposal site is called a **sanitary landfill.** The major issues involved in the design of a sanitary landfill are control of leachate migration, selection of an appropriate site, engineering of the facility, and monitoring. Let's briefly examine each of these concerns.

Leachate When rainwater percolates through a pile of solid waste, it picks up soluble materials. The resulting contaminated water, called **leachate,** represents one of the most significant environmental problems in the design and maintenance of landfills. Depending on the composition of the waste, the leachate can be highly toxic or may contain infectious agents. One study of heavy metals in leachate from a municipal landfill found arsenic, cadmium, copper, lead, manganese, zinc, and others, in some cases at concentrations 100 times the maximum levels permitted in drinking water. In other words, leachate is a hazardous liquid waste derived from the interaction of rainwater with solid waste in landfills. If leachate is allowed to migrate out and come into contact with groundwater, surface water bodies, or any part of the biosphere, it can cause environmental problems.

Siting Constraints Perhaps the most important constraint on the design of a sanitary landfill is the selection of an appropriate site, one that will prevent the leachate from being able to migrate very far. The level of the water table, characteristics of the rock or regolith surrounding the landfill, the amount of precipitation, and characteristics of the groundwater flow system thus are crucial factors in site selection.

In a dry environment, where the water table is low and precipitation is minimal, the level of leachate production will be low. Moreover, when groundwater flow rates are very slow, rates of leachate migration will also be slow. The leachate will have little opportunity to migrate into the groundwater system before becoming neutralized or dispersed by natural filtration, adsorption, and biodegradation. Thus, a dry environment is preferable to a moist one in siting a solid waste disposal facility (Fig. 16.5A).

In a wetter environment, with more precipitation and/or a higher water table, leachate production and migration will be more effective (Fig. 16.5B). Therefore the landfill must be sited on an impermeable or relatively impermeable rock or sediment—that is, an *aquiclude* or an *aquitard*—in order to impede the flow of leachate. An impermeable layer

A. Favorable site for landfill–dry environment.

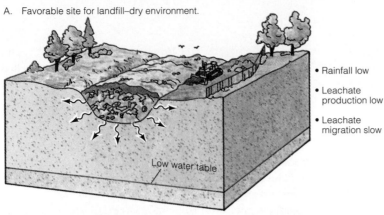

- Rainfall low
- Leachate production low
- Leachate migration slow

Low water table

B. Favorable site for landfill–presence of clay-rich attenuation layer.

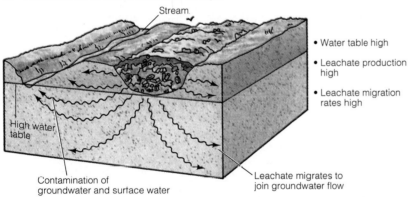

- Water table and leachate production relatively high, but leachable migration slowed by presence of impermeable clay attenuation layer

Clay attenuation layer

Adsorption of metals

Water table

C. Unavorable site for landfill–water table too high.

Stream

High water table

Contamination of groundwater and surface water

Leachate migrates to join groundwater flow

- Water table high
- Leachate production high
- Leachate migration rates high

D. Unfavorable site for landfill–underlying materials too porous and permeable.

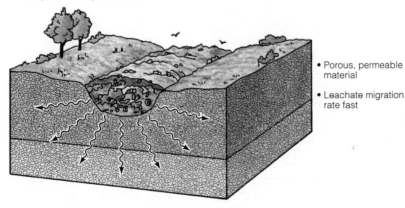

- Porous, permeable material
- Leachate migration rate fast

◄ F I G U R E 16.5
A series of diagrams showing favorable and unfavorable sites for landfills. A. Favorable; dry conditions with low water table and minimal leachate migration. B. Favorable; presence of an attenuation layer underlying the site. C. Unfavorable; moist environment with a high water table. D. Unfavorable; underlying materials too porous and permeable, permitting migration of leachate.

of Earth material that underlies a waste disposal site and slows the migration of leachate is called an **attenuation layer.** Clay-rich aquitards are particularly effective as attenuation layers; they are porous, so they can contain liquids, but they are usually relatively impermeable (the pore spaces are small and not highly interconnected), so the liquids cannot flow easily. Particles of clay also carry a negative electrical charge, so they tend to attract and hold positively charged particles, such as metals, that may be carried in the leachate solution.

An unfavorable site for a landfill would be one in which the water table is too high (such as the Fresh Kills site), or the underlying units are porous and permeable, thereby facilitating the flow and migration of leachate (Fig. 16.5C and 16.5D). Landfill siting can be particularly challenging in coastal areas and arctic ecosystems. Very humid, marshy, or perennially cold conditions can severely inhibit the natural breakdown of materials in the refuse, and as a result the impacts of waste disposal on these fragile ecosystems may be especially severe.

Engineering and Monitoring In addition to careful siting and natural attenuation, modern landfills typically include a variety of devices for leachate control. The most commonly employed devices are illustrated in Fig. 16.6. These include impermeable caps and liners to prevent the formation and migration of leachate, drainage pipes and leachate collection systems, leachate treatment systems, and wells for monitoring groundwater in the vicinity of the site.

Another engineering feature that is quite common in modern landfills is a system for collecting the gas generated by the decay of materials in the refuse pile. This gas is typi-

cally dominated by CH_4 (methane) but may also contain an appreciable amount of CO (carbon monoxide) as well as other gases. The gas has a disagreeable odor; in addition, methane is explosive when mixed with air in certain proportions, and CO is toxic. Some modern landfills collect and sell methane gas as an energy source, thereby reducing air pollution, recycling resources, and generating extra income at the same time.

The engineering of landfills has evolved considerably in recent years. Landfills have become more expensive to build, operate, and maintain, but they are also much more effective at minimizing negative environmental impacts. In some older landfills, poorly designed leachate control systems can actually cause more problems than they solve, as shown in Fig. 16.7. In landfills with impermeable bottom liners where no drainage or leachate collection pipes were installed, or where the pipes have collapsed, leachate can collect in the landfill and eventually overflow; this problem is known as the *"bathtub effect."*

Incinerators

Incineration—the burning of refuse, often in a specially designed facility—is another widely used method of solid waste disposal. Incineration methods are as varied as dumps and landfill types, ranging from the burning of paper, leaves, and other yard waste in the backyard to modern, highly engineered, high-temperature municipal incinerators. Many apartment buildings, hospitals, and businesses have small private incinerators. Modern municipal incinerators are designed to burn garbage at a very high temperature. They are most effective if the waste has been previously sorted, and noncombustible materials have been

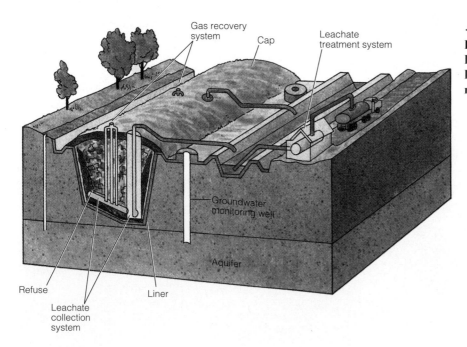

◄ F I G U R E 16.6
Engineered features of modern landfills, including liners, leachate collection pipes, and monitoring wells.

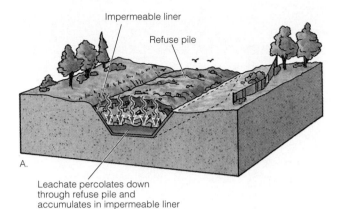

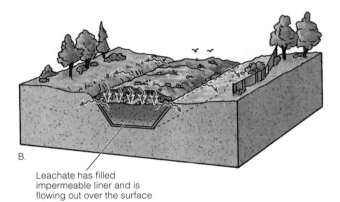

▲ **F I G U R E 16.7**
A. The "bathtub effect" occurs when leachate collects within an impermeable liner at the bottom of a landfill. B. The leachate builds up until it overflows the edges of the impermeable "tub" and flows out over the surface. This can happen in poorly sited or poorly engineered facilities or where features such as drainage pipes at the bottom of the refuse pile have collapsed.

removed. Incinerators can emit CO_2 (a greenhouse warming gas) and small amounts of acidic gases, but the newer facilities are fairly effective at controlling emissions of other toxic and particulate air pollutants (Table 16.2A).

All incinerators, even the most modern ones, generate ash, which often contains toxic heavy metals and organic chemicals such as dioxins, furans, and polyaromatic hydrocarbons (PAHs) (Table 16.2B). Incinerator ash is usually disposed of in landfills, but sometimes it requires landfills that are specially designed and licensed for the disposal of hazardous wastes. Whether the ash is classified as hazardous or nonhazardous waste can be critical in determining the ultimate cost of incineration, since the cost of disposing of hazardous ash adds significantly to the cost of incinerating municipal solid waste.

T A B L E 16.2A • **Typical Emissions of 15 Contaminants from Incinerators With and Without Pollution Control Devices[a]**

Contaminant	Typical Emissions[b]	Emissions from Obsolete Incinerator with No Air Pollution Controls
Hydrogen chloride	75 mg/m^3 (50 ppmdv)	430 ppmdv
Carbon monoxide	57 mg/m^3 (50 ppmdv)	150 ppmdv
Dioxins and furans	0.5 ng/m^3	250 ng/m^3
Particulate matter	20 mg/m^3	6,300 mg/m^3
Sulfur dioxide	260 mg/m^3	260 mg/m^3
Oxides of nitrogen	400 mg/m^3	400 mg/m^3
PAHs	5 µg/m^3	70 µg/m^3
PCBs	1 µg/m^3	3 µg/m^3
Pentachlorophenol	1 µg/m^3	2.7 µg/m^3
Polychlorobenzene	1 µg/m^3	12 µg/m^3
Lead	50 µg/m^3	34,000 µg/m^3
Cadmium	100 µg/m^3	1,500 µg/m^3
Mercury	200 µg/m^3	320 µg/m^3
Arsenic	1 µg/m^3	130 µg/m^3
Chromium	10 µg/m^3	2,000 µg/m^3

T A B L E 16.2B • **Contaminants Found in Bottom Ash and Fly Ash from a Municipal Incinerator**

Compounds	Range of Concentrations	
	Bottom Ash	Fly Ash
Organic compounds		
Dioxins	ND[c] - 0.16 ng/g	0.7 - 1,040 ng/g
Furans	ND	1.4 - 373 ng/g
PAHs	0.23 - 968 ng/g	18 - 5,640 ng/g
Inorganic compounds		
Cadmium	ND - 18 µg/g	23 - 1,080 µg/g
Chromium	984 - 3,170 µg/g	86 - 1,070 µg/g
Lead	1,000 - 9,900 µg/g	1,400 - 26,000 µg/g
Mercury	2.1 - 3.4 µg/g	8.0 - 54 µg/g
Zinc	1,300 - 5,210 µg/g	4,700 - 70,000 µg/g

[a]Based on Tables 24.4, 24.5, and 24.6 of *The State of Canada's Environment,* Supply and Services Canada, 1991, Ottawa.
[b]Note: ppmdv = parts per million dry volume.
[c]ND = not detected.

Many municipalities have started programs to recapture energy from waste. In some cases this takes the form of energy derived from incineration. Approximately 100 such plants are now in operation. The most modern plants can burn 90 percent of the solid waste, extracting energy and recovering metals. This form of power generation is sometimes referred to as *refuse-derived fuel* (or *RDF*). Some landfill operations also harness the energy from methane gas generated by the decomposition of materials in the landfill. This energy can be used to meet the energy needs of the landfill operation itself, or it can be sold to local communities.

One of the arguments commonly levied against incinerators in general, and refuse-derived fuel plants in particular, is that the incineration of waste will discourage residents from recycling. However, studies have shown that communities with plants that recapture energy from waste also tend to have above-average rates of recycling.

Composting and Recycling

Since organic waste is an important component of municipal solid waste, it makes sense to try to reduce its volume as much as possible. One way of doing this is to encourage **composting,** that is, facilitating the breakdown of organic refuse such as food waste, leaves, and yard clippings by the action of bacteria and other naturally occurring microorganisms. In the composting of household wastes, decomposition in *aerobic* conditions (i.e., in the presence of oxygen) is most efficient. The material left after decomposition, the *compost,* is a clean, soil-like material that is rich in organic matter. Some communities now have composting programs in which food and yard wastes are collected from special curbside bins and composted in a municipal facility.

Recycling, which has gained popularity in recent years, is another way of reducing the volume of solid waste (Fig. 16.8). The materials that are most commonly collected in curbside recycling programs are newspapers, glass bottles,

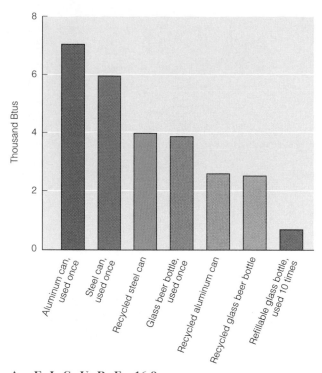

▲ F I G U R E 16.9
Energy consumption per use for a 12-ounce beverage container.

corrugated cardboard, cans (aluminum and steel), and plastics of various types. Recycling can occur in industrial settings as well as in municipal programs. Some industries, such as the steel and soft plastics industries, have a long tradition of recycling materials.

Recycling can contribute to significant reductions in solid waste, as well as saving both materials and energy. However, it is not as effective as reusing a product (Fig. 16.9). Sometimes recycled materials are made into products that are essentially the same as the original; for example, glass bottles are crushed and melted, then formed into new glass bottles. Another type of recycling, called *secondary recycling,* involves using recycled materials to make a completely different product; for example, plastic bottles can be recycled into park benches or lawn furniture. For recycling programs to be effective, there has to be a market for products made from recycled materials.

LIQUID WASTE

There are two major categories of *liquid waste:* (1) sewage (what we flush down the toilet and wash down the drain every day) and (2) industrial effluents. The latter are usually characterized by hazardous concentrations of metals and toxic chemicals that require special handling; the methods used to dispose of this type of waste are discussed

▲ F I G U R E 16.8
Recycling begins at home. Containers holding recyclable household materials await curbside collection.

later in the chapter. In this section we concentrate on problems associated with the disposal of sewage.

Sewage represents a major waste disposal challenge, if only because the volume of sewage generated each day is so great. On average, each person in North America generates about 550 liters of sewage per day. Sewage is primarily a liquid waste composed of water, urine, detergent, grease, and other liquids, although it contains significant amounts of solid materials—paper, feces, dirt, food waste, and other garbage. In rural and suburban areas, domestic sewage is often managed on an individual household basis by *septic systems*. In areas with higher population densities, sewage is commonly disposed of through municipal sewer systems combined with various kinds of facilities for treating the effluents and solid residues.

Domestic Sewage

A **septic tank** is a holding tank designed to receive domestic sewage from a single household. In principle, a septic tank contains the waste long enough to allow the organic material in it to partially decompose and the solids to settle. The clear liquid component of the waste is then slowly released into the surrounding sediment, either through a perforated tank called a *cesspool* or through a system of porous pipes called a *leach field* (Fig. 16.10). In passing through the sediment, the effluent is purified by natural filtration and biochemical processes and by the *adsorption* (clinging of an ion to a solid surface) of contaminants to particles.

If the geologic setting is appropriate, a septic tank can be an effective way of dealing with sewage effluent from households. The surrounding soil or sediment must be permeable enough to permit the liquid effluent to pass through, but not so permeable that it passes through quickly and reaches either the ground surface or the saturated zone before being sufficiently purified. Also, the water table must be well below the level of the septic tank to avoid groundwater contamination. If these conditions

are not met, or if septic tanks are too closely spaced or sited too close to a domestic well, contamination of sediment or groundwater may occur.

Municipal Sewage

Municipal sewage is a significant source of water pollution, especially in heavily urbanized areas. In many communities sewage is released directly into rivers, lakes, bays, and the open ocean without undergoing any treatment to remove the impurities. Although this is an unhealthy way of handling sewage effluents, it is still practiced in many parts of the world, including North America. The release of untreated sewage may not pose significant problems in very small communities; however, when a community grows, surface water bodies can no longer absorb the waste as efficiently as they did when the amount of effluent was smaller.

For example, until recently the city of Halifax, Nova Scotia, released 200 million liters of untreated raw sewage a day into the city's harbor from 40 different sewage outlets. This practice resulted in unpleasant smells and algal blooms on the water as well as high levels of fecal coliform bacteria (a standard measure of biological and microbial pollution) and toxic chemicals in the water and bottom sediments. In Massachusetts, clam harvests were halved between 1985 and 1990 as a result of algal blooms and contamination by sewage effluents. In 1992 there were more than 2600 beach closures in the United States, caused mainly by high levels of fecal bacteria from raw human sewage.

Sewage Treatment

In most municipalities wastewater is treated in a sewage plant before being discharged. *Primary treatment* is a physical process involving mechanical removal of solid wastes such as trash and fine sediment from the effluent. *Secondary treatment* is a biological process in which bacteria

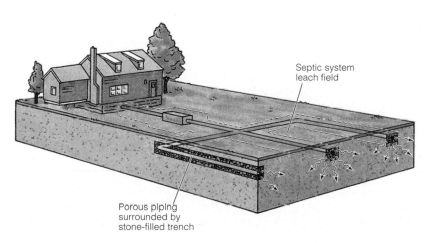

◀ F I G U R E 16.10
Septic system leach field, showing the stone-filled trench and porous piping system through which liquid effluents are released into the surroundings.

Septic system leach field

Porous piping surrounded by stone-filled trench

degrade the bulk of the dissolved organic material in the effluent. Sometimes the water is also chlorinated at this stage to disinfect it. Finally, *tertiary treatment* is a chemical process designed to remove any remaining contaminants, such as excess organic material (e.g., nitrogen and phosphate), heavy metals, and inorganic dissolved solids. Few communities currently offer tertiary sewage treatment.

The quality of the water returned to the system depends on the level of treatment. In many communities where freshwater supplies are under stress, wastewater that has undergone secondary or tertiary treatment is reused for other purposes, such as watering of lawns or city landscaping. To remove disease-causing bacteria from the effluents, secondary treatment is necessary. In principle, it is possible to return sewage effluents to drinkable standards, although many people find it hard to accept that concept. Also, it has not been proved that even tertiary treatment is completely effective in removing trace amounts of toxic chemicals from the effluents.

Sewage treatment and wastewater reclamation generates its own solid waste in the form of *sludge,* the residual material that remains after the effluents have been treated. Sewage sludge is usually incinerated, placed in landfills, or dumped into a large body of water. In some cases, because of its high organic content, sewage sludge can be recycled as a fertilizer for nonagricultural uses (e.g., parks, golf courses,

or urban landscaping). However, sludge sometimes contains high levels of toxic chemicals or heavy metals; in such cases the sludge itself can actually constitute a hazardous contaminant that requires special handling.

HAZARDOUS WASTE

Hazardous wastes are wastes that pose a present or potential danger to humans or wildlife (Table 16.3). They include toxic (poisonous or carcinogenic), caustic, acidic, explosive, infectious, and radioactive materials. Such materials are sometimes called *special wastes* because they require special handling and treatment. Hazardous waste represents one of the leading environmental concerns of industrialized countries and is a growing problem in many countries of the developing world as well.

Causes for Concern

One reason for the widespread concern about hazardous wastes is that they have been handled badly in the past. Perhaps the most famous incident of mismanagement occurred at Love Canal, a small community near Niagara Falls, New York (Fig. 16.11). In the 1940s more than 21,000 tons of toxic chemicals were disposed of in a large ditch (ac-

T A B L E 16.3 • Some Illustrative Examples of Hazardous Wastes[a]

Sector	*Source*	*Hazardous waste*
Commerce, services, and agriculture	Vehicle servicing	Waste oils
	Airports	Oils, hydraulic fluids, etc.
	Dry cleaning	Halogenated solvents
	Electrical transformers	PCBs
	Hospitals	Pathogenic/infectious wastes
	Farms/municipal parks, etc.	Unused pesticides, "empty" containers
Small-scale industry	Metal treating (electroplating etching anodizing, galvanizing)	Acids, heavy metals
	Photofinishing	Solvents, acids, silver
	Textile processing	Cadmium, mineral acids
	Printing	Solvents, inks, and dyes
	Leather tanning	Solvents, chromium
Large-scale industry	Bauxite processing	Red muds
	Oil refining (petrochemical manufacture)	Spent catalysts Oily wastes
	Chemical/pharmaceutical manufacture	Tarry residues, solvents
	Chlorine production	Mercury

[a]Based on M. Tolba, editor, *The World Environment, 1972–1992* Chapman & Hall, 1992, London.

▲ F I G U R E 16.11
**Toxic waste site, Love Canal, New York. Houses that
stood on the now empty swath were removed because
they were erected on contaminated soil.**

heavily monitored. The homes that suffered the worst contamination were removed, and others in the vicinity have
been cleaned up and were recently declared safe for resale.
But what about other hazardous waste sites? In the United
States there are between 30,000 and 50,000 contaminated
sites, many of them untreated and unmanaged, and in
Canada there are approximately 1000. Many of these are
"orphaned," with no known or traceable owners. Merely assessing and quantifying the hazard from these sites is an
enormous task, and the costs of cleaning them up will be
staggering.

Sources of Hazardous Waste

A large proportion of the hazardous waste that eventually
enters the environment comes from leaks at old, inappro-
priately managed, or badly designed disposal sites like Love
Canal. These sources are important because they represent
significant concentrations of hazardous materials. But another very important entryway for hazardous materials into
the environment, and one that is potentially much more
difficult to control, is through the release of wastes in the
normal course of industrial, agricultural, or household operation. For example, hazardous wastes may be released by
oil spills and other types of chemical spills, leaking underground petroleum storage tanks, legal discharge of liquid
wastes into surface water bodies, atmospheric emissions,
contaminated leachate migrating from landfills, and acid
mine drainage.

Industry

Industry is by far the largest generator of hazardous wastes
(Fig. 16.12), which may take the form of liquids, solids, or
atmospheric emissions. Common examples of industrial

tually part of a canal) by the Hooker Electrochemical Company. The ditch was eventually filled in and the surface was
restored, but beneath the surface a toxic chemical stew was
brewing. In 1953 the land was sold to the local school
board for $1. (The deed clearly stated that the land had
previously been used for the disposal of chemical waste.) In
the ensuing years a school, a playground, and homes were
constructed on and adjacent to the site. Toxic liquids eventually began coming to the surface, and residents of the
community experienced health problems, such as spontaneous abortions and birth defects. Although studies were
carried out, it was not until 1978 that New York State officially acknowledged the hazard and intiated a cleanup effort. In 1980 the site was declared a national state of emergency (the first state of emergency in the United States not
related to a natural hazard), and more than 1000 Love
Canal families were eventually forced to leave their homes.

Since then, the Love Canal site has been extensively re-
mediated. Much of the waste remains underground, but
impermeable caps have been placed over it and the area is

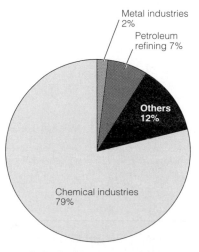

▲ F I G U R E 16.12
**Hazardous waste generated in the United States by
source (in percentage).**

hazardous wastes are acids from metallurgical processes, caustics from pulp and paper mills, and residues from refineries. Hazardous atmospheric emissions include volatile organic compounds (VOCs), particulates (soot), and lead. Some industrial hazardous wastes consist of materials that break down fairly quickly in the environment, such as oil and grease, phenols, acids, or ammonia. Others last a long time and may accumulate in the environment or in organisms. The most problematic of the persistent toxic chemicals are heavy metals (especially lead, mercury, cadmium, and arsenic); chlorinated organic compounds (PCBs, dioxins, and furans); and polyaromatic hydrocarbons (PAHs).

Households

As mentioned earlier, households generate hazardous wastes through the use of such products as batteries, paints, solvents, and pesticides. Even though the amount of hazardous waste generated by households is only a fraction of that generated by industry, household hazardous waste is a growing concern because these materials are turning up in ordinary landfills with alarming frequency. Some communities now have mobile units that make regular rounds or will come to your house by appointment to pick up and dispose of hazardous wastes that may require special handling.

Agriculture

Agriculture contributes hazardous materials to the environment primarily through runoff of toxic chemicals that have been applied to fields for the control of pests. Pesticides, which include insecticides, herbicides, and fungicides, are a major source of pollution in waterways near agricultural areas. This category of hazardous waste is difficult to deal with because the chemicals come from nonpoint sources and they tend to be very persistent. Many types of pesticides accumulate and become concentrated in organisms, and they are commonly harmful to *nontarget species* (i.e., they harm plants and wildlife that were not the intended targets of the control). Also, toxic organic compounds can be highly mobile in soil, water, and air. (Controls on the behavior of such contaminants in the environment are discussed in more detail in Chapter 17.)

Disposing of Hazardous Waste

When a company produces hazardous waste in small quantities, it is common practice to allow it to dispose of the waste by burial or incineration on-site, emission into the air (if it is in gaseous form), or discharge into an adjacent body of water, provided that the level of contaminants in the effluent is below a legally set limit. For small companies that can't afford on-site treatment or disposal, or for companies that generate very large quantities of hazardous wastes, off-site disposal is required.

Experience has demonstrated that uncontrolled surface dumping of hazardous wastes quickly leads to contamination of surface and subsurface water and hence to the potential for serious health problems. Some types of hazardous materials can be incinerated at high temperatures or stabilized by chemical treatments. Many hazardous materials need to be treated and then disposed of in specially designed landfills at the surface. For highly toxic materials and high-volume liquid wastes, it is common practice to dispose of the waste by isolating it deep underground.

Surface Disposal

Many of the problematic disposal sites mentioned earlier are surface sites where containers of hazardous waste have been piled up and abandoned, often in completely inappropriate locations such as along lake shores or river banks. Drums, barrels, and canisters disposed of in this way are prone to corrosion and leakage (Fig. 16.13).

Another form of surface disposal, often used in the past but less common now, is *surface impoundment.* In this method of disposal a small depression, either natural or excavated, is used to contain liquid hazardous wastes, often near the site where they were generated. The depression, called an *aeration pit* or a *lagoon,* is sometimes lined with clay or plastic. The lagoons are designed to hold the waste and allow it slowly to aerate and (hopefully) degrade. Although surface impoundment has traditionally been used in treating hazardous waste at many facilities, it has fallen out of favor because the lagoons are highly susceptible to leaks, which may result in soil and groundwater contamination in surrounding areas.

Secure Landfills

Some landfills are specifically designed and engineered to contain hazardous materials. Such facilities are called **secure landfills.** The engineering features of secure landfills are similar to those of sanitary landfills (Fig. 16.6), but the leachate drainage and collection systems are more sophisticated, the impermeable liners and caps are more substantial, and the concentration of monitoring wells around the perimeter may be greater. In addition, there is commonly an on-site treatment facility where hazardous materials are incinerated, neutralized, or otherwise stabilized before being placed in the landfill. Particularly careful attention must be paid to the siting of a secure landfill. It is usually assumed that a leak will occur in spite of the facility's engineering features; therefore, the geologic environment must provide as much protection from contamination as possible.

Geologic Isolation

Most studies of toxic and radioactive waste disposal have concluded that underground storage is an acceptable disposal method, provided appropriate sites can be found. Disposal of hazardous wastes underground in a geologic

▲ F I G U R E 16.13
Toxic wastes leak from rusting containers and soak into the groundwater system beneath this unsupervised waste disposal dump.

repository—a rock unit with specific waste-absorbing and containment characteristics—is referred to as **geologic isolation.**

The placement of hazardous materials underground, even far underground, immediately raises concerns about groundwater. Water is a nearly universal solvent, and the weakly acidic character of most groundwater means that any container of toxic or radioactive substances is likely to corrode, whereupon the contents will be dissolved and transported away from the storage site. Water is present in crustal rocks to depths of many kilometers, and it can circulate at rates of up to 50 m/year. Over thousands of years, even such slow rates can move dissolved substances over

great distances and introduce them to more rapidly flowing parts of the hydrologic system.

Deep-Well Disposal

One approach to geologic isolation is to inject liquid wastes deep into the ground in a specially designed well. This method, called **deep-well disposal,** is used when liquid wastes are highly toxic or the volume of the waste is very large. For example, brines generated in oil fields are commonly disposed of in deep disposal wells in quantities up to several *billions* of liters per day. There are thousands of such disposal wells in the oil-producing regions of North America.

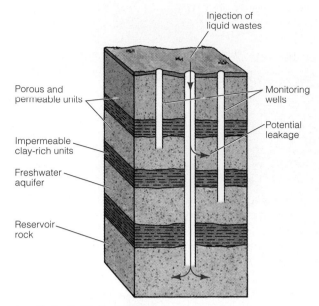

▲ F I G U R E 16.14
Deep-well disposal of liquid hazardous wastes. The reservoir rock is a porous and permeable rock, such as a sandstone or limestone, separated from freshwater aquifers by several intervening aquitard layers, that is, impermeable units.

In deep-well disposal (Fig. 16.14), a concrete-encased well is drilled to a depth of 700 m or more. In principle, the reservoir rock should be porous and permeable in order to permit the flow and storage of the liquid waste. Most commonly, therefore, the reservoir rock is a sandstone or a limestone. The most important geologic characteristic of the site, however, is that the reservoir rock must be isolated from any aquifers by one or more aquicludes—impermeable rock layers—such as shale or salt. A properly managed deep disposal well will also be monitored very closely for any signs of leakage or contamination.

Deep-well disposal poses some potentially serious problems. First, the number of sites with the ideal geologic characteristics for deep-well disposal is limited. The well itself may break or leak, releasing waste directly into rock or sediment layers above the reservoir unit. It is possible that the waste will be chemically incompatible with the reservoir rock or that there will not be enough pore space in the reservoir rock to handle all of the waste. More serious problems have occurred when fluids pumped into deep wells under considerable pressure have reactivated old, deep-seated faults and fractures in the underlying rocks. A striking example of such reactivation occurred at the Rocky Mountain Arsenal in Colorado in the early 1960s. In 1962 the U.S. Army began (legally) disposing of liquid toxic wastes in a deep disposal well. At about the same time, residents of nearby Denver were troubled by swarms of small earthquakes. The epicenters of the earthquakes were traced

to the vicinity of the well, and it was eventually demonstrated that the earthquakes occurred when the pressure and volume of waste being injected into the well were highest. Similar incidents have been reported at other locations in the United States.

In the final analysis, the main concern in selecting an effective method and an appropriate site for hazardous waste disposal is to isolate the waste and prevent any potential leaks or byproducts from interacting with the soil, groundwater, surface waters, or the biosphere. We end this chapter by considering the problems associated with the disposal and, in particular, the geologic isolation of a special category of hazardous material: radioactive waste.

RADIOACTIVE WASTE

Radioactive waste, or *radwaste* as it is sometimes called, consists of leftover radioactive material or equipment for which there is no further economic use. Some radwaste is so highly radioactive that exposure to even minute quantities can prove fatal. Many of these waste materials are extraordinarily persistent, and some are toxic as well. Radioactive wastes are closely regulated, although this does not always ensure proper disposal. The transport and disposal of radioactive wastes are also subject to international standards and agreements.

Radioactive Decay

Before we examine the problems of radioactive waste disposal, we need some background information. Let's briefly look at the process and products of radioactive decay.

Nuclear Fission
Radioactive materials undergo a process of spontaneous decay in which atoms, or *nuclides,* disintegrate to form another nuclide (or nuclides) that may, in turn, be unstable and decay radioactively. The decay process continues until a stable nuclide is eventually formed. A nuclide that is unstable and undergoing radioactive decay is referred to as a *radionuclide.* (Note that *nuclide* is essentially synonymous with *atom,* but *nuclide* is the preferred term when the nuclear characteristics of an element are being discussed.)

Radioactive decay occurs naturally in any Earth material (soil, rock, or sediment) that contains a radioactive material, such as uranium or thorium. We take advantage of this process to generate power in a *nuclear reactor* by inducing large, unstable atoms—usually ^{235}U—to disintegrate by splitting into two smaller nuclei. The smaller nuclei are called *daughters* or *fission products,* and the splitting process itself is nuclear *fission.* The masses of the two daughter nuclei add up to slightly less than the mass of the original nuclide; the "missing" mass is converted into energy, which is harvested in a nuclear reactor (see Chapter 12).

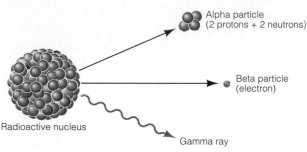

▲ FIGURE 16.15
Radioactive decay of an unstable isotope with the three common types of radioactive emission: alpha particles, beta particles, and gamma rays.

Radiation

During the process of radioactive decay, the disintegrating nuclide emits one or more forms of **radiation,** which may include *alpha particles, beta particles,* and *gamma rays* (Fig. 16.15). Of these three, alpha particles are the most massive; they consist of two protons and two neutrons. Beta particles, or electrons, are much less massive. Gamma rays have no mass at all but are a high-energy form of electromagnetic radiation similar to x-rays. It is the release of radiation and the interaction of the radiation with biologic tissue that can cause genetic and somatic damage or even death, depending on the dose received.

Units of Radiation Measurement   A bewildering array of units of measurement are used to quantify radiation and doses to living organisms. One of the most commonly used units is the *curie,* a measure of the number of radioactive disintegrations per unit of time. Specifically, 1 curie (Ci) is equivalent to 37 billion radioactive disintegrations per second. This amount, which is approximately equal to the rate of decay of 1 gram of pure radium, represents a *very* large

amount of radioactivity; smaller amounts of radiation are commonly measured in *microcuries* (10^{-6}, or one millionth Ci) or *picocuries* (10^{-12}, or one billionth Ci).

Another unit that is often used to quantify radiation is the **rad,** which measures radiation in terms of the amount of energy absorbed by biological tissue. A rad is equivalent to 0.01 joule of energy absorbed by 1 kg of tissue. Recall, however, that there are different types of radiation; because they are more massive, alpha particles are much more damaging to biologic tissue than are beta particles or gamma rays. This difference is taken into account by another unit of measurement, the **rem.** An organism can receive the same number of rads in the form of alpha particles, beta particles, and gamma rays, but the damage done (rems) by the alpha particle radiation will be 10 times greater. Note that the curie is not directly convertible into rads or rems; the former is a pure number, while the latter are stated in energy units.

What do these doses mean? Exposure to 500 rems will be lethal to about 50 percent of people exposed to it. Doses of 100 to 200 rems are enough to cause vomiting, spontaneous abortions, temporary sterility in males, and other effects. The maximum allowable dose of radiation per year for people exposed to radiation in the workplace is 5 rem (5000 mrem, or millirems).

Penetrating Power   Another feature of radiation that is important, both in assessing the potential damage to living organisms and in designing effective disposal technologies, is its penetrating power. Alpha radiation does not have much penetrating power; that is, it can't pass easily through thicknesses of materials (Fig. 16.16). Beta particles have somewhat more penetrating power, and in order to stop gamma rays you must surround the source with a 1-meter thickness of concrete or an equivalent volume of water. Alpha radiation may not be able to penetrate very far from the outside, but if a source of alpha radiation were inhaled

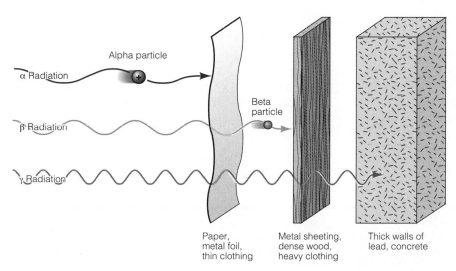

◀ FIGURE 16.16
The penetrating power of different types of radiation. Alpha radiation is massive; it is therefore particularly damaging to organisms but has low penetrating power. Beta radiation (electrons) and gamma radiation are more penetrating.

BOX 16.2
•
THE HUMAN PERSPECTIVE

THE INTERNATIONAL WASTE TRADE

*E*ach year more than 55,000 tons of hazardous waste cross the United States-Mexico border legally. However, environmental agencies in both countries are concerned that far more may be crossing the border illegally. With the high cost of liquid waste disposal in the United States, there is reason to suspect that wastes are being dumped illegally in sewers, landfills, or the desert. High levels of lead, mercury, cyanide, and other toxic materials have been detected in Mexican sewers and in unlined ditches near the Rio Grande River.

In 1992 the U.S. Environmental Protection Agency (EPA) established a computer database to track the movement of hazardous wastes across the border and determine whether they have been handled properly. Truckers are required to place an EPA identification number on the records they show at the border. Those who do not comply may face fines of up to $25,000 per day. In addition, Mexico's environmental agency, Secretaria de Desarrollo Social (Sedesol), and the EPA plan to beef up inspections of both generators and transporters of hazardous wastes.

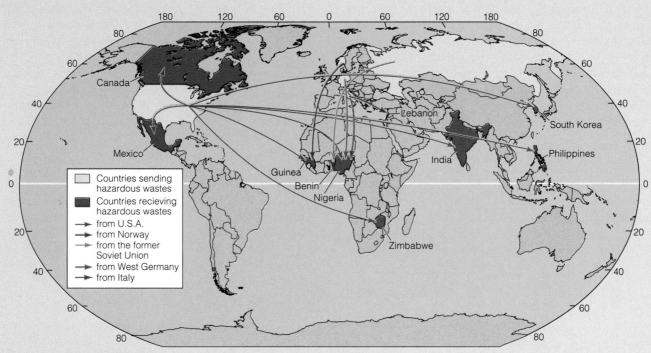

▲ F I G U R E B2.1
Some major transfers of hazardous waste across international borders.

it would be particularly damaging to the lining of the lungs.

Half-Life

Radioactive materials decay at different rates. For each radioisotope there is a characteristic rate at which it spontaneously decays, regardless of the surrounding environment

In other words, whether it is in rock, soil, water, air, or a living organism, that particular radioisotope will always decay at the same characteristic rate. The time it takes for one-half of the atoms in a given sample to decay is called a **half-life.** The half-lives of the different radioisotopes range from microseconds to billions of years. For example, ^{147}Sm (samarium-147, a naturally occurring radioisotope that is

The situation on the United States-Mexico border represents only a portion of the transfer of hazardous wastes—both legal and illegal—across international boundaries (Fig. B2.1). Aside from the safety issues arising from the long-distance transport of hazardous materials, some environmental and ethical issues also arise when international borders are involved. For example, an importing country may have much less stringent regulations regarding the disposal of hazardous materials than the country in which the material originated. The importing country may not have the technical or geologic capability to dispose of the wastes in an environmentally sound manner. Questions have also arisen as to whether it is ethical to pay poor countries to accept hazardous wastes or whether this type of trade is essentially a form of imperialism by industrialized nations.

In 1987 a working group of the United Nations Environment Programme produced a set of *Guidelines and Principles for Environmentally Sound Management of Hazardous Wastes*. These guidelines paved the way for the drafting, in 1992, of the *Basel Convention on the Control of Transboundary Movement of Hazardous Wastes and Their Disposal*, which focused specifically on the problems associated with international transfers (Table B2.1).

T A B L E B2.1 • The *Basel Convention on the Control of Transboundary Movement of Hazardous Wastes and Their Disposal* Was Adopted by 116 Countries and the European Community on March 22, 1989, and Took Effect on May 5, 1992[a]

The main points in the *Basel Convention* are as follows:

- A signatory state cannot send hazardous waste to another signatory state that bans its import or to any other country that has not signed the treaty.

- Every country has the sovereign right to refuse to accept a shipment of hazardous waste.

- The exporting country must first provide detailed information on the intended export to the importing country.

- Shipments of hazardous waste must be packaged, labeled, and transported in accordance with international rules and standards.

- The consent of the importing country must be obtained before shipment.

- Should the importing country be unable to dispose of the imported waste in an environmentally sound manner, the exporting country has a duty either to take it back or to find another way to dispose of the shipment safely.

- Illegal traffic in hazardous waste is criminal.

- A secretariat was set up to supervise and facilitate implementation of the treaty.

[a]From *The World Environment 1972–1992*, M. Tolba, ed., Chapman & Hall, 1992, London; p. 272, Box 7, Chapter 10.

not a problem in radwaste) has a half-life of 106 billion years. This means that it takes 106 billion years for *half* of the samarium atoms in a given sample to decay, another 106 billion years for half of the *remaining* samarium atoms to decay, and so on. By contrast, ^{214}Po (polonium-214, another naturally occurring radioisotope) has a half-life of just 0.2 microseconds.

One of the most difficult problems in the disposal of radioactive waste is that some of the radionuclides in the waste—a small but very significant fraction—have exceptionally long half-lives, which range up to tens of thousands of years. This material must be kept in isolation from the hydrosphere and the biosphere until its radiation has dissipated to a reasonably safe level.

Sources of Radiation Exposure

Environmental Radiation

There is a very distinct difference between radioactive waste and so-called **environmental radiation,** the dose of radiation each of us receives in our normal daily lives (Fig. 16.17). Sources of environmental radiation include terrestrial sources (i.e., radiation that comes from the decay of naturally occurring radioisotopes in rocks and soils); cosmic radiation; medical sources (e.g., diagnostic x-rays); nuclear fallout (airborne radioactive materials resulting from accidents at nuclear power plants, the testing of nuclear weapons, and other atmospheric sources); and home and occupational sources. The normal operation of nuclear power plants and the disposal of radioactive waste also contribute to environmental radiation, but these are relatively minor sources. On a fuel-equivalent basis, the burning of coal actually releases more radioactive material—as well as more carcinogens—than does normal nuclear power generation.

A figure that is widely quoted as representative of the average annual dose of environmental radiation received by each person per year is 106 mrem (millirems). (Note that this figure does *not* include the average annual dose of radiation from radon, a naturally occurring radioactive gas found in soil, which can pose a serious indoor air pollution problem if it becomes concentrated in buildings; radon is discussed in more detail in Chapter 18.)

Sources and Types of Radwaste

In this discussion we will focus not on environmental radiation but on radiation that is highly concentrated in the form of radioactive waste materials, or radwaste. The primary source of radwaste is nuclear power plants. Other sources include the decommissioning of nuclear weapons, scientific and medical laboratories, and rock wastes and tailings from uranium mining.

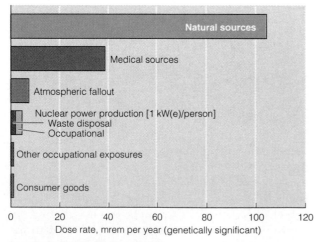

▲ **F I G U R E   16.17**
Major sources of environmental radiation.

Radioactive wastes are classified as *high level, intermediate level,* or *low level,* depending on the amount and type of radioactivity they emit and how long the harmful radiation persists.

Low-level Wastes Low-level wastes include a variety of materials that are generated in the normal operation of a nuclear reactor, such as mildly contaminated clothing and equipment (solid waste); cooling water (liquid waste); and ventilation exhausts (gaseous waste). Low-level liquid and gaseous wastes are generally released directly into a nearby body of water or into the atmosphere through a smokestack. The radioactivity in such materials must be below strictly regulated levels before their release. A problem with low-level gaseous emissions is *thermal inversion,* in which topographic or climatic factors cause the emissions to become concentrated in the vicinity of the release site. The release of low-level liquid wastes, especially water used as a coolant in the reactor, can also cause *thermal pollution* (excessive heat) in the receiving water body. Low-level solid wastes are generally put in metal canisters and buried in shallow disposal pits, either at the site where they were generated or at a licensed disposal facility.

Intermediate-level Wastes Wastes with levels of radioactivity that are too high to permit release directly into the environment are either chemically processed into their high-level and low-level components or held on-site until their radioactivity has dissipated enough for them to be disposed of as low-level wastes.

High-level Wastes Virtually all high-level radwaste comes from spent fuel. The fuel in nuclear reactors is in the form of bundles of metal alloy fuel rods containing pelletized uranium. The fission process in a reactor is initiated by bombarding the fissionable uranium with a neutron (refer to Fig. 12.10). The bombardment causes the uranium nuclide to split and release more neutrons (as well as alpha, beta, and gamma radiation). These neutrons go on to bombard neighboring uranium nuclei, which then split and release still more neutrons, and a fission chain reaction is initiated.

When the uranium nuclei split, daughter nuclei are created. Some of the daughter nuclei are also radioactive and will go on to decay spontaneously. Ultimately a very wide and unpredictable range of fission products is created. When the fission process has been going on for a while and a large number of daughter nuclei have been generated, the chain reaction will eventually begin to slow down. This happens because the daughter nuclei are absorbing so many of the fission-inducing neutrons that they are unable to bombard the remaining fissionable material effectively.

At this point the fuel rods must be removed from the reactor core. There will still be some fissionable material (i.e., uranium fuel) left in the fuel bundle. In principle, it is possible to recycle this material for use in new fuel rods. The

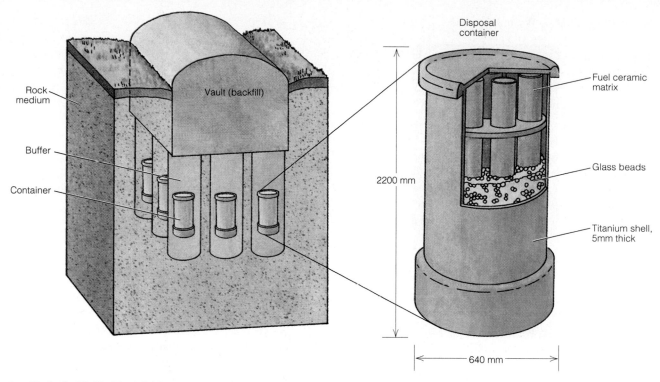

▲ FIGURE 16.18

A system of multiple barriers for preventing leakage of radioactive waste. Waste is first immobilized in glass or ceramic pellets and then surrounded by an absorptive buffering material and placed in corrosion-resistant canisters. The canisters are disposed of in a permanent repository.

remaining material is high-level radwaste, a highly radioactive, chemically diverse mix of fission products. Among the waste products of particular concern are ^{239}Pu (plutonium-239), which has a half-life of 24,000 years and is highly toxic, and ^{90}Sr (strontium-90), a beta-emitter that acts much like calcium (Ca) and can substitute for it in bone structures, making it exceptionally hazardous to human health.

Options for the Disposal of High-Level Radwaste

High-level radwaste is currently stored in water pools, large concrete bays, or underground storage tanks. These temporary storage facilities are not desirable for long-term storage. For permanent disposal, radioactive wastes must be isolated from any contact with the hydrosphere or biosphere for at least 10 half-lives. This means that many of the common fission products require isolation for hundreds or even hundreds of thousands of years. Isolation for at least 10,000 years is widely viewed as the minimum standard.

It is extraordinarily difficult to find a disposal method whereby these wastes can remain absolutely isolated for such a long time. This is why there are few operating permanent disposal facilities for high-level radioactive waste anywhere in the world, and none in North America. The search for an appropriate permanent disposal site has been under way for years in both Canada and the United States.

Several design, siting, and engineering criteria are crucial in establishing a permanent disposal facility. For example, in addition to being radioactive, high-level wastes boil continuously because they are constantly emitting heat. This places special constraints on the design of containers, which must be highly resistant to corrosion, and on the choice of materials with which to surround the waste. The facility must also be designed so that it cannot be easily entered or sabotaged, and it must be large enough to handle a sufficient volume of waste. It must be economically feasible, and it must be remote from current or potential human activities.

All the options for permanent disposal currently under consideration involve immobilizing high-level waste into the most inert form possible (usually a pelletized glass or ceramic), surrounding it with a buffer material, placing it in corrosion-resistant canisters, and disposing of the canisters in a permanent repository (Fig. 16.18). Some repositories that have received serious consideration are outer space, subduction zones, and ice caps; however, proposals involving geologic isolation on land are the most widely accepted.

Outer Space

Disposal of wastes in outer space is theoretically possible, but not technologically feasible at present. Under the best of circumstances it would be an expensive and risky undertaking. For example, the possibility exists that a waste repository might crash to Earth, with potentially devastating consequences, or that a spacecraft might collide with orbiting waste.

Subduction Zones

Subduction zones occur along plate boundaries where oceanic lithosphere is colliding with either continental or oceanic lithosphere (refer to Fig. 1.17). Proposals for disposal in subduction zones are based on the fact that deep-sea trenches, where subduction is actively occurring, have very high sedimentation rates. Any waste canisters deposited in such a trench would theoretically be covered within a relatively short period. We also know that some ocean floor sediments are subducted into the mantle along with the downgoing lithospheric plate; it is possible that waste canisters could also be subducted into the mantle.

The problem with this proposal is that we don't have a very detailed understanding of the processes that occur in deep-ocean trenches. For wastes to be buried quickly, the trench sediments would have to liquefy, allowing the canisters to sink to the bottom. It is theorized that trench sediments do occasionally liquefy, but it is not at all clear that this always happens. The technical challenges in transporting waste canisters to the bottom of the deepest parts of the ocean, and the political challenges in obtaining international agreements to do so, are also daunting.

Ice Caps

It has been calculated that a canister of high-level waste would melt its way to the bottom of the Greenland or Antarctic ice cap within a few years. The ice would close in behind the sinking canister, and the waste would wind up several kilometers down at the bottom of the ice sheet. The ice would be very effective at dissipating the heat generated by the waste, and the site would be remote from any current or likely future human activity. However, this option requires that wastes be transported over long distances, a risky proposition. It would also require international agreements to dispose of wastes in these remote areas. And scientists don't have a very thorough understanding of what might happen to the waste once it reached the bottom of the ice sheet.

Geologic Isolation on Land

The proposals that have received the most serious attention are those calling for land-based geologic disposal. There is general agreement among geologists that the ideal underground storage site for radioactive wastes should possess the following characteristics:

1. The enclosing rock should have few fractures and low permeability.
2. The enclosing rock should be effective at dissipating heat and chemically absorbing any potential leaks.
3. The enclosing rock should contain no minerals with present or future economic potential.

4. Local groundwater flow should be minimal and in a direction away from the biosphere.
5. Only very long paths of groundwater flow should be directed toward places accessible to humans.
6. The area should have low rainfall.
7. The zone of aeration should be thick.
8. The rate of erosion should be very low.
9. The probability of earthquakes or volcanic activity should be very low.
10. Future changes of climate in the region should be unlikely to affect groundwater conditions substantially.

The types of reservoir rock that are being investigated for land-based geologic isolation of high level radioactive wastes include shale, salt, tuffs, and crystalline rocks.

Shale Disposal of high-level wastes in shale would probably involve the use of deep wells like those described earlier for toxic wastes. This sounds like a rather risky plan until one considers the absorptive properties and relative impermeability of shale as a host for the wastes. A naturally occurring deposit of radioactive materials in shale is located in the African country of Gabon. The site has served as a "natural laboratory" where scientists have been able to observe, model, and test the migration of radioactive materials through shale. Their findings show that migration rates are very low and that shale could serve as an effective barrier in a repository for high-level wastes.

Salt Vaults You may recall from Chapter 11 that salt occurs in bedded layers within marine sedimentary sequences and also (as a result of flow) in upside-down teardrop-shaped deposits called *domes.* The mining of salt domes often leaves a cavity, called a *vault,* in the middle of the dome. Salt vaults are well suited for the disposal of radioactive waste for a number of reasons. The presence of a salt body implies that it has not come into contact with groundwater since it originally formed—otherwise the salt would have been dissolved and carried away in solution. Salt deforms plastically—that is, it tends to bend or flow rather than break or fracture. This means it is unlikely that any fractures would form and allow leaking wastes to enter the environment. Because salt is a good conductor, heat from the wastes would be dissipated effectively. Salt has been considered as a potential host rock for high-level wastes in both Canada and the United States, and an operating radwaste disposal facility in Germany is sited in an old salt vault.

Volcanic Tuffs The disposal site that is currently favored by many scientists in the United States is Yucca Mountain, Nevada (Fig. 16.19). It is situated in a thick sequence of volcanic tuffs (pyroclastics). The site was chosen because of the thickness and lateral continuity of the rock unit; its

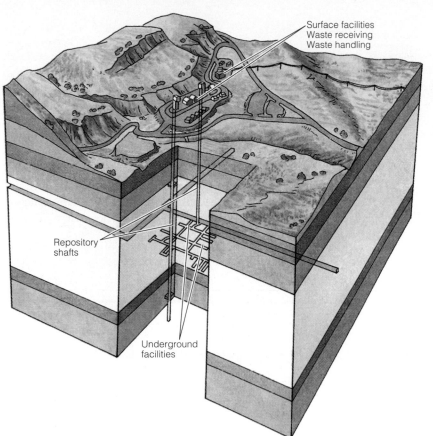

◀ F I G U R E 16.19
The proposed radioactive waste disposal site at Yucca Mountain, Nevada, would have extensive surface facilities for the handling of wastes, in addition to the permanent repository 300 m below the mountain.

highly impermeable, welded character; and its location in the unsaturated zone about 250 m above the water table. The rock contains some minerals, called *zeolites,* that may prove to be effective chemical absorbers should a leak ever occur. However, proposals for siting a facility at Yucca Mountain are highly controversial. Much of the debate has centered on rates of groundwater flow, the likelihood of fracturing, the potential for volcanic activity, erosion rates, and the future natural resource potential of the site.

Crystalline Rock Cavities   The preferred disposal option in Canada (Fig. 16.20), and one that has been looked at se-

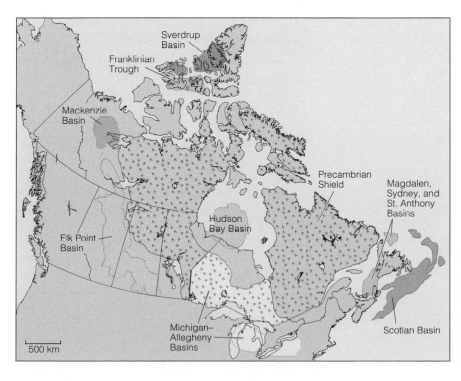

◀ F I G U R E 16.20
Areas of hard, crystalline rock in the Canadian Precambrian Shield, which are under consideration as potential host rocks for permanent storage of high-level radioactive wastes. The other colors denote basins that host salt deposits.

riously in the United States as well, is to excavate a large vault in very deep-seated, stable plutonic rock bodies like the granitic rocks of the Canadian Precambrian Shield. These old crystalline rock units have been tectonically stable for very long periods—hundreds of millions of years. There is a lot of this type of rock in Canada, much of it located in areas that are remote from human activity and potential resource interests. However, crystalline rocks have a tendency to fracture, especially when subjected to heat. At an underground research facility located in Lac du Bonnet, Manitoba, scientists are studying the fracturing behavior of granitic rocks and the potential for groundwater migration into these deep-seated areas.

Future Challenges

Historically, geologists have studied past events; when considering plans for the disposal of toxic and radioactive waste they are asked to predict future events. To do so with confidence, they must have considerable knowledge of local and regional groundwater conditions. They also need to understand how complex groundwater systems might respond to crustal movements, climatic change, and other natural factors that can affect the stability of a storage site. Finding a solution for the problem of long-term storage of toxic and nuclear wastes presents a major challenge for geologists as well as for policy makers.

SUMMARY

1. An understanding of geology is important in the management of waste, for three reasons: (1) wastes are generated through the use of Earth materials, and we need to understand the properties of the materials we are dealing with; (2) wastes and contaminants occur in the geologic environment, and we need to understand the nature of this environment and its transport processes; and (3) most waste management schemes involve the use of geologic reservoirs for containment or the use of geologic processes for dilution and dispersal of wastes.

2. The main sources of solid waste are agriculture and mining, followed by industrial and municipal sources. Agricultural solid waste consists primarily of organic materials and eroded sediment; the sediment often carries agricultural chemicals as well. Solid wastes from mining and processing of minerals consists primarily of piles of waste rock (gangue) and tailings.

3. Municipal solid waste is a mixture of various kinds of wastes produced by commercial, residential, and industrial sources. Municipal waste is a small but rapidly growing category of solid waste that is especially problematic because of its concentration in areas with high population densities where land for disposal is not readily available.

4. Today it is more common for municipal solid waste to be placed in sanitary landfills. Another disposal option is incineration. Community, industrial, and household composting and recycling programs can significantly reduce the volume of municipal solid waste.

5. In designing and siting a landfill, the primary concern is to prevent leachate from coming into contact with groundwater. To minimize production and migration of leachate, attention must be paid to the geologic characteristics of the site, such as depth to the water table, the porosity and permeability of surrounding materials, and the presence of an attenuation layer. Engineering features such as leachate collection systems, impermeable caps and liners, and monitoring wells can help prevent contamination of the surrounding environment.

6. Domestic and municipal sewage represents one of the most challenging problems of liquid waste disposal. Household sewage is often managed with a septic tank in which solid components settle out and clear effluents are slowly released into the surrounding regolith. Municipal sewage treatment can include a number of processes. *Primary treatment* is the mechanical removal of solid wastes such as trash and fine sediment. *Secondary treatment* is a biologic process, in which bacteria degrade dissolved organics in the effluent. *Tertiary treatment* is a chemical process that removes remaining contaminants, such as excess organic material, heavy metals, and inorganic dissolved solids.

7. Hazardous waste encompasses materials that pose a present or potential danger to humans or wildlife, including toxic (poisonous or carcinogenic), caustic, acidic, explosive, infectious, and radioactive materials. One of the biggest problems in dealing with hazardous waste is undoing the effects of past mismanagement of such waste.

8. Industry is by far the largest source of hazardous waste. Common examples of industrial hazardous wastes are acids, caustics, airborne particulates, ammonia, and persistent toxic chemicals such as heavy metals, chlorinated organic compounds, and polyaromatic hydrocarbons. Households also generate hazardous wastes through the use of such products as batteries, paints, solvents, and pesticides, which turn up in ordinary landfills with alarming frequency. The agricultural sector contributes hazardous materials to the environment primarily through runoff of toxic chemicals, including insecticides, herbicides, and fungicides.

9. In the past, hazardous materials have been disposed of by means of surface impoundment in aeration pits or lagoons. Now it is common practice to treat the materials (usually by incineration or chemical neutralization or stabilization) and then dispose of them in secure landfills. Geologic isolation in deep wells is another method for disposing of liquid hazardous wastes.

10. Everyone on the Earth is constantly exposed to environmental radiation from a variety of sources, including cosmic rays; natural decay of radioactive nuclides in rocks, minerals, and soils; medical sources; nuclear fallout; and other home and occupational sources. The normal operation of nuclear power plants contributes only a small portion of this radiation, but it generates high-level radioactive waste that represents an extraordinarily difficult disposal problem. Radioactive waste requires special treatment because some of the materials it contains can release harmful doses of radioactivity for a very long time.

11. High-level radioactive waste must be isolated from contact with the hydrosphere or biosphere. Therefore, a disposal site must have certain characteristics, including minimal fracturing and low permeability of the host rock; effective dissipation of heat and chemical absorbtion of potential leaks; no minerals with present or future economic potential; minimal goundwater flow; low rainfall and a thick zone of acration; a low rate of erosion; and a very low probability of earthquakes or volcanic activity.

12. Suggested repositories for the permanent disposal of high-level radioactive wastes include outer space, subduction zones, and ice caps. The most serious proposals involve land-based geologic isolation, which would entail stabilization of high-level waste in the form of glass or ceramic pellets, containment in corrosion-resistant canisters, and disposal in a permanent repository. Potential host rocks include shale, salt, volcanic tuff, and crystalline plutonic rocks.

IMPORTANT TERMS TO REMEMBER

attenuation layer (p. 416)
composting (p. 418)
deep-well disposal (p. 423)
environmental radiation (p. 428)

geologic isolation (p. 423)
half-life (p. 426)
hazardous waste (p. 420)
incineration (p. 416)

leachate (p. 414)
rad (p. 425)
radiation (p. 425)
radioactive waste (p. 424)

rem (p. 425)
sanitary landfill (p. 414)
secure landfill (p. 422)
septic tank (p. 419)

QUESTIONS AND ACTIVITIES

1. About 400,000 tons of municipal waste is generated each day in the United States. How much does this amount to per person per week? Try to calculate how much space might be taken up by 400,000 tons of waste. This is trickier than you might think; before you can do the calculation, you will need to find out the approximate proportions of different materials in the waste and the densities of those materials.

2. Where does your household waste go after it is picked up by the sanitation department on garbage day? Visit the local landfill and find out what kinds of engineering features it has. Is there a leachate collection system or monitoring wells? What is the underlying rock or sediment type? How high is the water table in the area?

3. If you live near a medium-sized or large city, the chances are good that your municipality is in the process of looking for a new landfill site (or has just completed such a search, or is about to start one). Find out what the criteria for site selection are (or were). In particular, what are the geologic criteria? Have certain environments been excluded from consideration because of their geologic or hydrologic characteristics? Is there a specific rock unit that has been identified as a desirable host for a landfill site? If so, what characteristics make it a potentially good host rock?

4. If you live near a nuclear power plant, try to arrange to visit it. Find out what happens to the various levels of wastes generated at the plant. Do any of the wastes remain on-site, or are they all shipped to off-site disposal facilities?

5. A visit to your local municipal sewage treatment plant can be very educational. Find out what level of sewage treatment is offered in your area. What happens to the treated effluents? How is the sewage sludge disposed of?

6. The controversy over the proposed nuclear waste disposal site at Yucca Mountain, Nevada, has been ongoing for many years. Find out more about the site's characteristics. What are the major pros and cons for its use for the permanent disposal of nuclear wastes?

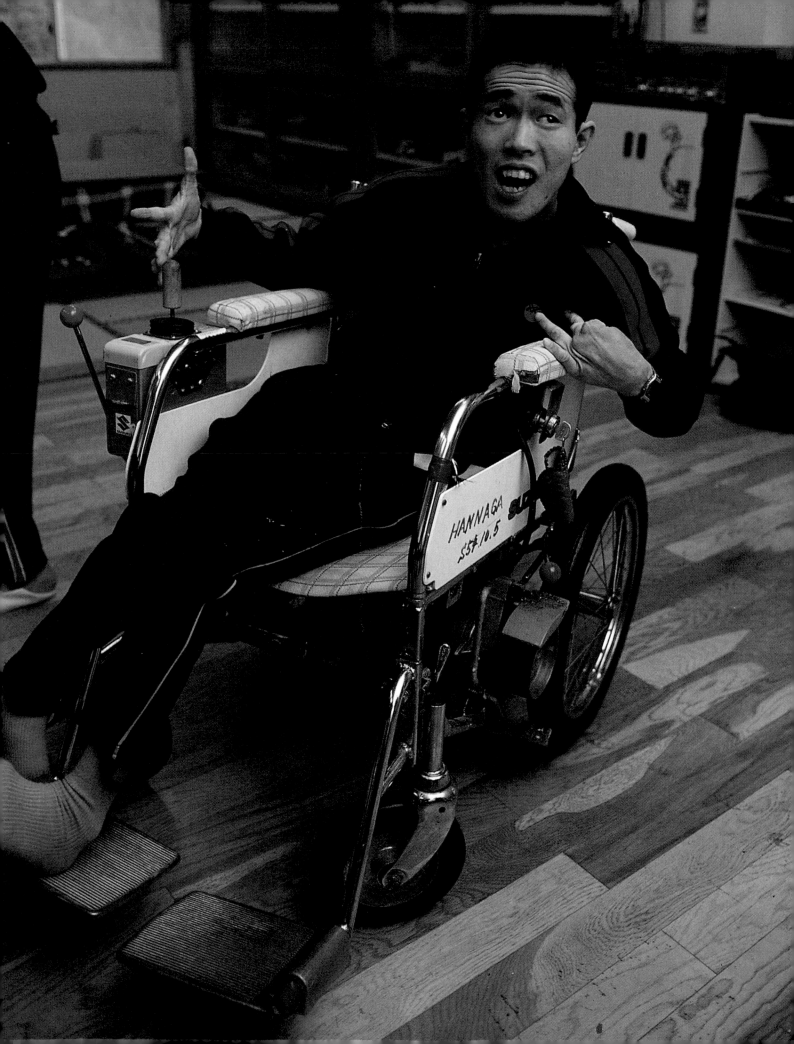

CHAPTER 17

CONTAMINANTS IN THE GEOLOGIC ENVIRONMENT

Don't look back. Something or somebody might be gaining on you.

• Satchel Paige, 1950

*I*n the 1950s the fishing village of Minamata, Japan, suffered a tragedy that came to be symbolic of the enormous human costs of pollution, and painfully instructive about the unforeseen problems associated with the disposal of hazardous materials. A chemical plant located in the village used mercuric oxide as a catalyst in its manufacturing process. The plant routinely (and legally) dumped mercury-laden liquid wastes into the adjacent bay. Local people depended on the waters of the bay for fish, a staple of their diet. In 1953 many of the villagers began showing symptoms of neurological disorders, such as blurred vision and speech and numbness in the lips and limbs. Children were born with serious congenital defects. People began to die.

Within a few years the problems had reached epidemic proportions, and their cause was a subject of intense debate. By 1959 it was evident that the villagers were suffering from mercury poisoning produced by the wastes from the chemical plant. Water-soluble forms of mercury had entered the marine food chain and become highly concentrated in the fatty tissues of fish, which were subsequently eaten by the villagers. Mercury levels in some of the fish were measured at 50 parts per *million* (mercury contamination is commonly measured in parts per *billion*). Astonishingly, the dumping of mercury-laden wastes into the bay was allowed to continue until 1971. Forty-three people eventually died from the effects of mercury poisoning, and it has been suggested that as many as 1000 others were affected.

In the 1950s, not only in Japan but throughout the world, the cardinal rule of hazardous waste disposal was "out of sight, out of mind." Chances are that, in the beginning at least, the managers of the chemical company had no idea that the mercury in the waste they were dumping would behave in such a manner or cause such horrific effects. Even if they had foreseen the problems, they might have felt no corporate responsibility since they were following legal and widely accepted procedures. These days we have a different perspective on the handling of contaminants and a very different attitude concerning the personal accountability of those whose actions produce situations like the one at Minamata.

WASTE AND POLLUTION

Waste and pollution are different issues, but they are so closely interrelated that it is difficult to discuss them in isolation from each other. Human-generated waste is not the only factor that has an impact on the natural environment; there are many processes and materials in the natural environment that can contaminate soils or degrade water quality. But by far the most serious occurrences of pollution involving soils, surface water, and groundwater are caused, either directly or indirectly, by anthropogenic wastes and emissions.

Remember, too, that waste isn't found only in garbage dumps, and waste management is not restricted to the organized disposal of large masses of material in landfills. Waste is generated by virtually every human activity. The

A victim of pollution. Physical deformity arising from mercury poisoning as a result of eating contaminated fish from Minamata Bay, Japan.

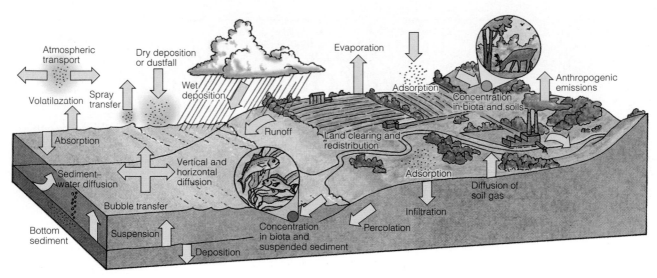

▲ F I G U R E 17.1
Some of the innumerable ways in which contaminants enter and are transported in the geologic environment.

harmful, contaminating, polluting components of waste are released into the environment from a *very* wide range of sources. Some of the possible entryways and pathways for contaminants in the geologic environment are shown in Figure 17.1. Hazardous materials are often released into the environment in small quantities over long periods, eventually accumulating to dangerous levels. They are transported through the environment in different forms along an almost infinite variety of pathways in water, in the air, in living organisms, and in rocks and soils.

In this chapter we begin by looking at the factors that control the transport, distribution, and concentration of contaminants in the environment. We then look in some detail at problems associated with sediment, surface water, and groundwater pollution, and at the processes—both natural and anthropogenic—that influence water quality. The chapter concludes by focusing on pollution in the marine environment.

BEHAVIOR OF CONTAMINANTS IN THE ENVIRONMENT

Wastes *can* be treated and disposed of properly, as we learned in Chapter 16. However, wastes and, more importantly, the harmful contaminants associated with some types of waste often evade proper treatment. A contaminant may enter the natural environment from any of a vast number of sources. Once released, it will be influenced by a variety of factors, including its rate of decay or decomposition; the length of time it remains in a given reservoir; the

mechanisms by which it is transported; and the way it interacts with other contaminants and with materials in the natural environment. Let's examine each of these factors in turn.

Point and Nonpoint Sources

Before a polluting substance can be dealt with properly, it is necessary to determine the nature of the source from which the material is emanating. **Point sources** are confined locations that can be represented by a single point on a map. Examples of point sources are a municipal sewage disposal pipe, a factory releasing contaminated liquids, a chemical spill, a landfill site leaking leachate, a mine, a slaughterhouse, and a drilling operation (Fig. 17.2A). Many point sources emit harmful substances into the environment at low concentrations; these emissions may be legal, but problems ensue when the substances accumulate in the environment. Other point sources are a result of accidental or illegal releases of contaminants.

Nonpoint sources of pollutants are not discrete locations but broad areas. An example of a nonpoint source is runoff carrying chemicals from agricultural land, a large urbanized area, or a suburban area (Fig. 17.2B). Air masses can also represent significant nonpoint sources of pollution. Even if a contaminating substance is released from a single smokestack (point source), the chances are good that it will be widely dispersed by atmospheric processes, eventually settling out over a broad area (nonpoint source). For example, it is estimated that more than 90 percent of the lead in Lake Superior has come from atmospheric deposition.

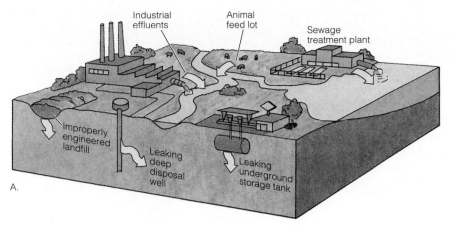

A. Some common point sources of pollution. B. Some common non-point sources of pollution.

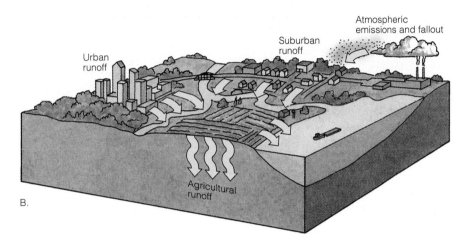

Point-source pollution tends to be more obvious than nonpoint-source pollution; a pipe spewing waste is hard to ignore, for example (Fig. 17.3). But it is inherently easier to manage point sources because they can be identified, tested, and monitored. Nonpoint-source pollution may originate from countless individual sources, each of which contributes only a very small amount of the contaminating substance. When taken together over a large area, small individual sources can add up to a very large and potentially uncontrollable pollution problem.

▶ F I G U R E 17.3
Chemically polluted waste water, Merseyside, England.

Decay and Decomposition

An important determinant of the behavior of a contaminant is its chemical stability. Many harmful substances naturally break down or decompose into other substances, either spontaneously or through interactions with other components of the environment. This can happen through **biodegradation,** the decomposition of a substance through the action of microorganisms; through radioactive decay, the spontaneous breakdown of unstable nuclides (discussed in Chapter 16); or through any of a wide range of chemical reactions through which compounds are altered and broken down.

Contaminants that take a very long time to decompose or to dissipate their harmful properties are referred to as **persistent.** Some of the substances that are of most concern in the disposal of radioactive waste are very persistent. For example, plutonium-239 (^{239}Pu) has a *half-life* of 24,000 years; that is, it takes that long for half of the ^{239}Pu atoms present to decay radioactively.

Another example of a persistent contaminant is dichlorodiphenyltrichloroethane, commonly known as DDT, a toxic chlorinated organic compound that is used as a pesticide. DDT does decompose in the environment, but not very quickly. It tends to remain in water or sediment for years. DDT and its breakdown products are retained in the tissues of living organisms through the process of *bioaccumulation,* and they increase in concentration in successively higher levels of the food chain through the process of *biomagnification* (Fig. 17.4). Together, these biologic processes leading to the buildup of contaminants in organisms are called **bioconcentration.** Even though the use of DDT has been restricted in North America since the early 1970s, the levels of DDT detected in organisms continued to increase until well into the 1980s.

Residence Time

Materials in the environment tend to cycle from one reservoir to another. As an example, consider lead, which is commonly added to gasoline to reduce engine knock. The lead used to produce the additive is derived from a geologic reservoir, most likely from a lead sulfide mineral, such as PbS (galena), in a hydrothermal ore deposit. When emitted in automobile exhaust, the lead enters the atmospheric reservoir, where it may remain for a couple of weeks. Eventually the airborne lead will settle and enter a hydrologic reservoir, finding its way into river water or the ocean via surface runoff. It may accumulate in bottom sediment and remain there for many years, or it may accumulate in the tissues of a plant or animal.

The average length of time a substance remains in a particular reservoir is commonly referred to as its **residence time.** For example, molecular nitrogen (N_2) has an average residence time in the atmosphere of about 44 million years.

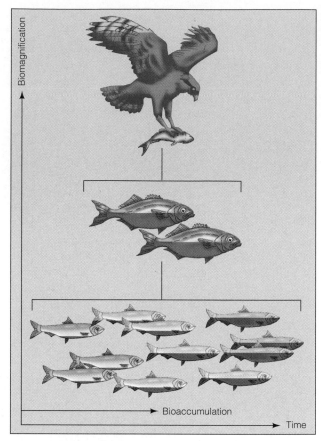

▲ FIGURE 17.4
Bioconcentration of toxic contaminants in the food chain. Through bioaccumulation, contaminants are retained and concentrated in the tissues of organisms. Through biomagnification, contaminants increase in concentration at successively higher levels of the food chain.

In other words, if you could identify and follow a number of individual nitrogen molecules and measure the amount of time they remain in the atmosphere before being recycled into water, soil, or some other reservoir, the average time measured would be about 44 million years. Water, by contrast, has a much shorter atmospheric residence time; the average water molecule remains in the (lower) atmosphere only 10 days before moving to another reservoir in the hydrologic cycle. The average residence time for lead in the atmosphere is also quite short—about 2 weeks—because, among other reasons, lead is heavy and therefore is not easily transported by air.

The concept of residence time can be applied to naturally occurring chemicals, such as water in the ocean or nitrogen in the atmosphere, and also to anthropogenic contaminants, such as lead from automobile emissions or DDT from agricultural runoff. Residence times are difficult to determine because they depend on a wide range of factors, including the physical and chemical characteristics of the material (such as its tendency to decay or biodegrade); the specific characteristics of the surrounding medium (air, water, soil, rock, or biota); the interactions between the contaminant and other chemicals present in the environment; and the capacity of the reservoir to hold the contaminant.

The size or holding capacity of the reservoir is particularly important in determining residence times. The average residence time of a substance, also called its *turnover time,* represents the amount of time it takes for the entire stock of that substance to cycle through a particular reservoir. For example, the air in your house periodically cycles through and is replaced with new air. How quickly the air exchange takes place depends on how big your house is and how leaky or airtight your doors and windows are. The same is true of natural reservoirs; residence time or turnover time—the amount of time required for the complete cycling of a material through the reservoir—depends on the size of the reservoir and the balance between rates of inflow and outflow of the material.

Transport Mechanisms

Once they have entered the environment, contaminants can be carried long distances by winds and water currents, as well as by living things that have absorbed them. If a substance is highly persistent, it may spread far from its point of origin. For example, samples of tissue from wildlife in the Arctic have been found to contain traces of PCBs and pesticides that originated in the industrial and agricultural regions of Europe, Asia, and North America (Fig. 17.5).

The mobility of some materials means that pollution from local sources can be augmented by contaminants originating in other areas. A situation of this kind occurs in the

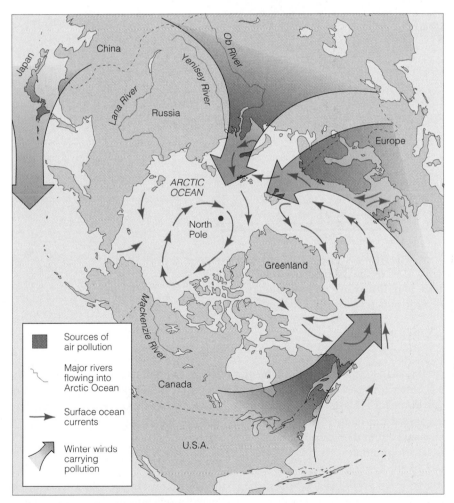

▲ F I G U R E 17.5
Transport of pollutants into the Arctic. Persistent contaminants may be transported long distances by winds, rivers, and ocean currents, sometimes accumulating far from their original sources.

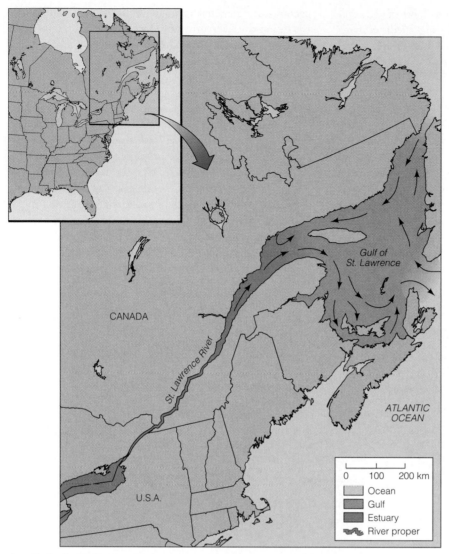

▲ F I G U R E 17.6
The St. Lawrence River, estuary, and gulf. The St. Lawrence is a 1500-km-long waterway that flows from the Great Lakes to the Atlantic Ocean. The estuary, on the northeast side of the system, is polluted by contaminants transported from the Great Lakes, hundreds of kilometers away at the southwest end of the system.

estuary of the St. Lawrence River, a 1500-km-long river and estuary system that flows from Lake Ontario to the Atlantic Ocean (Fig. 17.6). This fragile environment is affected not only by pollution from industry, agriculture, and shipping along the river, but also by contaminants from the Great Lakes basin. The pesticide Mirex, for example, reaches the estuary (in the northeast part of the system) from the Great Lakes (hundreds of kilometers away in the southwest part of the system), rather than from local sources.

Another consequence of the transport and dispersal of contaminants is that national or regional bans on the use of specific chemicals cannot be completely effective in con-

trolling them if the substances remain in use in other parts of the world. For example, although most applications of DDT have been banned in North America, the pesticide is still used in many parts of the world.

Interactions and Synergistic Effects

When contaminants are transported in the environment, they do not travel in isolation. Instead, they mix and interact chemically with the surrounding or transporting medium, as well as with other chemicals and contaminants. These interactions can directly affect the behavior of contaminants. For example, studies have shown that some chemicals that by themselves are only mildly toxic to wildlife can become many times more toxic when mixed with two or three other chemicals. This effect, in which the overall toxicity of the mixture is greater than the sum of the individual toxicities of the compounds, is called *synergism*. In a system such as the Great Lakes, in which at least 362 chemical contaminants have been identified, the potential for dangerous synergistic effects is very great.

Sometimes interactions between a contaminant and the surrounding medium serve to *decrease* the concentration of the contaminant. For example, negatively charged particles of clay in the silty bottoms of streams attract dissolved metal ions, which are positively charged. This is a natural mechanism through which both surface water and groundwater can be purified of some contaminants. Over time, however, the process can lead to significant accumulations of those contaminants in sediments. This is not necessarily a problem as long as the sediment remains undisturbed, but if it is disturbed—by dredging or excavation, for example—it can liberate potentially hazardous substances.

SEDIMENT POLLUTION

Sediment pollution is a term with two very different meanings. It can refer to the contamination of sediment by hazardous substances or to situations in which the sediment itself acts as a pollutant.

Siltation

Pollution in which the sediment itself is the pollutant is one of the negative impacts of accelerated soil erosion (see Chapter 14). The topsoil that washes away from agricultural land has to end up somewhere, and most often it ends up in adjacent waterways. Problems caused by eroded soil washing into waterways include reduced channel capacity because of silt accumulation; destruction of fish habitat; acceleration of algal growth; buildup of heavy metals, pesticides, and other toxic substances; loss of recreational value; and cleanup costs. (It has been estimated that the economic impacts of topsoil loss from agricultural land may be as great *off* the farm as *on* the farm.) Rates of erosion and resulting siltation also tend to be very high in areas with high rates of deforestation, such as Malaysia. Most of the estuaries in Malaysia are heavily silted; partly as a result of the siltation, the number of fish in coastal waters has decreased dramatically and coral reefs have died.

Contamination of Sediment

Where sedimentation and siltation are a problem, it is likely that the buildup of toxic substances will also be a source of concern. Eroding sediments represent one of the primary mechanisms by which contaminants are transported from land into surface water bodies. For example, phosphorus, heavy metals, pesticides, and other organic compounds can bind themselves to soil particles. The eroding soil carries the substances along to the surface water body. Contaminated sediment at the bottom of a body of water is of particular concern because it can act as a reservoir, holding toxic substances and slowly releasing them into the water. Bottom-dwelling organisms may be affected, leading to bioconcentration of the contaminant.

Sources of chemical contamination in sediment include agricultural fallout; runoff from urban, suburban, and agricultural areas; runoff from mining and logging operations; and infilling along shorelines. It is common practice to build up shorelines and create more land area by infilling with rubble, bricks, concrete, and soil, materials that together are referred to as *fill*. If some of the materials in the fill are contaminated, both water and sediment quality may eventually be affected.

When sediment contamination becomes a problem, it may be necessary to either remove the contaminated sediment pile or attempt to clean it up. One of the concerns arising from contaminated sediment is that if it is disturbed (e.g., by dredging) it will be stirred up and may release its contaminants into the water more readily than if it had been left undisturbed. If the sediment is severely contaminated, or contaminated with an acutely toxic substance, it may be more desirable to leave the sediment in place and isolate it physically. This is usually accomplished by capping the contaminated sediment with a layer of clean sediment or an impermeable plastic liner.

WATER QUALITY

Before we examine pollution in surface water and groundwater, let's look briefly at some of the factors that influence water quality. Citizens of modern industrialized nations take it for granted that when they turn on a faucet, water that is safe and drinkable will flow from the tap. However, throughout much of the world tapwater is unfit for human consumption. Not only do natural dissolved substances make the water undrinkable in some places, but many municipal water supplies are contaminated by human and industrial wastes.

Quantifying Water Quality

The specific, legal definition of "good" water quality varies according to the intended use of the water—that is, whether it is intended for drinking or swimming or as a home for fish and wildlife. As well, different jurisdictions have different legal standards for water quality control. Water quality is usually measured by scientists or technicians, who take samples of well water, groundwater, or surface water in the field. Portable instruments are used to measure such characteristics as temperature, turbidity (i.e., how cloudy the water is), dissolved oxygen, alkalinity, and pH. Then the water samples are brought to a lab, where other instruments (such as gas chromatographs, mass spectrometers, and plasma emission spectrometers) are used to detect and quantify the abundance of a wide variety of possible contaminants.

Many of the substances that are of concern in water quality control are present in minute quantities. For example, 1 gram of 2,4-D (a common household herbicide) can render 10 million liters of water undrinkable, and 1 gram of PCB can make a billion liters of water unsuitable for aquatic life. Sophisticated equipment must be used to detect very low concentrations of these substances. The testing and lab work required to characterize one sample completely with respect to a compete range of possible contaminant substances may cost hundreds of dollars.

Natural Influences

Much of the discussion in this chapter focuses on contamination resulting from wastes generated by human activities; however, a variety of natural processes can also influence water quality. The quality of water varies naturally with climate, season, and the geology of the local bedrock. Natural variations can have a negative impact on the suitability of water for human use. In some instances, natural characteristics of a water body can also influence its ability to resist acidification or its ability to absorb or dilute contaminants.

Analyses of many wells, springs, and surface water bodies show that the compounds dissolved in natural waters are mainly chlorides, sulfates, and bicarbonates of calcium, magnesium, sodium, potassium, and iron. These substances are derived from minerals in the rocks from which they have been weathered. As might be expected, the composition of water varies from place to place according to the kind of rock in which the water occurs. Where limestones are abundant, for instance, the water is likely to be rich in calcium and magnesium bicarbonates dissolved from the local bedrock. Taking a bath in such water, called *hard water*, can be frustrating because soap does not lather easily and a crustlike ring forms in the tub. Hard water also deposits a scaly crust in pipes, eventually restricting the flow of water. By contrast, water that contains little dissolved matter and no appreciable calcium is called *soft water*; it is commonly found in regions underlain by volcanic rocks and sandstones.

Water sometimes dissolves noxious elements from rocks through which it flows, making the water unsuitable for consumption. This is especially true of groundwater, which flows slowly and remains in contact with surrounding rocks or sediments for long periods. Water circulating through sulfur-rich rocks may contain dissolved hydrogen sulfide (H_2S) which, although harmless, has the disagreeable odor of rotten eggs. Dissolved iron can also affect the taste and color of water. In some arid regions, the concentration of dissolved materials, especially sulfates and chlorides, is so great that the groundwater is unusually noxious.

Toxic materials, such as mercury, arsenic, fluoride, and uranium, also can occur naturally in both groundwater and surface water. For example, mercury has been found to occur at very high levels in fish harvested from isolated lakes in the Arctic. For mercury to become concentrated in these systems, it must either be released gradually from underlying rocks or sediments or be transported through the atmosphere from a distant source. Studies have shown that the locations of some of these lakes correlate closely with known geologic sources of mercury but do not show a clear relationship to distance from smelters or other industrial sources of airborne mercury. This suggests that the gradual release of mercury from underlying geologic sources may be a source of mercury contamination of some surface water bodies.

Rates of flow can affect a stream's ability to dissolve or absorb contaminants. When the rate of flow is low, water quality tends to decline because there is less water available to dilute the materials entering the water. Low rates of flow also create poorer conditions for aquatic life because the water is generally warmer. The opposite extreme—an exceptionally high rate of flow—is not necessarily better. During periods of heavy rain or snowmelt, runoff carries an increased amount of dissolved and suspended material to streams. Water pollution and the contamination of domestic water supplies are of particular concern during floods. Thus, both high and low rates of flow can result in a reduction in water quality.

POLLUTION OF SURFACE WATER

Contaminants in surface water come primarily from urban, suburban, and agricultural runoff. Industrial effluents, particularly those produced by the pulp and paper industries and chemical manufacturing, are significant contributors. So are discharges related to resource extraction, among which mining, logging, and the petroleum industry are important. Other sources of contamination are airborne acid-forming substances (acid precipitation), sewage, and landfill leachates. In this section we examine some of the ways in which these effluents act to contaminate surface waters.

Organic Contamination

Many types of waste contain abundant organic material. We tend to think of anything "organic" as being "environmentally friendly." However, too much of a good thing can have some very negative effects on surface water bodies.

BOD

When organic matter biodegrades in an *aerobic* (oxygen-bearing) environment, oxygen is used up in the process. (Other types of biodegradation, generally much slower, can

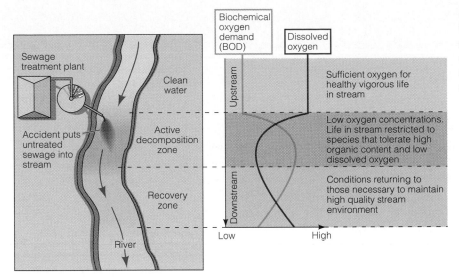

occur in *anaerobic* environments. Anaerobic biodegradation is carried out by different microorganisms that do not require oxygen.) Oxygen, usually measured in terms of the amount of dissolved oxygen in the water, is critical for the maintenance of most aquatic life forms. Effluents that cause the depletion of dissolved oxygen in a water body are said to place a **biochemical oxygen demand, or BOD,** on the system.

The greater the amount of organic matter discharged into the system, the higher the BOD. For example, the lower the level of treatment given to sewage effluents, the greater the BOD on the body of water into which the effluents are discharged; dissolved oxygen in the water responds by becoming depleted. Figure 17.7 shows the increase in BOD and the corresponding decrease in dissolved oxygen

that typically occur just downstream from a source of organic-rich effluents.

Eutrophication

One of the most common forms of surface water contamination results from an excess of organic material or *nutrients,* especially phosphorus and nitrogen. Nutrients are, of course, necessary to sustain plant growth. But when a system is overloaded with nutrients, plant growth runs out of control. In an aquatic system that has been overfertilized, excessive growth of aquatic weeds, plankton, and algae commonly occurs. When the growth of plankton and algae is uncontrolled (a situation called an *algal bloom*), oxygen in the water becomes depleted (Fig. 17.8). Other aquatic organisms die; the water usually turns green and mucky

▼ F I G U R E 17.8
Advanced case of algal growth as a result of high phosphorus levels in the water.

RESTORING DEGRADED WATERWAYS:
THE KISSIMMEE RIVER PROJECT

*T*he Kissimmee River begins in a chain of lakes just south of Orlando, Florida, and meanders southward toward Lake Okeechobee. At the southern edge of this very broad, shallow lake the water periodically spills over, flowing through and replenishing the swampy Everglades ecosystem. Beginning in the early 1900s the whole system underwent extensive engineering modifications. Many kilometers of dams and canals were constructed, and agriculture and urban development expanded onto the newly drained land. The expansion brought new hazards to land owners from flood-prone Lake Okeechobee, and the U.S. Army Corps of Engineers eventually encircled the lake with a massive dam to prevent further flooding.

The Kissimmee River, which originally carried the water to feed the entire Okeechobee–Everglades system, was particularly flood-prone. In 1961 the Corps of Engineers undertook a program of intensive channel modifications on the river, designed to control flooding. The channelization involved straightening and deepening the meandering river. Ten years and nearly $30 million later, a major stretch of the Kissimmee River, formerly 160 km long, had been turned into an 83-km-long canal cut straight through the floodplain.

Almost immediately it became apparent that the channelization was having unanticipated effects on the entire Everglades watershed. With water in the upper parts of the system increasingly being diverted for agricultural, municipal, and industrial uses, the Everglades—where water is the lifeblood of the ecosystem—began to run dry (refer to Box 15.3). In the altered flow regime, nutrient enrichment and eutrophication became serious problems; Lake Okeechobee was declared a "dying" lake. Along the river itself, flooding was still not under control. In addition, oxbow meanders isolated from the main flow had become stagnant pools, and 40,000 acres of wetlands had been destroyed.

As early as 1971—only months after the channel modifications had been completed—environmentalists began pressuring water managers to return the river to its original condition, and in 1976 the State of Florida passed the Kissimmee Restoration Act. However, restoring the river to its natural state will be a complicated endeavor. Lengthy experimentation will be needed to discover whether it will even be possible to rehabilitate the river. So far the consensus is that restoration is possible, and some test segments have already been reflooded. This represents the first time in history that an Army Corps of Engineers project has been reversed, and the costs will be high. It is estimated that it may cost as much as $300 million to undo the damage caused by the Kissimmee River channelization.

looking; and the entire system stagnates. This process is called **eutrophication.**

Eutrophication occurs naturally during the life cycle of a lake; over long periods of geologic time the lake will eventually be turned into a swamp or bog. Problems arise, however, when the natural process of eutrophication is tremendously accelerated by the addition of excess nutrients from anthropogenic sources. Runoff from agricultural land tends to be very rich in fertilizers, which are chemical nutrients. Untreated sewage discharged into a surface water body may also contain excessive quantities of nutrients. Phosphate-bearing detergents and effluents from food-processing plants can also contribute organic-rich material to surface waters.

Lake Erie provides an example of the problems associated with eutrophication. In the 1960s the lake began to turn an unhealthy-looking green color. This was caused by a bloom of bluegreen algae resulting from an excess of nutrients (especially phosphorus) in the water. Beaches were covered with green, slimy masses of rotting algae. Oxygen in the lake water was depleted (a condition called *anoxia*), and many aquatic species began to die off. Although oxygen depletion in deep waters is still a concern, Lake Erie has largely recovered from the effects of eutrophication. The recovery was an outcome of the Great Lakes Water Quality Agreement (1972), an agreement between Canada and the United States that limited the discharge of phosphorus from point sources (Box 17.2).

Infectious Agents

The presence of infectious agents is another form of surface water contamination associated with organic-rich effluents. Bacteria, viruses, and other disease-causing organisms may be found wherever human or animal wastes are among the substances contained in effluents. Measurements of the level of harmful microorganisms in water are usually based on the abundance of a particular fecal coliform bacterium, *Escherichia coli* (or *E. coli*), which lives in the intestines of humans and some animals. In many parts of the world, consumption of water contaminated with infectious agents is linked with outbreaks of diseases such as cholera. In North America, cholera and other waterborne diseases are less of a problem than they were 100 years ago; still, high levels of *E. coli* routinely cause beach closures in many coastal areas.

Toxic Contamination

Many of the pollutants found in surface water bodies are toxic substances with a wide range of physical and chemical characteristics. About 10 million chemicals are known, and 100,000 of them are used commercially. Many of these are routinely released into surface waters in small quantities, yet complete toxicity information is available for only a very small percentage of them. Some highly toxic substances, such as chlorine and cyanide, have been used in industrial processes since the 1800s and even earlier. In some respects these substances are relatively easy to control, since it is obvious that they require special handling. Chemicals that are toxic only after prolonged exposure at very low concentrations, such as DDT and some other pesticides, some heavy metals, and some PCBs, are more problematic because their presence and toxicity are less obvious.

Defining Toxicity

So far we have used the words *toxic* and *toxicity* without defining them precisely. In a general sense, **toxicity** is the capacity of a substance to cause adverse effects in a living organism. According to this very broad definition, any characteristic that enables a substance to damage a living organism would constitute toxicity. Such characteristics might include radioactivity or excess heat, for example, although in discussing toxicity we usually think in terms of chemical "poisoning" by toxic substances. A substance is *acutely toxic* if it acts quickly (within 96 hours) to produce lethal or damaging effects. If it takes a longer time to cause changes in growth or reproduction or to cause genetic mutations or death, the substance is said to exhibit *chronic toxicity*. Some substances have both acute and chronic effects.

Several problems arise when one attempts to determine the level of toxicity of a particular substance in the environment. A substance may be toxic to one type of organism but not to another. Or it may be toxic only if it is present above a certain concentration. Some forms of a substance may be toxic whereas other chemical forms of the same substance may be harmless. For example, inorganic mercury and phenyl mercury—common constituents of mercury-rich industrial effluents—are harmless and tend to accumulate in bottom sediments. However, bacteria can convert these forms of mercury into toxic methyl mercury compounds, which enter the food chain and become concentrated in the tissues of organisms.

Toxic Substances

Both persistent and nonpersistent contaminants can be toxic. Nonpersistent, acutely toxic substances can "shock" river and lake ecosystems, suddenly killing or severely stressing aquatic life. Persistent substances have the potential to cause both acute and chronic problems and can affect an ecosystem long after the source of the pollution has ceased to exist. Several major groups of substances occur as anthropogenic toxic contaminants in both surface and groundwater. The chemical names of these substances look very complicated, but many of them will be familiar to you from news reports about incidences of contamination.

Chlorinated Organic Compounds Some of the most significant toxic contaminants are chlorinated organic pesticides such as DDT and Mirex. Dioxins and furans, some of which are highly toxic, are byproducts of chemical processes using chlorine. Another group of contaminants consists of polychlorinated biphenyls (PCBs), some of which are highly toxic, persistent, and easily dispersed in the environment. PCBs are used as lubricants and heat exchangers in electrical transformers and capacitors, among other applications.

Heavy Metals Heavy metals, including cadmium, lead, tin, plutonium, and mercury, are released during mining and ore processing, metallurgical procedures (such as electroplating), paper manufacture, petroleum refining, and many other industrial processes. Of these, lead and mercury have received the most attention because they are so ubiquitous and have such devastating effects on living organisms. Long-term exposure to even very low levels of these substances can cause neurologic damage (especially in children), kidney damage, and a variety of other severe problems. As shown in Figure 17.4, heavy metals tend to bioaccumulate in the fatty tissues of organisms that eat them, eventually becoming concentrated at the higher levels of the food chain.

Hydrocarbons Many hydrocarbon compounds are toxic. One of the most common is benzene, which comes from industrial effluents and oil spills as well as from incomplete combustion of hydrocarbon fuel in incinerators and vehi-

BOX 17.2
•
THE HUMAN PERSPECTIVE

THE GREAT LAKES: A COMMON CONCERN

The Great Lakes represent a valuable shared resource and a common concern of the United States and Canada. The five lakes are a reservoir for the world's largest supply of fresh water, although there is a wide range in volume between Lake Superior's 12,100 km^3 and Lake Erie's 484 km^3. During the twentieth century, however, the quality of Great Lakes water has changed from pristine to polluted. Runoff carrying sediments and dissolved nutrients from farms in the Great Lakes Basin has accelerated the eutrophication process, and chemical dumping by industries located on the lake shores has threatened fish and other species in the lake waters.

Although all the Great Lakes are polluted, Lake Erie was actually declared "dead" in the 1960s. Because of its small size and shallow depth, Lake Erie did not contain the volume of water needed to dilute the wastes created by dense populations and industrial activities on its shores. In addition, the lake received extremely high levels of agricultural phosphorus, which overfertilized the water and caused algal blooms, depletion of the oxygen supply in the water, and other problems. As a consequence, some of Lake Erie's beaches were closed, the water was covered by a thick algal slime, and several native fish species became extinct or endangered. Lake Erie was the early warning signal for the whole Great Lakes Basin.

Lake Erie's condition has improved since the 1970s. In 1972 the United States banned the use of phosphates in laundry detergents. This act alone reduced 40 percent of the phosphorus entering Lake Erie and improved the water quality downstream in Lake Ontario as well. Lakes Michigan, Superior, and Huron also benefitted from the phosphorus ban. A more serious problem for all the lakes is contamination by hundreds of toxic chemicals and heavy metals. These contaminants, which are lethal to organisms, are found in effluents from sewage treatment and industrial plants, runoff from farmlands, contaminated fill, and emissions from fossil fuel combustion. Once in the ecosystem, they become concentrated at higher levels of the food chain through biomagnification and bioaccumulation.

In 1972 the United States and Canada signed the Great Lakes Water Quality Agreement to address these problems.

Since then it has become obvious to both countries that it will not be possible to eliminate all pollutants; instead, priority should be given to the most contaminated sites (Fig. B2.1) and limits should be placed on the most dangerous contaminants. Separately and together, both countries are taking steps to clean up the Great Lakes and ensure the long-term viability of the ecosystem.

Several examples of such actions illustrate the trend toward more stringent water quality policies. Through the International Joint Commission (IJC), the United States and Canadian governments initiated the Remedial Action Plan with the goal of cleaning up the most badly contaminated sites around the Great Lakes. Canada's federal government and the provincial government of Ontario signed an agreement committing $1.1 billion over a 6-year period to restore and protect the Great Lakes Basin ecosystem. The province of Ontario also initiated the Municipal/Industrial Strategy for Abatement program, which is aimed at reducing or, in some cases, eliminating the discharge of contaminants into the lakes and feeder streams. In a broad overhaul of the 1972 Clean Water Act, the United States proposed the Great Lakes Water Quality Initiative, the first effort to standardize water quality control for the eight Great Lakes states. Finally, the North American Free Trade Agreement (NAFTA) established the Commission on Environmental Cooperation to oversee conservation efforts and settle environmental disputes arising from trade activities. NAFTA is the first commercial agreement to explicitly address environmental issues.

On the other hand, some collaborative efforts have not been successful. For example, the IJC has called for a phaseout of the production and use of chlorine and a number of other chemicals; 23 of the 42 chemicals (or classes of chemicals) known to have damaging effects on reproductive systems and fetal development contain chlorine. Many of these chemicals are already present in Great Lakes water and sediment. The IJC suggested a managed transition away from chlorine, to be funded by a bilateral tax to mitigate economic losses. The recommendations were rejected by both the United States and the Canadian governments.

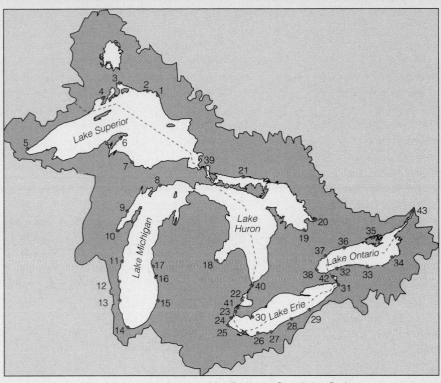

Forty–three areas of concern identified in the Great Lakes Basin

Lake Superior

1. Peninsula Harbour
2. Jackfish Bay
3. Nipigon Bay
4. Thunder Bay
5. St. Louis Bay/River
6. Torch Lake
7. Deer Lake–
   Carp Creek/River

Lake Michigan

8. Manistique River
9. Menominee River
10. Fox River/Southern Green Bay
11. Sheboygan River
12. Milwaukee Estuary
13. Waukegan Harbor
14. Grand Calumet River/
    Indiana Harbor Canal
15. Kalamazoo River
16. Muskegon Lake
17. White Lake

Lake Huron

18. Saginaw River/Saginaw Bay
19. Collingwood Harbour
20. Severn Sound
21. Spanish River Mouth

Lake Erie

22. Clinton River
23. Rouge River
24. River Raisin
25. Maumee River
26. Black River
27. Cuyahoga River
28. Ashtabula River
29. Presque Isle Bay
30. Wheatley Harbour

Lake Ontario

31. Buffalo River
32. Eighteen Mile Creek
33. Rochester Embayment
34. Oswego River
35. Bay of Quinte
36. Port Hope
37. Metro Toronto
38. Hamilton Harbour

Connecting Channels

39. St. Mary's River
40. St. Clair River
41. Detroit River
42. Niagara River
43. St. Lawrence River
    (Cornwall/Massena)

▲ F I G U R E B2.1

The Great Lakes' most contaminated "hot spots," identified by the International Joint Commission.

cles and even from tobacco smoke. Another group of potentially toxic hydrocarbon compounds that has received a lot of attention is the polycyclic aromatic hydrocarbons (PAHs), which also come from incomplete combustion of gasoline, wood, and coal.

Acidic and Caustic Effluents

Some hazardous chemicals are harmful to the environment because they make water more *acidic* (e.g., battery acid or sulfuric acid) or more *alkaline* (e.g., lye, a caustic substance). Either of these chemical changes, if substantial, can render an ecosystem inhospitable to most of its indigenous life forms. One problem resulting from the release of such substances is the acidification of lakes through the deposition of airborne acids. (We discuss acid precipitation in Chapter 18.) *Acid mine drainage* also causes acidification of surface water bodies.

Acid Mine Drainage

When piles of waste rock or tailings are left exposed, the waste will eventually interact with rainwater, groundwater, or surface water. If the waste pile contains coal or sulfide minerals, this interaction is likely to produce sulfuric acid (H_2SO_4). The acidified runoff from the waste pile is called **acid mine drainage** or *acid runoff.* Acid mine drainage was first officially noted in 1698 in Pennsylvania coal mines. In addition to acids, mine runoff can contain toxic contaminants such as heavy metals or cyanide. Other problem substances, such as asbestos fibers, salt, and radioactive materials, have also been found in runoff from mine waste.

Buffering of Surface Water Acidification

Bedrock geology and soil chemistry are the main factors that determine whether or not a surface water body will be vulnerable to acidification (Fig. 17.9). As noted earlier, the chemistry of a surface water body is controlled to a great extent by the chemistry of the materials in which it occurs. Sedimentary rocks, which are soft by comparison with most igneous and metamorphic rocks, tend to have rela-

tively high rates of weathering, erosion, and soil formation. This means that soluble salts that are capable of counteracting or *buffering* the effects of acidification will be released into surface waters at a high rate. Lakes that form on calcareous sedimentary rocks such as limestones, which contain large amounts of acid-neutralizing $CaCO_3$ (calcite) and $CaMg(CO_3)_2$ (dolomite), therefore are resistant to acidification. In contrast, igneous and metamorphic rocks tend to weather and erode more slowly, and their mineral constituents tend to be less water soluble. Lakes that occur on hard, crystalline rocks, especially siliceous rocks such as granites, tend to be highly vulnerable to acidification. Sometimes lake acidification can be counteracted by the addition of crushed or powdered limestone.

Alkalinization and Salinization

Anthropogenic activity can also cause increases in the alkalinity of a surface water body, although this is much less common than acidification. The sudden release of a caustic substance, such as ammonia or lye, would have this effect. Certain types of land use can also contribute to increases in alkalinity. For example, accelerated soil erosion releases mineral salts that dissolve in water to form bases that can raise the alkalinity of the water. An increase in overall salinity, that is, the total amount of dissolved salts, can also result from runoff from road salting or improperly designed irrigation programs.

Thermal Pollution

Many industrial and energy-producing processes release effluents that are warm or hot compared to the receiving environment. Excess heat, or **thermal pollution,** can harm or kill both plants and animals. When the temperature of water rises, the amount of dissolved oxygen also decreases. Thus, another negative impact of thermal pollution is to lower the oxygen content of surface waters into which the heated effluent is released. It is usually possible to control the effects of heat by allowing effluents to cool before they are released into the environment, but this is not very effective at raising the dissolved oxygen content of the effluents.

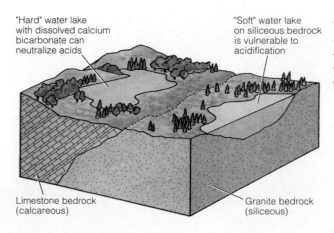

"Hard" water lake with dissolved calcium bicarbonate can neutralize acids

"Soft" water lake on siliceous bedrock is vulnerable to acidification

Limestone bedrock (calcareous)

Granite bedrock (siliceous)

◀ F I G U R E 17.9
A lake situated in granitic bedrock is more susceptible to acidification than a lake situated in limestone. Calcium carbonate in the limestone helps neutralize or buffer the effects of acids added to the lake water.

Suspended Solids

Not all contaminants occur as dissolved constituents in the water. Particles ranging from sand grains to submicroscopic particles called *colloids* can be carried in suspension by surface waters. As discussed earlier, any type of sediment, including ordinary sand, silt, and clay, can act as a pollutant if it accumulates in harmful concentrations. Some industrial processes also discharge suspended solids that pollute surface waters. Pulp and paper mills discharge hundreds of tons of wood waste every day. Since wood pulp is an organic material, it places a BOD on the water into which it is discharged. A more serious problem with suspended solids, however, is that they tend to settle and accumulate on the bottom of the water body, where they can literally suffocate bottom-dwelling organisms.

POLLUTION OF GROUNDWATER

Many of the types and sources of groundwater contamination overlap with those that cause surface water and sediment contamination. However, although many of the same substances cause pollution in both surface and subsurface waters, the processes whereby these substances enter the system, their interactions with the surrounding medium, and the processes whereby they are transported through the environment are very different. In addition, because of its hidden nature, groundwater contamination is inherently harder to detect, control, and clean up than surface water or sediment contamination.

Sources and Types of Groundwater Contamination

A potentially serious contaminant of groundwater has four basic characteristics: (1) it is water soluble, (2) it is resistant to biodegradation, (3) it is used in large quantities, and (4) it is toxic or noxious to humans. Dioxins, for example, although they can be highly toxic in very small doses, are not particularly water soluble, so they typically represent more of a soil contamination problem than a groundwater contamination problem. The most widespread of all groundwater pollutants—although not as toxic as many others—is nitrate, which comes from fertilizers, sewage, and landfills, among other sources. Twenty of the top 25 groundwater contaminants are volatile organic compounds, like benzene, toluene, ethylene, and xylene (together referred to as *BTEX*), all of which are constituents of gasoline. Chlorinated organic solvents such as trichloroethylene (TCE), commonly known as dry-cleaning fluid, also pose a significant threat to groundwater quality in many sites.

Pollution by Sewage

The most common source of water pollution in wells and springs is untreated sewage. Drainage from septic tanks and cesspools, broken sewers, and barnyards contaminates groundwater. If water contaminated with sewage passes through sediment or rock with large pores, such as coarse gravel or cavernous limestone, it can travel long distances and remain polluted (Fig. 17.10). On the other hand, if the contaminated water percolates through sand or permeable sandstone, it can become purified within short distances, in

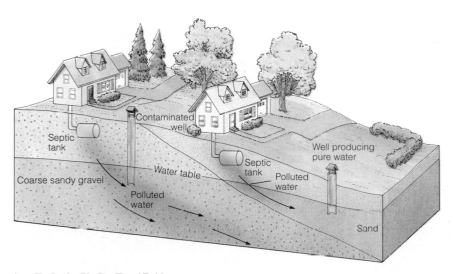

▲ **FIGURE 17.10**
Purification of groundwater contaminated by sewage. Pollutants percolating through a highly permeable sandy gravel contaminate the groundwater and enter a well downslope from the source of contamination. Similar pollutants moving through permeable fine sand higher up are removed after traveling a relatively short distance, and do not reach a well downslope.

▲ F I G U R E 17.11
Removing a leaking tank at a gasoline station, Silver Spring, Maryland.

some cases less than 30 m from where the pollution occurred (Fig. 17.10). Sand is an especially effective cleansing agent because it promotes purification in three ways: (1) by mechanically filtering out bacteria (water gets through but most of the bacteria do not); (2) by oxidizing bacteria so that they are rendered harmless; and (3) by bringing bacteria into contact with other organisms that consume them.

Toxic Substances

Harmful chemicals leaking from waste disposal facilities can slowly leach into groundwater reservoirs and contaminate them. Onsite disposal facilities for hazardous wastes, such as aeration pits or lagoons at chemical and metallurgical plants, can contribute substantial amounts of harmful contaminants to groundwater. Leaking underground storage tanks (commonly referred to by the acronym "*LUST*") at gas stations, refineries, and other industrial settings represent one of the most significant groundwater contamination problems in Canada and the United States. It has been estimated that at least 25 percent of all underground storage tanks currently in operation are leaking. This means that at any intersection with a gas station on each corner the chances are high that one of the four gas stations will have a leaking underground storage tank (Fig. 17.11). The implications become more evident when one realizes that one cup of gasoline can contaminate a volume of groundwater sufficient to fill an olympic-sized swimming pool.

Agricultural Chemicals

Agricultural pesticides and fertilizers, or *agrochemicals,* which are significant sources of surface water pollution, are also common contaminants in groundwater. Because of the manner in which they are applied, such nonpoint-source chemicals can invade the groundwater over wide areas as precipitation flushes them into the soil.

Many interrelated factors determine whether or not the groundwater in a particular area is vulnerable to contamination by agrochemicals. For example, physical and chemical properties of the soil, such as its permeability and moisture content, control the rate at which the chemicals infiltrate the soil. Characteristics of the chemicals themselves, such as how readily they break down or vaporize, determine whether they are able to reach the water table. Other factors, such as climate and the timing of pesticide application, also play a role. By combining information about all of these factors it is possible to identify areas that are vulnerable to groundwater contamination by agrochemicals (Fig. 17.12). Problem areas can be targeted for research, monitoring, or regulatory attention.

Contamination by Seawater

A different type of groundwater contamination occurs along coasts, where fresh groundwater is separated from seawater by a thin transition zone of brackish water (Fig.

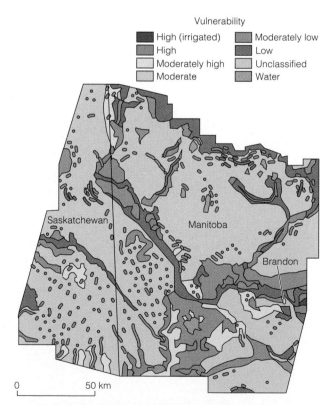

▲ F I G U R E 17.12
Vulnerability to groundwater contamination from agrochemicals in an area of Manitoba. This map was based on a combination of information about land use, soil types, climate, and agrochemical use.

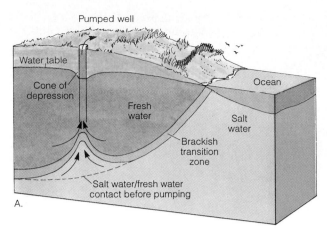

Saltwater intrusion contaminates a pumped well in a coastal area. A. Near the coast, a body of fresh groundwater overlies salty marine water. If pumping is not excessive, the well will draw only fresh water from the aquifer. B. Heavy pumping of groundwater forms a pronounced cone of depression both at the top and at the base of the groundwater body, eventually permitting salty water to enter and contaminate the well.

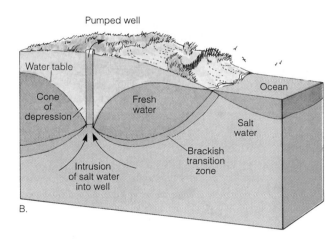

17.13A). Pumping groundwater from an aquifer near the coast will reduce the flow of fresh groundwater toward the sea, allowing saltwater to move landward through permeable strata. This condition is very difficult to reverse. Pumping that exceeds the natural flow of fresh groundwater toward the sea may eventually permit saline water to encroach far inland and reach major pumping centers. The resulting *saltwater intrusion* can contaminate water supplies (Fig. 17.13B).

Behavior of Contaminants in the Subsurface

Contaminants may be present in the subsurface in many different forms. If a contaminant vaporizes easily, it is likely that much of it will be present in the form of a gas. Contaminants can also occur as dissolved constituents or colloidal suspensions in groundwater, or as separate liquid phases. Physical and chemical interactions between the contaminants, the water, and the surrounding rock or sediment combine to determine the eventual fate of the contaminating substances.

Groundwater Flow

Unlike streamflow, which is measured in kilometers per hour, groundwater moves so slowly that rates of flow are expressed in centimeters per day or meters per year. This is because stream water flows in open channels, whereas groundwater must move through small, constricted passages between grains in rock or regolith. Therefore, the flow of groundwater depends to a great extent on the nature of the rock or sediment through which the water moves, especially the *porosity* and *permeability* of the material.

The mechanisms of groundwater movement are discussed in some detail in Chapter 15. Recall that the *water table* (the boundary between the *saturated zone* and the *zone of aeration*) tends to mimic the topography of the surface: the water table is high under hills and lower under valleys (refer to Fig. 15.2). Responding to gravity, groundwater percolates from areas where the water table is high toward areas where it is lower; in other words, it generally percolates toward surface water bodies. Part of the groundwater moves directly down the slope of the water table; the rest flows along curving paths that go deeper through the

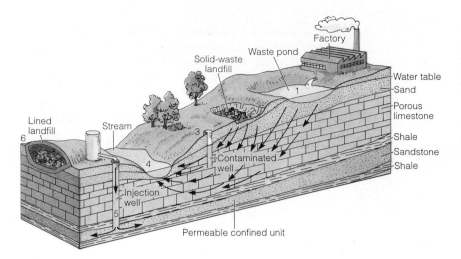

A groundwater system contaminated by toxic wastes. Toxic chemicals in an open aeration pit (1) and an unlined landfill (2) percolate downward and contaminate an underlying aquifer. Also contaminated are a well downslope (3) and a stream (4) at the base of the hill.

ground, eventually turning upward to enter the surface water body from beneath. Whenever development or waste disposal occurs in a groundwater recharge zone there is the potential for groundwater contamination that may eventually affect municipal, industrial, or agricultural water supplies (Fig. 17.14). When the contaminated groundwater reaches a discharge zone, the contamination may enter the adjacent surface water body as well.

Transport of Contaminants

Pollutants typically travel from their source in the form of a spreading mass, or *plume,* of contaminated groundwater; the direction of travel depends on the regional groundwater flow pattern. The normal flow of groundwater and its dissolved constituents through a permeable medium is called *advection* or *advective flow* (Fig. 17.15A). If a contaminant travels by advective flow at approximately the same rate as the groundwater in which it is being transported, it is referred to as being *unretarded;* in other words, the contaminant has not been retarded or slowed down relative to the rate of movement of the groundwater. **Retardation,** therefore, is a measure of how much a contaminant has been slowed down relative to the groundwater flow. It can be defined as follows:

$$R_f = \frac{V_{gw}}{V_c}$$

where R_f is the retardation factor, V_{gw} is the velocity of groundwater flow, and V_c is the velocity of movement of the contaminant.

There are three main mechanisms whereby contaminants can become retarded relative to groundwater flow. They are (1) sorption, (2) dispersion, and (3) biodegradation. In *sorption,* compounds adhere or "stick" to particles in the soil (Fig. 17.15C). *Dispersion* alters the simple advective flow of

the contaminant by forcing contaminants to take a less direct route (Fig. 17.15B). When a contaminant is subject to dispersion, its molecules follow a winding, indirect pathway around and between the sediment particles. Finally, as time goes by, some compounds break down chemically or biodegrade in the subsurface environment, often with the help of bacterial action. Biodegradation in a contaminant plume can occur either aerobically or anaerobically, depending on the conditions within the plume and the types of contaminants present. Sorption, dispersion, and biodegradation typically act in concert to retard the movement of a contaminant, spread it out over a broad area, and eventually reduce its concentration in groundwater.

Monitoring an unretarded contaminant is a good way to trace the movement of a contaminant plume and determine the maximum rate of groundwater flow within the plume. The most frequently used tracer is the chloride ion (Cl⁻). Chloride is a common constituent of many contaminant plumes, and it is essentially unretarded. If the concentration of chloride ions in a "clean" well suddenly increases, it may indicate that the rest of the contaminant plume (the retarded portion) is following some distance behind, moving more slowly than both the groundwater and the unretarded chloride tracer. Chloride ion thus is a harbinger of bad news; it represents the fastest moving constituents—the "front line"—of the contaminant plume. If the progress of the chloride tracer is continually monitored, the direction and rate of travel of the plume itself can be determined.

DNAPLs and LNAPLs

Among the most important physical characteristics used in determining the behavior and predicting the flow of a groundwater contaminant plume is the density of the contaminant relative to the groundwater itself. If a liquid contaminant is less dense (i.e., lighter) than water, it is referred to

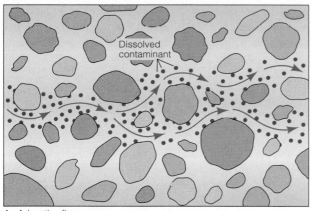

A. Advective flow

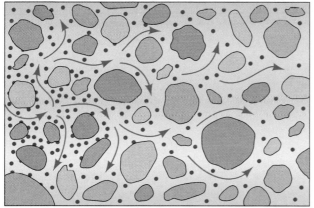

B. Dispersion

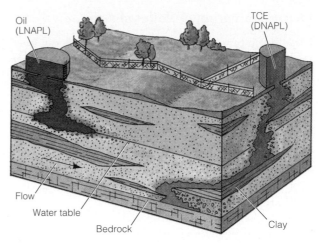

▲ F I G U R E 17.16
The flow of contaminants in the subsurface environment. Oil, a light nonaqueous phase liquid (LNAPL) percolates downward and then floats in a pool on top of the saturated zone. Trichloroethylene (TCE), a dense nonaqueous phase liquid (DNAPL), percolates downward through the zone of aeration and continues to sink past the water table and into the saturated zone.

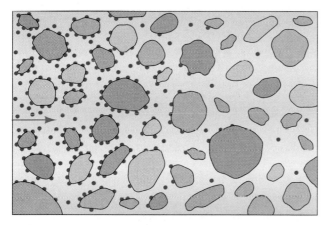

▲ F I G U R E 17.15
Groundwater flow processes that affect the distribution of contaminants in the subsurface. A. Advective flow. The contaminant moves along an essentially straight path at the same rate as the groundwater. B. Dispersion. The contaminant is spread out over a large area and becomes diluted. C. Sorption. The contaminant is electrically attracted to a particle in the surrounding medium. In advective flow, the contaminant plume moves at the same rate as the groundwater, but when dispersion and sorption occur, the flow of the contaminant may be retarded relative to that of the groundwater.

as a *light nonaqueous phase liquid* or *LNAPL*. Petroleum is the most common LNAPL groundwater contaminant. In contrast, *dense nonaqueous phase liquids,* or *DNAPLS,* are denser (i.e., heavier) than groundwater. Trichloroethylene (TCE) is a common example of a DNAPL contaminant.

A contaminant's behavior in the subsurface environment depends on whether it is lighter or denser than the groundwater (Fig. 17.16). For example, a contaminant plume composed of petroleum will infiltrate the soil layers, and upon reaching the saturated zone, will float on top of the groundwater at the level of the water table, where it will tend to spread out in a lens-shaped body. On the other hand, a plume of TCE will sink through the zone of aeration and then through the saturated zone to the bedrock below, becoming mixed to a certain extent with the groundwater as it passes through. DNAPL plumes can be extraordinarily difficult to monitor and control. Because they are not always affected by lateral flow in the groundwater, their movement is not very predictable. They pass easily through fractures and may contaminate deep-seated aquifers or seep between layers of bedrock. It is often difficult or impossible to locate the original source of a DNAPL plume.

Decontamination of Polluted Groundwater

Cleaning up contaminated groundwater plumes can be exceedingly expensive. Estimates of the cost of cleaning up sites identified by the U.S. Superfund program run as high as $1.5 *trillion*. In some cases, the cost of cleaning up a particular site may not be warranted because the risk to humans is very low. In other cases it simply may not be possible to carry out a completely effective cleanup program. This often happens, for example, in situations where the contaminant plume has passed all the way through the regolith and saturated zone and seeped into bedrock layers.

Treatment of groundwater contamination carries with it a host of problems. The technical challenges alone are daunting, but many legal and financial questions arise as well. For example, it may be impossible to identify the original source of the contamination. Even if the source can be located, there may no longer be an owner, or the present owners may have had nothing to do with the activities that caused the contamination. Moreover, legal jurisdictions vary in the way they handle groundwater contamination. In many places, for example, it is not actually illegal to contaminate groundwater until the plume crosses under the property boundary line; at that point the owner of the source property is responsible for dealing with the contaminant plume.

There are two basic approaches to the cleanup of contaminated groundwater plumes: *active* and *passive remediation*. **Passive remediation** is a fancy way of saying "let nature take its course." As mentioned earlier, sorption, dispersion, and biodegradation are natural processes that occur in the subsurface environment. Biodegradation is an especially important contributor to the eventual dilution of contaminants. The aerobic bacteria present in groundwater degrade such chemicals as benzene, toluene, ethylene, and xylene. Chlorinated organic solvents, on the other hand, are degraded by anaerobic bacteria, which are also present in groundwater. However, biodegradation may operate very slowly in the subsurface, especially in regions where the groundwater is cold. Bacteria function best at about room temperature (25°C), and it is not uncommon for groundwater to be many degrees colder than this, slowing bacterial action considerably.

In contrast to passive remediation, **active remediation** involves a range of intervention technologies. Following the identification and delineation of a plume, one of the first tasks undertaken is to contain the plume if possible, that is, to prevent it from traveling any farther. Sometimes this can be accomplished by physical means, such as digging a trench or placing an underground containing wall in the path of the plume. The next step is usually to clean up any residual materials remaining at the source in order to prevent any more contaminants from entering the groundwater.

It is sometimes possible to leave the contaminant plume in place and intervene by stabilizing the contaminant or by speeding up the natural processes of degradation and dilution. This can be accomplished by adding oxygen to stimulate aerobic biodegradation, or by adding specific types of anaerobic bacteria to the system. This type of intervention is called *biostimulation*. In the case of inorganic contaminants, which will not biodegrade, it may be possible to inject chemicals that will cause the substances to precipitate, thereby taking them out of solution and immobilizing them.

In some cases it may be more effective or efficient to remove the contaminant physically. For example, a floating pool of LNAPL contaminants may be removed from the surface of the groundwater by pumping. For some types of DNAPL contaminants it may be necessary to inject chemicals, such as solvents or surfactants, to facilitate the flow of the contaminant. Removal of any contaminant that is present as a gas in surrounding soils is a necessary part of this step in the remediation process. It is often necessary to remove a large quantity of mildly contaminated groundwater as well; this can sometimes be treated and reinjected into the ground. This approach to active remediation is called *pump-and-treat*.

POLLUTION IN THE MARINE ENVIRONMENT

Many of the pollution problems discussed in this chapter come together in a unique and vitally important system: the oceans. The United Nations periodically assesses the status of the world's oceans through meetings of the Group of Experts on the Scientific Aspects of Marine Pollution (GESAMP). GESAMP reports indicate that pollution is not yet a critical concern in the open oceans, but that most of the world's coastal areas are polluted, some of them severely, and that overall marine pollution is increasing. Because the world's oceans are a vital resource that is shared and must be managed by all coastal communities, the types and sources of ocean contamination deserve special attention.

Types of Marine Pollutants

The most important types of marine pollutants will be familiar to you from previous discussions. They include sewage, siltation, and hazardous substances, as well as marine litter. Many of these contaminants enter the marine environment as a result of human activities (Table 17.1).

Sewage is a major cause of ocean pollution, as it is for other surface water bodies. Even after secondary treatment, sewage effluents still contain suspended solids, nitrates, phosphates, pathogenic viruses (some of which can survive for up to 17 months in ocean water), heavy metals, pesticides, and even radioactive materials. Because these wastes are rich in organic matter, they place a heavy BOD on

T A B L E 17.1 • **Potential Effects on the Marine Environment of Various Activities and Sources of Contamination[a]**

Activity or Source of Contamination	Bacterial/Viral Contamination	Oxygen Depletion	Toxicity	Bioaccumulation	Habitat Degradation	Depletion of Biota	Degradation of Aesthetic Values
Exploration for, and exploitation of, oil and gas reserves		X	X	X	X	X	X
Ocean dumping	X	X	X	X	X	X	
Coastal developments	X				X	X	X
Discharges of municipal wastewater	X	X	X	X	X	X	X
Discharges from pulp and paper mills	X	X	X	X	X	X	X
Food and beverage processing		X			X		X
Oil refineries			X	X			X
Chlor-alkali plants			X	X			
Mining wastes			X	X	X	X	X
Chemical spills and leaks		X	X	X			X
Urban and agricultural runoff	X	X	X	X	X		
Litter					X	X	X
Agriculture	X	X	X	X	X		X
Pesticides			X	X			
Atmospheric emissions	X		X	X			

Source: *The State of Canada's Environment*, Supply & Services Canada, 1991; from Waldichuk (1988); T.R. Parsons, University of British Columbia, personal communication; D.J. Thomas, Seakem Oceanography Ltd., personal communication.

[a]Source: *The State of Canada's Environment*, Supply & Services Canada, 1991; from Waldichuk (1988); T.R. Parsons, University of British Columbia, personal communication; D.J. Thomas, Seakem Oceanography Ltd., personal communication.

coastal waters, contributing to eutrophication and algal blooms. Shellfish poisoning and contamination of recreational beaches are the most common problems associated with sewage pollution in areas near the shore.

Wherever there is human activity in shallow waters, siltation and sediment pollution are likely to result. Most discharges of liquid wastes carry suspended solids, which eventually settle out. Logging, farming, mining, and dredging also contribute to accelerated sedimentation in coastal zones. Excess sediment can be a source of concentrated contaminants, and it can also "suffocate" some types of marine organisms.

Chlorinated organic compounds (including pesticides and industrial chemicals), heavy metals, and radioactive substances contaminate ocean waters via many of the same pathways by which they reach other surface water bodies: through direct industrial discharges, chemical spills, runoff from land, and atmospheric deposition, as well as ocean dumping. Oil is a significant contaminant, both because of its toxicity and because it is so widespread; oil spills are discussed in detail later in the chapter.

Population growth in coastal zones and increased shipping activity are also contributing to increases in the amount of persistent litter and debris in coastal waters and along heavily traveled shipping lanes. Plastic, which is both durable and buoyant, accounts for by far the greatest portion of floating debris that accumulates in ocean waters. Abandoned fishing nets and oil drums are other significant sources of marine litter. The main problems associated with persistent litter are the death of marine mammals, fish, and birds, usually as a result of entanglement, and the degradation of beaches.

Sources of Marine Pollution

Most ocean contaminants come from coastal regions, which tend to be heavily populated. Nearshore ocean waters become polluted as a result of ocean dumping, discharge of liquid wastes, and contaminated runoff from the land. When wastes are discharged or dumped in nearshore waters, wave action and longshore currents in the littoral cell (Chapter 9) often combine to transport the contaminants downcurrent, concentrating them along the shoreline instead of dispersing them into the open ocean (Fig. 17.17). The result can be a cell of heavily contaminated water situated just offshore (Fig. 17.18), often leading to beach contamination and closures.

Effects on Coastal Environments

Pollution in coastal waters is especially damaging because shallow continental shelves host a great abundance of marine life. Moreover, coastal areas are characterized by delicately balanced ecosystems—wetlands, estuaries, mangrove swamps, coral reefs, and barrier island beaches. When these sensitive ecosystems are altered or contaminated, the nat-

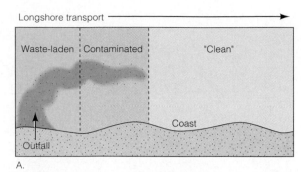

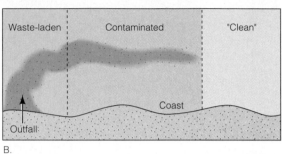

▲ F I G U R E 17.17
Contamination of a coastal circulation cell. A. Liquid wastes are released offshore. B. Continued injection of wastes saturates the coastal littoral cell, and the saturated water moves progressively along the shore with the longshore current. C. The result is a strip of nearshore water, parallel to the coast, that is contaminated by liquid waste.

ural services they perform—buffering against storm erosion, storing groundwater, providing habitat for marine life—are impaired.

Coral reefs, which are found in shallow coastal waters in tropical and subtropical areas, are particularly sensitive to the effects of excess sedimentation and other types of coastal contamination. Reefs are built up over thousands of years by the activity of very small organisms called *polyps*. Each polyp deposits a protective layer of calcareous material; over time the material deposited by millions of these organisms develops into a complex reef structure (Fig. 17.19). Coral reefs are highly productive ecosystems, inhabited by a wealth of life—fish, mollusks, crustaceans, and a multitude of less familiar creatures. Reefs perform an important role in the recycling of nutrients in shallow coastal environments. They provide physical barriers that

◄ F I G U R E 17.18
Coastal strip of water contaminated by liquid wastes being
transported along the coast by
longshore currents. Oil spill,
Galveston Bay, Texas, 1990.

◄ F I G U R E 17.19
The abundant life of a coral reef.
The Rainbow Reef, Fiji.

dissipate the force of high-energy waves, thereby protecting the ports, mangrove swamps, and beaches behind them. They are also an important aesthetic resource.

Because they have very specific water temperature and light level requirements, coral reefs are particularly susceptible to damage from both human activities and natural causes, such as cyclones. Because of very slow rates of growth, it may take hundreds to thousands of years for a damaged reef to repair itself, if it does at all. Siltation is one of the most serious threats to the long-term health of reefs. Even sediment that is too light to actually suffocate the polyps can affect a colony, since corals cannot reproduce in murky waters where light levels are too low. Changes in water quality, such as thermal pollution by industrial waste-waters, decreased salinity caused by freshwater runoff, or contamination from oil or industrial wastes, can also affect coral colonies.

Pollution in the Open Ocean

In contrast to coastal zones, the open ocean is relatively clean. Most contaminants settle out and become increasingly dilute in waters that are far from the source of the pollution. Still, anthropogenic contaminants can be detected even in deep ocean waters. Some of this comes from atmospheric deposition, but pollution in the open ocean is of greatest concern along heavily traveled shipping lanes.

Ships leave a wide variety of contaminants in their wakes, including bilge water, litter and debris, and chemicals. In addition, ballast water (water that is taken on to provide weight after the ship unloads its cargo and released when the ship reloads) may contain exotic organisms. For example, it is believed that zebra mussels were introduced into the Great Lakes by the discharge of ballast water from a European vessel in the late 1980s. The mussels have created an enormous and costly problem since then. They clog industrial and municipal water intake pipes, and strip the plankton from large volumes of water, thereby endangering the food supplies of fish and other indigenous aquatic species.

Oil Spills

As a source of marine pollution, oil spills have received a great deal of attention. However, it should be noted that the petroleum industry has a good record for safe transportation of its product; only a small fraction of the total oil transported is accidentally spilled. In terms of volume, the contamination resulting from oil spills in the oceans is much less significant than that resulting from careless handling of oil, industrial mishaps, and LUSTs on land. In terms of overall environmental impact, sewage is considerably more significant than oil as a marine pollutant.

Nevertheless, oil spills do occur in the marine environment each year—not just occasionally, but in the thousands. Although only the most catastrophic spills make headlines, the many small spills add up: Each year small spills release an amount of oil equivalent to 17 *Exxon Valdez* spills into the Mediterranean Sea alone. Most oil spills occur in coastal zones, where they have the greatest impact on living marine resources. (Fig. 17.20).

Although estimates vary widely, the total amount of oil entering the world's oceans each year is on the order of 6 million tons. The main source of this oil is accidental spills and discharges that occur in the course of transport (Fig. 17.21). Other sources that contribute to the overall flow of oil into ocean water include municipal and industrial discharges, urban runoff, atmospheric fallout (redeposition of oil that has evaporated into the atmosphere), and offshore exploration and production of oil.

Although it received a great deal of publicity and was the largest oil spill ever in North American waters, the spill resulting from the wreck of the tanker *Exxon Valdez* in March 1989 was not the largest accidental spill in history; that distinction belongs to the blowout of the offshore oil platform *Ixtoc I* in the Gulf of Mexico in 1979. The *Exxon Valdez* ran aground off the shore of Valdez, Alaska, spilling 10 million gallons of crude oil into the water. The oil contaminated more than 5000 kilometers of Alaska's shoreline. (The largest *intentional* oil spill in history happened as a result of the Persian Gulf War.)

According to John Vandermeulen, an expert on oil spills, "There are three myths about oil spills that need clearing up. First—that oil spills can be controlled. Not so.

▲ F I G U R E 17.20
Sea birds covered with oil as a result of the spill by the *Christos Bitas*, October 1978, on the coast of Wales. Birds are wrapped prior to being taken to a cleaning center.

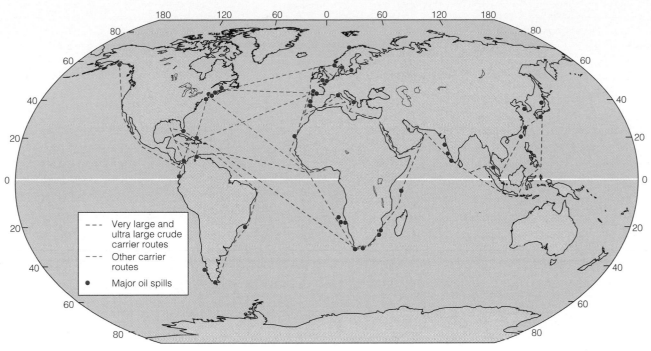

▲ F I G U R E 17.21
Main tanker routes and locations of major oil spills.

Second—that oil spills can be cleaned up. Not so. Third—that damaged environments are doomed. Wrong again." A large oil spill caused by the wreck of an oil tanker or an accident on an offshore drilling rig can be extremely damaging to the local environment. Each spill is unique: the behavior of a spill, the effectiveness of natural dilution processes, and the type of cleanup interventions required are determined by a variety of factors, including temperature, wind, ocean currents, wave action, and the specific chemical composition of the oil.

The first response to a spill is usually to attempt to contain it (e.g., with floating "booms") and to deal with the source of the spill. Oil floats on seawater; if it is contained, much of it can be skimmed off the water surface. The remaining oil is exposed to air and will eventually either biodegrade or sink to the bottom. The oil that is not contained eventually reaches the shore, where it creates a different set of problems. For example, shorebirds are often among the wildlife most obviously affected by a major oil spill; their feathers can become coated with sticky black oil, causing the birds to freeze or drown. Usually the cleanup of contaminated shore areas requires a combination of biostimulation and mechanical cleaning—spraying with high-pressure hot water, scrubbing with brushes, or manually wiping rocks and boulders clean.

One of the problems associated with oil spills is that within a few days the oil begins to mix with the water, forming a messy material called *mousse*. In the case of the *Exxon Valdez* spill, cold conditions and heavy wave action

in the Alaskan waters rendered the mousse particularly intractable. Exxon called off the cleanup effort on September 15, 1989, out of fear that winter storms would jeopardize the safety of the workers and because the cold weather was increasingly hampering the success of the cleanup (Fig. 17.22).

▲ F I G U R E 17.22
Steam-cleaning the rock shore of Latouche Island, Alaska, after the oil spill from the *Exxon Valdez*.

BOX 17.3

•

THE HUMAN PERSPECTIVE

GEOLOGY AND HEALTH

Medical geology is a subfield of geology that studies the effects of chemical elements in the environment, especially trace elements, on the health of humans and animals. This involves an understanding of the biologic effects of exposure to these elements, as well as knowledge of their distribution in the geologic environment. *Trace elements* occur in very low concentrations (usually defined as less than 1 percent) in the rocks and minerals of the Earth's crust. Sometimes they become concentrated in soil, water, or air and are taken up by plants and ingested by humans or animals. In addition to abundant elements such as iron and calcium, many trace elements are essential for human nutrition; these include zinc, magnesium, iodine, and cobalt. Other trace elements are biologically active but not essential for nutrition; some are harmful to organisms.

For some trace elements there is a very narrow line between beneficial and toxic levels of exposure. An example is selenium, which is essential to both human and animal nutrition, but only in very small quantities. If ingested in quantities greater than 5 mg per day, selenium can be toxic to humans. Selenium is strongly bioconcentrated in "locoweed," a common weed in the southwestern United States. When animals graze on this plant, they develop a selenium-related disease called blind staggers.

The relationship between the presence of fluorine in groundwater and resistance to tooth decay was one of the earliest correlations made by medical geologists. Studies carried out in the 1940s and 1950s demonstrated that people who lived in areas where groundwater had a high fluorine content were less likely to develop dental caries (cavities). In parts of North America where natural fluorine concentrations are low, many municipalities now fluoridate their drinking water. Like selenium, fluorine can cause some problems, such as brown mottling of the teeth, when taken in high doses. But the toxic dose of fluorine is so high that it is not a cause for concern. There may also be some synergistic effects between fluorine and other trace elements present in groundwater. For example, boron, molybdenum, and lithium appear to reinforce the effects of fluorine, whereas high selenium intake may have the opposite effect.

Medical geologists have also identified a correlation between the concentration of iodine in soils and the prevalence of goiter. Goiter is a disease that causes enlargement of the thyroid gland. It occurs primarily in regions where the iodine content of the soil is very low, as it is across much of the northern United States and southern Canada, the so-called "goiter belt." The lack of iodine in the soils of the goiter belt was probably caused when soluble iodine salts were washed away during the melting of glaciers following the last glaciation. Because commercial table salt is now iodized, goiter is no longer a major health problem in North America.

Other studies have attempted to correlate the composition of soil, rock, and groundwater with the distribution of diseases such as osteoporosis, heart disease, and various cancers. Although such studies often reveal intriguing correlations, it is difficult to draw definitive conclusions from the findings. One reason for this is that most people no longer eat food grown only in their own area. This means that they are exposed, through food intake, to whatever trace elements have become concentrated in crops from different parts of the world. Moreover, people are exposed in their daily lives to a wide variety of other materials that have an impact on health. For example, radon, a radioactive gas present in soil, is thought to contribute to the incidence of lung cancer. However, it is very difficult to separate the effects of exposure to radon from other potential causes of lung cancer, such as smoking. Similarly, studies conducted in Georgia have suggested a correlation between local geology and deaths from cardiovascular disease. The reason for the correlation is as yet unexplained, and at any rate it would be exceedingly difficult to distinguish the geologic effects from the effects of diet or lifestyle.

The *Exxon Valdez* oil spill highlighted a number of problems, including the fact that we simply do not know enough about how oil is absorbed into beaches and transported along shorelines. One of the main concerns has been subsurface oil, that is, oil that penetrates into gravel and sand, both on the beach and underwater, and therefore cannot be removed using surface washing techniques. Some traditional cleanup techniques, such as biostimulation, were less effective than usual in dealing with this spill because of the cold temperatures. In addition, the company's response to the spill, the role of human error in the accident, and the company's liability for the environmental damage have given rise to further concerns and much criticism.

SUMMARY

1. Waste and pollution are distinct, but closely related; it is difficult to discuss them in isolation from each other. Many natural processes can contribute to the degradation of soil or water quality, but the worst contamination comes from anthropogenic sources.

2. Contaminants can be released into the environment either from point sources or from nonpoint sources. Once a contaminant has been released, its behavior is controlled by a range of factors, including the characteristics of the material itself, such as its chemical stability or its tendency to biodegrade. It is also common for contaminants to interact with the surrounding medium, with organisms, or with other contaminants. Sometimes the effects of these interactions are mutually reinforcing or synergistic.

3. Persistent contaminants take a long time to break down, and their harmful effects are dissipated very slowly. Some persistent chemicals build up in the tissues of organisms and accumulate in the food chain through the process of bioconcentration.

4. Contaminants can be transported long distances by wind or water or by organisms in which they have become concentrated. The average length of time a contaminant remains in a particular reservoir—its residence time—depends on the chemical and physical characteristics of the contaminant, the characteristics of the surrounding medium, and the holding capacity of the reservoir.

5. In sediment pollution, the sediment itself is sometimes the pollutant. The effects of excess sediment from erosion include siltation and clogging of waterways, destruction of habitat, acceleration of algal growth, loss of recreational value, and cleanup costs. In other cases, heavy metals, pesticides, or other toxic substances accumulate in sediment. Contaminated sediment in a surface water body can be especially difficult to clean up because any attempt to do so will accelerate the release of contaminants into the water.

6. The definition of "good" water quality varies from place to place and according to the potential use of the water. A variety of natural processes, including climate, season, and the geology of the local bedrock, can affect water quality. In some instances the natural characteristics of a body of water can determine its ability to absorb, dilute, or neutralize contaminants.

7. Surface water bodies are often contaminated by excess nutrients from organic materials, such as sewage and fertilizers. Biodegradation of organic material places a high biochemical oxygen demand (BOD) on the water, which may lead to oxygen depletion and, in some cases, eutrophication. In some cases organic contaminants may carry infectious agents such as bacteria or viruses.

8. Many of the pollutants found in surface water bodies are toxic. Acutely toxic substances cause harm to organisms within 96 hours; substances that exhibit chronic toxicity cause harm over an extended period of exposure, sometimes at very low levels. Among the types of toxic contaminants found in surface water are chlorinated organic compounds, heavy metals, and hydrocarbons.

9. Acidic effluents can also contaminate surface water, as in the case of acid mine drainage. The chemistry of the underlying soil and bedrock help determine the vulnerability of a surface water body to acidification. Other, less common types of surface water contamination include alkalinization, salinization, and thermal pollution.

10. A potentially serious groundwater contaminant is soluble in water, resistant to biodegradation, used in large quantities, and toxic or noxious to humans. The most widespread type of groundwater contaminant is nitrate from sewage, agricultural chemicals, and landfill leachate. Contamination is also caused by toxic substances released from leaking disposal sites or underground storage tanks. Saltwater intrusion can be a problem in coastal zones.

11. The transport of contaminants in the subsurface is governed by many of the same processes that control the flow of groundwater. An unretarded contaminant

travels by advective flow at the same rate as the ambient groundwater flow. The flow of contaminants can be retarded relative to that of groundwater by the processes of dispersion and sorption. Biodegradation, both aerobic and anaerobic, also contributes to the dilution of contaminants in plumes. Different types of contaminants behave differently: LNAPLs, such as petroleum, tend to form a pool on top of the water table, whereas DNAPLs, such as TCE, sink through the saturated zone.

12. Passive remediation of polluted groundwater relies on the natural processes of biodegradation, dispersion, and sorption to dilute contaminants. Active remediation, on the other hand, can involve a range of interventions such as containment, biostimulation, chemical stabilization or neutralization, and pump-and-treat.

13. Ocean pollution is most common in coastal zones, where human activities such as industry, mining, dredging, logging, and tourism create contaminated runoff and effluents. The two most common types of ocean pollutants are sewage and excess sediment, followed by hazardous substances (including oil) and marine litter. Pollution is particularly damaging in coastal zones because of the high concentration of marine organisms and the delicately balanced ecosystems that characterize these areas.

14. Large oil spills from tankers and offshore drilling rigs can be extremely damaging to local marine environments and coastal zones. However, most of the oil flowing into the marine environment comes from oil spills and leaks on land, as well as from the thousands of small spills and accidental discharges that occur in the oceans. Each oil spill requires different cleanup techniques, which may include containment, skimming, mechanical cleanup, or biostimulation.

IMPORTANT TERMS TO REMEMBER

acid mine drainage (p. 448)
active remediation (p. 454)
biochemical oxygen demand (BOD) (p. 443)
bioconcentration (p. 438)
biodegradation (p. 438)

eutrophication (p. 444)
medical geology (p. 460)
nonpoint source (p. 436)
passive remediation (p. 454)
persistent contaminants (p. 438)
point source (p. 436)

residence time (p. 438)
retardation (p. 452)
sediment pollution (p. 441)
thermal pollution (p. 448)
toxicity (p. 445)

QUESTIONS AND ACTIVITIES

1. Could an event like the Minamata mercury poisoning happen today? Probably not in North America, where water quality and waste disposal standards are very high; however, the negative impacts of pollution on health are just beginning to be felt in many parts of the world. For example, the breakup of the Soviet Union has left a legacy of pollution and improperly handled wastes. As a research project, investigate the status of pollution in one of the countries of Eastern Europe or the former Soviet Union. In what ways are the environmental and health impacts of past mismanagement becoming evident?

2. Find out about the trace elements that affect your health. Do you live in the "goiter belt"? Is the water in your area hard or soft? Is your drinking water naturally high in fluoride? If not, does your community fluoridate the water supply? Is there evidence of unusual concentrations of any trace elements in rocks, soils, or groundwater in your area?

3. The largest release of oil into the environment occurred during the Persian Gulf War. Find out how much oil was spilled during the war and what impacts it had on land and water. What is the current status of the areas in which the worst spills occurred?

4. Take a field trip to identify point sources and nonpoint sources of pollution in your area. How many can you find? See if you can document the amount of pollution contributed by each source.

5. Most urban and suburban areas have sites that are undergoing remediation, such as former petroleum refineries or chemical plants. Take a field trip to one of these sites and find out as much as you can about the underlying geology, the type and extent of contamination that has occurred, and the remediation methods being used.

6. Investigate the current status of pollution and cleanup efforts in the Great Lakes region. How many of the "hot spots" identified by the International Joint Commission are cases of water pollution and how many are cases of sediment pollution? Choose a site to investigate in detail. What is the nature of the contamination? How is the cleanup being approached? Have agreements between the United States and Canada facilitated the cleanup effort?

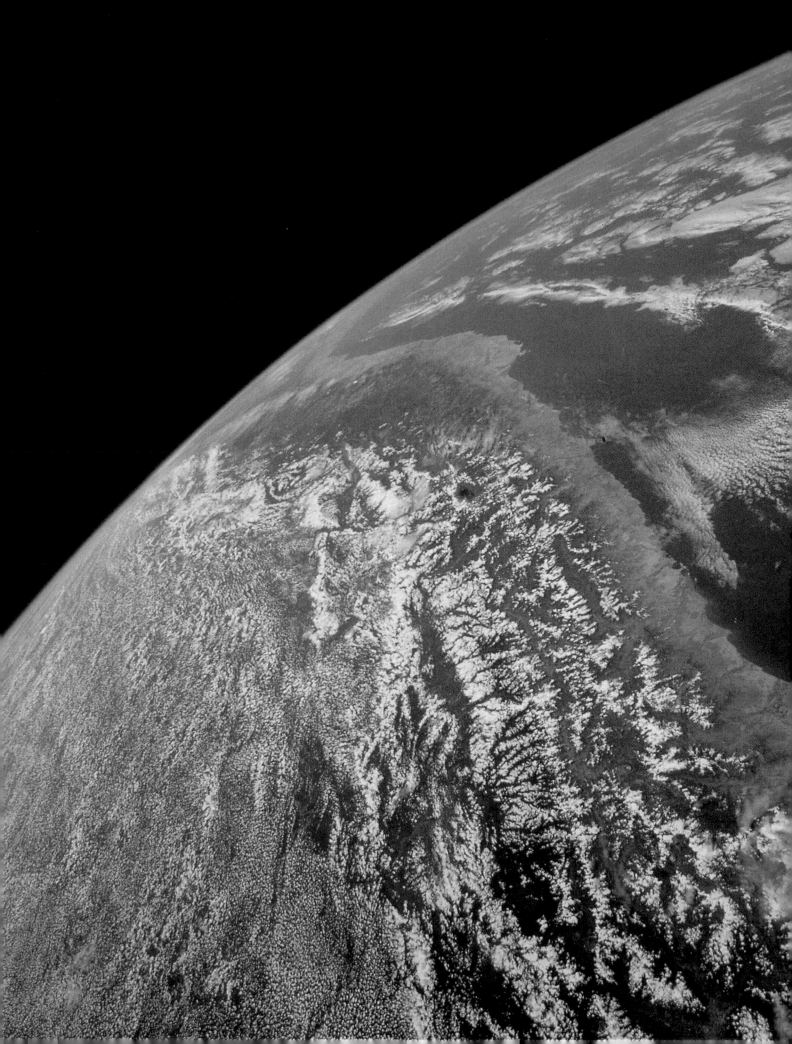

CHAPTER 18

ATMOSPHERIC CHANGE

Some say the world will end in fire,
Some say in ice.
From what I've tasted of desire
I hold with those who favor fire.
But if I had to perish twice,
I think I know enough of hate
To say that for destruction ice
Is also great
And would suffice.

• **Robert Frost**

In 1958, in response to mandates that emerged from the International Geophysical Year (1957), a small project was undertaken at Mauna Loa Observatory in Hawaii. The original goal of the project was to measure seasonal fluctuations in the amount of carbon dioxide (CO_2) in the atmosphere. The equipment was scheduled to be turned off when this had been accomplished; luckily, it wasn't. The information gathered in this study was largely responsible for initiating the intense debate, which is still going on in both scientific and political arenas, concerning the possibility and consequences of global climatic change.

Atmospheric CO_2 is still being measured at Mauna Loa. The measurements have revealed that since 1958, CO_2 levels in the atmosphere have risen exponentially (that is, not only is the *amount* of CO_2 increasing, but also the *rate of increase*). Other studies indicate that atmospheric CO_2 levels have been increasing exponentially ever since the beginning of the Industrial Revolution, when anthropogenic CO_2 emissions from the burning of fossil fuels first began to increase significantly.

◀ ——————————————————

Space shuttle view of the Earth through the thin, blue atmosphere. Looking south along the South American coast with the Pacific Ocean (right). The snow covered Andes Mountains are in the center; the Amazon River Basin is lower left.

Carbon dioxide is known to be an important contributor to the *greenhouse effect,* the complex set of atmospheric processes that keeps the surface of the Earth warm enough to sustain life. Thus, the rising curve of atmospheric CO_2 raises profound questions concerning the Earth's future climate: Does the increase in CO_2 in the atmosphere mean that the Earth's surface is warming? If so, by how much and at what rate? Can the trend be reversed? To what extent has human activity caused or contributed to atmospheric change? And finally, what are the implications of global warming for ecosystems and human interests?

THE THIN BLUE LINE

Without the unique characteristics of the atmosphere, the Earth would be like other planets in the solar system: barren and lifeless. We take for granted the comfortable environment provided by the atmosphere, as we take for granted its stability, its resilience, and its apparently limitless extent. Yet in many respects nothing could be more fragile than the thin wisp of gases held to the Earth by gravity. Astronauts on their first missions often comment on this, as did West German astronaut Ulf Merbold in 1990:

For the first time in my life, I saw the horizon as a curved line. It was accentuated by a thin seam of dark blue light—our atmosphere. Obviously, this was not the "ocean" of air I had been told it was so many times in my life. I was terrified by its fragile appearance.

For all its apparent fragility, the atmosphere gives us much more than just the air we breathe. A protective layer of ozone in the atmosphere shields us from harmful ultraviolet radiation; gases and water vapor trap the Sun's heat near the surface of the Earth, warming it just enough to sustain life; interactions among the atmosphere, oceans, and land masses create weather and precipitation patterns and facilitate the movement of elements that are critical for life-sustaining processes. All of these functions are essential for the continued existence of life on Earth; all depend on the integrity and stability of the atmosphere.

Unfortunately, it is becoming increasingly clear that what once seemed to be a vast, stable, endlessly replenishable reservoir is not only finite and delicately balanced but subject to a variety of changes resulting from both natural processes and human activities. Because of the dynamic nature of the atmosphere, these changes occur on scales ranging from local to regional to global. The most significant anthropogenic impacts on the atmosphere are (1) smog and other forms of air pollution that are primarily local in their impacts; (2) acid precipitation, which is regional in its effects; (3) depletion of the ozone layer, an essentially global phenomenon; and (4) global climatic warming. In this chapter we look in detail at these four categories of atmospheric change. We start with a brief look at the structure and composition of the atmosphere, how it has evolved, and how it supports life on the Earth.

THE STRUCTURE AND EVOLUTION OF THE ATMOSPHERE

The atmosphere is a mixture of gases dominated by nitrogen (about 78 percent), oxygen (21 percent), and argon (just under 1 percent). Small amounts of other gases, some water vapor, and clouds composed of water droplets constitute the remainder (Figure 18.1). When we talk about "oxygen" in the atmosphere we are usually referring to the dominant form, molecular oxygen (O_2). However, *free oxygen,* that is, oxygen that has not combined with other elements in compounds such as carbon dioxide or water, can also occur in the atmosphere in other forms, primarily ozone (O_3) and atomic oxygen (O). Carbon dioxide, a source of concern because of its role in global warming, is a trace gas, accounting for only 0.035 percent. Some other trace constituents of the atmosphere are methane (CH_4), nitrous oxide (N_2O), hydrogen (H_2), and carbon monoxide (CO).

The composition of the atmosphere is relatively constant; that is, the relative proportions of its constituents do not vary substantially from one place to another. The notable exception is water vapor, which is unevenly distributed because of its short atmospheric residence time and its active role in the hydrologic cycle.

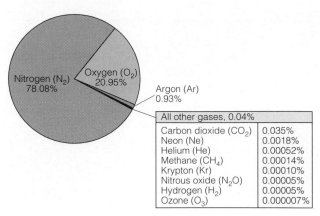

▲ F I G U R E 18.1
The composition of dry, aerosol-free air in percentage of total volume. Three gases—nitrogen, oxygen, and argon—make up 99.96 percent of the air.

Vertical Structure

Atmospheric scientists divide the atmosphere into layers, each with its own physical, chemical, and temperature characteristics (refer to Fig. 9.1). The bottom layer, the **troposphere,** extends from the ground to an altitude of 6 to 17 km, depending upon season, latitude, and other factors. Extending from the *tropopause,* the top of the troposphere, to an altitude of 50 km is the **stratosphere.** The *mesosphere* extends from the top of the stratosphere (the *stratopause*) to an altitude of 90 km, and the *thermosphere* extends from the top of the mesosphere (the *mesopause*) to an altitude of 120 km. Many scientifically interesting processes occur in the upper layers of the atmosphere, but from an environmental point of view the most important layers are the troposphere and the stratosphere.

The Troposphere

The troposphere contains most of the atmosphere's actual mass, about 80 percent, and virtually all of the water vapor. The troposphere is extremely dynamic and well mixed. However, except during violent weather events or volcanic eruptions, little mixing occurs between the troposphere and the layers overlying it. Most weather-related phenomena originate in the troposphere. The uneven distribution of solar heating (more intense in the equatorial regions than at the poles) creates wind systems in the troposphere, which eventually distribute heat and moisture—as well as pollutants—to all parts of the globe. (The processes of weather formation are discussed in Chapter 9.)

The atmosphere is a vast reservoir of elements that are essential to life. The troposphere is in a state of dynamic equilibrium with the oceans and soils; this facilitates the transfer of biochemical components such as carbon, nitrogen, hydrogen, and oxygen from one reservoir to another

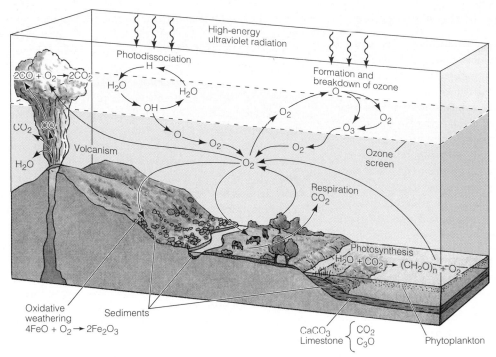

▲ F I G U R E 18.2
The oxygen cycle is complicated because oxygen appears in so many forms and combinations. Oxygen occurs primarily as molecular oxygen (O_2), in water, and in organic and inorganic compounds. Some global pathways of oxygen are shown here in simplified form.

(Fig. 18.2). The troposphere also contains most of the heat-trapping gases responsible for insulating the Earth's surface and making it warm enough to be hospitable to life.

The Stratosphere

The stratosphere contains another 19 percent of the atmosphere's total mass. (Note that this leaves only about 1 percent to reside in the upper 250 km or so of the atmosphere.) One of the most important features of the stratosphere is that it contains a concentration of the gaseous chemical ozone (O_3), which protects life by absorbing harmful short-wavelength ultraviolet solar radiation.

The stratosphere also plays an important role in the global redistribution of materials in the atmosphere. If a pollutant is injected into the atmosphere but does not enter the stratosphere, chances are that its most significant impacts will be local or regional rather than global in extent. This was the case with smoke from burning oil wells during the Persian Gulf War of 1991. At the beginning of the war there were fears of global atmospheric impacts, but the smoke never rose above 6 km and thus did not go high enough to be distributed globally by stratospheric circulation processes (Fig. 18.3).

The first time scientists were able to study upper atmospheric circulation directly was during the 1883 eruption of Krakatau in Indonesia. This was the first major volcanic eruption to occur after the establishment of global communications, and scientists had a unique opportunity to monitor the ash cloud as it spread across the globe, causing strangely colored skies and spectacular sunsets for the next 3 years.

▲ F I G U R E 18.3
Two Kuwaitis view the burning oil wells left behind by the retreating Iraqis during the Gulf War. Note the dense black clouds of smoke.

The Chemical Evolution of the Atmosphere

Like other Earth systems, the atmosphere has evolved through time. More than 4 billion years ago the Earth's atmosphere was very thin, almost nonexistent. The envelope of gases or *primary atmosphere* that surrounded the Earth at the time of its formation was stripped away by the solar wind, and the planet generated a new, *secondary atmosphere* by degassing volatile constituents from its interior. The early atmosphere was chemically very different from the present atmosphere—it contained no free oxygen and most likely would have been highly inhospitable to any life forms other than some algae and bacteria. The atmosphere developed and changed chemically over geologic time, just as the systems with which it is interconnected—the hydrosphere, the biosphere, and the lithosphere—developed and changed.

The Role of Volcanism

Throughout geologic history, the main process by which volatile materials have been released from the Earth has been volcanism. Gas is still a major component of volcanic emissions; clearly, therefore, the degassing process must still be going on. As you may recall from Chapter 4, the main chemical constituent of volcanic gases (as much as 97 percent by volume) is water vapor, with varying amounts of nitrogen, carbon dioxide, hydrogen, sulfur dioxide, chlorine, hydrogen sulfide, methane, ammonia, carbon monoxide, and other gases. Overall, the volume of volcanic gases released over the past 4 billion years or so is thought to be in approximately the right proportions and sufficient amounts to account for the entire volume of the oceans and atmosphere, with one extremely important exception: oxygen. Where has the abundant free oxygen in the atmosphere come from?

Some oxygen was almost certainly generated in the early atmosphere by the breakdown of water molecules into hydrogen and oxygen as a result of interactions with ultraviolet light, called *photodissociation* (Fig. 18.2). This is an important process, but it doesn't come close to accounting for the present high level of oxygen in the atmosphere. Another oxygen-producing process was required, but it had to wait for the appearance of life on the planet.

The Role of Life

Practically all organisms in the biosphere today utilize **photosynthesis,** either directly or indirectly, to obtain food. In photosynthesis, light energy is used to cause carbon dioxide to react with water, synthesizing organic substances (carbohydrates) and releasing oxygen in the process. The basic photosynthetic reaction is as follows:

$$CO_2 + H_2O + \text{light energy} \rightarrow (CH_2O)_n + O_2$$

▼ FIGURE 18.4
Towering limestone spires near Guilin in the karst region of Guanxi Province, China. White limestone can be seen underneath vegetation on some of the vertical cliff faces.

Most of the oxygen currently in the atmosphere originated through photosynthesis. But oxygen is toxic to most carbon-based life forms unless oxygen-mediating enzymes are present. These enzymes are themselves a result of biologic processing; yet without them or some other receptor to carry away the free oxygen, early life forms would have had difficulty dealing with the oxygen released during photosynthesis. Therefore, it seems likely that the earliest life forms—*prokaryotes*—created their food from nonbiologic sources through the anaerobic process of *fermentation*.

Another possibility is that the toxic oxygen waste produced during photosynthesis was carried away by iron in ocean water and deposited in the form of iron-bearing minerals. In fact, evidence of the gradual transition from oxygen-poor to oxygen-rich ocean water is preserved in the ancient record of seafloor sediments. The minerals in seafloor sedimentary rocks older than about 2 billion years are characterized by the presence of reduced (oxygen-poor) iron compounds; in rocks younger than about 1.8 billion years, oxidized compounds predominate. The chemical sediments precipitated on the ocean floor during this period reflect this transition through alternating bands of red (oxidized iron) and black (reduced iron) minerals. These rocks are called *banded iron formations;* their significance as a source of iron ore is discussed in Chapter 13. Since ocean water is in constant contact with the atmosphere, and the two reservoirs maintain a dynamic state of chemical equilibrium, the transition to an oxygen-rich atmosphere must also have occurred during this period.

Eventually enough oxygen was created through photosynthesis by early oxygen-producing life forms (*eukaryotes,* or bluegreen algae) to permit free oxygen to build up in the atmosphere. With the buildup of molecular oxygen came a parallel increase in ozone. When the level of oxygen in the atmosphere reached just 1 percent of its present level, the ozone began functioning as a screen, filtering out harmful ultraviolet radiation. This permitted life forms to survive and flourish in shallow waters and, eventually, on land. This critical stage in the evolution of the atmosphere was probably reached around 600 million years ago; sure enough, the fossil record shows that there was an explosion of life forms at that time.

Thus, life itself was a major player in the chemical transformation of the atmosphere. But the role of life in the complex story of the atmosphere didn't stop there. Shelled marine organisms utilize carbon dioxide to make their shells, which are composed primarily of calcium carbonate ($CaCO_3$). When these organisms die, their shells are buried and eventually are transformed into the sedimentary rock called limestone (Fig. 18.4). If all the carbon dioxide now stored in great thicknesses of limestone were to be released, there would be as much CO_2 in the Earth's atmosphere as there is in the atmosphere of Venus, where the greenhouse effect runs rampant and the surface temperature averages about 480°C!

ATMOSPHERIC CHANGE AS A NATURAL PROCESS

The atmosphere is a dynamic reservoir, constantly changing on time scales ranging from less than a day (in the formation of weather systems) to billions of years (in the chemical evolution of the atmosphere). Different processes are responsible for initiating atmospheric changes on different time scales. Scientists have direct measurements of changes going back about a century, but the evidence for earlier changes, changes that predate the main impacts of human activity, is preserved in records that range from tree rings to ice caps, layered sediments, and ancient rock strata (Fig. 18.5).

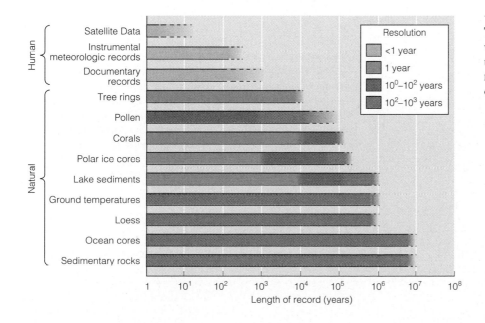

◀ F I G U R E 18.5
Time range and resolution of various sources of information used to reconstruct or document past environmental changes.

The Role of Geology in Understanding Atmospheric Change

Geology plays an important role in understanding atmospheric change, for geologists have access to a record of the Earth's changing climate that extends into the remote past. It is impossible to understand current patterns of atmospheric change and to appreciate the impact of human activities unless we view atmospheric change in a long-term context.

Atmospheric changes can be "read" from the geologic record in many ways, as shown in Fig. 18.5. *Paleontologists* (geologists who study fossils) infer past climates from assemblages of fossil plants and animals. Fossilized pollen spores in old bogs and lake bottom sediments have been particularly useful in reconstructing past climatic changes on a relatively fine scale. *Sedimentologists* and *stratigraphers* (geologists who study the processes of sedimentation and the formation of layered sedimentary rocks) can infer many things about past climates from the nature of the sediments they study. The sedimentary rock record reveals the past distribution of sediments, their mineralogy, and the mechanisms of weathering, transport, and deposition.

Sediments and sedimentary rocks can also preserve a record of short-term changes in the chemical composition of water and air, such as changes in acidity (pH). Ancient soil horizons, called *paleosols,* which represent former land surfaces, can provide a great deal of information about climate and weather in ancient environments. Geochemists can learn a lot about the environments in which weathering has occurred by examining the minerals present in paleosols; these enable them to determine the chemical compositions of the ambient air and water at the time that the soil was formed. Former ground surface temperatures can be determined by studying variations in temperature profiles

preserved in sediment layers. Isotope geologists can determine past temperatures from studies of terrestrial and marine sediments and core samples from polar ice. Layered lake bottom sediments and corals can reveal fine seasonal variations. The amount of dust in the atmosphere in earlier periods can be deduced from studies of the distribution of *aeolian* (wind-deposited) sediments such as loess. Scientists have even obtained actual samples of ancient air, up to 160,000 years old, trapped in bubbles deep within the Greenland and Antarctic ice sheets.

By reconstructing past climates, geologists can determine the range of climatic variation on different time scales and use the data to test the accuracy of computer models that simulate climatic change. One important reason for these studies is to learn how the atmosphere and climate system behaves, what controls it, and how it is likely to change in the future. We know that it *will* change, for the geologic record of atmospheric change is abundantly clear. What we still lack is a clear view of *how* it will change and at what rate. The answers to these questions are important not only scientifically but socially, politically, and economically as well. In the process of extracting and burning the Earth's immense supply of fossil fuels, humans have unwittingly begun a great geochemical "experiment" that is likely to have a significant impact on this planet and its inhabitants.

Atmospheric Change and Tectonism

On a time scale of billions of years, changes in the atmosphere have been governed primarily by changes within the planet itself, such as the release of volatile materials through volcanic degassing. It is possible that global tectonic activity contributed to and, to a certain extent, controlled the buildup of free oxygen in the atmosphere. It has been proposed that major episodes of plate tectonic activity at dif-

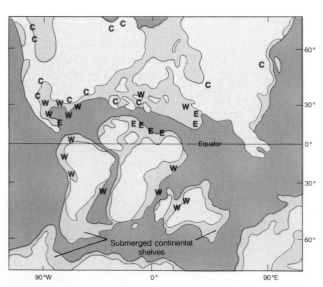

◀ F I G U R E 18.6
During the Middle Cretaceous Period the global climate was much warmer and wetter than it is now. Sea level was 100 to 200 m higher, and the oceans flooded large areas of the continents, producing shallow seas. Warm-water fossil assemblages (W) and evaporite deposits (E) were present at low to middle latitudes, while coal deposits (C) developed in northern latitudes, implying warmer year-round temperatures.

ferent times in the Earth's history caused surges of oxygen to be released into the atmosphere. How might this have happened?

Oxygen and carbohydrates are created during photosynthesis, but most of the oxygen is recaptured almost immediately when bacteria or other organisms digest available organic carbon in the process called *respiration.* A small amount of oxygen may escape, however, if some of the organic carbon is covered over by seafloor sediments. In other words, the locking away of organic carbon in seafloor sediments would allow free oxygen to build up in the environment. At times of intense geologic upheaval and tectonic activity, sedimentation rates (and thus the rate of burial of organic carbon) tend to increase. Episodes such as this may have influenced the chemical evolution of the atmosphere by allowing oxygen to accumulate in the atmosphere.

On a somewhat shorter time scale—tens to hundreds of millions of years—climatic changes have been controlled mainly by the shifting of continental land masses. For example, 40 million years ago the Western Australian Shield, which now has a hot, arid climate, was about 2000 km south of its present position. There it lay within a rainier, more temperate climatic zone much like that of southern Chile today. These wetter climatic conditions resulted in the deeply weathered paleosol horizons preserved in the rocks of the Western Australian Shield.

Aside from the actual movement of continents from one climatic zone to another, the distribution of land and water masses and the growth of high mountain chains can play an important role in regulating climate. The height and areal extent of land masses influence air circulation and the distribution of moisture; for example, there is generally a rain "shadow" on the leeward side of a high mountain chain, like the Andes. During the Middle Cretaceous Period (100 million years ago) the climate was particularly warm and wet (Fig. 18.6). The distribution of land and water masses at that time—low-lying continents grouped closely together in the middle of a great expanse of ocean—influenced atmospheric and oceanic circulation and helped create a tropical climate.

The Glacial Ages

In the past few million years numerous cycles of cooling and warming have been superimposed on the long-term cooling trend in the Earth's overall climate (Fig. 18.7A). Periods during which the average surface temperature dropped by several degrees and stayed cool long enough for the polar ice sheets to grow larger are referred to as **glaciations,** also known as *glacial periods* or *ice ages.* Periods between the glaciations, when temperatures were relatively warm, the glaciers retreated, and sea levels rose, are called *interglaciations* or *interglacial periods.* During the past 1.6 million years—a time period known as the *Pleistocene Epoch*—the Earth has experienced no fewer than 20

glacial–interglacial cycles. The timing of these cycles has varied from 20,000 to 40,000 years, with extreme temperature minima (low points) occurring roughly every 100,000 years.

Glaciation was especially prevalent in the Pleistocene, but it is not a new phenomenon; the rock record contains evidence of glacial ages occurring as long ago as 2.3 billion years. On a global scale, the areas of former glaciation add up to a total of more than 44 million km², or about 29 percent of the Earth's present land area. Today, by comparison, only about 10 percent of the world's land area is covered with glacier ice; of this area, 84 percent lies in the Antarctic region.

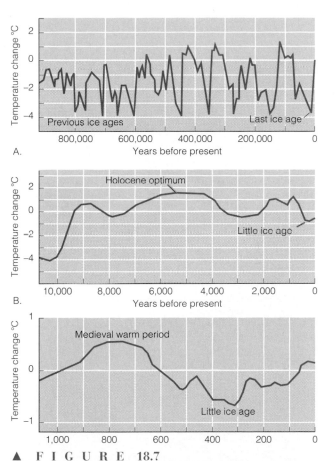

▲ **F I G U R E   18.7**
Global temperature variations over the past 1 million years. A. Temperature changes from 1 million years ago to the present. Note the cyclical variation between temperature minima—the glacial periods—and temperature maxima—the interglacial periods. B. Temperature changes from 11,000 years ago to the present, showing the Holocene optimum and the Little Ice Age. C. Temperature changes from 1100 years ago to the present, showing the Little Ice Age in comparison to the medieval warm period.

The Last Ice Age

The most recent glaciation began about 30,000 years ago when an extensive ice sheet that had formed over eastern Canada spread south toward the United States and west toward the Rocky Mountains (Fig. 18.8). Simultaneously, another great ice sheet that originated in Scandinavia spread southward across northwestern Europe. Other large ice sheets grew over arctic regions of North America and Eurasia, including some areas that are now submerged by shallow polar seas, and over the mountain ranges of western Canada. The ice sheets in Greenland and Antarctica grew larger and advanced across areas of the surrounding continental shelves. Glaciers also developed in major mountain ranges, including the Alps, Andes, Himalaya, and Rockies, as well as in numerous smaller ranges and on isolated peaks. About 10,000 years ago the Earth emerged from the last ice age. We have now passed the time of maximum warmth in

the cycle and—in principle—should soon begin the slow decline into the next glacial age.

Causes of Glaciation

The factors responsible for initiating and terminating periods of glaciation are complex. Geographic changes resulting from tectonism—the shifting of continents, the uplift of continental crust and creation of large mountain chains, and the opening or closing of ocean basins—have a significant impact on climate and atmospheric circulation. The astronomic characteristics of the Earth's orbit and rotation also play an important role in controlling the cyclicity of climatic variations. The *eccentricity* (departure from circularity) of the Earth's orbit, the tilt of the planet's axis of rotation, and the *precession* (wobbling) of the axis all have an impact on the amount of solar radiation reaching the Earth's surface at any given time (Fig. 18.9). The tilt and

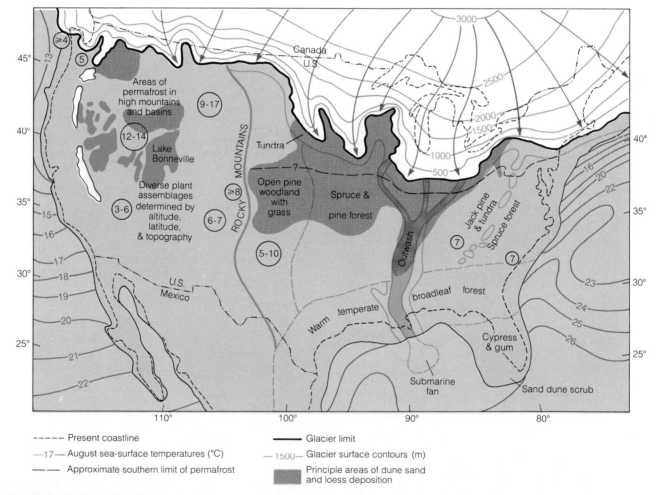

▲ F I G U R E 18.8

Map of central North America during the last glaciation, about 20,000 years ago. Coastlines lie farther seaward owing to a fall in sea level of about 100 m. Sea-surface temperatures are based on analysis of microfossils obtained from deep-sea core samples. Circled numbers show estimated temperature lowering, relative to present temperatures, at selected sites.

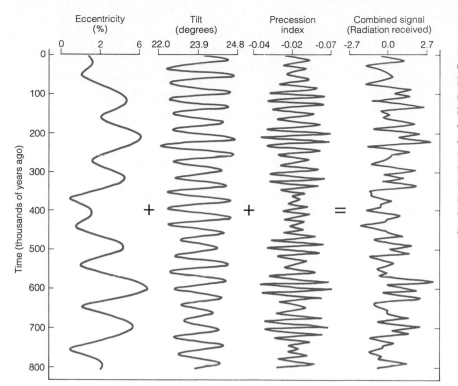

◄ F I G U R E 18.9
Curves showing variations in orbital eccentricity, tilt, and precession during the last 800,000 years. Summing these factors produces a combined signal that shows the amount of radiation received on the Earth at a particular latitude through time. This type of variation is thought to be partially responsible for variations in climate exhibited in the glacial–interglacial cycles.

precession of the axis may account for the 20,000- to 40,000-year glacial–interglacial cycles, whereas the eccentricity may contribute to the 100,000-year intervals between extreme temperature minima.

Variations in the composition of the Earth's atmosphere, especially "greenhouse" gases like carbon dioxide and methane, also influence the warming or cooling of the Earth's surface, and thus may reinforce the onset or termination of a glacial period. The concentrations of both methane and carbon dioxide in the atmosphere were low during glacial periods, when air temperatures were cool, and high during interglacial periods, when conditions were warmer overall (Fig. 18.10). Establishing this correlation between concentrations of greenhouse gases and prevailing surface temperatures represents a critical step toward understanding the processes involved in global warming.

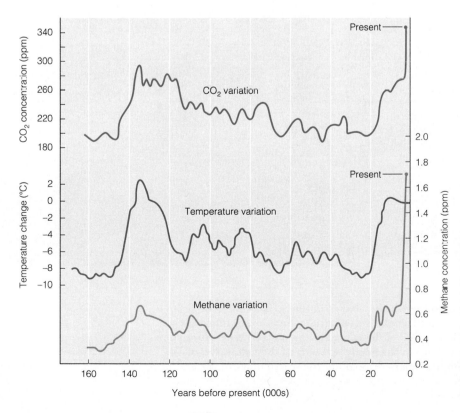

◄ F I G U R E 18.10
Variation of temperature over Antarctica and of global atmospheric carbon dioxide and methane during the last 160,000 years, as inferred from air found in core samples taken from Antarctic ice. The establishment of a firm correlation between temperature and atmospheric composition was a crucial step toward understanding the factors that control global climatic change.

Short-Term Fluctuations

Since the end of the last ice age (Fig. 18.7A) the Earth's climate has been more or less constant, with average surface temperatures varying within a degree or two of their present levels (Fig. 18.7B). But superimposed on this relative constancy has been a succession of shorter term fluctuations that have had geologic, biologic, and even socioeconomic impacts.

The Little Ice Age and the Holocene Optimum

As noted earlier, the Earth is now in a period of interglacial warmth that started 10,000 years ago. About 6000–7000 years ago temperatures peaked in a warm period (about 1°C warmer than now, on average) called the *Holocene optimum.* Since then temperatures have been gradually cooling, with several distinctly cooler fluctuations superimposed on the overall trend (Fig. 18.7B). The latest of these small fluctuations (Fig. 18.7C), characterized by an average global surface temperature just a little bit (about 1°C) lower than today's, occurred between about 1300 A.D. and 1900 A.D. During this *Little Ice Age,* as it is commonly known, climates were generally cool and glaciers expanded more than they had at any time since the end of the last glacial period 10,000 years ago. Documents, lithographs, and paintings dating from the sixteenth and seventeenth centuries describe glaciers that advanced over small villages

and became larger than at any other time in human memory (Fig. 18.11).

Volcanism and Atmospheric Change

A combination of global tectonism, astronomic influences, and fluctuations in atmospheric composition is probably responsible for the relatively long-term cyclicities of the glacial–interglacial ages. But what causes such shorter term fluctuations as the Little Ice Age, the Holocene optimum, and the small temperature variations that occur from year to year and from decade to decade? A number of interacting climatic processes play a role; El Niño is particularly influential in altering short-term temperature and precipitation patterns (Chapter 9). However, the factor that correlates best with recent fluctuations in temperature is volcanic emissions.

When a major volcanic eruption occurs, dust and gases are emitted. Volcanic gases react with water vapor in the atmosphere to form **aerosols,** extremely fine particles or droplets carried in suspension. An important component of most volcanic emissions is sulfur dioxide (SO_2). The aerosols formed from volcanic SO_2 consist primarily of dispersed sulfuric acid (H_2SO_4) droplets, which have the effect of reflecting sunlight back into outer space. It has been observed that the level of transmission of sunlight through the atmosphere tends to drop following major eruptions such as that of El Chichón in 1982 (Fig. 18.12).

A.

B.

▲ F I G U R E 18.11
A. A lithograph made in the 1850s shows the Rhône Glacier in the Swiss Alps close to its maximum extent during the Little Ice Age. The glacier terminates in a broad lobe that crosses the floor of the valley. B. A photograph of the same area taken 110 years later shows the terminus of the glacier perched high up in the headward part of the valley. Extensive thinning and recession of this and other glaciers throughout the world marked the end of the Little Ice Age.

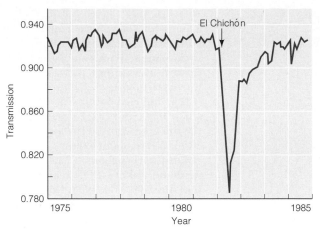

Effect on light transmission of volcanic aerosols from the eruption of El Chichón in 1982.

Eruptions that eject large amounts of aerosols high into the stratosphere tend to be associated with periods of global cooling, which may continue for several years following the eruption. Any process that controls or influences climate is called *climate forcing;* therefore, this effect is called *volcanic aerosol climate forcing.* For example, the eruption of Krakatau in 1883 led to a 0.5°C drop in average global temperature. The even larger Tambora eruption of 1815 was followed by the "year without a summer," during which midsummer snow and frost caused severe hardships in Europe and New England. The huge eruption of Mount Pinatubo in June 1991 introduced a vast quantity of fine ash and sulfurous gas into the stratosphere, where it quickly spread into the Northern and Southern hemispheres. As a result, average surface temperatures during the following year were lowered by about 0.5°C throughout the world.

In contrast to the global cooling effect caused by volcanic aerosols in the stratosphere, aerosols in the troposphere may actually cause warming on a local or regional scale. Under certain circumstances tropospheric aerosols absorb solar radiation, trapping it close to the ground and warming the surface. Volcanic emissions can also have other local and regional effects as well. For example, when volcanic SO_2 combines with water vapor to form H_2SO_4, acidic precipitation may result. Volcanoes with alkaline magma compositions have also been known to generate precipitation with a high concentration of NaCl (sodium chloride, or salt), which can be devastating to agriculture.

ANTHROPOGENIC IMPACTS ON THE ATMOSPHERE

In this brief introduction to a very complex topic, we have taken a broad view of the natural causes of change in the atmosphere and climate. Just as the atmosphere has changed on a wide variety of time scales, the spatial scales of change are variable as well. Let's look closely at the most recent trends in atmospheric change, focusing in particular on the anthropogenic causes of atmospheric change on scales ranging from local to regional, and ending with a look at the global impacts of human activities.

Recent Atmospheric Trends

The most reliable information concerning past atmospheric change comes from scientific observations and data collected over the past century. These records contain abundant evidence of fluctuations in surface temperature and atmospheric composition. The records have been maintained for varying lengths of time; for example, reasonably accurate information about temperatures has been available for more than a century, but observations of atmospheric CO_2 concentrations have been made on a systematic basis only since 1958. Some other constituents, such as ozone and chlorofluorocarbons (CFCs), were not measured systematically until very recently.

As shown in Fig. 18.13, the atmospheric concentrations of some anthropogenic gases have been increasing in recent years. The relatively long-term records (150 to 200 years) that exist for some gases suggest that many of the trends began in the early 1800s, when the burning of fossil fuels first occurred on a large scale. Ozone, a naturally occurring atmospheric trace gas, is the only atmospheric constituent to show a decreasing trend.

These recent trends in atmospheric composition are interrelated. They can be traced directly to greatly increased industrial activity and the resulting emissions of gases. Through population growth and industrialization, humans have become a force capable of altering the composition of the atmosphere, perhaps permanently. Although there is a wide variety of sources for these emissions, one stands out above all others: burning of fossil fuels is the single most prolific cause of anthropogenic emissions into the atmosphere.

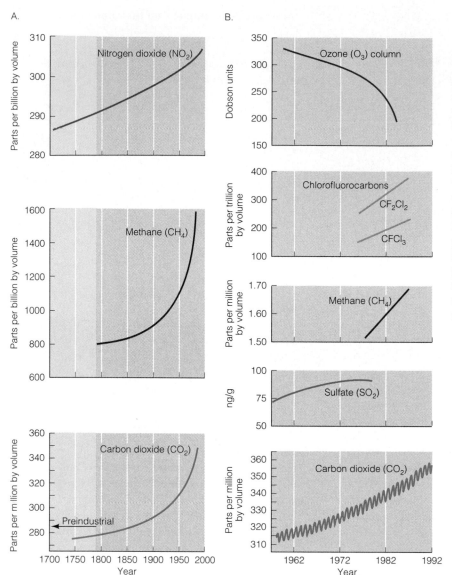

A.

B.

▶ FIGURE 18.13
Trends in atmospheric constituents that have been affected by human activities since the Industrial Revolution. A. Trends in nitrogen dioxide (N_2O), methane (CH_4), and carbon dioxide (CO_2) since 1800, and CO_2 released from the burning of fossil fuels since 1850. B. Trends in ozone (O_3), chlorofluorocarbons (CFCs), methane, sulfate (SO_4^{2-}), and carbon dioxide since 1958. Units: ppmv stands for parts per million by volume; ppbv stands for parts per billion by volume; ng/g stands for nanograms per gram; DU stands for Dobson units (see text for explanation).

Local Impacts

The local impacts of atmospheric change are often the most obvious. You are aware of those impacts when you see a yellowish haze hanging over the city, when the weather report refers to a "smog watch," or when you pass a truck belching black smoke. In addition to adding contaminants, humans alter the atmosphere by changing the landscape, primarily through urban development.

Changing the Landscape

In the process of urbanization, large expanses of asphalt, concrete, and other artificial surfaces are created. These contribute to the warming of the city by absorbing and reradiating solar energy to the air above them. The warming effect is compounded by the release of heat from sources such as vehicles and heated or air-conditioned buildings. In addition, some airborne contaminants, which are typically found in higher concentrations in urban environments, directly absorb solar energy and reradiate heat. The net result is that air temperatures in urban areas tend to be higher than those in the surrounding countryside. This is called the *urban heat island* effect (Fig. 18.14). Long-term meteorologic records from pairs of stations—one within and the other outside a city—show how this effect increases with urban growth.

Urban effects on local climate can encompass more than air temperature. Tall buildings promote turbulence and the convection of heat above the city. This can lead to increased cloudiness and greater frequency of precipitation, especially in the summer and downwind of the urban core. Some pollutants provide nuclei for the formation of water droplets in clouds. The pattern of air circulation that typically develops in an urban heat island also tends to trap and recirculate contaminants, especially fine dust particles, contributing to poor air quality.

Smog and Tropospheric Ozone

A wide variety of airborne contaminants cause pollution at lower levels in the atmosphere, especially in urban centers. The behavior of an airborne contaminant depends on a variety of factors, including its own chemical and physical

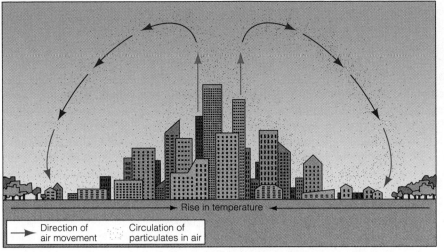

◄ F I G U R E 18.14
The urban heat island effect. Ground covers such as concrete and asphalt absorb solar energy and reradiate it as heat. Heat from buildings and vehicles adds to the effect. Some atmospheric contaminants also absorb heat, trapping it close to the surface. The result is a concentration of heat and contaminants over the city.

characteristics, the local topography, and prevailing weather conditions. The characteristics of the pollutant determine its atmospheric residence time, which controls how long it can remain in suspension in the atmosphere and, therefore, how far it will be transported. Winds and precipitation also play an important role in the transport of airborne contaminants; when the air is very still, pollutants are not trans-ported as far and tend to become concentrated over the source area.

Sometimes weather and topography join forces to exacerbate local air pollution. Under normal circumstances, temperatures in the lower part of the atmosphere decrease steadily with increasing height (refer to Figure 9.1). However, if a warm air mass moves into the

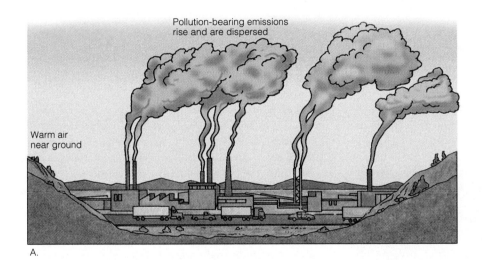

◄ F I G U R E 18.15
Thermal inversion and air pollution. A. The normal case, with temperature decreasing upward. Polluted air rises from cars and factories and is dispersed at a high level. B. A warm air mass has trapped cool air near the ground. Polluted air rising from the city is trapped beneath the layer of warm air, concentrating pollutants over the city. In this example the effect is especially pronounced because of the basinlike topography of the setting.

area it may trap a pocket of cool air underneath it, a phenomenon known as a **thermal inversion.** When warm, contaminant-laden gases are released from factories or vehicles, the plume of polluted air rises because it is less dense (more buoyant) than the cooler air around it (Fig. 18.15A). In a thermal inversion, however, the plume of polluted air will rise until it hits the warm air mass and then spread laterally (Fig. 18.15B). Most episodes of acute air pollution are associated with thermal inversions.

Sometimes local topography can intensify the effect of a thermal inversion. In the Los Angeles area, for example, air pollution and thermal inversions are a particular problem because the city and its surroundings are built on a series of basins that tend to trap cool air and concentrate pollutants.

When airborne contaminants interact with sunlight, they undergo a variety of complex photochemical reactions. The end product is a yellowish-brown haze called **smog** (a combination of *smoke* and *fog*). Smog is usually at its worst in urban areas (near the source of the contaminants), on sunny days (when smog-forming reactions are facilitated), and during a thermal inversion (when the smog is trapped near the ground).

Airborne pollutants occur either as fine suspended particulate matter or as gases. In either state they can have negative effects on health, particularly for people with respiratory problems. Some types of air pollution can also affect the health of plants and animals and the durability of materials.

Suspended Particulate Matter **Particulates** are pollutants that are carried in suspension as extremely fine, dispersed solid particles or liquid droplets. Suspended particulates are produced by some natural processes, such as

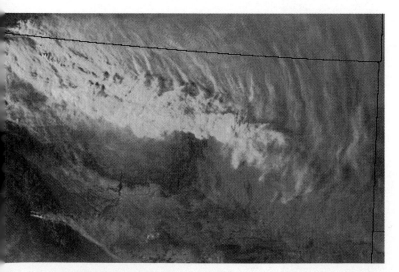

▲ F I G U R E 18.16
A satellite image reveals vast smoke plumes from fires in Yellowstone National Park, July, 1988. The black lines mark state boundaries. The source of the smoke is the northwest corner of Montana.

volcanic eruptions; smoke from forest fires; and blowing sea salt, dust, and pollen. Most anthropogenic particulates originate from industrial activities, transportation, and the generation of electrical power. Refuse incineration and the burning of wood and agricultural wastes also contribute particulates to the atmosphere. There is growing concern about the contribution of biomass burning, especially slash-and-burn agriculture, to both particulate and gaseous air pollution. In recent years astronauts have noticed a dramatic increase in the amount of smoke from biomass burning observable from space (Fig. 18.16).

Heavy Metals In the atmosphere, lead is the most abundant of the *heavy metals* (which include mercury, zinc, arsenic, cadmium, and manganese). Airborne lead has traditionally come mainly from automobile exhaust, but now the main sources are industrial (e.g., smelting, mining). Lead-based paint, which is no longer used but is found in older housing, is of particular concern because ingesting or breathing in lead particles can cause brain damage. The other heavy metals are derived mainly from emissions from smelting operations and, to a lesser extent, from waste incinerators. Heavy metal particulates eventually settle out, but once they have settled they may become concentrated in soils in the area surrounding the source.

Carbon Monoxide The oxides of carbon, sulfur, and nitrogen are among the most dangerous gaseous air pollutants. Carbon monoxide (CO), which occurs naturally in the atmosphere in minute quantities, is toxic and sometimes fatal in high doses; chronic exposure even to low doses can be fatal. Carbon monoxide is generated by the combustion of carbon-bearing materials (i.e., fossil fuels) in a low-oxygen environment. Cars and small trucks are the main source of CO pollution, with some contribution from space heating, industry, and the burning of wood and wood wastes.

Sulfur Dioxide Emissions of sulfur dioxide (SO_2) come mainly from industrial processes, especially the burning of coal. Coal-fired power plants are a particular culprit. Sulfur dioxide can cause respiratory problems, eye irritation, and other health problems. When it combines with water vapor in the atmosphere, it can cause acidic precipitation (discussed later in the chapter). Sulfur dioxide emissions can be controlled through the burning of low-sulfur fuels and the imposition of stricter emissions standards.

Nitrogen Oxides Nitrogen and oxygen, both of which are essential components of the atmosphere, combine to form a variety of gaseous compounds (referred to as NO_x). Human activities involving the burning of fossil fuels typically generate nitric oxide (NO), which combines with oxygen in the atmosphere to form nitrogen dioxide (NO_2). Nitrogen dioxide is one of the main constituents of photochemical smog; it causes respiratory problems, and it can inhibit plant growth and corrode or degrade a variety of materials. Nitrogen dioxide can also combine with water vapor to form nitric acid (HNO_3), leading to acidic precipitation.

Volatile Organic Compounds   Easily volatilized chemicals, which come from incomplete combustion of fossil fuels as well as from solvents and paints, are another source of respiratory problems. Volatile organic compounds (VOCs) also play a role in the formation of photochemical smog.

Ground-level Ozone   One of the main products of the photochemical reactions that create smog is ozone (O_3). Most of the ozone in the atmosphere occurs in the stratosphere, where it performs the important function of shielding organisms from harmful radiation. When it occurs in the lower part of the atmosphere, it is referred to as **ground-level ozone** or *tropospheric ozone*. Ozone is an eye irritant and can cause respiratory problems as well as damaging crops and vegetation. Ground-level ozone is not generated directly by human activities, but occurs as a product of photochemical reactions in air contaminated with VOCs or NO_x.

Indoor Air Pollution

Air pollution problems can occur indoors as well as outdoors. Some types of indoor air pollution, such as smoke from wood fires, have been a problem ever since people began living and cooking indoors. However, the buildup of contaminants inside buildings has been a particular problem since the 1970s, when concerns about energy consumption led to the construction of very tightly sealed, energy-efficient homes. The old, "leaky" type of construction required more energy for heating and cooling but facilitated the exchange of air, which prevented the buildup of airborne contaminants inside the building. Among the most common indoor pollutants are cigarette smoke, carbon monoxide (mainly from furnaces), and formaldehyde (a VOC used in blown foam insulation). But perhaps the best known and most controversial pollutants found in indoor environments are asbestos and radon.

Asbestos   Asbestos is a mineral with fire-retardant characteristics that was used for many years in ceiling materials and pipe casings. **Asbestos** is actually an industrial term referring to a group of minerals that grow with a fibrous or *asbestiform* crystal habit. Thus, there are several commonly used asbestos minerals. The most important of these are *chrysotile* (the most common), *amosite,* and *crocidolite.* Some scientific evidence has linked chrysotile with diseases such as lung cancer and mesothelioma, giving rise to widespread concern about its use in public buildings. In many parts of North America ceiling materials containing chrysotile have been removed from such buildings at great

expense. However, new evidence implicates the other varieties of asbestos, rather than the commonly used chrysotile, as carcinogens. This has prompted some researchers to recommend the removal of all asbestiform materials.

Radon   **Radon** is a colorless, odorless radioactive gas produced by the radioactive decay of uranium in soils derived from uranium-bearing rocks (Fig. 18.17). As a soil gas, radon diffuses naturally through cracks and openings in permeable soils. Once it enters the atmosphere, it is normally diluted and dispersed; the concentration of outdoor radon therefore is quite low. However, problems occur when radon enters the lower levels of buildings. Soil gas can enter a building via a number of pathways, including cracked foundations, pipes or windows with loose fittings, sump pumps, well water, construction materials, and even soil in potted plants (Fig. 18.18). Because radon is very heavy (it is the densest known gas), it tends to become concentrated in the basements.

Once inside a building, radon continues to decay radioactively. Some of its daughter products are also radioactive, and they emit harmful radiation as they decay (Fig. 18.17). Some of the daughter products attach themselves to dust particles and thus become more concentrated in the lower levels of the buildings. The real problem occurs when air contaminated by radon or one of its daughter products is inhaled. The radioactive decay of radon emits *alpha particles,* which, as you may recall from Chapter 16, are particularly damaging to biologic tissue. Direct doses of alpha radiation to the lining of the lung are thought to cause more deaths from lung cancer than any other cause except smoking.

▲   F I G U R E   18.17
The radioactive decay series leading to the formation of radon and its daughter products. A. Radioactive uranium occurs naturally in soils derived from uranium-bearing rocks. In the soil, uranium undergoes radioactive decay, forming radium and then radon, a gas, as its daughter products. B. In the gaseous state (i.e., as a soil gas or in the atmosphere), radon decays through a series of short-lived radioactive daughter products. The dose of alpha radiation to the lining of the lungs resulting from inhalation of these products can cause lung cancer.

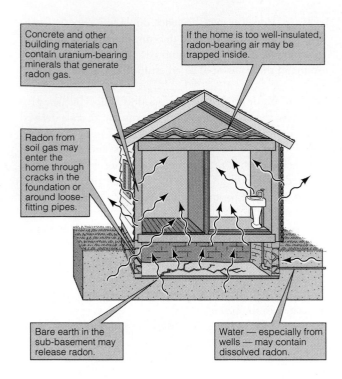

Concrete and other building materials can contain uranium-bearing minerals that generate radon gas.

If the home is too well-insulated, radon-bearing air may be trapped inside.

Radon from soil gas may enter the home through cracks in the foundation or around loose-fitting pipes.

Bare earth in the sub-basement may release radon.

Water — especially from wells — may contain dissolved radon.

◄ F I G U R E 18.18
Entryways for radon into a building. Once it escapes as a soil gas, radon can enter a building through a number of pathways, including fractures in the foundation, poorly sealed windows or pipes, well water, soil in potted plants, construction materials, and sump pumps.

It is exceedingly difficult to predict the level of radon in a building. If the building is constructed on a known uranium-bearing rock formation, chances are that it will have an elevated level of radon. But many other factors come into play, including the construction (style and materials); weather (the problem tends to be worse in the winter, when buildings are shut tight); chemical and physical characteristics of the underlying soil or bedrock; and a host of other factors. One approach to prediction is to fly airborne spectrometers over large areas in an attempt to measure uranium concentrations at the surface and radon levels in soil gas. These measurements are used to identify areas of potential concern, where more detailed measurements should be undertaken.

Regional Impacts

The regional impacts of airborne contaminants arise mainly from the transport of materials over long distances, referred to as *long-range transport of airborne pollutants.* Long-range transport is facilitated if a pollutant attaches itself to water droplets or dust particles or undergoes chemical reactions to form new pollutants in the atmosphere.

Materials transported in suspension will eventually settle out either as precipitation, called **wet deposition,** or as dry matter, called **dry deposition.** Typically the concentrations of contaminants deposited in this way will be low because the contaminants have been diluted and dispersed by atmospheric processes. However, continued deposition over a long period can lead to the buildup of contaminants in

soils, sediments, water, and organisms. Deposition of contaminants can occur very far from the source of the contaminants. For example, after the accident at the Chernobyl nuclear power plant in 1986, radioactive particles derived from the facility encircled the globe within 2 weeks and eventually settled out as both dry and wet deposition (Fig. 18.19).

One problem associated with the long-range transport of contaminants is *arctic haze.* This reddish-brown haze, composed of fine particles of sulphates, soot, dust, hydrocarbons, toxic metals, and pesticides, occurs in the lower 1 to 2 km of the atmosphere during the arctic winter. It is caused by long-range transport of pollutants, primarily from Europe and Asia, by prevailing winds (as shown in Fig. 17.5), and can have negative impacts on sensitive arctic ecosystems. It is possible that arctic haze, which absorbs heat, may intensify global warming.

Acid Precipitation

Perhaps the most significant regional effects of long-range transport come from the transport of acid-forming compounds, which eventually settle out in the form of *acid precipitation.* Acidity is measured on the **pH** scale (Fig. 18.20), which ranges from 0 (extremely acidic) to 14 (extremely alkaline). A pH of 7, at the midpoint of the scale, represents a neutral material, such as distilled water. The pH scale is a logarithmic scale, which means that each point on the scale represents a tenfold change in acidity. Ordinary rainfall (i.e., rainfall that does not contain anthropogenic pollutants) is slightly acidic, with a pH of about 5.6 to 5.0.

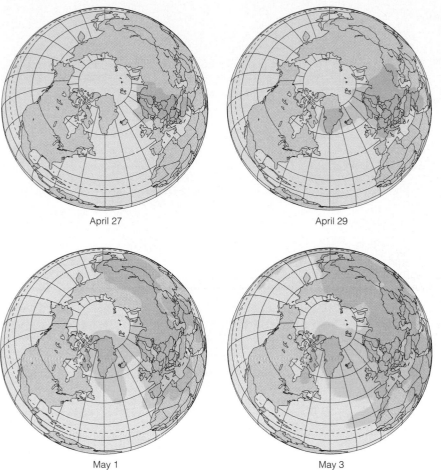

April 27

April 29

May 1

May 3

◀ F I G U R E 18.19
The Chernobyl disaster occurred
on April 26, 1986. By May 3 radioactive contamination from the
damaged reactor had spread over
half the Northern Hemisphere.

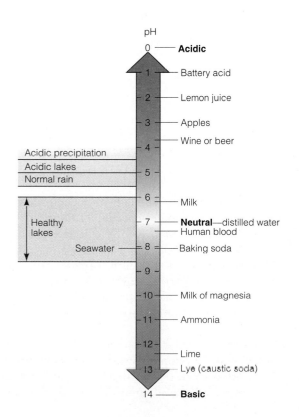

pH

0 — **Acidic**

1 — Battery acid

2 — Lemon juice

3 — Apples

4 — Wine or beer

Acidic precipitation

Acidic lakes

Normal rain

5

6 — Milk

Healthy
lakes

7 — **Neutral**—distilled water
Human blood

Seawater — 8 — Baking soda

9

10 — Milk of magnesia

11 — Ammonia

12 — Lime

13 — Lye (caustic soda)

14 — **Basic**

◀ F I G U R E 18.20
The pH scale. A pH of 7 is neutral. Higher pHs are indicative of alkaline or basic substances such as ammonia
or lye. Lower pHs are indicative of acids such as tomato
juice or battery acid. Acid rain has a pH of less than 5.0.
The scale is logarithmic, which means that each increase
of 1 on the scale indicates a tenfold increase in acidity.

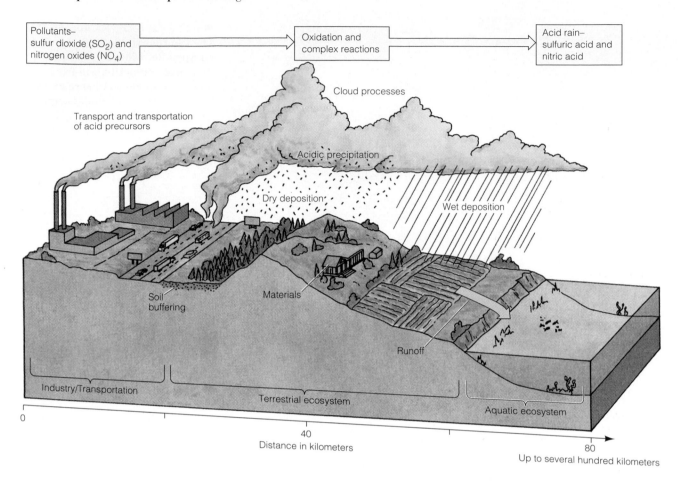

Pollutants–
sulfur dioxide (SO_2) and
nitrogen oxides (NO_4)

Oxidation and
complex reactions

Acid rain–
sulfuric acid and
nitric acid

Cloud processes

Transport and transportation
of acid precursors

Acidic precipitation

Dry deposition

Wet deposition

Soil
buffering

Materials

Runoff

Industry/Transportation

Terrestrial ecosystem

Aquatic ecosystem

0

40
Distance in kilometers

80
Up to several hundred kilometers

▲ F I G U R E 18.21

The sources, transport, formation, and deposition of acid precipitation. The main sources of acid precursors (mainly SO_2 and NO_x) are human activities, primarily those involving the burning of fossil fuels. The acid precursors are transported by atmospheric circulation. In the atmosphere they interact chemically with water, forming sulfuric and nitric acids. The acids can fall to the surface through wet deposition (acid rain, snow, or fog) or dry deposition. Acidification affects vegetation, aquatic ecosystems, and the resilience of materials. Certain soils and rocks can buffer or neutralize the effects of acidification.

Acid precipitation is defined as any precipitation with a pH lower than about 5.0.

Acid precipitation is often called *acid rain,* although it can also occur as snow. Airborne acidic contaminants can also occur in the form of clouds and fog or settle to the ground as dry deposition, either as gases or associated with dust particles. The term *acid deposition* is sometimes used to encompass all of the processes that cause wet and dry deposition of acids (Fig. 18.21).

The main chemical compounds that cause acid deposition are sulfur dioxide (SO_2) and nitrogen oxides (NO_x), both of which are generated by the burning of fossil fuels. The major sources are industries, such as smelting and refining operations, and power generation. In the air, SO_2 and NO_x combine with water vapor to form sulfuric acid (H_2SO_4) and nitric acid (HNO_3), which eventually fall as

acid precipitation or in dry deposition. Acid deposition is a problem over much of Europe and eastern North America (Fig. 18.22).

Acid deposition has a wide range of negative impacts, including the corrosion and degradation of materials (Fig. 18.23) and the acidification of soils. The most obvious problem, however, is the effect of acidification on aquatic ecosystems, especially lakes. When the pH of a freshwater lake drops below 6.0, some species may be negatively affected; if the pH drops below 4.5, no fish can survive. Many lakes in eastern North America have "died" as a result of acidification.

An important constraint on an ecosystem's vulnerability to acidification is the composition of the underlying bedrock. If the bedrock and soils derived from it are alkaline, such as calcium carbonate–bearing limestones, they can

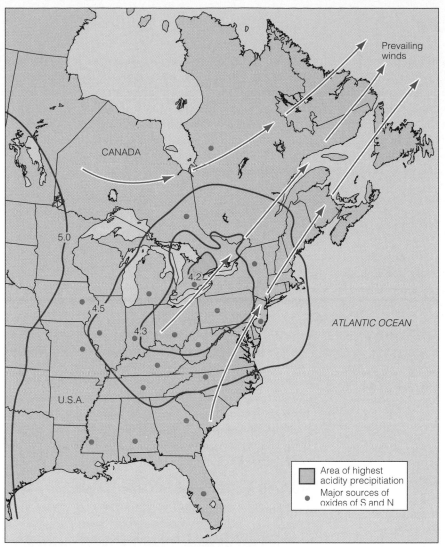

◄ F I G U R E 18.22
Mean acidity of precipitation in eastern North America, 1982–1987. The circles indicate some of the principal sources of acid precursors, oxides of sulfur and nitrogen. Arrows indicate prevailing winds in winter and summer.

Prevailing winds

CANADA

ATLANTIC OCEAN

5.0

4.2

4.5

4.3

U.S.A.

Area of highest acidity precipitation
Major sources of oxides of S and N

◄ F I G U R E 18.23
Limestone and sandstone are especially vulnerable to the effects of acidic pollution. Acid rain has severely disfigured this limestone statue on the facade of the cathedral at Reims, France.

neutralize or *buffer* much of the acid deposited in the water. On the other hand, if the rock or soil is siliceous, like granites and some sandstones, its acid-buffering capacity will be much lower, and the lake will be much more vulnerable to acidification. In some cases the addition of alkaline material—most commonly powdered limestone distributed by helicopter—can increase the acid-buffering capacity of a lake.

DEPLETION OF STRATOSPHERIC OZONE

So far we have looked at human-generated atmospheric change on the local and regional scales. We now turn to a problem that is essentially global in extent. As mentioned earlier, one of the atmosphere's main life-supporting functions is the screening out of harmful short-wavelength ultraviolet radiation (*UV-B* and *UV-C*) by a layer of ozone gas in the stratosphere. In recent decades scientists have warned that some anthropogenic gaseous emissions may be depleting the ozone layer.

The Ozone Layer

The **ozone layer** actually is not a layer but a zone in the stratosphere where ozone occurs in higher concentrations than elsewhere in the atmosphere (Fig. 18.24). The total amount of ozone in the ozone layer is surprisingly small. Ozone concentration is often measured in terms of the amount of ozone in a vertical column above a given point on the Earth's surface. The measurements are given in *Dob-*

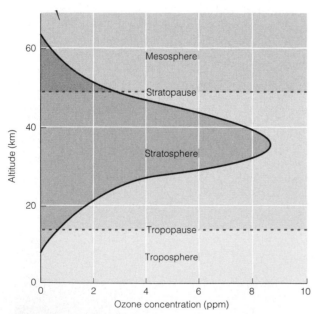

▲ F I G U R E 18.24
The vertical structure of the atmosphere and the distribution of ozone.

son units (DU), where 100 DU is equivalent to a layer of pure ozone 1 mm thick at standard sea level temperature and pressure. The average concentration of ozone at any point above the Earth is 350 DU, the equivalent of a layer of pure ozone just 3.5 mm thick! The greatest concentration of stratospheric ozone, which occurs at an altitude of about 25–35 km, is never more than 10 ppmv (parts per million by volume).

The Chemistry of the Ozone Layer

Ozone is created naturally in the upper atmosphere when high-energy ultraviolet solar radiation strikes an oxygen molecule and breaks it down into two highly reactive oxygen atoms. One of the freed oxygen atoms combines with an intact oxygen molecule to form ozone:

$$O + O_2 = O_3$$

Ozone can also be broken down by interaction with solar radiation, as well as by interactions with various chemicals in the stratosphere. Some of the naturally occurring chemicals that play a role in the destruction of ozone are oxides of hydrogen and nitrogen. Under normal circumstances, the processes of ozone formation and destruction balance each other and the concentration of ozone in the stratosphere remains relatively stable.

Anthropogenic Causes of Ozone Depletion

Concerns about the impact of anthropogenic gases on the ozone layer emerged in the 1960s and 1970s, when it was suggested that airplanes, rockets, and supersonic jets flying through the stratosphere might cause ozone depletion because their exhaust contains nitrogen oxides. Shortly thereafter, with the development of the space shuttle, an additional threat materialized. It was feared that chlorine—a trace gas in the shuttle's exhaust gases—could initiate ozone depletion. It was quickly determined that exhaust from the space shuttle was not a major concern, but chlorine itself became a subject of much interest and research.

Anthropogenic chlorine (Cl) in the atmosphere is derived mainly from a group of chemicals known as *chlorofluorocarbons,* or *CFCs.* CFCs are used in refrigerators, aerosol propellants, industrial solvents, cleaning fluids, foam-blowing agents, fire retardants, and a variety of other applications. Because they are relatively stable chemically, they survive in the atmosphere long enough to migrate up to the stratosphere. Once there, they are subjected to intense solar radiation and break down, releasing an atom of chlorine in the process (Fig. 18.25A). The chlorine acts as a chemical catalyst to cause the breakdown or *dissociation* of ozone molecules by combining with one of the oxygen atoms in the ozone to form a molecule of chlorine monoxide (ClO) (Fig. 18.25B). The chlorine is later rereleased when the ClO reacts with oxygen.

Because the chlorine is a catalyst and is not, itself, altered in the process of ozone dissociation, it can be recycled

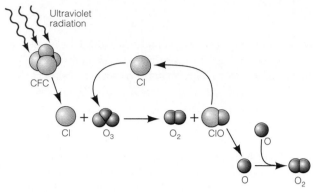

▲ FIGURE 18.25
CFCs act as a chemical catalyst, facilitating the breakdown of ozone in the stratosphere. A. Chlorine is released when CFC molecules break down as a result of interaction with ultraviolet radiation. B. The chlorine atom reacts with ozone, breaking down the ozone to form molecular oxygen (O_2) and chlorine monoxide (ClO). C. Eventually the chlorine monoxide breaks down. The freed oxygen combines with another oxygen atom to form O_2 and the chlorine is recycled, causing the breakdown of other ozone molecules.

to break down more ozone molecules (Fig. 18.25C). One atom of chlorine may be able to break down as many as 100,000 ozone molecules. Bromine, another element released from the breakdown of anthropogenic gases in the upper atmosphere, is less abundant but even more efficient than chlorine in causing the dissociation of ozone molecules. (Chlorine and bromine belong to a group of elements called *halogens*. When combined with carbon, as in CFCs, they are called *halocarbons*. Other chlorine- and bromine-bearing chemicals that have been implicated in ozone depletion are halons, carbon tetrachloride, methyl chloroform, and hydrochlorofluorocarbons, all of which have very long atmospheric residence times.)

The Ozone Hole

Although it was known that elements like chlorine and bromine had the potential to cause the breakdown of ozone molecules, concern increased when ozone depletion was actually measured and documented. In 1985 it was announced that ozone levels in Antarctica had declined by as much as 50 percent since the 1970s; these measurements and the continuing depletion of ozone have since been confirmed. Ozone depletion is both seasonal and spatially controlled; it is greatest in the spring (September to November in the Southern Hemisphere) and most pronounced over the poles (especially the South Pole). The phenomenon of ozone depletion centered over the south polar region has come to be known as the **ozone hole**. The depletion continues to become more se-

▼ FIGURE 18.26
Total atmospheric ozone over the Southern and Northern Hemispheres on September 30, 1992. Reddish and yellowish colors indicate high ozone concentrations, whereas bluish colors denote the abnormally low concentrations that define the ozone hole.

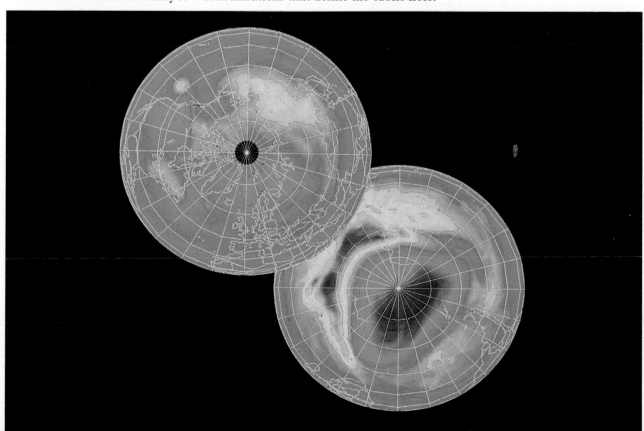

vere, though it fluctuates somewhat from year to year. In 1987 ozone concentrations over Antarctica dropped to 121 DU. In 1988 the depletion was less intense, but record lows were measured in the 3 following years. In 1992 the areal extent of depletion over the South Pole was more than 23 million km², and new concerns were raised about the depletion of stratospheric ozone in the Northern Hemisphere.

It is not absolutely clear why ozone depletion is most intense over the poles (Fig. 18.26). Ice-mediated reactions in polar clouds during the cold winter months may enhance the ozone-depleting capacities of chlorine and bromine, converting them into more reactive forms. In the spring, when warmer weather and more sunlight return, the ozone-destroying reactions begin, with the chlorine and bromine acting as catalysts. The dynamics of stratospheric wind systems over the poles—the *polar vortex*—also play a role in determining the extent and intensity of ozone depletion by isolating polar air masses and allowing the reactions to proceed without influxes of new, warmer air.

Effects of Ozone Depletion

Ozone depletion allows more ultraviolet radiation—especially UV-B rays, which are selectively absorbed by ozone—to penetrate the atmosphere and reach the Earth's surface. The resulting higher exposure to UV-B radiation is expected to lead to an increase in the incidence of skin cancer, which is correlated with intensity of UV-B exposure. Within continental North America exposure to ultraviolet radiation increases by approximately 1 percent for each 2–3° of latitude as one moves toward the south. However, it is difficult to predict the extent of the increase in skin cancer; it depends not only on the extent of ozone depletion at a given latitude but also on factors such as people's willingness to protect themselves with sunscreens. Ultraviolet radiation can also affect other aspects of human health, such as the general effectiveness of the immune system, and the rate of occurrence of eye diseases, such as cataracts.

Exposure to UV radiation is also harmful to some plants, animals, and materials. For example, phytoplankton production, a measure of the degree of photosynthetic activity, has been shown to decrease significantly in areas undergoing ozone depletion. This could have an effect on the availability of food for animals higher in the food chain. Some studies have shown that exposure to UV radiation causes reduced yields, mutations, and chemical changes in the leaves of some crops. Even some inanimate materials, such as rubber, paints, and plastics, are more susceptible to deterioration when they are exposed to increased levels of ultraviolet radiation.

T A B L E 18.1 • Control Measures Agreed upon in the *Montreal Protocol on Substances That Deplete the Ozone Layer,* Which Came into Force in 1989 and Was Revised in 1990[a] (values represent levels of production as a percentage of production in the base year).

Year	CFCs identified in protocol	Other CFCs[b]	Halons identified in protocol[c]	Carbon tetrachloride	Methyl chloroform
1992	freeze	—	freeze	—	—
1993	20%	20%	—	—	freeze
1994	—	—	—	—	—
1995	50%	—	50%	85%	30%
1996	—	—	—	—	—
1997	85%	85%	—	—	—
1998	—	—	—	—	—
1999	—	—	—	—	—
2000	100%	100%	100%	100%	70%
2005					100%
Base year	1986	1989	1986	1989	1989

[a]Source: Environment Canada, Atmospheric Environment Service.

[b]HCFCs are to be phased out between 2020 and 2040. In the interim, their usage is to be confined to CFC substitutes.

[c]Other halons are not used much commercially. Studies are under way through the Montreal Protocol to determine whether—and, if so, how—they can be included in the protocol.

Responding to Ozone Depletion

In 1985 concerns about ozone depletion led political leaders from around the world to sign the *Vienna Convention,* which called for the establishment of an action plan to deal with the issue. The agreement also called for scientists to report in detail on the latest findings and probable future trends in the condition of the ozone layer. At the time there was still some uncertainty about whether ozone levels had really been declining, but scientists warned that there was no time to waste; once the ozone-depleting chemicals began to work, damage to the ozone layer would continue for decades even if production of CFCs were completely halted.

The Montreal Protocol

By the time of the next international meeting, which took place in Montreal in 1987, the existence of the ozone hole over Antarctica had been confirmed and the connection of ozone depletion with anthropogenic gases was well established. Twenty-four nations signed the original *Montreal Protocol on Substances That Deplete the Ozone Layer,* which called for a phaseout of CFC production by 2000. One of the features that makes the Montreal Protocol unique is that it provides for regular assessments of the status of scientific understanding of ozone depletion and the effectiveness of the control measures undertaken. The agreement went into effect in 1989 and was revised in 1990 to include additional ozone-depleting substances (Table 18.1). Signing the Montreal Protocol was an exceedingly difficult political decision for countries that are major producers of CFCs; in the United States alone, CFCs represent a multibillion dollar industry.

There is a lot more to be learned about the processes involved in ozone depletion and its possible effects. Even with reductions in CFC use, ozone depletion will probably continue until at least 2030. But the positive side of this story is that scientific understanding led to a political consensus and an international agreement to act on behalf of all the Earth's inhabitants. As of the end of 1992, 86 countries, including all of the major producers of ozone-depleting substances, had ratified the Montreal Protocol and its amendments.

GLOBAL WARMING

As we have seen, short-term climatic fluctuations are a normal part of the evolution of the atmosphere. During the past decade or so, however, the possibility of accelerated global warming and the role of anthropogenic emissions in the warming process have occupied an important position in both scientific and political debates.

Historical records of surface temperatures suggest that over the past 100 years or so, average global temperatures do, indeed, appear to have increased by about 0.5° to 0.6° (Fig. 18.27). But is the warming trend real? It is extremely difficult to measure average surface temperature changes conclusively on a time scale as short as this because the magnitude of the changes—0.5° or 0.6°C—is small by comparison with seasonal changes and variations from one climatic zone to another. Therefore, studies of this type tend to be somewhat controversial. If the warming trend is real, will it continue? How warm will it get? What is the relationship between global warming and the chemical composition of the atmosphere? Is the warming trend really caused by human activities? What will be its effects? Can it be reversed? Let's try to answer these questions by looking first at the natural processes whereby the atmosphere warms the surface of the Earth.

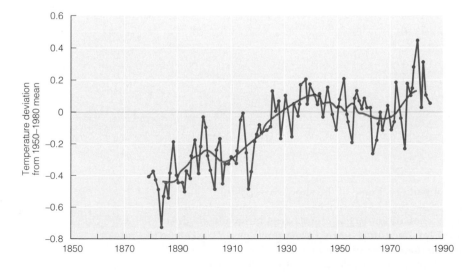

◄ F I G U R E 18.27
Change in global temperature over the past century. The mean annual temperature (spiky curve) and the 5-year running mean (smooth curve) are plotted.

The Greenhouse Effect

The atmosphere is the engine that drives the Earth's climate system, and the Sun provides the energy that allows the engine to work (Fig. 18.28). Recall from Chapter 1 that some of the solar radiation reaching the atmosphere is reflected from clouds and dust and returns to outer space. Of the radiation that reaches the planet's surface, some is absorbed by the land and oceans and some is reflected into space by water, snow, ice, and other reflective surfaces. This visible reflected solar radiation has a short wavelength. The Earth also absorbs some radiation and eventually reradiates it as long-wavelength infrared radiation (i.e., heat).

However, a portion of the outgoing long-wavelength radiation encounters gases in the atmosphere with chemical properties that prevent this energy from escaping. Instead, the energy is retained in the lower atmosphere, causing the temperature at the Earth's surface to rise. Clouds can also trap infrared radiation near the surface. (Thus, clouds play a dual role in regulating surface temperature: cooling the surface by reflecting incoming radiation and warming it by trapping heat underneath.) A comparable effect explains why the temperature of air in a glass greenhouse is warmer than that of the air outside: the glass, acting in much the same way as the clouds and atmospheric gases, prevents the

escape of radiant energy. Hence, this phenomenon is referred to as the **greenhouse effect.**

It is the greenhouse effect that makes the Earth habitable. Without it, the surface of the Earth would be as inhospitable as those of the other planets in the solar system. If the Earth lacked an atmosphere, its surface environment might be like that of the Moon. On the sunlit side of the Moon temperatures are close to the boiling point of water, whereas on the dark side temperatures are far below freezing. By contrast, on Venus the runaway greenhouse effect generates surface temperatures hot enough to melt lead.

Greenhouse Gases

Atmospheric gases that are efficient absorbers of infrared radiation are referred to in technical terminology as **radiatively active gases** and are commonly known as *greenhouse gases.* Water vapor naturally present in the atmosphere is by far the most important "greenhouse gas," accounting for about 80 percent of natural greenhouse warming. The remaining 20 percent is due to other gases that are present in very small amounts. Despite their low concentrations, which are measured in parts per billion by volume (ppbv) of air, some of these trace gases contribute significantly to the greenhouse effect (Table 18.2).

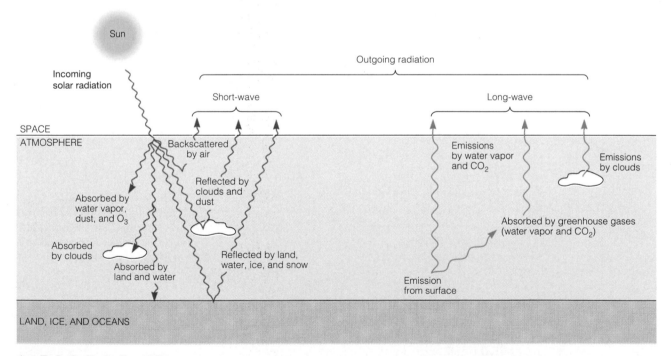

▲ F I G U R E 18.28
The greenhouse effect in the Earth's atmosphere. Some of the shortwave (visible) solar radiation reaching the Earth is absorbed by the land and oceans, while some is bounced back into space from reflective surfaces such as snow, ice, clouds, and atmospheric dust. The Earth also radiates longwave radiation back into space. Greenhouse gases trap some of the outgoing longwave radiation, causing the atmosphere to heat up. With increasing concentrations of these trace gases, the temperature of the lower atmosphere rises.

T A B L E 18.2 • **Atmospheric Trace Gases Involved in the Greenhouse Effect**[a]

	Carbon Dioxide (CO₂)	Methane (CH₄)	Nitrous Oxide (N₂O)	Chlorofluoro-carbons (CFCs)	Tropospheric Ozone (O₃)	Water Vapor (H₂O)
Greenhouse role	Heating	Heating	Heating	Heating	Heating	Heats in air, cools in clouds
Effect on stratospheric ozone	Can increase or decrease	Can increase or decrease	Can increase or decrease	Decrease	None	Decrease
Principal anthropogenic sources	Fossil fuels; deforestation	Rice culture, cattle, fossil fuels, biomass burning	Fertilizer, land-use conversion	Refrigerants; aerosols; industrial processes	Hydrocarbons (with NOₓ); biomass burning	Land conversion, irrigation
Principal natural sources	Balanced in nature	Wetlands	Soils, tropical forests	None	Hydrocarbons	Evapotrans-piration
Atmospheric lifetime	50–200 yr	10 yr	150 yr	60–100 yr	Weeks to months	Days
Present atmospheric concentration at surface (ppbv)	353,000	1720	310	CFC-11: 0.28 CFC-12: 0.48	20–40	3000–6000 in stratosphere
Preindustrial (1750–1800) concentration at survace (ppbv)	280,000	790	288	0	10	Unknown
Present annual rate of increase	0.5%	1.1%	0.3%	5%	0.5–2.0%	Unknown
Relative contribution to the anthropogenic greenhouse effect	60%	15%	5%	12%	8%	Unknown

[a]Source: Earthquest, vol. 5, No. 1, 1991.

Chief among the anthropogenic gases is carbon dioxide, which has an average concentration of about 350,000 ppbv in surface air. Other significant radiatively active gases are methane, nitrogen oxides, and ozone, as well as commercially produced CFCs. During the past decade the anthropogenic greenhouse gases have received increasing scientific and public attention as it has become clear that the atmospheric concentrations of all of these gases is rising.

Carbon Dioxide In 1958 scientists measured the carbon dioxide concentration in the atmosphere near the top of Mauna Loa in Hawaii. This site was chosen because of its altitude and its location far from sources of atmospheric pollution. The measurements, which are still being made, show two remarkable things (Fig. 18.29). First, the amount of CO_2 fluctuates regularly throughout the year. In effect, the Earth—actually, the biosphere—is breathing. During the growing season, CO_2 is absorbed by vegetation, and the

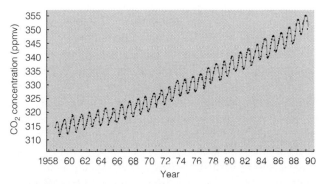

▲ F I G U R E 18.29

Concentration of carbon dioxide in dry air since 1958 as measured at the Mauna Loa Observatory in Hawaii, in ppmv (parts per million by volume). Annual fluctuations reflect seasonal changes in biologic uptake of CO_2, whereas the long-term trend shows a persistent increase in atmospheric concentrations of this greenhouse gas.

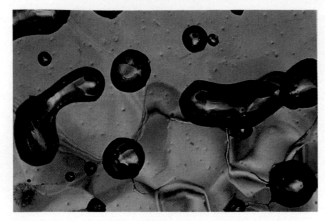

▲ F I G U R E 18.30
Air bubbles trapped in glacier ice. By melting the ice in a laboratory and collecting the gas, the content of CO_2 and other trace gases in these ancient samples of the atmosphere can be measured.

atmospheric concentration falls; during the winter, more CO_2 enters the atmosphere than is removed by vegetation and the concentration rises. Second, the long-term trend is unmistakable: Since 1958 the CO_2 concentration has risen from 315,000 to 355,000 ppbv at a rate that is not linear but exponential.

The increase in CO_2 in the atmosphere raises two questions: Is the observed rise in CO_2 unusual, and how can it be explained? To answer the first question we turn to the geologic record. Core samples obtained from the Antarctic and Greenland ice sheets contain tiny bubbles of air that were trapped when snow falling on the glaciers was slowly compressed and transformed into glacier ice (Fig. 18.30). If the ice is melted in a laboratory and the air is collected and analyzed, the amounts of CO_2 and other trace gases present in the ancient atmosphere can be measured. Ice from Vostok Station in Antarctica contains a record of atmospheric composition extending back 160,000 years (Fig. 18.10). The measurements show that during glacial ages the atmosphere typically contained about 200,000 ppbv of CO_2, whereas during interglacial times the concentration rose to about 280,000 ppbv. In addition, levels of CO_2 in preindustrial times were close to 280,000 ppbv, the typical value for an interglacial age. The subsequent rapid increase to 355,000 ppbv during the last 200 years is unprecedented in the ice-core record and implies that something very unusual is taking place.

No known natural mechanism can explain such a rapid increase in CO_2, but a possible explanation becomes apparent if we examine the rate at which CO_2 has been added to the atmosphere since the beginning of the Industrial Revolution (Fig. 18.13). The curve tracking the increase in CO_2 closely resembles the curve showing the increase in carbon dioxide emissions into the atmosphere by the burning of fossil fuels (Fig. 18.13A). A reasonable conclusion, there-

fore, is that the burning of fossil fuels must be a primary reason for the observed increase in atmospheric CO_2. Additional contributing factors include widespread deforestation, with its attendant burning and decay of cleared vegetation, and the use of wood as a fuel in many developing countries with rapidly growing populations. Human activities liberate CO_2 that was previously held in long-term geologic reservoirs, releasing it back into the carbon cycle (Fig. 18.31). (Refer to Chapter 1 for a more detailed discussion of the carbon cycle.)

Methane Methane (CH_4) absorbs infrared radiation 25 times more effectively than CO_2, making it an important greenhouse gas despite its relatively low concentration (1720 ppbv). Since the late 1960s, when scientists began measuring atmospheric methane, the concentration has increased at a rate of about 1 percent per year (although since 1984 the rate has decreased slightly, for unknown reasons). Methane levels for earlier times have been obtained from ice-core studies; these show an increase that essentially parallels the growth of the human population. This relationship is not surprising, for much of the methane now entering the atmosphere is generated by biologic activity related to rice cultivation, leaks in domestic and industrial gaslines, and the digestive processes of domestic livestock, especially cattle (Fig. 18.31). The global livestock population increased greatly in the past century, and the total acreage under rice cultivation has increased more than 40 percent since 1950. In prehistoric times, methane levels, like CO_2 levels, increased and decreased with the glacial–interglacial cycles (Fig. 18.10).

Other Trace Gases CFCs, which as noted earlier are implicated in the depletion of the stratospheric ozone layer, can also act as greenhouse gases. For example, CFC-12 has 20,000 times the capacity of carbon dioxide to trap infrared radiation. CFC-11, which is widely used in making plastic foams and as an aerosol propellant, has 17,500 times the greenhouse warming capacity of carbon dioxide. The concentrations of both compounds in the atmosphere are increasing at an annual rate of about 5 percent.

Ozone in the upper atmosphere is beneficial because it traps harmful ultraviolet solar radiation. However, when ozone builds up in the troposphere it contributes to the greenhouse effect as well as to the production of photochemical smog. Together, tropospheric ozone and nitrogen oxides account for about 13 percent of the anthropogenic greenhouse effect. Atmospheric concentrations of these substances are increasing at annual rates of 0.5 to 2 percent and 0.3 percent, respectively. Nitrogen oxides, which are released by microbial activity in soil, the burning of timber and fossil fuels, and the decay of agricultural residues, have a long residence time in the atmosphere; accordingly, atmospheric concentrations are likely to remain well above preindustrial levels even if emission rates are stabilized.

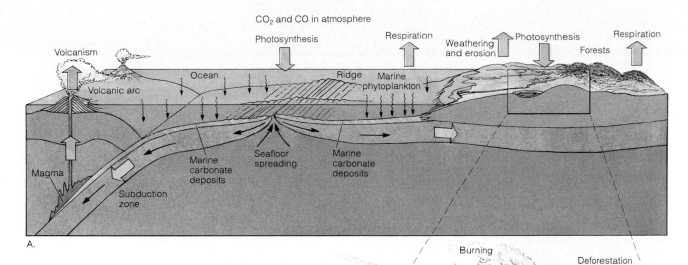

▲ F I G U R E 18.31
The carbon cycle. A. Natural fluxes of carbon through the atmosphere, hydrosphere, biosphere, and lithosphere. Carbon enters the atmosphere through volcanism, weathering, biologic respiration, and decay of organic matter in soils. Photosynthesis incorporates carbon in the biosphere, from which it can become part of the lithosphere if it is buried with accumulating sediment. B. Human activities release carbon to the atmosphere, primarily through the burning of forests and fossil fuels.

Modeling Global Warming

If atmospheric concentrations of greenhouse gases are rising, what does this portend for the Earth's future climate? Does it mean that surface temperatures will continue to increase, and, if so, by how much and at what rate? To answer these questions let's look briefly at how scientists model future climatic changes using the most powerful tool available: the supercomputer.

General Circulation Models

Global climate models are three-dimensional mathematical models of the Earth's climate system; they are an outgrowth of efforts to forecast the weather. The most sophisticated climate models are **general circulation models (GCMs),** which portray interconnected processes in the atmosphere, the hydrosphere, and the biosphere. The sheer complexity of these natural systems means that such models are, of necessity, greatly simplified representations of the real world. Moreover, many of the linkages and processes in the climate system are poorly understood and hence difficult to

model (see Box 18.1). Existing models do not adequately portray the dynamics of ocean circulation or cloud formation, two of the most important elements of the climate system. Also absent are many of the complex biogeochemical processes that link climate to the biosphere. Despite these limitations, GCMs have been very successful in simulating the general character of present climates and have greatly improved weather forecasting.

To run a modeling experiment that simulates the climate, a set of *boundary conditions* must be specified. Boundary conditions are mathematical expressions of the physical state of the Earth's climate system during the period under study. Thus, an experiment designed to simulate the present climate would prescribe as boundary conditions present values for the orbitally determined solar radiation reaching the Earth; the geographic distribution of land and ocean; the positions and heights of mountains and plateaus; the concentrations of atmospheric trace gases; sea-surface temperatures; the limits of sea ice; the amounts of snow and ice cover on land; the albedo (reflectivity) of land, ice, and water surfaces; and the effective soil moisture (the sum of water input and water loss).

BOX   18.1

•

FOCUS ON . . .

GLOBAL CLIMATE MODELS: HOW GOOD ARE THEY?

Global climate models, also known as *general circulation models (GCMs),* are computerized, mathematical simulations of complex, large-scale atmospheric dynamics. The purpose of a GCM is to predict future climate by calculating the behavior of different components of the atmosphere. A GCM is divided into submodels, each covering a specific atmospheric or oceanographic phenomenon, such as heat exchange between the atmosphere and the oceans. In a GCM the global atmosphere is divided into vertical columns and calculations are made at several altitudes within each column. To run the model, the calculations are repeated at specified intervals and predictions are made for global temperature and precipitation 10, 20, or 100 years in the future.

The development of GCMs began in 1975 with the publication of a research paper by Syukoro Manabe and Richard Wetherald. Using a computer model they developed at the Geophysical Fluid Dynamics Laboratory in Princeton, New Jersey, they projected global warming of 4°C resulting from a doubling of atmospheric carbon dioxide. Their paper was intended to focus a scientific discussion of the limitations of GCMs; instead, by the mid-1980s the spinoffs from their paper included at least four additional GCMs and many congressional hearings. Today GCMs are a widely accepted approach to the prediction of climatic change.

Like the first GCM, later models were developed at major universities and research laboratories in the United States and England. The models differ in the assumptions made in their submodel components; because of the complexity of the processes being simulated, some of the as-

sumptions are overly simplistic. For example, early GCMs were based on the assumption of shallow oceans without currents, an Earth with constant daylight, and thunderstorms that raged uniformly within huge cells as large as the diameter of the central United States.

Because a considerable amount of computer time is required to solve the mathematical equations of a GCM, the three-dimensional grid spacing (the distance between points on and above the globe for which the solutions are calculated) in the model experiments is large in order to keep costs at a manageable level. As a result, the resolution of such models is rather coarse: grid points commonly are separated by 4 or 5 degrees of latitude, or 450–550 km (Fig. B1.1). These models therefore can generate a reasonable picture of global and hemispheric climatic conditions, but they are poor at resolving conditions at smaller scales. Until more powerful computers are built and/or the cost of running a model experiment decreases, the spacial resolution of GCMs is likely to remain relatively coarse. In theory, a smaller GCM grid network is possible and would be more reliable in detecting regional variations in climate. Using current computer technology, however, reducing the grid spacings would mean increasing the required computer time from months to years.

Scientists have been revising and refining the assumptions on which their submodels are based to include such factors as interactions between the oceans and the atmosphere. The resulting temperature increases are several degrees lower than earlier predictions. For example, the United Kingdom Meteorological Office decreased its earlier estimate of the expected increase in average global tem-

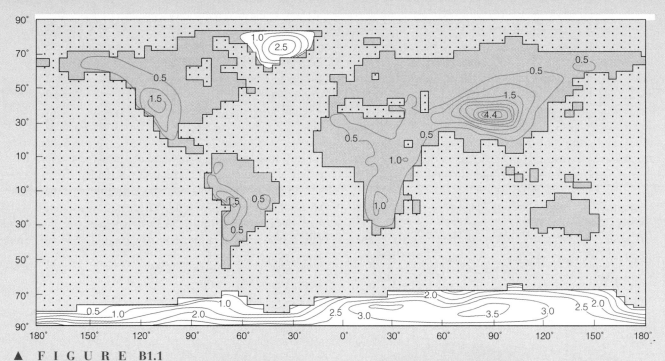

▲ FIGURE B1.1
View of the Earth as depicted by a global climate model with a grid spacing of 4° of latitude by 5° of longitude. Spacing of grid points is shown over the oceans but not over land. The altitudes of mountain ranges and plateaus are generalized and expressed in thousands of meters. Because of the map projection, Antarctica appears at the bottom as a long, narrow continent.

perature from 5.2°C to 1.9°C. The latest GCMs have more realistic submodels that reveal differential temperature changes within regions of the Southern and Northern hemispheres. International agencies are supporting the efforts of research programs to produce GCMs with fewer uncertainties. However, it is likely that another decade will pass before GCMs can provide truly reliable information about how climate will vary seasonally and regionally in response to increases in atmospheric concentrations of greenhouse gases.

Predictions of Climatic Change

Predictions of climatic change related to greenhouse warming are based mainly on the results of GCMs. The models differ in degree of detail as well as in the assumptions they employ. Nevertheless, they all predict that the anthropogenically generated greenhouse gases already in the atmosphere will lead to an average increase in global temperature of *at least* 0.5 to 1.5°C over the next century. This prediction is consistent with the 0.5°C rise in temperature inferred from historical records. The models further predict that if the greenhouse gases continue to build up until their combined effect is equivalent to a doubling of the preindustrial CO_2 concentration, then average global temperatures will rise between 1 and 5°C. This does not mean, however, that temperatures will increase uniformly all over the Earth. Instead, the projected temperature change varies geographically, with the greatest change occurring in the polar regions (Fig. 18.32).

Because of uncertainties about the functioning of the climate system, scientists tend to be cautious in their predictions. Nevertheless, they generally agree on the following points: (1) Human activities have led to increasing atmospheric concentrations of carbon dioxide and other trace gases, which have enhanced the greenhouse effect; (2) the global mean surface air temperature has increased by 0.3 to 0.6°C during the last 100 years, and this increase may be a direct result of the enhanced greenhouse effect; (3) during the next century the global average temperature will likely increase at about 0.3°C per decade, assuming that emission rates do not change. This projected increase will lead to a global average temperature about 1°C warmer than the present average by 2025 and as much as 3°C warmer by the end of the next century. If governmental controls lead to lower emission rates, the average temperature may rise by only 0.1 to 0.2°C per decade. Nevertheless, the temperature increase related to the continued release of greenhouse gases will be larger and more rapid than any experienced in human history. In effect, we may be about to experience a "super interglaciation" that will be warmer than any interglaciation of the past 2 million years.

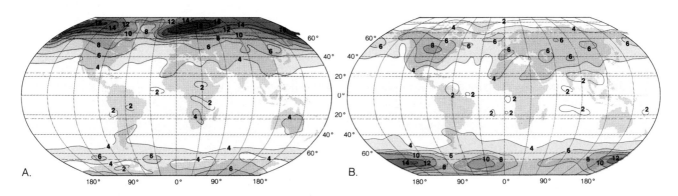

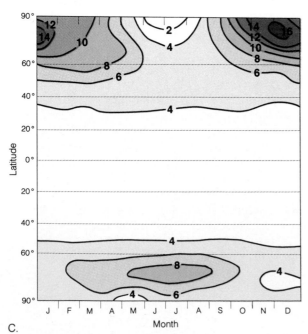

m **FIGURE 18.32**

A forecast of future changes in surface air temperature (in °C) that would result from an effective doubling of atmospheric CO_2 concentrations relative to present levels. A. Temperature increases for winter (December, January, February). For example, along the lines labeled 4, the projected temperature increase is 4°C. B. Temperature increases for summer (June, July, August). C. A latitudinal cross-section showing changes in zonal average air temperature through the year. This graph is a summary of the map patterns shown in A and B but includes the spring and autumn months as well. Greenhouse warming is greatest at high latitudes, where the model forecasts temperature increases as great as 16°C for winter months in the Northern Hemisphere.

Environmental Effects of Global Warming

An increase in average surface air temperature by a few degrees does not sound like much; surely people can put up with such an insignificant change. However, when one considers that the average global temperature is only 5°C higher today than it was during the coldest part of the last ice age, it is clear that even a temperature change of 1 or 2 degrees could have major consequences.

Global warming is just one of the results of human-induced climatic change. There are many other potential physical and biologic effects that are of considerable interest and concern, including changes in precipitation and vegetation patterns, increased storminess, changes in the areal extent of glaciers and sea ice, thawing of frozen ground, rise in sea level, changes in the hydrologic cycle, decomposition of organic matter in soil, and the breakdown of gas hydrates. Let's briefly look at each of these potential impacts.

Global Precipitation Changes A warmer atmosphere will lead to increased evaporation from oceans, lakes, and streams and to more precipitation. However, changes in precipitation will be unevenly distributed. Climate models suggest that the equatorial regions will receive more rainfall, partly because warmer temperatures will increase rates of evaporation over the tropical oceans and promote the formation of rainclouds. Some interior portions of large continents, which are distant from precipitation sources, will become both warmer and drier.

Changes in Vegetation Shifts in patterns of precipitation are likely to upset ecosystems, requiring both plants and animals to adjust to new conditions. Forest boundaries may shift in response to altered temperature and precipitation regimes. Some prime agricultural regions are likely to face increased droughts and substantially reduced soil moisture, which will have a negative impact on crops. Higher latitude regions with short, cool growing seasons may see increased agricultural production as summer temperatures rise.

Increased Storminess The shift to a warmer, wetter atmosphere could produce an increase in tropical storm activity. Regions that are prone to hurricanes and typhoons could experience these devastating storms with greater frequency, and the size of the storms could increase.

Melting (or Growing?) Glaciers Because warmer summers result in greater *ablation* (loss of mass from glaciers), low- and middle-latitude mountain glaciers are likely to retreat in a warmer world. On the other hand, warmer air in high latitudes can evaporate and transport more moisture from the oceans to adjacent ice sheets. This means that more precipitation will fall over high-latitude continental ice sheets and may cause them to grow larger.

Reduction of Sea Ice The enhanced heating projected for high northern latitudes in some models would cause shrinkage of sea ice. A reduction in polar sea ice, which has a high albedo, would enhance the greenhouse effect by reducing the amount of solar radiation reflected back into space. Models show less heating in high latitudes in the Southern Hemisphere, suggesting little change in sea ice there.

Thawing of Frozen Ground Rising summer air temperatures will begin to thaw vast regions of perennially frozen ground *(permafrost)* at high latitudes. The thawing is likely to affect natural ecosystems as well as cities and engineering works built on frozen ground.

Eustatic Rise in Sea Level As the temperature of ocean water rises, its volume will expand, causing sea level to rise. This **eustatic** (global) rise in sea level, supplemented by meltwater from shrinking mountain glaciers, is likely to increase calving along the margins of tidewater glaciers and ice sheets, thereby causing a further rise in sea level. The rising sea will inundate coastal regions and make tropical regions even more vulnerable to larger and more frequent cyclonic storms.

Changes in the Hydrologic Cycle Shifting patterns of precipitation and warmer temperatures will probably cause significant local and regional changes in stream runoff and groundwater levels.

Decomposition of Organic Matter in Soil As temperatures rise, the rate of decomposition of organic matter in soil will increase. Soil decomposition releases CO_2 to the atmosphere, thereby further enhancing the greenhouse effect. If average world temperatures rise by 0.3°C per decade, during the next 60 years, soils will release an amount of CO_2 equal to nearly 20 percent of the projected amount released by combustion of fossil fuels, assuming that the present rate of fuel consumption is maintained.

Breakdown of Gas Hydrates Gas hydrates are icelike solids in which molecules of gas, mainly methane, are locked in the structure of water. They are found in ocean sediments and beneath frozen ground. By one estimate, gas hydrates worldwide may hold a total of 10,000 billion metric tons of carbon, twice the amount contained in all the known coal, gas, and oil reserves on land. Gas hydrates accumulate in ocean sediments beneath water at least 500 m deep, where the temperature is low enough and the pressure high enough to permit their formation. They also accumulate beneath frozen ground, which acts as a seal to prevent the gas from migrating upward and escaping. When gas hydrates break down, they release methane. Global warming at high latitudes will result in the thawing of frozen ground and may destabilize the hydrates there, releasing large volumes of methane and thus amplifying the greenhouse effect.

Responding to Global Change

Why is it so difficult to predict the effects of global change and respond to them with policy initiatives? One reason is that the Earth's climate system is incredibly complex, and scientific uncertainties persist about the processes involved as well as the potential effects of global warming. In this final section we consider how scientists and policy makers deal with decision making in an atmosphere of uncertainty.

Dealing with Scientific Uncertainty

Present knowledge of how the Earth works, coupled with computer modeling, enables scientists to make educated projections of surface environmental changes that will result from global warming. However, most scientists are reluctant to make firm forecasts. Their caution emphasizes the gap between what we know about the Earth and what we would *like* to know, and points to the many challenges facing the study of global change.

The rate at which the projected warming will occur depends on a number of basic uncertainties: How rapidly will concentrations of the greenhouse gases increase? How rapidly will the oceans, a major reservoir of heat and a fundamental element in the climate system, respond to climatic changes? How will changing climates affect ice sheets and cloud cover? What is the range of natural variations in the climate system on a time scale of centuries? The potential complexity of these uncertainties is well illustrated by clouds. If the temperature of the lower atmosphere increases, more water will evaporate from the oceans. The increased atmospheric moisture will create more clouds, but clouds reflect solar energy back into space. This will have a cooling effect on the surface air, thereby having a result opposite to the greenhouse effect.

Added to the scientific uncertainties are various social and political uncertainties: Will nations reach a consensus on the appropriate response to the threat of global warming? Will it be possible to reduce the generation of greenhouse gases or even to freeze them at present levels? If so, can climatic changes be halted or reversed? And what would be the socioeconomic impacts of imposing limits on the production of greenhouse gases? Unfortunately—but understandably—policy makers generally don't like to deal with uncertainties; it is much easier to make decisions on the basis of facts and certainties. In deciding how to respond to the threat of global warming, political decision makers are faced with a conflict: potentially enormous social and economic costs in the present, versus future environmental changes that might be catastrophic—or might not materialize as predicted.

One approach to dealing with this conflict has been to use *risk assessment* as a planning tool. In risk assessment, the costs of taking action and the consequences of inaction are weighed against the possible outcomes of the projected changes. If the costs of taking action are very high or the

consequences of inaction are minimal, it may be desirable to postpone any action. Conversely, if the consequences of inaction are extremely high, it may be deemed best to act immediately even if the costs of taking action are also high. Using this approach, many policy makers have decided that the risks involved—even if there is only a small possibility that appreciable global warming will occur—are simply too great to justify inaction.

Toward an International Consensus

An international consensus seems to be emerging among both scientists and policy makers. It is generally agreed that (1) more scientific understanding is needed; (2) anthropogenic contributions to atmospheric change should be limited wherever possible; (3) any actions taken must be fair to both developing and industrialized nations, whose contributions to the problem of anthropogenic change differ vastly; and (4) any actions taken should be coordinated internationally.

Such actions are beginning to materialize. In 1979 the *World Climate Programme* was launched to promote international research on global climate processes and the impacts of climatic changes. The *International Geosphere–Biosphere Programme* and the *Human Response to Global Change* program are two other major international programs devoted to increasing scientists' understanding of the processes of global change. In 1989 the *World Meteorological Organization,* the *United Nations Environment Programme,* and the *International Council of Scientific Unions* cooperated in the establishment of an *Intergovernmental Panel on Climate Change (IPCC),* eventually involving policy advisers and scientists from 38 countries. Since then the IPCC has been the main coordinating body for research on climatic change, meeting periodically to report on the current status of scientific understanding of global warming and its implications.

In modeling global climatic change and predicting future climate scenarios, the scientific uncertainties are many and profound, yet the social, political, and economic risks are enormous. The global climate system depends on a complex, delicately balanced web of interconnected processes and subsystems involving the atmosphere, the hydrosphere, the solid Earth, and the biosphere; only with the advent of supercomputers have scientists begun to answer some of the basic questions about how this system works. It appears that the short-term prospect is for a warmer world. But if we stand back and look at these changes from a long-term geologic perspective, we will perceive that they represent only a brief, very rapid, nonrepeatable perturbation in the Earth's climatic history. The perturbation is nonrepeatable because once the Earth's store of easily extractable fossil fuels has been used up, as is likely within the next several hundred years, human impacts on the atmosphere will decline and the climate system should revert to a more natural state.

SUMMARY

1. The atmosphere performs many functions that are crucial for the maintenance of life on Earth. These include shielding us from harmful ultraviolet radiation with a layer of protective ozone; keeping the Earth's surface warm by trapping the Sun's heat; creating weather and precipitation patterns through interactions among atmosphere, oceans, and land masses; and facilitating the movement of elements in biogeochemical cycles.

2. The atmosphere consists of a mix of gases dominated by nitrogen and oxygen and including a number of less abundant gases such as argon, carbon dioxide, methane, and water vapor. It is divided into vertical layers, each with its own chemical, physical, and thermal characteristics. The lowest layer, the troposphere, is where most of the mass of the atmosphere resides and weather patterns originate. The next layer, the stratosphere, is where most of the ozone in the atmosphere resides and global circulation patterns occur.

3. Like other parts of the Earth system, the atmosphere has evolved chemically. The main source of new gases into the atmosphere is volcanic degassing. The evolution of photosynthetic life forms was crucial for the buildup of free oxygen in the atmosphere. Living organisms have also influenced the chemical evolution of the atmosphere by sequestering carbon dioxide in the form of limestones created from the remnants of calcium carbonate shells.

4. Geology plays an important role in understanding atmospheric change, for geologists have access to a record of the Earth's changing climate that extends into the remote past. Paleontologists infer past climates from the assemblages of fossil plants and animals they find. Sediments and sedimentary rocks can reveal the past distribution of sediments; the mechanisms of weathering, transport, and deposition; and changes in the chemical composition of water and air. Isotope geologists can determine past temperatures and atmospheric compositions from studies of terrestrial and marine sediments and sample cores of polar ice, and from samples of ancient air trapped in bubbles deep within ice sheets.

5. On long time scales, the Earth's atmospheric and climatic processes are controlled by changes within the planet and by interactions among the atmosphere, hydrosphere, biosphere, and lithosphere. Global tectonism influences climate by changing rates of sedimentation (and, thus, the rates of biogeochemical cycles) and by moving continents from one climatic zone to another. The distribution of land and water masses and the heights of land masses help determine atmospheric circulation patterns.

6. Periodic cyclicity from glacial to interglacial periods, on time scales from tens of thousands to 100,000 years, is controlled by a combination of global tectonism, astronomic influences, and long-term fluctuations in atmospheric composition. The last ice age started about 30,000 years ago and ended about 10,000 years ago; the Earth is now in a period of interglacial warming and is poised—in principle—to begin the descent into another ice age. Shorter fluctuations, such as the Little Ice Age and the Holocene optimum, are also caused by the complex interplay of a variety of factors, notably El Niño and volcanic aerosol climate forcing.

7. Recent trends show increases in the atmospheric concentrations of many anthropogenic gases, including carbon dioxide, methane, CFCs, and oxides of nitrogen and sulfur. Studies suggest that some of the increases began around the time of the Industrial Revolution, when industrial activity and, in particular, the burning of fossil fuels first began in earnest.

8. People affect the atmosphere on a local scale, not only by releasing contaminants but also by changing the landscape. Urbanization can affect local air circulation patterns, the distribution of heat, and the concentration of pollutants in a given area. Urban activity combined with a basinlike topography can lead to serious air pollution episodes associated with thermal inversions. Photochemical smog, of which tropospheric ozone is a major component, is formed when airborne contaminants interact with solar radiation at low levels in the atmosphere.

9. Another form of air pollution is caused by the buildup of contaminants inside buildings. Some of the most common indoor air pollution problems are associated with the presence of asbestos fibers or radioactive radon gas and its daughter products.

10. Long-range transport of airborne pollutants occurs only when contaminants reach the stratosphere. Perhaps the most serious regional effect of air pollution is acid precipitation, which is caused when acid precursors (mainly oxides of sulfur and nitrogen derived from the burning of fossil fuels) interact with water vapor in the atmosphere to form acids. The acids are transported by atmospheric circulation, eventually reaching the surface of the Earth through wet or dry deposition. Acid precipitation can cause acidification of aquatic ecosystems and soils and corrosion of materials.

11. Some anthropogenic gases, notably CFCs, facilitate the chemical breakdown of ozone in the stratosphere. Measurements have revealed a steady depletion in levels of stratospheric ozone, especially during the sum-

mer months over the South Pole. Ozone depletion allows more ultraviolet radiation to reach the surface of the Earth's surface, where it can have harmful effects on humans and other organisms. The *Montreal Protocol* is an international agreement that calls for limits on the production of ozone-depleting substances and for scientific monitoring and research on the causes and mechanisms of ozone depletion.

12. The greenhouse effect is a natural process that keeps the surface of the Earth warm enough for life to exist. The most important natural greenhouse gas is water vapor. Concerns have recently arisen because of sharp increases in some highly effective anthropogenic greenhouse gases, notably carbon dioxide, methane, nitrogen oxides, and CFCs.

13. Predictions of global climatic change are based on sophisticated computer models called general circulation models (GCMs), which attempt to quantify the inter-connections among the oceans and atmosphere on the basis of certain assumptions and boundary conditions. The amount of warming predicted by such models varies but is generally in the range of 0.5 to 1.5°C over the next century. The potential environmental effects of global warming include changes in patterns of precipitation and vegetation, increased storminess, changes in the areal extent of glaciers and sea ice, thawing of frozen ground, a global rise in sea level, changes in the hydrologic cycle, decomposition of organic matter in soil, and breakdown of gas hydrates.

14. It is difficult for policy makers to respond to global climatic change because of the fundamental scientific uncertainties and the high socioeconomic stakes that are involved. Such programs as the International Geosphere–Biosphere Programme and the establishment of the Intergovernmental Panel on Climate Change (IPCC) represent steps toward a global consensus.

IMPORTANT TERMS TO REMEMBER

acid precipitation (p. 482)
aerosols (p. 474)
asbestos (p. 479)
dry deposition (p. 480)
eustatic (sea level) (p. 495)
general circulation models (GCMs) (p. 491)
glaciation (p. 471)

greenhouse effect (p. 488)
ground-level ozone (p. 479)
ozone hole (p. 485)
ozone layer (p. 484)
particulates (p. 478)
pH (p. 480)
photosynthesis (p. 468)

radiatively active gases (p. 488)
radon (p. 479)
smog (p. 478)
stratosphere (p. 466)
thermal inversion (p. 478)
troposphere (p. 466)
wet deposition (p. 480)

QUESTIONS AND ACTIVITIES

1. Investigate the status of the agreements made in the *Montreal Protocol on Substances That Deplete the Ozone Layer*. When was the protocol last revised? Which chemicals are scheduled to be phased out or frozen at current levels, and by what year? How close are the signatory countries to meeting their agreements?

2. Have you ever thought about having your home (or school) tested for indoor air pollutants? Testing for some pollutants, such as asbestos fibers, can be very costly. For others, such as radon, home testing kits are available that may only cost $30 or $40. In most cases, you follow the instructions and then send the kit back to the laboratory to be analyzed, although in some cases you are given instructions for analyzing the results yourself. Find out the average indoor radon level for your region. If it is high, consider having your home tested. Is there an obvious reason for the high readings, such as proximity to uranium-bearing rock? Remember that it is a good idea to repeat the testing several times because of the potential impacts of weather and other factors on radon concentrations.

3. Policy makers in most developing countries differ fundamentally from those in industrialized countries in their attitudes toward global atmospheric change (i.e., ozone depletion and global warming). This difference has been one of the most difficult political stumbling blocks in the path of an international consensus on global change. Find out more about this difference of opinion. Why are developing countries reluctant to participate wholeheartedly in solving the problems of global warming and ozone depletion? Do you agree with their opinion (or can you understand their point of view)? How did the difference of opinion surface at the 1992 Earth Summit in Rio de Janeiro? What can be done to foster international cooperation on these issues?

4. What are the major sources of gaseous emissions in your area? To what extent do industrial activities such as manufacturing or smelting contribute to air pollution? Is transportation a major contributor? What are the main local impacts of air pollution? Are thermal inversions common? If so, are they caused by topography? Do airborne contaminants in your area have any regional effects, such as lake acidification?

5. Investigate the likely effects of global warming in your region. For example, if you live in a coastal zone, what effects would a 50-m rise in sea level have? How would coastlines change? How would a change in precipitation (wetter or drier) or a change in the length of the growing season (longer or shorter) affect your area?

UNITS AND THEIR CONVERSIONS

ABOUT SI UNITS

Regardless of the field of specialization, all scientists use the same units and scales of measurement. They do so to avoid confusion and the possibility that mistakes can creep in when data are converted from one system of units, or one scale, to another. By international agreement the SI units are used by all, and they are the units used in this text. SI is the abbreviation of Système International d'Unités (in English, the International System of Units).

Some of the units are likely to be familiar, some unfamiliar. The SI unit of length is the meter (m), of area the square meter (m^2), and of volume the cubic meter (m^3). The SI unit of mass is the kilogram (kg), and of time the second (s). The other SI units used in this book can be defined in terms of these basic units. Three important ones are:

- The newton (N), a unit of force defined as that force needed to accelerate a mass of 1 kg by 1 m/s^2; hence 1 N = 1 kg·m/s^2. (The period between kg and m indicates multiplication.)

- The joule (J), a unit of energy or work, defined as the work done when a force of 1 newton is displaced a distance of 1 meter; hence 1 J = 1 N·m. One important form of energy so far as the Earth is concerned is heat. The outward flow of the Earth's internal heat is measured in terms of the number of joules flowing outward from each square centimeter each second; thus, the unit of heat flow is J/cm^2/s.

- The pascal (Pa), a unit of pressure defined as a force of 1 newton applied across an area of 1 square meter; hence 1 Pa = 1 N/m^2. The pascal is a numerically small unit. Atmospheric pressure, for example (15 lb/in^2), is 101,300 Pa. Pressure within the Earth reaches millions or billions of pascals. For convenience, earth scientists sometimes use 1 million pascals (megapascal, or MPa) as a unit.

Temperature is a measure of the internal kinetic energy (expressed as movement) of the atoms and molecules in a body. In the SI system, temperature is measured on the Kelvin scale (K). The temperature intervals on the Kelvin scale are arbitrary, and they are the same as the intervals on the more familiar Celsius scale (°C). The difference between the two scales is that the Celsius scale selects 100°C as the temperature at which water boils at sea level, and 0°C as the freezing temperature of water at sea level. Zero degrees Kelvin, on the other hand, is absolute zero, the temperature at which all atomic and molecular motions cease. Thus, 0°C is equal to 273.15 K, and 100°C is 373.15 K. The temperatures of processes on and within the Earth tend to be at or above 273.15 K. Despite the inconsistency, earth scientists still use the Celsius scale when geological processes are discussed.

Appendix A provides a table of conversion from older units to Standard International (SI) units.

PREFIXES FOR MULTIPLES AND SUBMULTIPLES

When very large or very small numbers have to be expressed, a standard set of prefixes is used in conjunction with the SI units. Some prefixes are probably already familiar; an example is the centimeter (which is one hundredth of a meter, or 10^{-2} m). The standard prefixes are

giga	1,000,000,000	=	10^9
mega	1,000,000	=	10^6
kilo	1,000	=	10^3
hecto	100	=	10^2
deka	10	=	10
deci	0.1	=	10^{-1}
centi	0.01	=	10^{-2}
milli	0.001	=	10^{-3}
micro	0.000001	=	10^{-6}
nano	0.000000001	=	10^{-9}
pico	0.000000000001	=	10^{-12}

One measure used commonly in geology is the nanometer (nm), a unit by which the sizes of atoms are measured; 1 nanometer is equal to 10^{-9} meter.

COMMONLY USED UNITS OF MEASURE

Length

Metric Measure

1 kilometer (km)	= 1000 meters (m)
1 meter (m)	= 100 centimeters (cm)
1 centimeter (cm)	= 10 millimeters (mm)
1 millimeter (mm)	= 1000 micrometers (μm) (formerly called microns)
1 micrometer (μm)	= 0.001 millimeter (mm)
1 angstrom (Å)	= 10^{-8} centimeters (cm)

Nonmetric Measure

1 mile (mi)	= 5280 feet (ft) = 1760 yards (yd)
1 yard (yd)	= 3 feet (ft)
1 fathom (fath)	= 6 feet (ft)

Conversions

1 kilometer (km)	= 0.6214 mile (mi)
1 meter (m)	= 1.094 yards (yd) = 3.281 feet (ft)
1 centimeter (cm)	= 0.3937 inch (in)
1 millimeter (mm)	= 0.0394 inch (in)
1 mile (mi)	= 1.609 kilometers (km)
1 yard (yd)	= 0.9144 meter (m)
1 foot (ft)	= 0.3048 meter (m)
1 inch (in)	= 2.54 centimeters (cm)
1 inch (in)	= 25.4 millimeters (mm)
1 fathom (fath)	= 1.8288 meters (m)

Area

Metric Measure

1 square kilometers (km²)	= 1,000,000 square meters (m²)
	= 100 hectares (ha)
1 square meter (m²)	= 10,000 square centimeters (cm²)
1 hectare (ha)	= 10,000 square meters (m²)

Nonmetric Measure

1 square mile (mi²)	= 640 acres (ac)
1 acre (ac)	= 4840 square yards (yd²)
1 square foot (ft²)	= 144 square inches (in²)

Conversions

1 square kilometer (km²)	= 0.386 square mile (mi²)
1 hectare (ha)	= 2.471 acres (ac)
1 square meter (m²)	= 1.196 square yards (yd²)
	= 10.764 square feet (ft²)
1 square centimeter (cm²)	= 0.155 square inch (in²)
1 square mile (mi²)	= 2.59 square kilometers (km²)
1 acre (ac)	= 0.4047 hectare (ha)
1 square yard (yd²)	= 0.836 square meter (m²)
1 square foot (ft²)	= 0.0929 square meter (m²)
1 square inch (in²)	= 6.4516 square centimeter (cm²)

Volume

Metric Measure

1 cubic meter (m³)	= 1,000,000 cubic centimeters (cm³)
1 liter (l)	= 1000 milliliters (ml)
	= 0.001 cubic meter (m³)
1 centiliter (cl)	= 10 milliliters (ml)
1 milliliter (ml)	= 1 cubic centimeter (cm³)

Nonmetric Measure

1 cubic yard (yd³)	= 27 cubic feet (ft³)
1 cubic foot (ft³)	= 1728 cubic inches (in³)
1 barrel (oil) (bbl)	= 42 gallons (U.S.) (gal)

Conversions

1 cubic kilometer (km³)	= 0.24 cubic miles (mi³)
1 cubic meter (m³)	= 264.2 gallons (U.S.) (gal)
	= 35.314 cubic feet (ft³)
1 liter (l)	= 1.057 quarts (U.S.) (qt)
	= 33.815 ounces (U.S. fluid) (fl. oz.)
1 cubic centimeter (cm³)	= 0.0610 cubic inch (in³)
1 cubic mile (mi³)	= 4.168 cubic kilometers (km³)

1 acre-foot (ac-ft)	=	1233.46 cubic meters (m³)
1 cubic yard (yd³)	=	0.7646 cubic meter (m³)
1 cubic foot (ft³)	=	0.0283 cubic meter (m³)
1 cubic inch (in³)	=	16.39 cubic centimeters (cm³)
1 gallon (gal)	=	3.784 liters (l)

Mass

Metric Measure

1000 kilograms (kg)	=	1 metric ton (also called a tonne) (m.t)
1 kilogram (kg)	=	1000 grams (g)

Nonmetric Measure

1 short ton (sh.t)	=	2000 pounds (lb)
1 long ton (l.t)	=	2240 pounds (lb)
1 pound (avoirdupois) (lb)	=	16 ounces (avoirdupois) (oz) = 7000 grains (gr)
1 ounce (avoirdupois) (oz)	=	437.5 grains (gr)
1 pound (Troy) (Tr. lb)	=	12 ounces (Troy) (Tr. oz)
1 ounce (Troy) (Tr. oz)	=	20 pennyweight (dwt)

Conversions

1 metric ton (m.t)	=	2205 pounds (avoirdupois) (lb)
1 kilogram (kg)	=	2.205 pounds (avoirdupois) (lb)
1 gram (g)	=	0.03527 ounce (avoirdupois) (oz) = 0.03215 ounce (Troy) (Tr. oz) = 15,432 grains (gr)
1 pound (lb)	=	0.4536 kilogram (kg)
1 ounce (avoirdupois) (oz)	=	28.35 grams (g)
1 ounce (avoirdupois) (oz)	=	1.097 ounces (Troy) (Tr. oz)

Pressure

1 pascal (Pa)	=	1 newton/square meter (N/m²)

1 kilogram/square centimeter (kg/cm²)	=	0.96784 atmosphere (atm) = 14.2233 pounds/square inch (lb/in²) = 0.98067 bar
1 bar	=	0.98692 atmosphere (atm) = 10^5 pascals (Pa) = 1.02 kilograms/square centimeter (kg/cm²)

Energy and Power

Energy

1 joule (J)	=	1 newton meter (N.m)
	=	2.390×10^{-1} calorie (cal)
	=	9.47×10^{-4} British thermal unit (Btu)
	=	2.78×10^{-7} kilowatt-hour (kWh)
1 calorie (cal)	=	4.184 joule (J)
	=	3.968×10^{-3} British thermal unit (Btu)
	=	1.16×10^{-6} kilowatt-hour (kWh)
1 British thermal unit (Btu)	=	1055.87 joules (J)
	=	252.19 calories (cal)
	=	2.928×10^{-4} kilowatt-hour (kWh)
1 kilowatt hour	=	3.6×10^6 joules (J)
	=	8.60×10^5 calories (cal)
	=	3.41×10^3 British thermal units (Btu)

Power (energy per unit time)

1 watt (W)	=	1 joule per second (J/s)
	=	3.4129 Btu/h
	=	1.341×10^{-3} horsepower (hp)
	=	14.34 calories per minute (cal/min)
1 horsepower (hp)	=	7.46×10^2 watts (W)

Temperature

To change from Fahrenheit (F) to Celsius (C)

$$°C = \frac{(°F - 32°)}{1.8}$$

To change from Celsius (C) to Fahrenheit (F)

$$°F = (°C \times 1.8) + 32°$$

To change from Celsius (C) to Kelvin (K)

$$K = °C + 273.15$$

To change from Fahrenheit (F) to Kelvin (K)

$$K = \frac{(°F - 32°)}{1.8} + 273.15$$

Radioactivity

1 rad = the unit of absorbed radiation, which corresponds to 0.01 J of energy per kg of material

1 rem = the dose of ionizing radiation that gives the same biologic effect as 1 rad of 250 kvp x-rays

A rem is the product of the dose in rads and a quality factor that depends on the nature of the radiation; alpha particles and neutrons have factors of 10, whereas beta particles and gamma rays are weighted as 1; hence, 1 rad of gamma radiation = 1 rem, but 1 rad of alpha particles = 10 rem.

1 curie (Ci) = a measure of the number of radioactive disintegrations per second; specifically, 37 billion disintegrations per second; (more commonly used in measuring radiation is the picocurie (pCi) = 10^{-12} Ci or .037 disintegrations per second).

1 sievert (Sv) = 1 Sv = 1J/kg = 100 rem

1 gray (Gy) = 1 J/kg = 100 rads (the SI unit of dose equivalent is the sievert; the gray is the SI unit of absorbed dose)

1 working level (WL) = the concentration of radioactive gas with a potential alpha energy release of 1.3×10^5 MeV per liter of air (approximately equivalent to 200 pCi per liter of air).

TABLE OF THE CHEMICAL ELEMENTS

Element	Symbol	Atomic Number	Crustal Abundance, Weight Percent	Element	Symbol	Atomic Number	Crustal Abundance, Weight Percent
Actinium	Ac	89	Human-made	Iron	Fe	26	5.80
Aluminum	Al	13	8.00	Krypton	Kr	36	Not known
Americium	Am	95	Human-made	Lanthanum	La	57	0.0050
Antimony	Sb	51	0.00002	Lawrencium	Lw	103	Human-made
Argon	Ar	18	Not known	Lead	Pb	82	0.0010
Arsenic	As	33	0.00020	Lithium	Li	3	0.0020
Astatine	At	85	Human-made	Lutetium	Lu	71	0.000080
Barium	Ba	56	0.0380	Magnesium	Mg	12	2.77
Berkelium	Bk	97	Human-made	Manganese	Mn	25	0.100
Beryllium	Be	4	0.00020	Mendelevium	Md	101	Human-made
Bismuth	Bi	83	0.0000004	Mercury	Hg	80	0.000002
Boron	B	5	0.0007	Molybdenum	Mo	42	0.00012
Bromine	Br	35	0.00040	Neodymium	Nd	60	0.0044
Cadmium	Cd	48	0.000018	Neon	Ne	10	Not known
Calcium	Ca	20	5.06	Neptunium	Np	93	Human-made
Californium	Cf	98	Human-made	Nickel	Ni	28	0.0072
Carbon[a]	C	6	0.02	Niobium	Nb	41	0.0020
Cerium	Ce	58	0.0083	Nitrogen	N	7	0.0020
Cesium	Cs	55	0.00016	Nobelium	No	102	Human-made
Chlorine	Cl	17	0.0190	Osmium	Os	76	0.00000002
Chromium	Cr	24	0.0096	Oxygen[b]	O	8	45.2
Cobalt	Co	27	0.0028	Palladium	Pd	46	0.0000003
Copper	Cu	29	0.0058	Phosphorus	P	15	0.1010
Curium	Cm	96	Human-made	Platinum	Pt	78	0.0000005
Dysprosium	Dy	66	0.00085	Plutonium	Pu	94	Human-made
Einsteinium	Es	99	Human-made	Polonium	Po	84	Footnote[d]
Erbium	Er	68	0.00036	Potassium	K	19	1.68
Europium	Eu	63	0.00022	Praseodymium	Pr	59	0.0013
Fermium	Fm	100	Human-made	Promethium	Pm	61	Human-made
Fluorine	F	9	0.0460	Protactinium	Pa	91	Footnote[d]
Francium	Fr	87	Human-made	Radium	Ra	88	Footnote[d]
Gadolinium	Gd	64	0.00063	Radon	Rn	86	Footnote[d]
Gallium	Ga	31	0.0017	Rhenium	Re	75	0.00000004
Germanium	Ge	32	0.00013	Rhodium[c]	Rh	45	0.00000001
Gold	Au	79	0.0000002	Rubidium	Rb	37	0.0070
Hafnium	Hf	72	0.0004	Ruthenium[c]	Ru	44	0.00000001
Helium	He	2	Not known	Samarium	Sm	62	0.00077
Holmium	Ho	67	0.00016	Scandium	Sc	21	0.0022
Hydrogen[b]	H	1	0.14	Selenium	Se	34	0.000005
Indium	In	49	0.00002	Silicon	Si	14	27.20
Iodine	I	53	0.00005	Silver	Ag	47	0.000008
Iridium	Ir	77	0.00000002	Sodium	Na	11	2.32

Element	Symbol	Atomic Number	Crustal Abundance, Weight Percent	Element	Symbol	Atomic Number	Crustal Abundance, Weight Percent
Srontium	Sr	38	0.0450	Unnilhexium	Unh	106	Human-made
Sulfur	S	16	0.030	Unniloctium	Uno	108	Human-made
Tantalum	Ta	73	0.00024	Unnilpentium	Unp	105	Human-made
Technetium	Tc	43	Human-made	Unnilquadium	Unq	104	Human-made
Tellurium[c]	Te	52	0.000001	Unnilseptium	Uns	107	Human-made
Terbium	Tb	65	0.00010	Uranium	U	92	0.00016
Thallium	Tl	81	0.000047	Vanadium	V	23	0.0170
Thorium	Th	90	0.00058	Xenon	Xe	54	Not known
Thulium	Tm	69	0.000052	Ytterbium	Yb	70	0.00034
Tin	Sn	50	0.00015	Yttrium	Y	39	0.0035
Titanium	Ti	22	0.86	Zinc	Zn	30	0.0082
Tungsten	W	74	0.00010	Zirconium	Zr	40	0.0140
Unnilennium	Une	109	Human-made				

Source: After K. K. Turekian, 1969.

[a]Estimate from S. R. Taylor (1964).

[b]Analyses of crustal rocks do not usually include separate determinations for hydrogen and oxygen. Both combine in essentially constant proportions with other elements, so abundances can be calculated.

[c]Estimates are uncertain and have a very low reliability.

[d]Elements formed by decay of uranium and thorium. The daughter products are radioactive with such short half-lives that crustal accumulations are too low to be measured accurately.

THE GEOLOGIC TIME SCALE

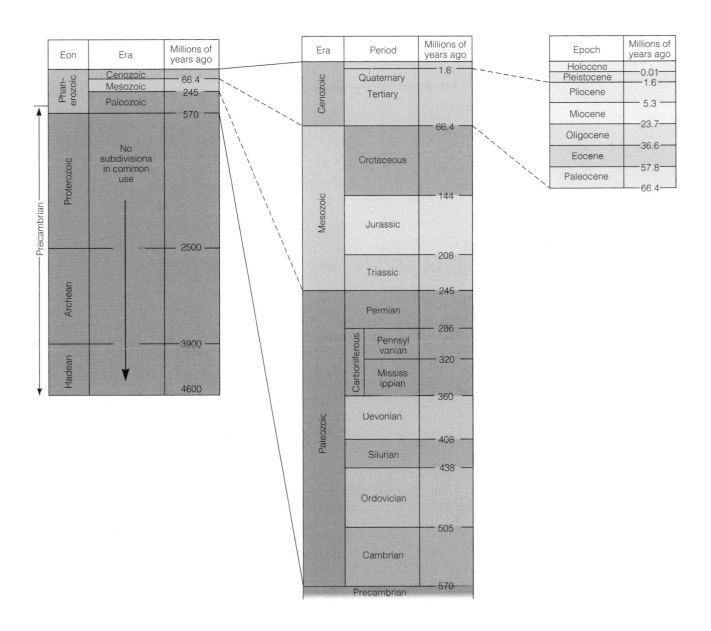

Eon	Era	Millions of years ago
Phanerozoic	Cenozoic	66.4
	Mesozoic	245
	Paleozoic	570
Proterozoic	No subdivisions in common use	2500
Archean		3900
Hadean		4600

Era	Period	Millions of years ago
Cenozoic	Quaternary	1.6
	Tertiary	
Mesozoic	Cretaceous	66.4
	Jurassic	144
	Triassic	208
		245
Paleozoic	Permian	286
	Carboniferous — Pennsylvanian	
	Carboniferous — Mississippian	320
	Devonian	360
	Silurian	408
	Ordovician	438
	Cambrian	505
	Precambrian	570

Epoch	Millions of years ago
Holocene	0.01
Pleistocene	1.6
Pliocene	5.3
Miocene	23.7
Oligocene	36.6
Eocene	57.8
Paleocene	66.4

The geologic time scale. Absolute ages obtained from radiometric dates. Note that the Pennsylvanian and Mississippian Periods are equivalent to the Carboniferous Period of Europe. The time boundary between the Archean and Hadean is uncertain as no rocks of the Hadean Eon are known on the Earth. Hadean rocks are known to exist on other planets in the solar system.

PHOTO CREDITS

Introduction
Opener: ©Earth Satellite Corp./SPL/Photo Researchers. Figure In-1: Robert Caputo/Stock, Boston. Figure In-2: Simon Jauncey/Tony Stone Images/New York, Inc. Figure In-3: Nani Gois/Abril Imagens. Figure In-4: ©Tom Bean/DRK Photo. Figure In-6: ©Steve Kaufman/Peter Arnold, Inc. Figure In-7: ©Blair Seitz/Photo Researchers. Figure In-8: ©Miles Ertman/Masterfile. Figure In-9: Dallas & John Heaton/Stock, Boston.

Part 1
Opener: ©Tsado/Tom Stack & Associates. Figure I-1: NRSC Ltd./Science Photo Library/Photo Researchers. Figure I-3: COMSTOCK, Inc. Figure I-5: E. R. Degginger/Bruce Coleman, Inc. Figure I-6: ©John S. Shelton. Figure I-7: ©Tom Van Sant/The Geosphere Project. Figure I-8: Earth Satellite Corporation. Figure I-9: Spencer Swanger/Tom Stack & Associates.

Chapter 1
Opener: ©David Robert Austen. Figure 1-9: Claus Meyer/Black Star. Figure 1-15: National Geographic Society. Figure 1-16: Courtesy NASA/Lyndon B. Johnson Space Center. Figure 1-20a: ©John S. Shelton. Figure 1-20b: Francois Gohier/Ardea London.

Chapter 2
Opener: ©Robert Chasin/Black Star. Figure 2-3: William E. Ferguson. Figure 2-7: Michael Hochella. Figure 2-10a: Brian J. Skinner. Figure 2-10b: Brian Parker/Tom Stack & Associates. Figure 2-11: Brian J. Skinner. Box 2-2-1a: Barry L. Runk/Grant Heilman Photography. Box 2-2-1b: Runk/Schoenberger/Grant Heilman Photography. Box 2-2-1c: Brian J. Skinner. Figure 2-12: Brian J. Skinner.

Part 2
Opener: Les Stone/Sygma.

Chapter 3
Opener: Iwasa/Sipa Press. Figure 3-1: ©John S. Shelton. Figure 3-4a: William C. Gaustein. Figure 3-4b: Brian J. Skinner. Figure 3-14: ©Steve Lehman/SABA. Figure 3-15: Paul X. Scott/Sygma Photo News. Figure 3-16: A. Nogues/Sygma Photo News. Figure 3-21: ©Mark Downey/Gamma Liaison. Figure 3-22: Kevin Schafer/Tom Stack & Associates. Figure 3-23: Lysaght/Gamma Liaison. Figure 3-24: Steve McCutcheon. Figure 3-26: Courtesy USGS Photo Library, Denver, Colorado. Figure 3-33: Centre National d'Etudes Spatiales.

Chapter 4
Opener: ©Alberto Garcia/SABA. Figure 4-1 & 4-2: ©Krafft Explorer/Photo Researchers. Figure 4-3 & 4-4: J.D. Griggs/USGS. Figure 4-5: Brian J. Skinner. Figure 4-6a: William E. Ferguson. Figure 4-6b: J. D. Griggs/USGS. Figure 4-6c: Schofield/Gamma Liaison. Figure 4-7: G. J. Orme, Dept. of Army, Fort Bragg, N.C.. Figure 4-9: Roger Werth/Woodfin Camp & Associates. Figure 4-10: S.C. Porter. Figure 4-11: Steve Vidler/Leo de Wys, Inc. Figure 4-12a: ©Krafft Explorer/Photo Researchers. Figure 4-12b: Tom Bean/DRK Photo. Figure 4-13: Brian J. Skinner. Figure 4-14: S.C. Porter. Figure 4-15: Greg Vaughn/Tom Stack & Associates. Figure 4-16: Brian J. Skinner. Figure 4-18: Lyn Topinka/USGS. Figure 4-19: J. D. Griggs/USGS. Figure 4-23: G. Brad Lewis/Gamma Liaison. Figure 4-24: David Hiser/The Image Bank. Box 4-1-3a: ©T. Orban/Sygma. Box 4-1-3b: ©Peter Turnley/Black Star. Figure 4-27: David Robert Austen. Figure 4-29: NRSC Ltd./Science Photo Library/Photo Researchers.

Chapter 5
Opener: ©Sankei Shimbun/Symga Photo News. Figure 5-2: Bishop Museum. Figure 5-6: ©Steve McCutcheon.

Chapter 6
Opener: El Tiempo/©Sipa Press. Figure 6-2: Jacques Jangoux/Tony Stone Images/New York, Inc. Figure 6-5: S.C.

Porter. Figure 6-6: Alex Soto/ Health Sciences Center for Educational Resources, University of Washington. Figure 6-7: USGS/ Vancouver. Figure 6-10: A. Lincoln Washburn. Figure 6-12: © George Herben Photography. Figure 6-19: Steve McCutcheon/Tony Stone Images/ Seattle. Figure 6-20: George Plafker, U.S Geological Survey. Figure 6-23: F. O. Jones/U.S. Geological Survey. Figure 6-24: Hulton Deutsch. Figure 6-25: S.C. Porter. Box 6-2-1a: Los Angeles County Department of Public Works. Box 6-2-1b: Perry Ehlig. Figure 6-26: ©John Elk III. Figure 6-28: ©Bill Brooks/Masterfile. Figure 6-29: S.C. Porter.

Chapter 7
Opener: ©Alex S. MacLean/Landslides. Figure 7-1: Michael Nichols/Magnum Photos, Inc. Figure 7-2: Courtesy NASA. Figure 7-3: Hiroji Kubota/Magnum Photos, Inc. Figure 7-4: Sam C. Pierson, Jr./Photo Researchers. Figure 7-6: ©Robert Baumgardner. Figure 7-10: C.R. Dunrud/Courtesy USGS. Figure 7-14: Courtesy City of Long Beach. Figure 7-15: ©Jim Pickerell/Gamma Liaison. Figure 7-16: ©Stevan Stefanovic/OKAPIA/Photo Researchers. Figure 7-17: Mirco Toniolo/Gamma Liaison. Figure 7-18: ©James Fraser/Impact Photos.

Chapter 8
Opener: ©Frank Oberle/Tony Stone Images/New York, Inc. Figure 8-1: G. Hading/Photo Researchers. Figure 8-2: Chip Hires/Gamma Liaison. Figure 8-3: ©Wim Ruigrok/Sygma Photo News. Figure 8-4 & 8-7: S.C. Porter. Figure 8-8: Courtesy NASA. Figure 8-17: Hiroji Kubota/Magnum Photos, Inc. Figure 8-18: A. Holbrooke/Gamma Liaison. Figure 8-21: Earth Satellite Corporation.

Chapter 9
Opener: ©David R. Frazier Photolibrary/Photo Researchers. Figure 9-6: Greg Scott/Masterfile. Figure 9-10: Nicholas DeVore III/ Photographers Aspen. Figure 9-12: Vince Streano/Tony Stone Images/New York, Inc. Figure 9-13: G. R. Roberts. Figure 9-19: ©Mark Wexler/Woodfin Camp & Associates. Figure 9-20: /Spaceshots, Inc. Figure 9-21: Valerie Taylor/Ardea London. Figure 9-22b: James Stanfield/National Geographic Society. Figure 9-23b: Dick Davis/Photo Researchers. Figure 9-24: ©Tom Sobolink/Black Star. Figure 9-25: National Oceanic and Atmospheric Administration/ DLR—FRG/Roger Ressmeyer/©CORBIS. Figure 9-26: Otto/ Visions/Impact Photos. Figure 9-27: Courtesy NASA Goddard Space Flight Center/ A. F. Hasler, H. Pierce, K. Palaniappan, and M. Manyin. Figure 9-28: Stephen Krasemann/DRK Photo. Figure 9-30: Victor Englebert/Photo Researchers. Box 9-2-1: Dr. Richard Iegeckis/ Science Photo Library/Photo Researchers.

Chapter 10
Opener: Courtesy Canada Centre for Remote Sensing, Department of Natural Resources Canada. Figure 10-1: Pekka Parviainen/ Science Photo Library/Photo Researchers. Figure 10-2: Dr. Ursula Marvin, with permission of the Keeper of

Manuscripts at the Universitatsbibliothek/Smithsonian Astro Observatory. Figure 10-3: Nabeel Turner/Tony Stone Images/New York, Inc. Figure 10-4: Dr. Ursula Marvin/Smithsonian Astro Observatory. Figure 10-5: Wards Science/ Science Source/Photo Researchers. Figure 10-7: ©World Perspectives/ Explorer/Photo Researchers. Box 10-1-1: Dr. Ursula Marvin/Smithsonian Astro Observatory. Figure 10-8: ©William K. Hartmann. Figure 10-11: ©Ronald E. Royer/JPL/Photo Researchers. Figure 10-13: ©1986 Max-Planck-Institut fur Aeronomie, Courtesy Harold Reitsema & Alan Delamere, Ball Aerospace/Roger Ressmeyer ©CORBIS. Figure 10-14: Novosti Press Agency/Science PhotoLibrary /Photo Researchers. Box 10-2-1: AP/Wide World Photos. Figure 10-18a: John Sanford/Science Photo Library/Photo Researchers. Figure 10-18b: ©Tom Till. Figure 10-21: Jonathan Blair/National Geographic Image Collection. Figure 10-25: J. Baker/ D. Milon/Photo Researchers.

Part 3
Opener: ©Stephanie Maze/Woodfin Camp & Associates. Figure III-1: ©Jeremy Hartley/ Panos Pictures. Figure III-2: ©Steve McCutcheon. Figure III-3: Stephen Krasemann/DRK Photo. Figure III-8: Panos Pictures.

Chapter 11
Opener: Aramco World. Figure 11-1: Runk/Schoenberger/Grant Heilman Photography. Figure 11-6: Bettmann Archive. Box 11-1-2: Robert Azzi/Woodfin Camp & Associates. Figure 11-8: G. Whitely/Photo Researchers. Figure 11-10: ©Charlie Ott/Photo Researchers. Figure 11-11: Kristin Finnegan/Tony Stone Images/Seattle. Figure 11-12: Grant Heilman Photography.

Chapter 12
Opener: John Cancalosi/Tom Stack & Associates. Figure 12-1: ©Marty Cordano/DRK Photo. Figure 12-3: Grant Heilman Photography. Figure 12-5b: Peter Christopher/Masterfile. Figure 12-6 b&c: Brian J. Skinner. Figure 12-7: Tommaso Guicciardini/Science Photo Library /Photo Researchers. Figure 12-8: ©Bob Clarke/Gamma Liaison. Figure 12-9: Roger Ressmeyer@ CORBIS. Figure 12-11: Igor Kostin/ Imago/Sygma. Figure 12-12: Simon Fraser/Science Photo Library /Photo Researchers. Figure 12-14: Charles Krebs/Tony Stone Images/Seattle. Figure 12-15: Ron Sanford/Black Star. Box 12-1-1: ©1995 CNES, Provided by Spot Image Corporation. Box 12-1-2: Barbara Murck. Figure 12-17: Ronald Toms/ Oxford Scientific Films, Ltd.

Chapter 13
Opener: Dudley Foster/Woods Hole Oceanographic Institution. Figure 13-7, 13-8 & 13-10: Brian J. Skinner. Figure 13-11: William E. Ferguson. Figure 13-12: ©Tom McHugh/Photo Researchers. Figure 13-13 & 13-14: Brian J. Skinner. Figure 13-17: Dane Penland/ Smithsonian Institution. Figure 13-18: Martin Miller. Figure 13-20 & 13-21: Brian J. Skinner. Figure 13-22: William Sacco. Figure 13-24, 13-25, 13-26 & 13-27: Brian J. Skinner. Figure 13-28: Dr. Ursula Marvin/Smithsonian

Astro Observatory. Figure 13-30: ©David L. Brown/Tom Stack & Associates. Box 13-2-1: Richard K. Brummer P. Eng. Figure 13-31: ©Gary Brettnacher/Tony Stone Images/New York, Inc.

Chapter 14
Opener: ©Jeremy Hartley/Panos Pictures. Figure 14-1: S.C. Porter. Figure 14-2: Ole P. Rorvik Aune Forlag. Figure 14-3: S.C. Porter. Figure 14-4: Stan Osolinski/ Oxford Scientific Films Ltd. Figure 14-5: S.C. Porter. Figure 14-6: Alan H. Strahler. Figure 14-9: S.C. Porter. Figure 14-10: Henry D. Foth. Figure 14-12: Official U.S. Navy Photographs. Figure 14-13: Mark A. Melton. Figure 14-14: ©Georg Gerster/COMSTOCK, Inc. Figure 14-15: S.C. Porter. Figure 14-16: ©Jim Richardson/Woodfin Camp & Associates. Figure 14-17: ©Georg Gerster/COMSTOCK, Inc. Figure 14-19: Nani Gois/Abril Imagens. Figure 14-20: Grant Heilman Photography. Figure 14-22: S.C. Porter.

Chapter 15
Opener: Grant Heilman Photography. Figure 15-16: COMSTOCK, Inc. Figure 15-17: ©Peter Essick/Aurora. Figure 15-18: David Turnley/Black Star. Figure 15-19: ©John Edwards/Tony Stone Images/New York, Inc. Box 15-2-2: Ed Kashi. Box 15-3-1: Randy Wells/Tony Stone Images/New York, Inc.

Part 4
Opener: ©Jose Azel/Aurora. Figure IV-1: David Prichard/First Light, Toronto. Figure IV-3: ©Alex McLean/Landslides. Figure IV-4: ©Thomas Hartwell/Sygma. Figure IV-5: Martin Bond/Science Photo Library/Photo Researchers.

Chapter 16
Opener: ©Louie Psihoyos/Matrix International. Figure 16-1: ©Alex McLean/Landslides. Figure 16-2: ©David Hiser/Photographers Aspen. Figure 16-4 & 16-8: Alan Pitcairn/Grant Heilman Photography. Figure 16-11: ©Andy Levin/Photo Researchers. Figure 16-13: Heidi Bradner/Panos Pictures.

Chapter 17
Opener: ©Fred Ward/Black Star. Figure 17-3: David Woodfall/Tony Stone Images/New York, Inc. Figure 17-8: Scott Cazamine/Photo Researchers. Figure 17-11: U. S. Environmental Protection Agency. Figure 17-18: ©Jim Olive/Peter Arnold, Inc. Figure 17-19: Stuart Westmorland/Tony Stone Images/New York, Inc. Figure 17-20: RSPB/Bruce Coleman, Ltd. Figure 17-22: ©Vanessa Vick/Photo Researchers.

Chapter 18
Opener: World Perspectives/Tony Stone Images/New York, Inc. Figure 18-3: Noel Quidu/Gamma Liaison. Figure 18-4: Bruno Barbey/Magnum Photos, Inc. Figure 18-11: S.C. Porter. Figure 18-16: Courtesy NASA. Figure 18-23: William E. Ferguson. Figure 18-26: Courtesy of IBM. Figure 18-30: S.C. Porter.

GLOSSARY

A

Abrasion Mechanical erosion that results from the impact of wind-driven grains of sediment. (Ch. 14)

Acid mine drainage The acidified runoff from a mine, waste rock pile, or tailings pile. (Ch. 17)

Acid precipitation Any precipitation with a pH lower than 5.0. (Ch. 18)

Active remediation An approach to the cleanup of contaminated groundwater that involves active intervention. (Ch. 17)

Active volcano A volcano that has erupted within recorded history. (Ch. 4)

Aerosols Extremely fine particles or droplets that are carried in suspension; volcanic aerosols result from the reaction of volcanic gases with water vapor in the atmosphere. (Ch. 18)

Aftershock An earthquake that occurs shortly after a major quake. (Ch. 3)

Amplitude Half the height from the peak to the trough of a wave; or, the height of a water wave from the still-water line. (Chs. 3, 5)

Andesite A fine-grained igneous rock with intermediate silica content. (Ch. 4)

Anthropogenic hazards Human-generated hazards that arise from pollution and degradation of the natural environment (e.g., contamination of surface and underground water bodies, depletion of the ozone layer, and global climatic warming). (Essay II)

Apollo objects Extraterrestrial bodies with earth-approaching orbits. (Ch. 10)

Aquiclude A body of impermeable or distinctly less permeable rock adjacent to a permeable one. (Ch. 15)

Aquifer A water-bearing rock or sediment unit. (Chs. 7, 15)

Artesian aquifer An aquifer in which water is under hydraulic pressure; a confined aquifer. (Ch. 15)

Asbestos An industrial term that refers to a group of minerals that grow with a fibrous crystal habit. (Ch. 18)

Asteroid A small, irregularly shaped rocky body that orbits the sun. (Ch. 10)

Asteroid belt A swarm of at least 100,000 asteroids nestled between the orbits of Mars and Jupiter, at the boundary between the inner (terrestrial) and outer (Jovian) planets. (Ch. 10)

Atmosphere The mixture of gases, predominantly nitrogen, oxygen, argon, carbon dioxide, and water vapor, that surrounds the Earth. (Essay I)

Atom The smallest individual particle that retains all the properties of a given chemical element. (Ch. 2)

Attenuation layer An impermeable layer that underlies a waste-disposal site and slows the migration of leachate. (Ch. 16)

Aven An open, chimneylike passageway that connects a cavern to near-surface rocks and unconsolidated overburden. (Ch. 7)

B

Bankfull discharge The discharge measured when a river is at bankfull stage. (Ch. 8)

Bankfull stage A condition in which a river's channel fills completely, so that any further increase in discharge results in water overflowing the banks. (Ch. 8)

Barrier island A long, narrow, sandy island close to and parallel to the coast. (Ch. 9)

Basalt A fine-grained igneous rock with low silica content. (Ch. 4)

Base metals Metallic resources that include any of the common metals, such as copper or lead. (Ch. 13)

Beach Wave-washed sediment along a coast. (Ch. 9)

Beach drift The irregular movement of particles along a beach as they travel obliquely up the slope of a beach with the swash and directly down this slope with the backwash. (Ch. 9)

Biochemical oxygen demand (BOD) The depletion of dissolved oxygen in a body of water that results from an excess of aerobic microorganisms. (Ch. 17)

Bioconcentration Biologic processes leading to the buildup of contaminants in organisms. (Ch. 17)

Biodegradation The decomposition of a substance through the action of microorganisms. (Ch. 17)

Biogas Gas derived from decaying organic wastes, which is collected and used as a fuel. (Ch. 12)

Biogeochemical cycle A natural cycle that describes the movements and interactions through the Earth's spheres of the chemicals essential to life. (Ch. 1)

Biomass energy Any form of energy that is derived more or less directly from the Earth's plant life. (Ch. 12)

Biosphere The totality of the Earth's living matter, including organic matter that has not yet completely decomposed. (Essay I)

Body waves Seismic waves that travel outward from an earthquake's point of origin and pass through the Earth. (Ch. 3)

Bore A wall of water that results when an onrushing wave becomes concentrated in a long, narrow bay or river mouth. (Ch. 5)

C

Caldera A roughly circular, steep-walled volcanic collapse basin several kilometers or more in diameter. (Ch. 4)

Cap rock A rock with very low porosity and permeability that impedes migrating oil and gas and prevents them from going any farther. (Ch. 11)

Carrying capacity The maximum number of people an ecosystem can support, imposed by the limited resources of that ecosystem. (Essay III)

Catastrophism The concept that all of the Earth's major features, such as mountains, valleys, or oceans, have been produced by a few great catastrophic events. (Ch. 1)

Cave A natural underground open space, generally connected to the surface and large enough for a person to enter. (Ch. 7)

Cavern A large cave or system of interconnected cave chambers. (Ch. 7)

Channel The passageway in which a river flows. (Ch. 8)

Channelization Modifications to river channels, consisting of some combination of straightening, deepening, widening, clearing, or lining of the natural channel. (Ch. 8)

Chemical weathering The decomposition of rocks and minerals as chemical reactions transform them into new chemical combinations that are stable at or near the Earth's surface. (Ch. 14)

Chondrites Very old, stony meteorites that contain small, round, glassy-looking spheres called chondrules. (Ch. 10)

Climate The average weather conditions of a place or an area over a period of years. (Ch. 9)

Closed system A system whose boundaries allow the exchange of energy, but not matter, with the surrounding environment; a system in which the amount of matter is fixed and the contents cycle within the boundaries of the system. (Ch. 1)

Coal A black, combustible, sedimentary or metamorphic rock that consists chiefly of decomposed plant matter and contains more than 50 percent organic matter. (Ch. 11)

Coalification The process of coal formation, which involves the loss of volatile materials, including water, carbon dioxide, and methane, from peat. (Ch. 11)

Comet An icy extraterrestrial object that glows when it approaches the sun, producing a long, wispy tail that points away from the sun. (Ch. 10)

Compaction A decrease in the thickness and porosity of a sediment layer, due to the weight of overlying material and, sometimes, to withdrawal of fluids. (Ch. 7)

Composting Facilitating the breakdown of organic refuse such as food waste, leaves, or yard clippings by the action of bacteria and other naturally occurring microorganisms. (Ch. 16)

Compressional waves (*P* waves) Seismic body waves that consist of alternating pulses of compression and expansion acting in the direction in which the wave is traveling. (Ch. 3)

Cone of depression A conical depression in the water table immediately surrounding a pumped well. (Chs. 7, 15)

Confined aquifer An aquifer that is bounded by aquicludes. (Ch. 15)

Contaminants Materials that have harmful impacts and degrade the environment; see also *Pollutants*. (Essay IV)

Convergent margin A type of plate boundary where plates meet as they move toward each other. (Ch. 1)

Core The spherical mass, largely metallic iron, at the center of the Earth. (Ch. 2)

Coriolis effect An effect which causes any body that moves freely with respect to the rotating Earth to veer to the right in the Northern Hemisphere and to the left in the Southern Hemisphere. (Ch. 9)

Crater (volcanic) A funnel-shaped depression opening upward near the summit of a volcano, from which gases, tephra, and lava are ejected. (Ch. 4)

Crust The outermost and thinnest of the Earth's compositional layers, which consists of rocky matter that is less dense than the rocks of the mantle below. (Ch. 2)

Crystalline Any solid with a crystal structure; that is, an orderly, repeated geometric pattern of atoms. (Ch. 2)

Current A broad, slow drift of water in a particular direction. (Ch. 9)

Cyclone An atmospheric low-pressure system that gives rise to roughly circular, inward-spiraling wind motion, called vorticity. (Ch. 9)

D

Decommissioning The cleaning up and rehabilitation of a contaminated site after industrial use. (Ch. 13)

Deep-well disposal A waste-disposal procedure that involves injecting liquid wastes deep into the ground in a specially designed well. (Ch. 16)

Deflation Erosion that occurs when the wind picks up and removes loose rock fragments, sand, and dust. (Ch. 14)

Desert Land where annual rainfall is less than 250 mm or in which the rate of evaporation exceeds the rate of precipitation. (Ch. 9)

Desertification The spread of desert into nondesert areas. (Ch. 9)

Discharge The quantity of water that passes a given point on the bank of a river within a given interval of time; or, in the context of groundwater, the process whereby water leaves the subsurface to join a surface water body or to become surface runoff. (Chs. 8, 15)

Dissolution The chemical weathering process whereby minerals and rock materials pass directly into solution. (Ch. 7)

Divergent margin A type of plate boundary along which plates are moving apart from one another. (Ch. 1)

Divide The topographic high point that separates adjacent drainage basins. (Ch. 8)

Dormant volcano A volcano that has not erupted in recent memory and shows no signs of current activity, but is not deeply eroded. (Ch. 4)

Downstream flooding Flooding that continues for a long time and extends over an entire region. (Ch. 8)

Drainage basin The total area that contributes water to a river. (Ch. 8)

Drought An extended period of exceptionally low precipitation. (Ch. 9)

Dry deposition Materials transported in suspension in the air that eventually settle out as dry matter. (Ch. 18)

E

Early warning A public declaration that the normal routines of life should be altered for a period of time to deal with the danger posed by an imminent hazardous event. (Essay II)

Earth-crossing asteroids Asteroids whose orbits pass within the orbit of the Earth. (Ch. 10)

Earth resources Materials of value to humans that are extracted (or extractable) from the solid Earth. (Essay III)

Economic geology The field of study devoted to how and where the different kinds of valuable mineral deposits form. (Ch. 13)

Ejecta A jet of rock and dust that results as a high-speed meteorite strikes and penetrates the surface of a planet. (Ch. 9)

Elastic rebound theory The theory that earthquakes result from the release of stored elastic strain energy by slippage on faults. (Ch. 3)

Element The most fundamental substance into which matter can be separated by chemical means. (Ch. 2)

El Niño An anomalous warming of surface waters in the eastern equatorial Pacific Ocean. (Ch. 9)

Energy cycle A natural cycle describing the movement of energy (solar, geothermal, and tidal) through the Earth system and its interactions with materials and processes in the biosphere, hydrosphere, atmosphere, and solid Earth. (Ch. 1)

Energy intensity Energy consumption per unit of economic activity. (Ch. 11)

Enhanced recovery The introduction of fluids into an oil well to facilitate the flow of oil. (Ch. 11)

Enrichment factor The extent to which a material must be concentrated over and above its average abundance in the Earth's crust, in order to be economically recoverable. (Ch. 13)

Environment The outer biophysical system, in which people and other organisms exist; anything—living or nonliving—that surrounds and influences living organisms. (Intro)

Environmental geology The study of the functioning of Earth systems and how they affect and are affected by human activities. (Introduction)

Environmental radiation The dose of radiation each of us receives in our normal daily lives. (Ch. 16)

Eolian erosion Wind erosion. (Ch. 14)

Epicenter The point on the Earth's surface directly above the focus of an earthquake. (Ch. 3)

Eruption column A hot, turbulent mixture of gas and tephra that rises rapidly in the cooler air above the vent of an erupting volcano. (Ch. 4)

Eustatic Global (refers to sea-level change). (Ch. 18)

Eutrophication A process whereby the oxygen in water becomes depleted as a result of the uncontrolled growth of plankton and algae. (Ch. 17)

Evaporite deposits Layers of salts that precipitate as a consequence of the evaporation of seawater. (Ch. 13)

Expansive soils Soils that expand greatly when saturated with water. (Ch. 6)

Exploration geology The branch of geology concerned with discovering new supplies of usable minerals. (Ch. 13)

Exponential growth The increase in a population by a constant percentage per unit of time. (Essay III)

Extinct volcano A volcano that has not erupted within recorded history, is deeply eroded, and shows no signs of future activity. (Ch. 4)

F

Fall A sudden, vertical movement of Earth material—for example, from an overhanging cliff. (Ch. 6)

Fault A fracture in a rock along which movement occurs. (Ch. 3)

Fission The splitting of an atom into two smaller atoms, with an attendant release of heat energy. (Ch. 12)

Fissure A long, linear or arc-shaped fracture in the Earth's crust. (Ch. 4)

Fissure eruption A volcanic eruption that occurs when lava reaches the surface via elongated fissures. (Ch. 4)

Flash flood A flood in which the lag time is exceptionally short—hours or minutes. (Ch. 8)

Flood A discharge great enough to cause a body of water to overflow its natural or artificial confines and submerge surrounding land. (Ch. 8)

Flood-frequency curve Flood magnitudes that are plotted with respect to the recurrence interval calculated for a flood of that magnitude at a given location. (Ch. 8)

Floodplain The part of any stream valley that is inundated during floods. (Ch. 8)

Flood stage A condition in which a river's channel fills completely, so that any further increase in discharge results in water overflowing the banks. (Ch. 8)

Flow A mass-wasting process that involves the chaotic movement of mixtures of sediment and water (or sediment, water, and air). (Ch. 6)

Focus The place where energy is first released to cause an earthquake. (Ch. 3)

Forecast The short- or long-term prediction of the magnitude and time of occurrence of a hazardous event. (Essay II)

Foreshock A swarm of small earthquakes that precedes a big quake. (Ch. 3)

Fossil fuels Hydrocarbons, formed from the remains of plants or animals trapped in sediment, that may be used as fuel; primarily oil, natural gas, and coal. (Ch. 11)

Frequency The time interval between crests of adjacent waves. (Ch. 5)

Frost heaving The lifting of regolith by the freezing of water contained within it. (Ch. 6)

Fumarole A vent or fracture from which volcanic gases are emitted. (Ch. 4)

Fusion The joining together, or fusing, of two small atoms to create a single, larger atom, with an attendant release of heat energy. (Ch. 12)

G

Gangue The nonvaluable minerals of an ore. (Ch. 13)

General circulation models (GCMs) Climate models that portray interconnected processes in the atmosphere, hydrosphere, and biosphere. (Ch. 18)

Geologic hazards The wide range of geologic circumstances, materials, processes, and occurrences that are hazardous to humans, such as earthquakes, volcanic eruptions, floods, or landslides. (Essay II)

Geologic isolation The underground disposal of hazardous wastes in a geologic repository. (Ch. 16)

Geologists Scientists who study the Earth. (Introduction)

Geology The scientific study of the Earth. (Introduction)

Geopressurized zone A high-pressure, high-temperature environment that holds large quantities of natural gas at great depths in the pore spaces of basinlike sedimentary rock sequences thousands of meters thick. (Ch. 12)

Geothermal energy Energy derived from the Earth's internal energy. (Ch. 1, 12)

Geyser A thermal spring equipped with a natural system of plumbing and heating that causes intermittent eruptions of water and steam. (Ch. 4)

Glaciations Periods during which the average surface temperature dropped by several degrees and stayed cool long enough for the polar ice sheets to grow larger; also referred to as *glacial periods* or *ice ages*. (Ch. 18)

Glowing avalanche A pyroclastic flow that is dense, gas-charged, and so hot that it incandesces. (Ch. 4)

Grade A term for the level of concentration of a metal or a mineral in an ore. (Ch. 13)

Gradient A measure of the vertical drop over a given horizontal distance; in the cases of streams, the vertical distance that a channel falls between two points along its course. (Chs. 6, 8)

Greenhouse effect The property of the Earth's atmosphere by which long-wavelength heat rays from the Earth's surface are trapped or reflected back by the atmosphere. (Ch. 18)

Ground-level ozone Ozone that occurs in the lower part of the atmosphere; tropospheric ozone. (Ch. 18)

Groundwater The water contained in spaces within bedrock and regolith. (Ch. 15)

Gullies Distinct, narrow stream channels that are larger and deeper than rills and that result as erosion from running water progresses. (Ch. 14)

H

Half-life The time it takes for half of the atoms in a given sample to decay. (Ch. 16)

Hazard assessment The process of determining when and where hazardous events have occurred in the past, the severity of the physical effects of past events of a given magnitude, the frequency of events that are strong enough to generate physical effects, and what a particular event would be like if it were to occur now; and portraying all this information in a form that can be used by planners and decision makers. (Essay II)

Hazardous wastes Wastes that pose a present or potential threat to humans or wildlife. (Ch. 16)

Historical geology The study of the chronology of events, both physical and biologic, that have occurred in the past. (Introduction)

Humus Partially decomposed organic matter derived from the decay of dead plants and animals in soils. (Ch. 14)

Hurricane A tropical cyclonic storm having winds that exceed 120km/h. See *Typhoon*. (Ch. 9)

Hydraulic gradient The slope of the water table. (Ch. 15)

Hydrocarbon An organic compound (gaseous, liquid, or solid) that consists primarily of carbon and hydrogen. (Ch. 11)

Hydrocarbon trap The combination of a source rock, reservoir rock, and cap rock that provides a barrier to the upward migration of liquid or gaseous hydrocarbons. (Ch. 12)

Hydroelectric power Power that is generated from the potential energy of stream water as it flows to the sea. (Ch. 12)

Hydrograph A graph in which river discharge is plotted against time. (Ch. 8)

Hydrologic cycle A natural cycle that describes the movement of water through the Earth system. (Ch. 1)

Hydrosphere The totality of the Earth's water, comprising oceans, lakes, streams, underground water, and all snow and ice, including glaciers. (Essay I)

Hydrothermal solutions Hot, water-rich solutions that can concentrate metallic ore minerals. (Ch. 13)

I

Igneous rock Rock formed by the cooling and crystallization of magma. (Ch. 1)

Impact crater A bowl-shaped depression that occurs as the result of the transfer of kinetic energy from an object to the surface of the Earth upon impact. (Ch. 10)

Incineration The burning of refuse, often in a specially designed facility. (Ch. 16)

Inexhaustible resource A resource that can never be depleted or used up. (Essay III)

Infiltration The process by which water from precipitation soaks into the soil. (Ch. 15)

Ion An atom that has excess positive or negative charges caused by electron transfer. (Ch. 2)

Irons Meteorites composed mainly of alloys of metallic nickel and iron. (Ch. 10)

Isohyetal map A map that depicts the distribution of rainfall over the area affected by a storm. (Ch. 8)

J

Jovian planets The giant planets (Jupiter, Saturn, Uranus, and Neptune) in the outer regions of the solar system, which are characterized by great masses, low densities, and thick atmospheres consisting primarily of hydrogen and helium. (Essay I)

K

Karst topography Topography characterized by many small, closed basins and a disrupted drainage pattern whereby streams disappear into the ground and eventually appear elsewhere in large springs. (Ch. 7)

L

Lag time The time that elapses between the onset of precipitation and the peak flood stage. (Ch. 8)

Lahar A volcanic mudflow. (Ch. 4)

Landslide Any perceptible downslope movement of bedrock, regolith, or a mixture of the two. (Ch. 6)

Lateral blast A sideways volcanic eruption of pulverized rock and hot gases. (Ch. 4)

Lava Magma that reaches the Earth's surface through a volcanic vent and pours out over the landscape. (Ch. 4)

Lava dome A formation that results from the squeezing out of sticky and viscous lava after a major volcanic eruption. (Ch. 4)

Leachate Contaminated water that results when water percolates through a pile of solid waste, picking up soluble materials. (Ch. 16)

Liquefaction The rapid fluidization of sediment as the result of a disturbance or an abrupt shock, such as an earthquake. (Chs. 3, 6)

Lithosphere The outermost rocky layer of the Earth; more precisely, the crust and the outermost portion of the mantle. (Essay I)

Load   The particles of sediment and dissolved matter that are carried along by a river. (Ch. 8)

Longshore current   A current within the surf zone that flows parallel to the coast. (Ch. 9)

M

Magma   Molten rock, sometimes containing suspended mineral grains and dissolved gases, that forms when temperatures rise sufficiently for melting to occur in the Earth's crust or mantle. (Ch. 4)

Mantle   The thick shell of dense, rocky matter that surrounds the Earth's core. (Ch. 2)

Mass-wasting   The movement of Earth materials downslope as a result of the pull of gravity. (Ch. 6)

Maturation   A series of complex physical and chemical changes that break down organic material into simpler liquid and gaseous hydrocarbons. (Ch. 11)

Mechanical weathering   The disintegation or physical breakup of rocks with no change in their chemistry or mineralogy. (Ch. 14)

Medical geology   The subfield of geology that studies the effects of chemical elements in the environment on human and animal health. (Ch. 17)

Metallic mineral resources   Minerals mined specifically for the metals that can be extracted from them. (Ch. 13)

Metamorphic rock   Rock whose original compounds or textures, or both, have been transformed into new compounds and new textures by reactions in the solid state as a result of high temperature, high pressure, or both. (Ch. 1)

Meteor   A glowing fragment of extraterrestrial material passing through the atmosphere. (Ch. 10)

Meteorite   A fragment of extraterrestrial material that strikes the surface of the Earth. (Ch. 10)

Meteorology   The study of the Earth's atmosphere and weather processes. (Ch. 9)

Mineral   Any naturally formed, crystalline solid with a definite chemical composition and a characteristic crystal structure. (Ch. 2)

Mineral deposits   Volumes of rock that contain enrichments of one or more minerals. (Ch. 13)

Mineral resource   Any element, compound, mineral, or rock that can be extracted from the ground and is of potential value as a commodity. (Ch. 13)

Modified Mercalli Intensity Scale   A scale used to compare earthquakes based on the amount of vibration people feel during low-magnitude quakes and the extent of damage to buildings during high-magnitude quakes. (Ch. 3)

Molecule   The smallest unit that retains all the properties of a compound. (Ch. 2)

N

Natural gas   Naturally occurring hydrocarbon compounds that are gaseous at ordinary temperatures and pressures. (Ch. 11)

Natural hazards   The wide range of natural circumstances, materials, processes, and events that are hazardous to humans, such as locust infestations, wildfires, or tornadoes, in addition to strictly geologic hazards. (Essay II)

NIMBY   "Not in my backyard"; a social attitude toward waste disposal that reflects the fact that no one wants to live next door to a waste-disposal facility. (Essay IV)

Nonmetallic mineral resource   A mineral resource, such as salt, gypsum, or clay, that is not used for the metals it contains but for its other physical or chemical properties. (Ch. 13)

Nonpoint sources   Broad areas where pollutants originate and enter the natural environment. (Ch. 17)

Nonrenewable resource   A resource that cannot be replenished or regenerated on a human time scale. (Essay III)

Nor'easter   A cyclonic storm that originates in regions of atmospheric instability in middle latitudes over land or coastal regions; named for its wind, which blows from the northeast. (Ch. 9)

Normal fault   A fault that is caused by tension—movement that tends to pull crustal blocks apart. (Ch. 3)

Nuclear energy   The heat produced during the controlled transformation of suitable radioactive isotopes. (Ch. 12)

O

Oceanography   The study of the physical, chemical, biologic, and geologic aspects of the Earth's oceans. (Ch. 9)

Ocean thermal energy conversion (OTEC)   The use of heat pumps to withdraw energy from ocean water. (Ch. 12)

Oil   The liquid form of petroleum. (Ch. 11)

Oil shale   A shale containing waxlike substances that break down to liquid and gaseous hydrocarbons when heated. (Ch. 12)

Ore   An aggregate from which one or more minerals can be extracted profitably. (Ch. 13)

Ozone hole   The phenomenon of stratospheric ozone depletion centered over the south polar region. (Ch. 18)

Ozone layer   A zone in the stratosphere where ozone occurs in unusually high concentrations. (Ch. 18)

P

Paleoflood hydrology   The study of ancient floods. (Ch. 8)

Paleoseismology   The study of prehistoric earthquakes. (Ch. 3)

Parent material   The rock and mineral regolith from which soil develops. (Ch. 14)

Particulates   Pollutants that are carried in suspension in the air as extremely fine, solid particles. (Ch. 18)

Passive remediation   An approach to the cleanup of contaminated groundwater which says, "Let nature take its course." (Ch. 17)

Peak discharge   The point at which the maximum discharge for a particular flood is reached. (Ch. 8)

Peat   An unconsolidated deposit of plant remains with a carbon content of about 60 percent. (Ch. 11)

Percolation   The movement of groundwater in a saturated zone. (Ch. 15)

Permeability   A measure of how easily a solid allows fluids to pass through it. (Ch. 15)

Persistant contaminants   Contaminants that take a very long time to decompose or to dissipate their harmful properties. (Ch. 17)

Petroleum   Gaseous, liquid, and semi-solid naturally occurring substances that consist chiefly of hydrocarbon compounds. (Ch. 11)

Photosynthesis   A process whereby light energy is used to cause carbon dioxide to react with water, synthesizing organic substances (carbohydrates) and releasing oxygen in the process. (Ch. 18)

Photovoltaic cells Devices used to convert solar energy directly into electricity. (Ch. 12)

Physical geology The study of the processes that operate at or beneath the surface of the Earth and the materials on which those processes operate. (Introduction)

Placer A deposit of heavy minerals that have been concentrated by mechanical processes. (Ch. 13)

Plate tectonics The special branch of tectonics that deals with the processes by which the lithosphere is moved laterally over the aesthenosphere. (Essay I, Ch. 1)

Point sources Discrete sources of contaminants that can be represented by single points on a map. (Ch. 17)

Pollutants Materials that have harmful impacts and degrade the environment; see also *Contaminants*. (Essay IV)

Pollution Materials with harmful impacts on the natural environment, or the act of releasing such materials. (Essay IV)

Porosity The percentage of the total volume of a body of regolith or bedrock that consists of open spaces or pores. (Ch. 15)

Precious metals Gold, silver, and platinum group metals. (Ch. 13)

Precursor A small physical change that precedes a catastrophic event. (Essay II)

Prediction A statement of probability based on scientific observation. (Essay II)

Primary recovery When oil flows from a well under its own pressure or flows easily with pumping. (Ch. 11)

Prior appropriation A doctrine that grants water-use rights on the basis of historical precedent. (Ch. 15)

Pyroclast A fragment of rock ejected during a volcanic eruption. (Ch. 4)

Pyroclastic flow A hot, highly mobile flow of tephra that rushes down the flank of a volcano during a major eruption. (Ch. 4)

Pyroclastic rock A rock formed from pyroclasts. (Ch. 4)

P waves Compressional, or primary, waves. (Ch. 3)

R

Rad A unit used to quantify radiation in terms of the amount of energy absorbed by biologic tissue. (Ch. 16)

Radiation Alpha particles, beta particles, and gamma rays that are emitted from disintegrating nuclides during radioactive decay. (Ch. 16)

Radiatively active gases Atmospheric gases that are efficient absorbers of infrared radiation; also known as *greenhouse gases*. (Ch. 18)

Radioactive waste Leftover radioactive material or equipment for which there is no further economic use; also called *radwaste*. (Ch. 16)

Radon A colorless, odorless radioactive gas produced by the radioactive decay of uranium in soils derived from uranium-bearing rocks. (Ch. 18)

Rapid onset hazards Events that strike quickly and with little warning, such as earthquakes, flash floods, or sudden windstorms. (Essay II)

Raveling The process of sinkhole formation initiated by the gradual downward movement of unconsolidated material into the aven. (Ch. 7)

Recharge Replenishment of groundwater that occurs when rainfall and snowmelt enter the ground in recharge areas, where precipitation infiltrates and percolates downward through soil layers to reach the saturated zone. (Ch. 15)

Recurrence interval The average interval between occurrences of two events, such as floods, of equal magnitude. (Ch. 8)

Recycling The process of taking apart an old product and using the material it contains to make a new product. (Essay IV)

Reef A generally ridgelike structure composed chiefly of the calcareous remains of sedentary marine organisms (e.g., corals, algae). (Ch. 9)

Regolith The irregular blanket of loose rock debris that covers the Earth. (Essay I, Ch. 14)

Rem A unit that measures the damage done to organisms by alpha particles, beta particles, and gamma rays. (Ch. 16)

Renewable resource A resource that can be renewed, replenished, or regenerated. (Essay III)

Reserve That portion of a resource for which there is detailed information concerning the quality and quantity of material available. (Ch. 13)

Reservoir A natural storage place for materials or energy in Earth cycles. (Ch. 1)

Reservoir rock A rock with a high proportion of interconnected pore spaces in which petroleum accumulates. (Ch. 11)

Residence time The average length of time a substance remains in a particular reservoir. (Chs. 1, 18)

Resource Any potentially valuable mineral compound, including hypothetical or speculative deposits of the material. (Ch. 13)

Resource recovery Any means of extracting economically valuable components, whether materials or energy, from wastes. (Essay IV)

Resurgent dome A volcanic dome formed when magma reenters the chamber and causes uplifting of the collapsed floor of a caldera. (Ch. 4)

Retardation A measure of how much a contaminant has been slowed down relative to normal groundwater flow. (Ch. 17)

Reuse Using a product repeatedly instead of throwing it away. (Essay IV)

Reverse fault A fault that arises from compression; movement that tends to push crustal blocks together. (Ch. 3)

Rhyolite A fine-grained igneous rock with high silica content. (Ch. 4)

Richter magnitude scale A scale, based on the recorded amplitudes of seismic body waves, for comparing the amounts of energy released by earthquakes. (Ch. 3)

Rills Small, narrow channels that first begin to cut into the regolith during erosion due to running water. (Ch. 14)

Riparian rights A principle that views bodies of water as a common property resource, owned by all who have access to it. (Ch. 15)

Risk assessment The process of establishing the probability that a hazardous event of a particular magnitude will occur within a given period and estimating its impact, taking into account the locations of buildings, facilities, and emergency systems in the community, the potential exposure to the physical effects of the hazardous situation or event, and the community's vulnerability when subjected to those physical effects. (Essay II)

River A stream of considerable volume with a well-defined passageway. (Ch. 8)

Rock Any naturally formed, nonliving, firm, and coherent aggregate mass of mineral matter that constitutes part of a planet. (Ch. 2)

Rock cycle The cyclic movement of rock material, in the course of which rock is created, destroyed, and altered through the operation of internal and external Earth processes. (Ch. 1)

Runoff The portion of precipitation that flows over the surface of the land. (Ch. 8)

Run-up The water level achieved by a tsunami once it hits the shore; expressed as height in meters above normal high tide. (Ch. 5)

S

Safety factor The ratio of shear strength to shear stress, which determines the tendency of a slope to fail. (Ch. 6)

Salinization Contamination of groundwater by salt water. (Ch. 17)

Sanitary landfill A disposal site for waste in which a layer of soil isolates the waste from birds, insects, and rodents and minimizes the amount of precipitation that can infiltrate the refuse pile. (Ch. 16)

Saturated zone The groundwater zone in which all pore spaces are filled with water. (Ch. 15)

Scarp A cliff that is formed where displaced material has moved away from undisturbed ground upslope. (Ch. 6)

Secure landfill A landfill that is specially designed and engineered to contain hazardous materials. (Ch. 16)

Sediment Regolith that has been transported by any of the external Earth processes. (Ch. 1)

Sedimentary rock Any rock formed by chemical precipitation or by sedimentation and cementation of mineral grains transported to a site of deposition by water, wind, ice, or gravity. (Ch. 1)

Sediment pollution The contamination of sediment by hazardous substances; or situations in which sediment acts as a pollutant. (Ch. 17)

Sediment yield The amount of sediment eroded from the land by runoff and transported by streams. (Ch. 14)

Seiche A periodic standing-wave oscillation of the water surface in an enclosed water body such as a lake. (Ch. 5)

Seismic belts Large tracts of the Earth's surface that are subject to frequent earthquakes. (Ch. 3)

Seismic gaps Places along a fault where earthquakes have not occurred for a long time even though tectonic stresses are still active and elastic energy is steadily building up. (Ch. 3)

Seismic sea waves (also called *tsunamis*) Long-wavelength ocean waves produced by sudden movement of the seafloor. (Ch. 3)

Seismic waves Elastic disturbances that spread out in all directions from an earthquake's point of origin. (Ch. 3)

Seismograms The characteristic signatures and arrival times of each seismic wave of an earthquake, as recorded by seismographs. (Ch. 3)

Seismograph A device used to record the shocks and vibrations caused by earthquakes. (Ch. 3)

Seismology The study of earthquakes. (Ch. 3)

Sensitive soils Materials that lose shear strength as a result of remolding. (Ch. 6)

Septic tank A holding tank designed to receive domestic sewage from a single household. (Ch. 16)

Shear strength The internal resistance of a body to movement. (Ch. 6)

Shear stress The force acting on a body that causes movement of the body parallel to a slope. (Ch. 6)

Shear waves (*S* waves) Seismic body waves that consist of alternating series of sideways movements, each particle in the deformed solid being displaced in a direction perpendicular to the direction of wave travel. (Ch. 3)

Sheet wash Erosion by water through overland flow during heavy rains; also called *sheet erosion*. (Ch. 14)

Shield volcano A volcano that emits fluid lava and builds a broad, dome-shaped edifice. (Ch. 4)

Silicate mineral A mineral that contains the silicate anion, $(SiO_4)^{-4}$. (Ch. 2)

Sinkhole A large dissolution cavity that is open to the sky. (Ch. 7)

Sinkhole karst A landscape dotted with closely spaced, circular basins of various sizes and shapes. (Ch. 7)

Slide The rapid displacement of rock or sediment in one direction, with no rotation. (Ch. 6)

Slump A type of slope failure involving the downward and outward movement of rock or regolith along a curved, concave-up surface. (Ch. 6)

Smog A yellowish-brownish haze that results when airborne contaminants interact with sunlight and undergo a variety of complex photochemical reactions. (Ch. 18)

Soil Earth material that has been broken down and altered in such a manner that it can support rooted plant life. (Ch. 14)

Soil fertility The ability of a soil to provide the nutrients needed for plant growth. (Ch. 14)

Soil horizons The subhorizontal weathered zones formed as a soil develops. (Ch. 14)

Soil profile The succession of soil horizons between the surface and the underlying parent material. (Ch. 14)

Solar energy Energy that reaches the Earth from the sun. (Ch. 12)

Solar system The sun and the group of objects in orbit around it. (Essay I)

Source reduction The process of cutting back on waste generation by reducing the volume of products, increasing the useful lifetime of products, reducing the amount of packaging, and decreasing consumption. (Essay IV)

Source rock Rock containing organic material that is eventually converted into oil and natural gas. (Ch. 11)

Splash erosion The spattering of small particles of loose soils dislodged as raindrops strike bare ground. (Ch. 14)

Spring A flow of groundwater that emerges naturally at the ground surface. (Ch. 15)

Stage The height of a body of water above a locally defined reference surface. (Ch. 8)

Stones Meteorites made primarily of silicate minerals, including many of the same minerals that are abundant in the Earth's crust, which resemble terrestrial igneous rocks. (Ch. 10)

Stony-iron Meteorites composed of mixtures of stony (silicate) and metallic material. (Ch. 10)

Stoping The process of sinkhole formation initiated by the sudden wholesale collapse of the bedrock "roof" into the underground void; the process of roof collapse can also occur in magma chambers and in underground mines. (Ch. 7)

Storm surge An abnormal, temporary rise in water levels in oceans or lakes due to low atmospheric pressure associated with storms. (Ch. 9)

Stratosphere The layer of the atmosphere that extends from the top of the troposphere (the tropopause) to an altitude of 50 km. (Ch. 18)

Stratovolcano A volcano that emits both tephra and viscous lava and that builds up steep, conical mounds of interlayered lava and pyroclastic deposits. (Ch. 4)

Stream A body of water that flows downslope along a clearly defined natural passageway, transporting particles and dissolved substances. (Ch. 8)

Streamflow The flow of surface water in a well-defined channel. (Ch. 8)

Strike-slip fault A fault that arises from stresses that lead to horizontal, or translational, motions of the fault blocks. (Ch. 3)

Subsidence The sinking or collapse of a portion of the land surface. (Ch. 7)

Surf Wave activity between the line of breakers and the shore. (Ch. 9)

Surface waves Seismic waves that are guided by and restricted to the Earth's surface. (Ch. 3)

S waves Shear, or secondary, waves. (Ch. 3)

System A portion of the universe that can be isolated from the rest of the universe for the purpose of observing changes. (Ch. 1)

T

Tar An oil that is viscous and does not flow. (Ch. 11)

Tar sands Deposits of dense, viscous, asphaltlike oil that are found in a variety of sedimentary rocks and unconsolidated sediments. (Ch. 12)

Technological hazards Hazards associated with everyday exposure to hazardous substances, such as radon, mercury, asbestos fibers, or coal dust, usually through some aspect of the use of these substances in our built environment. (Essay II)

Tephra A loose assemblage of pyroclasts. (Ch. 4)

Tephra cone A buildup of tephra around the vent of a volcano. (Ch. 4)

Terrestrial planets The innermost planets of the solar system (Mercury, Venus, Earth, and Mars), which have high densities and rocky compositions. (Essay I)

Thermal inversion A phenomenon whereby a warm air mass moves into an area and traps a pocket of cool air underneath it. (Ch. 18)

Thermal pollution Excess heat that can harm or kill plants or animals. (Ch. 17)

Thermal spring Descending groundwater that comes into contact with the hot rock near the magma chamber of a volcano, is heated, and rises to the surface along fractures. (Ch. 4)

Tidal bore A large, turbulent, wall-like wave of water caused by the meeting of two tides or by the rush of tide up a narrow inlet, river, estuary, or bay. (Ch. 9)

Tidal energy Energy derived from gravitational interactions among the Earth, the moon, and the sun. (Ch. 12)

Tide The twice-daily rise and fall of the ocean's surface that results from the gravitational attraction of the moon and sun. (Ch. 9)

Topsoil A layer of fertile soil near or at the surface of the uppermost horizon in a soil profile. (Ch. 14)

Tornado A cyclonic storm with a very intense low-pressure center. (Ch. 9)

Toxicity The capacity of a substance to cause adverse effects in a living organism. (Ch. 17)

Trade winds A globe-encircling belt of winds in the low latitudes, which blow from the northeast in the Northern Hemisphere and the southeast in the Southern Hemisphere. (Ch. 9)

Transform fault margin A type of plate margin along which two plates slide past one another. (Ch. 1)

Translational slide The rapid displacement of masses of rock or sediment involving the movement of relatively coherent blocks of material along well-defined, inclined surfaces such as faults, foliation planes, or layering. (Ch. 6)

Trap A geologic situation that includes a source rock (to contribute organic material), a reservoir rock (to allow for the migration of oil), and a cap rock (to stop the migration). (Ch. 11)

Troposphere One of the four thermal layers of the atmosphere, which extends from the surface of the Earth to an altitude of 10 to 16 km. (Chs. 9, 18)

Tsunami A very long wavelength ocean wave that is generated by a sudden displacement of the sea floor; also called *seismic sea wave*. (Ch. 5)

Tsunami earthquake An earthquake which generates a tsunami that is anomalously large with respect to the magnitude of the quake. (Ch. 5)

Tsunamigenic earthquake An earthquake that generates a tsunami. (Ch. 5)

T-value The tolerable annual rate of soil loss; losses that are balanced or offset by natural regeneration of the soil. (Ch. 14)

Typhoon A term used to describe a tropical cyclonic storm that originates in the western Pacific Ocean. See *Hurricane*. (Ch. 9)

U

Unconfined aquifer An aquifer whose upper surface coincides with the water table and therefore is in contact with the atmosphere (through overlying soil layers). (Ch. 15)

Unconventional hydrocarbon Any hydrocarbon fuel that occurs in an atypical reservoir, such as a tight sandstone or a geopressurized zone, or in an unconventional form. (Ch. 12)

Uniformitarianism The principle that the same external and internal processes we recognize in action today have been operating unchanged, though at different rates, throughout most of the Earth's history. (Ch. 1)

Upstream flooding Severe but local flooding caused by intense, infrequent storms of short duration. (Ch. 8)

V

Viscosity The internal property of a substance that offers resistance to flow. (Ch. 4)

Volcano The vent from which magma, solid rock debris, and gases erupt. (Ch. 4)

Vulnerability A concept that encompasses the physical effects of a natural hazard as well as the status of people and property in the area. (Introduction)

W

Waste   The residual materials and by-products that are generated by human use of the Earth's resources and wind up unwanted and unused. (Essay IV)

Water table   The upper surface of the saturated zone of groundwater. (Ch. 15)

Wave base   The effective lower limit of wave motion, which is half of the wavelength. (Ch. 9)

Wave-cut cliff   A cliff cut by wave action at the base of a rocky coast. (Ch. 9)

Wave energy   A secondary expression of solar energy; used to ring bells and blow whistles that serve as navigational aids. (Ch. 12)

Wavelength   The distance between the crests of adjacent waves. (Ch. 5)

Wave trap   A situation in which the energy of a long section of a wave is concentrated on a particular stretch of coastline as a result of seafloor topography or shoreline configuration. (Ch. 5)

Weather   The state of the atmosphere at a given time and place. (Ch. 9)

Weathering   The chemical alteration and physical breakdown of rock and sediment when exposed to air, moisture, and organic matter. (Ch. 14)

Wet deposition   Materials transported in suspension in the air, which eventually settle out as precipitation. (Ch. 18)

Wind energy   A secondary expression of solar energy used to power ships and windmills. (Ch. 12)

Z

Zone of aeration   A zone in which open spaces in regolith or bedrock are partially filled with air; also called the *vadose zone* or the *unsaturated zone*. (Ch. 15)